LINEAR SYSTEMS

A State Variable Approach
With Numerical Implementation

Raymond A. DeCarlo

School of Electrical Engineering
Purdue University

PRENTICE HALL, Englewood Cliffs, New Jersey 07632

Library of Congress Cataloging-in-Publication Data

DeCarlo, Raymond A.,
    Linear systems.

    Bibliography: p.
    Includes index.
    1. Linear systems.   2. State-spacemethods.   I. Title.
QA402.D34 1989   003   88-32435
ISBN   0-13-536814-6

Editorial/production supervision and
    interior design: **John Fleming**
Cover design: **Ben Santora**
Manufacturing buyer: **Mary Noonan**

Printed in the United States of America

10 9 8 7 6 5 4 3 2 1

ISBN 0-13-536814-6

PRENTICE-HALL INTERNATIONAL (UK) LIMITED, *London*
PRENTICE-HALL OF AUSTRALIA PTY. LIMITED, *Sydney*
PRENTICE-HALL CANADA INC., *Toronto*
PRENTICE-HALL HISPANOAMERICANA, S.A., *Mexico*
PRENTICE-HALL OF INDIA PRIVATE LIMITED, *New Delhi*
PRENTICE-HALL OF JAPAN, INC., *Tokyo*
SIMON & SCHUSTER ASIA PTE. LTD., *Singapore*
EDITORA PRENTICE-HALL DO BRASIL, LTDA., *Rio de Janeiro*

This text is dedicated to all those
who pass the doors of Covenant House.

CONTENTS

21   OBSERVABILITY AND THE BASICS OF OBSERVER DESIGN   355

22   STABILITY OF LUMPED TIME-INVARIANT SYSTEMS   404

23   CONTROLLABILITY AND OBSERVABILITY: THE TIME-VARYING
CASE   436

PREFACE

This text is an outgrowth of lectures given in a one-semester graduate level course on linear systems taught at Purdue for the past 15 years or so. The course syllabus Fncludes both continuous and discrete time concepts with asides on numerical implementations such as simulation of the state model. Students taking the course have diverse backgrounds ranging from solid state to computers to systems. Thus a certain philosophical structure emerges to better meet the uneven background skills present in a class.

Loosely speaking, the text has three levels. The first third of the text (Chapters 1 through 9) covers vocabulary, basic definitions, and heuristic introductions to some of the more advanced notions, like controllability and observability. The presence of abundant examples, few theorems, and number crunching mark this as a senior-level exposition.

Chapters 10 through 19, the middle third, represent the basic meat and potatoes of the course. The level here is advanced senior and beginning graduate. Theorems and proofs take on an essential part of the treatment, demanding a certain rapidity in the intellectual maturation process of the students.

Students often ask at this juncture if understanding the proofs rather than knowing the proofs is sufficient. My response overflows with assurances that understanding is paramount, but that an inability to reproduce a proof on a test will lead me to conclude a definitive lack of comprehension. The motivation here comes about because theorems and their proofs reflect the viewpoint of the pioneering researchers who have contributed to the elegant structure of state variable theory. Just as musical students must learn to play classics, so too must our students learn to understand the thinking of the contributors to system theory.

The last third of the book deals with the more advanced topics of controllability, observability, realization, stability, and disturbance decoupling. Hobnobbing introductions to topics more or less disappear. The developments mirror a clear reliance on the language of linear algebra to convey the essential concepts. Once again the student undergoes an intellectual stretching while mastering these topics.

A typical one semester course includes all of Chapters 1 (basic vocabulary), 2 (motivation and review), 3 (basic state model definitions and constructions), 4 (canonical forms except for the observability and controllability forms), and 5 (a primer on observability and controllability based on the problem of determining and setting up initial conditions of a state model constructed from a differential equation). Chapter 6 (Gaussian elimination) is mentioned but generally skipped. Only the basic structure of the Newton-Raphson algorithm in the front part of Chapter 7 is covered. All of the simulation material of Chapter 8 is taught. This provides a natural motivation for the discrete time state model basics laid out in Chapter 9 and the theoretical aspects of existence and uniqueness set forth in Chapter 10. Typically, only the introduction to Chapter 10 and the global existence and uniqueness theorem in the last part of Chapter 10 are explained. After this, the notions of the state transition matrix, the fundamental matrix, their properties, and the complete analytic solution to the time-varying state dynamics take up all of Chapters 11, 12, and 13. Skipping over Chapter 14, which describes several nonstandard techniques for computing the matrix exponential, the eigenvalue-eigenvector factorization method of exp (At) is covered in detail in Chapter 15. At this point the development of the previous Chapters is condensed, retold, and expanded upon for the discrete time case as delineated in Chapter 16. Our usual course then journeys through all of Chapter 17, a decomposition of the state trajectory and the state transition matrix into directed modes, and all of Chapter 18, the basic impulse response and frequency domain ideas of state variables. The singular value decompostion of a matrix in Chapter 19 is stated but not developed. Most of Chapter 20 (time-invariant controllability), about half of Chapter 21 (time-invariant observability and realization basics), and all of Chapter 22 (time-invariant bounded-input-bounded-state and bounded-input-bounded-output stability) make up the end of the one semester course.

There is much more I would like to say, but my experience is that long-winded prefaces go unread. I would be happy to discuss my decisions to treat topics as I have. Any suggestions for improvement of the text are welcome.

Raymond A. DeCarlo
W. Lafayette, Indiana

ACKNOWLEDGMENTS

Many persons have contributed in untold ways to the compilation of this text. No doubt, there are some whose names and contributions have slipped through the cracks of my memory. With an apology to all those who go unmentioned, I would like to acknowledge those whose footprints have remained clear. First, there is my wife who supported me through this writing and many other writings as well. The seeds for the writing of the text were planted by my major professor, Dick Saeks, and the many other professors who first introduced and guided me through the alley ways of systems and control. An abiding thank you is due my colleagues, Bob Barmish, P.M. Lin, Stan Zak, Ed Coyle, Abe Haddad, Carl Cowen, and unknown reviewers whose readings corrected and improved the manuscript. All the former EE-602 students, especially Larry Rapisarda, Greg Matthews, Dale Sebok, Cathy Hudson, Tom Harris, Philip Peleties, Mark Wicks, and Rob Frohne, who have sharpened and deepened my understanding of state variable ideas and the written expression of those ideas as set forth in this text are likewise due some recognition. Finally I would like to thank the technical typists, especially Mary Schultz, Linda Stovall, and Nancy Lein, for typing my course notes into a readable manuscript not once but many times, and my secretaries, Cathy Tanner and Jill Comer, who have typed, retyped, proofread, and re-proofread, homework problems, various chapters, and the index. Without their help and the help of others the immensity of the task would have surpassed my skills.

1

BASIC CONCEPTS, VOCABULARY, AND NOTATION

NOTION OF A SYSTEM AND SYSTEM MODEL

Daily life contains innumerable action-reaction types of situations, from popping bread into a toaster to the complexities of car manufacturing. In all cases some stimulus or set of stimuli activate the operation of some process (toasting or car manufacturing, for example) to produce some result or response (toasted bread or an assembled car.) Typically, some person or electronic device monitors the resulting response and makes adjustments in the process to improve, say, quality or appearance or yield. Figure 1.1 provides a pictorial description.

The term *system* generically categorizes such processes. Generally speaking, a system has provision for input stimuli and some mechanism for modifying or reacting to these input stimuli to produce a finished product or generate some desired behavior. A manufacturing process clearly fits this description: raw materials, parts, and human labor stimulate an assembly line to produce a finished product (output) such as a car, stereo, or VCR. A business such as a furniture store illustrates an economic system: orders serve as input and goods shipped as output. A space shuttle launch, loosely speaking, is a system in which a stable orbit could be a desired response. The idle speed control on a car engine provides an example of an electromechanical system: temperature and the difference between the desired

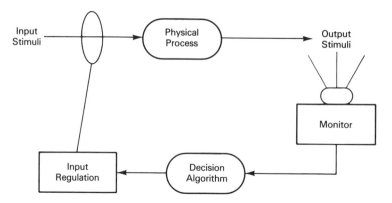

Figure 1.1   Illustration of a controlled process or system.

idle speed and the actual rpm are inputs to a device controlling the airflow into the carburetor. The consumer electronics industry produces a supermarket of electronic systems, most of which use energy (a battery or household current) and some information signal as inputs to produce visual images, sound, and/or music as outputs. Finally, the ubiquitous rectifier circuit serves as an example of a system where an AC signal is rectified and filtered to generate a DC signal.

In sum, a *system* refers to some physical process which generates outputs, finished products, desired behaviors, etc., in response to input stimuli. The input stimuli cause or initiate a set of events within the system. The system might be thought of as an "artificial intelligence" which reshapes, amplifies, or filters the input stimuli to obtain a desired result.

The input stimuli and related responses are typically represented by signals or functions. These functions will either depend on time in a continuous manner or be defined only at discrete-time instants. The former are *continuous* signals, the latter *discrete* signals. The physical process under study (analysis) or under design will be called the *physical system* and is represented by a mathematical model relating inputs to outputs.

The construction of a mathematical model often requires some strong assumptions about the nature of the physical process. The resulting model often reflects only the dominant behavior of the associated process. This is because, on the whole, most real world processes are sufficiently complex to evade understanding without the tools of a mathematical formalism. The inconsistencies of the standard weather prediction model serve as an example.

The importance of the mathematical model stems from its utility— e.g., in the analysis and design of information transmission systems and systems for the control and regulation of such things as oil refineries, electric power generation, commuter trains, airplanes, and the like. Once the physical process has a mathematical representation, computer simulations of the model can be used to predict the future behavior of the process, as, for example, in earthquake prediction. Differential equations are common models. Several good texts are available on constructing mathematical models of real systems [1,2,3].

For convenience, we shall use the term *system* to refer to the mathematical model of the physical system.

CONTINUOUS SIGNALS AND MATHEMATICAL MODELS

Figure 1.2 depicts a generic physical system. The figure contains three central features which sharpen our understanding of the idea of a system: an input space U containing the set of all admissible input stimuli or signals; an output space Y containing the set of all admissible output signals; and the system model, which defines a mathematical relationship between any admissible input signal and the resulting output signal. Typical mathematical relationships include differential equations, integral equations, and matrix equations, all relating inputs to outputs. These ideas suggest the block diagram representation of a system as depicted in Figure 1.3, where the notation N (usually called an *operator*) embodies the system in that it maps the admissible input $u(\cdot)$ to the admissible response $y(\cdot)$. Thus $N:U \rightarrow Y$ is defined as $y(\cdot)=Nu(\cdot)$—i.e., the signal $u(\cdot)$ defined over the infinite time interval, $(-\infty,\infty)$ maps into another signal $y(\cdot)$. For example the operator N could represent the convolution-like relationship between inputs $u(\cdot)$ and outputs $y(\cdot)$ given by

$$y(t) = \int_{-\infty}^{t} \exp[-(t-q)]u^2(q)dq \tag{1.1}$$

That is, each point $y(t)$ of the signal $y(\cdot)$ is computed by this convolution-integral relationship.

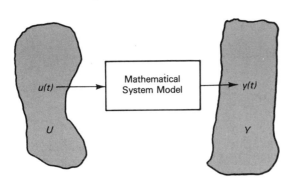

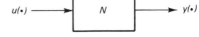

Figure 1.3 Block diagram of system mapping inputs $u(\cdot)$ to outputs $y(\cdot)$.

Figure 1.2 Generic system representation.

As a simple illustration of a circuit model consider the parallel RL circuit of Figure 1.4. Some simple calculations produce the differential equation model

$$\frac{di_L}{dt}(t) + \frac{R}{L}i_L(t) = \frac{R}{L}i_s(t) \tag{1.2}$$

where $i_L(t)$ is the circuit output and $i_s(t)$ the circuit input. Equivalent to this differential equation is the convolution-integral model

$$i_L(t) = \frac{R}{L}\int_{-\infty}^{t} \exp[-\frac{R}{L}(t-q)]i_s(q)dq \tag{1.3}$$

which, as expected, defines a kinship between the input $i_s(t)$ and the resulting output $i_L(t)$.

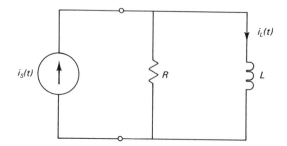

Figure 1.4   Parallel *RL* circuit.

A third type of model called a *frequency domain model* results by taking the two sided Laplace transform of the differential equation to obtain

$$i_L(s) = \frac{\frac{R}{L}}{s + \frac{R}{L}} i_s(s) \tag{1.4}$$

where the *s*-dependence of $i_L(s)$ and $i_s(s)$ indicates the Laplace transform of the corresponding time functions.

Each of these three models looks at the *RL* circuit differently. For example, the frequency-dependent circuit aspects displayed in equation 1.4 differ radically from the time-dependent characteristics of equations 1.2 and 1.3. The type of model that is best suited for analysis or design depends, of course, on the intended application. This text focuses on the so-called state model and its frequency domain equivalent.

The widespread use of Laplace transforms and frequency domain models in circuits and systems suggests that the input and output classes contain only those signals having well-defined two-sided Laplace transforms. Thus, it is natural as well as convenient to have the input and output spaces coincide.

Definition 1.1.   A continuous time signal is *admissible* if (*i*) it is piecewise continuous, (*ii*) there is a point t_0 such that for all $t \leqslant t_0$, the signal is zero, (*iii*) it is exponentially bounded, and (*iv*) it conforms to the appropriate dimensions imposed by the system model.

Condition (*i*) means that the signal is continuous except possibly at a finite number of points per finite interval of time. Condition (*ii*) implies that there exists a "turn-on" time before which the signal is identically zero. Also, it implies that at $t = -\infty$, the system is relaxed and that initial conditions at some t must result from the stimulus of an input. Loosely speaking, condition (*iii*) means that the signal will not approach infinity faster than an exponential signal. If the function is continuous and scalar-valued, this means that there exist constants c_1 and c_2 such that

$$|u(t)| \leqslant c_1 \exp(c_2 t)$$

Basically, these conditions guarantee the existence of a two-sided Laplace transform of an admissible signal [4,5,6]. As a consequence, many types of system models have frequency domain equivalents which prove useful in analysis and design.

Certain general kinships exist among the set of admissible signals. These kinships make certain system properties well posed. The first of these, arising from the definition of an admissible signal, is that a linear combination of admissible signals is admissible. The input space U and the output space Y are therefore said to be closed under addition, allowing us in turn to describe the system-theoretic notions of superposition and linearity. A second general kinship comes about because translations of signals in time produce admissible signals. This permits a discussion of the concept of time invariance to be well posed. Thirdly, under typical mathematical restrictions, limits of sequences of well behaved admissible signals are admissible. And finally, we note that when appropriate, derivatives and integrals of admissible signals are admissible. As a consequence of these properties, the set of admissible signals is sufficiently rich to permit the usual kinds of engineering analysis and design.

INPUT-OUTPUT LINEARITY

The first great division among the class of systems (system models) is the linear versus nonlinear distinction. Figure 1.5 provides a picture.

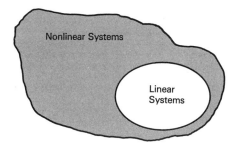

Figure 1.5 Linear and nonlinear systems as special subclasses of systems.

Definition 1.2.   A system described by the operator N (as in Figure 1.3) is *linear* if, for arbitrary complex (real) scalars c_1 and c_2 and admissible signals u_1 and u_2,

$$N(c_1 u_1 + c_2 u_2) = c_1 N u_1 + c_2 N u_2 \qquad (1.5)$$

In other words, the response to a scaled sum of admissible inputs is the scaled sum of the responses. The principle of *superposition* then falls immediately out of the definition: $N(u_1 + u_2) = N u_1 + N u_2$.

Proper interpretation of definition 1.2 critically depends on the fact that signals are defined over the infinite time interval $(-\infty, \infty)$ and that only an input signal can set up any internal or external initial conditions. Hence, at $t = -\infty$, the system is relaxed, with no initial conditions. Later, we shall modify the definition to take care of semi-infinite intervals, $[t_0, \infty)$, and the presence of initial conditions at t_0.

Some examples will help clarify these notions. Consider the low-pass filtering circuit of Figure 1.6. At $t = -\infty$, no initial conditions are present on the capacitors and inductor. Hence, the response of the circuit to a scaled sum of admissible input signals is the scaled sum of the individual responses. On the other hand, the

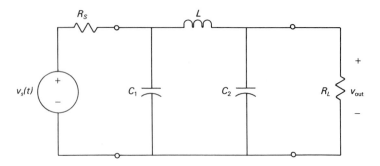

Figure 1.6   Linear low pass filtering circuit.

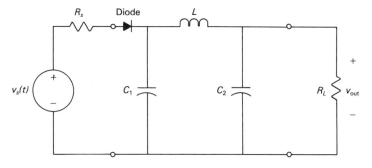

Figure 1.7   Nonlinear rectifier circuit.

simple insertion of a diode into the circuit (Figure 1.7) produces a half-wave rectifier with filter. The nonlinearity of the diode makes the circuit behavior nonlinear— i.e., the response to two arbitrary sinusoids fails to equal the sum of the responses except in very rare cases.

The class of linear systems also includes systems whose output is the derivative or exact integral of the input. Recall that the derivative of a scalar times a function is simply the scalar times the derivative of the function and that the derivative of the sum of any finite number of functions is the sum of the derivatives. Similar arguments hold for exact integration of signals. Hence, it follows that ordinary differential equations as the following are linear.

$$\ddot{y}(t) + a_1(t)\,\dot{y}(t) + a_2(t)\,y(t) = b_0(t)\,u(t) + b_1(t)\,\dot{u}(t) \qquad (1.6)$$

By contrast, some simple modifications to equation 1.6 produce the following nonlinear differential equation:

$$\ddot{y}(t) + a_1(t)\,\dot{y}(t)\,u(t) + a_2(t)\,y(t)\,u^2(t) = b_0(t)\,\dot{u}(t) + b_1(t)\,u(t) \qquad (1.7)$$

Finally, the small-displacement dashpot-spring combinations typical of elementary mechanical engineering also illustrate linear systems, since such systems can be characterized by linear ordinary differential equation models.

Although the real world is predominately nonlinear, a number of compelling reasons motivate a serious investigation of linear systems. First, linear models can often be used to approximate nonlinear systems with reasonable accuracy. Indeed, many control systems and other real world devices are designed from linear approximations to real-world nonlinear processes. In addition, a well-defined body of knowledge surrounds linear systems, which is essential to an understanding of the much more intricate theory of nonlinear systems and thus in comprehension of the surrounding physical world.

TIME INVARIANCE

Some systems change their structure with time and some do not. Those whose structure remains fixed with time are called *time-invariant* and those whose structure or characterizing parameters change with time are called *time-varying*. Equation 1.6 is a differential equation whose coefficients change with time. These coefficients embody the characteristic parameters of the system; hence, the model and system are time-varying. A simple *RC* circuit containing a time-varying capacitor will spawn such a differential equation. To see this consider the circuit of Figure 1.8.

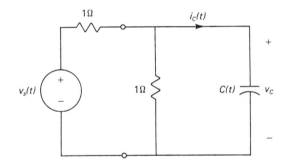

Figure 1.8 *RC* circuit.

Here, the capacitor current-voltage relationship satisfies the equation

$$\frac{d}{dt}[C(t)v_c(t)] = i_c(t) \tag{1.8}$$

which, after some arithmetic manipulations, produces the differential equation model of the circuit,

$$\dot{v}_c(t) + \left[\frac{\dot{C}(t) + 2}{C(t)}\right]v_c(t) = \left[\frac{1}{C(t)}\right]v_s(t) \tag{1.9}$$

On the other hand, the simple *RL* circuit of Figure 1.4 produces the constant-coefficient differential equation model of equation 1.2. Observe that the constant coefficients depend on the constant resistance and inductance parameters. Hence, the circuit is time-invariant.

To mathematically pin down the property of time invariance, recall that U and Y are closed under time translations; i.e., if $u(\cdot)$ is any function in U and $y(\cdot)$ any function in Y, then, for any real number T, the shifted or time-translated signals $u(\cdot - T)$ and $(\cdot - T)$ are in U and Y, respectively.

Definition 1.3. A possibly nonlinear system denoted by the operator $N{:}U \rightarrow Y$ is *time-invariant* if, whenever $y(\cdot) = Nu(\cdot)$, then for any real T, $y(\cdot - T) = Nu(\cdot - T)$.

Informally, if $y(\cdot)$ is the response $u(\cdot)$, then the response to the shifted signal $u(\cdot - T)$ must always be the shifted response $y(\cdot - T)$. An example of a time-invariant nonlinear circuit is given in Figure 1.9.

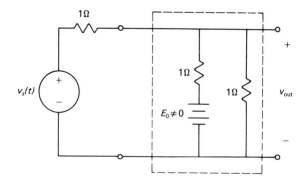

Figure 1.9   A nonlinear time-invariant circuit.

To illustrate the implications of time invariance and linearity, consider the hypothetical *RLC* (all linear elements) circuit of Figure 1.10, where $u(t)$ denotes the input signal and $y(t)$ the output signal. Let $u(t) = \exp(-t)\cos(2\pi t)1^+(t)$, and suppose that $y(t) = 0.5\exp(-t)\cos(2\pi t + 0.14\pi)1^+(t)$, where

$$1^+(t) = \begin{cases} 1 & t \geqslant 0 \\ 0 & t < 0 \end{cases} \tag{1.10}$$

is the so-called *step function*. Note that we assume that all initial conditions are zero at $t = 0$. If $u(t)$ is shifted by one time unit and scaled by a half to produce the new input $u_1(t) = 0.5u(t-1) = 0.5\exp(-(t-1))\cos(2\pi(t-1))1^+(t-1)$, then the new output is

$$y_1(t) = 0.5y(t-1) = 0.25\exp(-(t-1))\cos(2\pi t - 1.86\pi)1^+(t-1).$$

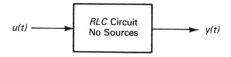

Figure 1.10   Hypothetical *RLC* circuit without internal sources.

CAUSALITY

Intuitively, causality means, "Nothing in, then nothing out," or "No cause, then no effect." The nonlinear circuit of Figure 1.9 provides a counterexample to these glib but common utterances. Because the battery voltage $E_0 \neq 0$, there is almost always a nonzero output ($v_{out} \neq 0$), even when the input $v_s(t) = 0$. That is, the "nothing in" still begets "something out." The following definition circumvents this subtlety:

Definition 1.4. A possibly nonlinear system modeled by some operator $N:U \rightarrow Y$ is *causal* [7,8] if and only if for any two inputs $u_1(\cdot)$ and $u_2(\cdot)$ with $u_1(t) = u_2(t)$ for all $t < T$, $Nu_1(t) = Nu_2(t)$ for all $t < T$.

To illustrate the meaning of this definition suppose a system has an output-input relationship given by $y(t) = u(t + 1)$. The graphs in Figure 1.11 illustrate the noncausal nature of this relationship. The relationship is noncausal because, although $u_1(t) = u_2(t)$ for $t \leqslant 1$, $y_1(t) \neq y_2(t)$ for $t \leqslant 1$. The importance of the definition will become evident at a later point in the text. Meanwhile, we can observe that causality is basic to the concepts of system stability, system invertibility, and estimation. Also it underlies the areas of filtering and controller realization [8,9].

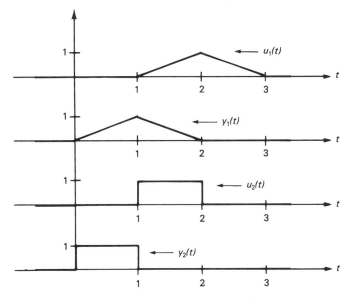

Figure 1.11 Input-output relationships for $y(t) = u(t + 1)$.

It is important to note as well that for linear systems, an identically zero input signal must produce a response which is identically zero. This fact can be used to show that a linear system is causal if and only if, for every signal $u(\cdot)$ such that $u(t) = 0$ for $t \leqslant T$, the output response $y(\cdot)$ satisfies $y(t) = 0$ for $t \leqslant T$. If the output were nonzero for some $t \leqslant T$, then the system would have had to predict that at a later time the input would be nonzero in order to produce the response. Thus, noncausal systems are *predictor* systems in that they use information from the future in producing a response in the present.

LUMPED AND DISTRIBUTED SYSTEMS

An inverse of a predictor system is a *delay* system. The response of such a system is a delayed modification of the input signal. If, in Figure 1.3, the operator N is defined according to the equation $y(t) = Nu(t) = u(t-1)$, then an ideal delay system of one second is defined— i.e., the response replicates the input delayed by one second. Systems containing delays— or predictors, for that matter— do not appear to be ordinary, i.e., they are not described by ordinary differential equations. The differential equations 1.6 and 1.7 are time-varying ordinary differential equations. However, the equation

$$\ddot{y}(t) + a_1(t)\,\dot{y}(t-1) + a_2(t)\,y(t-2) = b_0(t)\,u(t) + b_1(t)\,u(t+1) \qquad (1.11)$$

is not ordinary since the highest derivative of $y(\cdot)$ depends on output derivatives at more than one instant and on the input and its derivatives at various instants. Because of this, equation 1.11 and the system it models are said to be *distributed*, and in fact, the equation is noncausal but linear. (Why?)

Systems which do not contain delays and predictors and which are not described by partial differential equations are said to be *lumped*. Ideal resistors, capacitors, and inductors model lumped elements. In fact, virtually all of elementary circuit theory deals with lumped circuits which have ordinary differential equation models. This is because the wavelength of the signals involved is much larger than the size of the components. This motivates the following definition of a lumped system or model:

Definition 1.5. A system is *lumped* if its input-output variables are characterized by a set of ordinary differential equations; otherwise it is said to be *distributed*.

As an example, suppose the block diagram of Figure 1.3 has the differential equation model

$$\frac{d^n y}{dt^n} + a_1 \frac{d^{n-1}y}{dt^{n-1}} + \dots + a_n y = b_1 \frac{d^m u}{dt^m} + \dots + b_{m+1}u \qquad (1.12)$$

Then the system represented by N is lumped, time-invariant, and linear. The following two examples also illustrate the notions of lumped and distributed systems, as well as some of the other properties previously described.

EXAMPLE 1.1

Consider the *RLC* circuit of Figure 1.12. The differential equation governing the capacitor voltage is

$$\ddot{v}_c(t) + \frac{1}{RC}\,\dot{v}_c(t) + \frac{1}{LC}\,v_c(t) = \frac{1}{RC}\,\dot{v}_s(t) \qquad (1.13)$$

Hence, the circuit is lumped and again linear, time-invariant, and causal. More generally, all circuits containing ideal capacitors, inductors, resistors, transformers, and independent sources are lumped.

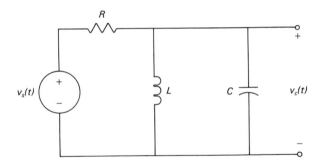

Figure 1.12 *RLC* circuit.

EXAMPLE 1.2

Consider the transmission line, which represents a distributed system, shown in Figure 1.13. The transmission-line equations are

$$\frac{\partial i(x,t)}{\partial x} = -[G\, v(x,t) + C\, \frac{\partial v(x,t)}{\partial t}] \tag{1.14}$$

and

$$\frac{\partial v(x,t)}{\partial x} = -[R\, i(x,t) + L\, \frac{\partial i(x,t)}{\partial t}]$$

where L, C, R, and G are the inductance, capacitance, resistance, and conductance, of the system, respectively, all per unit length of the transmission line. Note that L and R are series parameters whereas C and G are shunt parameters. The voltage and current distributions clearly depend on the spatial properties of the T-line. These *partial differential equations* are also linear, time-invariant, and causal.

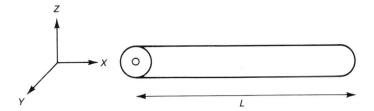

Figure 1.13 Diagram of transmission line.

The "lumped versus distributed" characterization comes about because the wavelengths of typical system signals are quite large relative to the physical dimensions of the devices contained in the circuits or systems. When this is true, the macroscopic behavior of the interaction of various devices satisfies Kirchhoff's laws. When the wavelengths become close to the physical dimensions of the system devices, Maxwell's equations must be used to describe the voltage and current distributions and thus the system is distributed.

CONTINUOUS SYSTEMS

Perhaps the best way to describe a *continuous system* is to contrast it with a specific example of a discontinuous system. Accordingly, in Figure 1.14, let

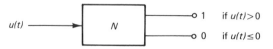

Figure 1.14   A discontinuous system.

$u_n(t) = \xi_n 1^+(t)$ for $n = 1,2,3,\ldots$ and $\xi_n = 1/n$, be a sequence of inputs which converge at each instant of time to the identically zero function. Then the response of each of these inputs is $1^+(t)$, whereas the response of the identically zero input is the identically zero output. In other words, the sequence of responses generated by the above input sequence does *not* converge to the identically zero output function, which is the response of the limit of the input sequence. Discontinuous systems are anomalies because Fourier analysis, common to system analysis, may break down. Hence, throughout the text, all systems will be assumed continuous unless stated otherwise. Mathematically, we will assume that if $\{u_i\}_{i=1}^{\infty}$ is a sequence of admissible input functions which converge to an admissible input u, i.e., if

$$\lim_{i \to \infty} u_i = u^* \quad (u_i \to u^*) \tag{1.15}$$

then

$$Nu^* = \lim_{i \to \infty} Nu_i \tag{1.16}$$

where "converge" is to be taken in some acceptable mathematical sense. In other words, the response of the limit is the limit of the responses. The reader should not confuse this mathematical notion of continuous system with that of a continuous time system.

SOME EXAMPLES

As a first example, let N in Figure 1.15 represent a linear, lumped, time-invariant causal system. The signals $u(\cdot)$ and $y(\cdot)$ are as defined. Figure 1.16 states the problem: find $y(t)$ if $u(t) = \delta(t - 2)$, where $\delta(t)$, the so called delta function (although not really a function [10]) is defined as

$$\delta(t) = \frac{d}{dt} 1^+(t) \tag{1.17a}$$

in which case $\delta(t) = 0$ for $t \neq 0$, $\delta(t) = \infty$ for $t = 0$, and

$$\int_{0^-}^{0^+} \delta(t)\, dt = 1 \tag{1.17b}$$

That is, the delta function has a well defined area of unity. Of course $\delta(t-2)$ is the shifted delta function. Since the delta function is the derivative of the step function and the derivative is a linear operator, the response to $\delta(t)$ is the derivative of $(1 - e^{-t})1^+(t)$ or $e^{-t}1^+(t)$. Since the system is time-invariant, the response to the shifted delta function is the shifted response. Hence, the answer to the problem is $y(t) = e^{-(t-2)}1^+(t-2)$.

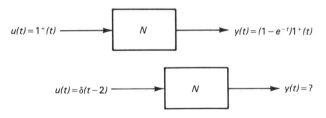

$u(t) = 1^+(t)$ N $y(t) = (1-e^{-t})1^+(t)$

Figure 1.15 A linear, lumped, time-invariant, causal system having the indicated step response.

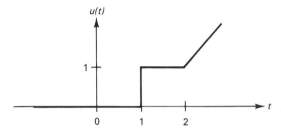

$u(t) = \delta(t-2)$ N $y(t) = ?$

Figure 1.16 Find $y(t)$ when $u(t) = \delta(t-2)$, given Figure 1.15.

As a second example, consider again Figure 1.15, except suppose now that it is known only that N is causal and time-invariant. That is, an assumption of linearity cannot be made. Suppose the new input shown in Figure 1.17 excites the system. Find $y(t)$ for $t \leqslant 2$.

$u(t)$

1

0 1 2 t

Figure 1.17 An input signal.

The key to the solution of this problem is the observation that the new input signal satisfies $u(t) = 1^+(t-1)$ for $t \leqslant 2$. Time invariance then implies that the response to $1^+(t-1)$ is $(1 - e^{-(t-1)})1^+(t-1)$. Since the system is causal we may conclude that the response to the input signal of Figure 1.16 is $(1 - e^{-(t-1)})1^+(t-1)$ for $t \leqslant 2$. If the system were not causal, this could not be ascertained.

As a third example consider the circuit of Figure 1.18. The goal here is to categorize the circuit as linear or nonlinear, time-invariant or time-varying, etc. After a little thought one observes that the circuit is linear; i.e., all devices are linear. The circuit is also time-varying due to the time dependence of the dependent-source voltage $\sin(t) v_1(t)$. By looking at the other dependent source, one notices that at $t=0$, the source voltage, $v_2(1-t)$ requires knowledge of v_2 at $t = 1$. Hence, the circuit is noncausal and distributed. Since all the circuit elements are continuous, the circuit is continuous.

DISCRETE-TIME SYSTEMS

As indicated earlier, the phrase "continuous time" refers to the fact that the time parameter takes values in the continuum of real numbers. Hence, the phrase "continuous time system" identifies a system whose underlying time structure is defined on the continuum of real numbers. In a similar manner, a discrete-time system is defined over discrete-time points, such as the set of points $t_0, t_1, t_2, \ldots$. Often the successive time instants differ by a fixed time step Δt, so that $t_k = t_0 + k\Delta t$, where $k = 0, 1, 2, 3, \ldots$.

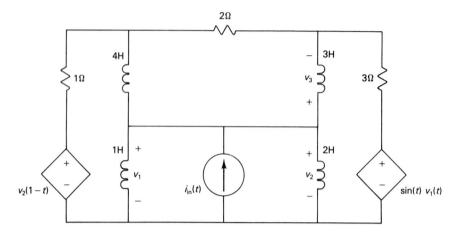

Figure 1.18   Sample circuit.

Definition 1.6.   Systems defined over sets of discrete-time instants (e.g., at $k\Delta t$ or simply k for short) as opposed to the continuum of real numbers are said to be *discrete-time systems*.

Examples of discrete-time systems abound [1,3]. For example, a dealership's car inventory has a discrete-model time formulation: daily customer demand in units and weekly restocking serve as inputs with customer purchases as an output. Similarly, based on census data, the U.S. Bureau of the Census generates population models of cities, states, regions, etc., all of which are discrete-time systems. Finally, the binary counter and digital filter are two examples of an electronic discrete-time system.

The input and output signals of a discrete-time system must of course be discrete, and hence the vehicle of a sequence best represents such signals. The designation for an *input sequence* will be $\{u(kT)\}_{-\infty}^{\infty}$, or more briefly, $\{u(k)\}_{-\infty}^{\infty}$, where $\pm\infty$ means that k takes integer values from $-\infty$ to ∞. For the sequence to be admissible we require that for each $\{u(k)\}$ there exist a k_0 such that $u(k)=0$ for all $k \leqslant k_0$. Furthermore, when necessary, we will assume that the sequence has a well-defined one sided Z-transform [11]. In a similar manner, we will let $\{y(kT)\}_{-\infty}^{\infty}$ designate the *output sequence*. Since notation becomes cumbersome, for convenience we will use $u(k)$ or $y(k)$ to denote both the sequence as well as the value of the sequence at point k. When it is unclear from the content, or when emphasis is necessary, the former notation will be used.

Although discrete-time signals occur quite naturally in the real world, one common method for obtaining them is via the sampling of a continuous time signal. Figure 1.19 illustrates the sampling of a hypothetical speech waveform. The successive samples generate a sequence of points which drives a digital filter or some other digital (discrete) signal processing device. With knowledge of the sampling time (the time between successive samples), these sampled values are quantized to binary numbers representing the amplitude of the signal at the particular sample instant. The binary numbers are then processed by a digital device (microproces-

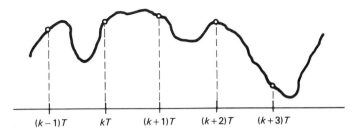

$(k-1)T$ kT $(k+1)T$ $(k+2)T$ $(k+3)T$

Figure 1.19 Sampled waveform.

sor) and passed through a D/A converter to produce a desired analog signal, for example, one which replicates the input signal except that background noise has been removed.

A discrete-time system therefore maps input sequences to output sequences. As a result, notions like linearity, time invariance, and the like have natural analogs. For example, a discrete-time system is *linear* if the response to a linear combination of input sequences is the same linear combination of the responses. In the same vein, *time invariance* means that whenever a discrete-time system produces an output sequence $\{y(k)\}$ in response to the input sequence $\{u(k)\}$, $\{y(k-n)\}$ must be the response to the shifted input sequence $\{u(k-n)\}$, where n is any finite integer. Causality, of course, is analogously defined.

In contrast to the differential equations of continuous time systems, difference equations are common models for discrete-time systems. Difference equations define a relationship between sets of terms in the output sequence and sets of terms in the input sequence. For example, an nth-order difference equation takes the form

$$y(k+n) + a_1(k)\,y(k+n-1) + \ldots + a_n(k)\,y(k)$$
$$= b_0(k)\,u(k) + b_1(k)u(k+1) + \ldots + b_m(k)\,u(k+m) \tag{1.18}$$

A little rewriting shows that

$$y(k+n) = b_0(k)\,u(k) + b_1(k)u(k+1) + \ldots + b_m(k)\,u(k+m)$$
$$- a_1(k)\,y(k+n-1) - \ldots - a_n(k)\,y(k) \tag{1.19}$$

which indicates that the $(k+n)$th term in the response sequence depends on past values of the output sequence and on values of various terms in the known input sequence. Thus, at the $(k+n)$th instant the value of the output depends on delayed values of the input and output signals. Consequently, by design, discrete systems have a distributed nature.

This brief introduction to discrete systems paves the way for their formal presentation in Chapter 9. A chapter on digital simulation where a continuous time model is converted to a discrete-time approximate one for simulation on a digital computer precedes Chapter 9. This is a more natural point for a formal development.

THE NOTION OF STATE

In basic linear circuits, independent sources can be viewed as inputs, various voltages and currents as outputs, and initial capacitor voltages and initial inductor currents as initial conditions. The initial conditions represent important internal variables which, in conjunction with the independent source waveforms, determine all voltage and current profiles for future time. Specifically, if the circuit is turned on at time t_0, the source voltages and currents and the initial capacitor and inductor voltages at t_0 uniquely determine the output voltages and currents. Thus, analysis focuses on the interval $[t_0,\infty)$, not the interval $(-\infty,\infty)$.

The central underlying assumption in this type of circuit analysis is that knowledge of the input signals over $[t_0,\infty)$ and knowledge of all internal initial conditions uniquely determine the future evolution of the output signals. Since the input is assumed to be known only over $[t_0,\infty)$, the initial voltages and currents embody the cumulative effect of all past excitation to the system. These important internal variables are called *states*.

States also appear in elementary dynamics. For example, initial velocity and position uniquely specify the motion of a projectile given the forces acting on the projectile, and similarly for fluid dynamics. In addition, it is well known that the initial conditions on well-behaved differential equations in conjunction with a known forcing function uniquely specify the solution of the equations. The ubiquitous differential equation model (such as equations 1.6 or 1.12) suggests that large classes of engineering systems have important internal variables, termed states. States represent the effect of all past excitations and are fundamental in determining the future evolution of the system.

A simple illustration drives the concept home. Consider the parallel *RL* circuit of Figure 1.4. The complete response for $t \geqslant 0$ to a step input $i_s(t) = 1^+(t)$, has the form

$$i_L(t) = i_L(0) \exp[-(R/L)t] + (1 - \exp[-(R/L)t]) \qquad (1.20)$$

Thus, a known value of $i_L(0)$ will uniquely specify the current $i_L(t)$; otherwise $i_L(t)$ is indeterminate.

Definition 1.7. The *state* of a system at a time t_0 is the minimum set of internal variables which is sufficient to uniquely specify the system outputs given the input signal over $[t_0,\infty)$.

Very often there are natural choices for states. Equally as often, however, the choice is arbitrary. For example, instead of voltages and currents in a circuit, one can choose scattering variables (various sums and differences of voltages and currents) to obtain an equivalent state of the system. More about this in Chapter 4. This flexibility leaves the designer with broad freedom.

Earlier, definitions of linearity, time invariance, causality, etc., were set forth for signals defined over the infinite time interval $(-\infty,\infty)$. Now, consideration of signals over $[t_0,\infty)$ requires some minor modifications of these definitions. A system is *linear* over $[t_0,\infty)$ if, whenever the system is relaxed at t_0 (no nonzero initial conditions), the response to a linear combination of input signals defined over $[t_0,\infty)$ is the same linear combination of the responses over that interval. An equivalent formulation of this linearity condition is as follows [10]:

Definition 1.8. Let $y = N[u,x(t_0)]$ be the response of a system to the input signal $u(\cdot)$ defined over $[t_0,\infty)$ with initial state $x(t_0)\epsilon\mathbf{R}^n$. Then the system is *linear* if and only if for any two admissible input signals u_1 and u_2 and any scalar k, $k(N[u_1,x(t_0)] - N[u_2,x(t_0)]) = N[k(u_1 - u_2),\theta(t_0)]$ for all initial state vectors $x(t_0)\epsilon\mathbf{R}^n$ where $\theta(t_0)$ is the zero vector.

It follows immediately from this definition that the response of a linear system is the sum of the response to the initial state, $x(t_0)$, and the response to the total input u. To see this, let $u_2 = 0$, and set $u_1 = u$, and $k = 1$. Then $N[u_1,x(t_0)] = N[0,x(t_0)] + N[u_1,\theta(t_0)]$.

The notion of time invariance needs a similar modification. Let $y(\cdot)$ be the response to $u(\cdot)$ over the interval $[t_0,\infty)$ when the initial state at t_0 is $\theta(t_0)$ — i.e., the system is relaxed. Then the system is *time-invariant* if the response to $u(\cdot - T)$ is $y(\cdot - T)$ over $[t_0 + T,\infty)$ given that $x(t_0 + T) = \theta(t_0)$ for every T.

If we define the delay system $N_T:U \to Y$ as

$$N_T[u(\cdot),\theta(t_0)] = u(\cdot - T) \tag{1.21}$$

so that N_T merely delays the input by T units, then, again following [10], a more formal definition is possible.

Definition 1.9. A system is *time-invariant* if for all $t \geqslant t_0$, there exists an initial state $x_1 \epsilon \mathbf{R}^n$ such that

$$N_T[N[u,x(t_0)],\theta(t_0)] = N[N_T[u,\theta(t_0)],x_1(t_0 + T)] \tag{1.22}$$

To understand this definition better, suppose that the step response to the *RL* circuit of Figure 1.4 is given by equation 1.20 for $t \geqslant 0$ as

$$N[1^+(\cdot),i_L(0)] = i_L(t) = K\exp[-(R/L)t]1^+(t)$$
$$+ (1 - \exp[-(R/L)t]1^+(t)$$

where K is the numerical value of $A = i_L(0)$. Then

$$N_T(N[1^+(\cdot),i_L(0)]) = K\exp[-(R/L)(t-T)]1^+(t-T)$$
$$+ (1 - \exp[-(R/L)(t-T)])1^+(t-T))$$

which is defined over the interval $[t_0 + T,\infty)$. It is straightforward to show that this expression coincides with $N[N_T[1^+(t),\theta(t_0)],i_L(t_0 + T) = A]$.

The notion of a state suggests that the models used to represent physical systems should explicitly reflect this internal behavior. In the early 1950s this awareness became compelling, and researchers formally introduced the state model representation of a physical system. A classic text is [12]. Since state models dominate the stage for this text, the various notations are now set forth.

The standard notation for the nonlinear state model is given by

$$\dot{x} = f(x,u,t)$$
$$y = g(x,u,t) \tag{1.23}$$

Here, the state vector, x, again denotes a vector of important physical variables such as currents and voltages in a circuit or position and velocity for the dynamics of projectile motion. Equations 1.23 says that the derivative $\dot{x}$ of the state vector (i.e., the change in position, velocity, or acceleration) is functionally dependent (via the function f) on the value of the state vector x, the input vector u, and the present time. Also the output of the system is functionally dependent (via g) on x, u, and t. The explicit writing of t in f and g is to emphasize the possibility that f and g have functional forms dependent on t. That is, f and g may be time-varying.

The nonlinear matrix pendulum equations

$$\begin{bmatrix} \dot{x}_1 \\ \dot{x}_2 \end{bmatrix} = \begin{bmatrix} x_2 \\ -4\sin(x_1) \end{bmatrix} + \begin{bmatrix} 0 \\ 1 \end{bmatrix} u \tag{1.24}$$

and

$$\begin{bmatrix} y_1 \\ y_2 \end{bmatrix} = \begin{bmatrix} x_1 \\ x_2 \end{bmatrix}$$

exemplify a nonlinear state model whose structure does not depend on time. Here, x_1 represents angular position and x_2 angular velocity.

Nonlinear system state models whose structure is time-independent have the form

$$\dot{x} = f(x,u)$$
$$y = g(x,u) \tag{1.25}$$

and the equations of motion of the pendulum fit more snugly into this form. The notation for the *linear time-invariant* state model is

$$\dot{x} = Ax + Bu$$
$$y = Cx + Du \tag{1.26}$$

where x, u, and y are vectors and A, B, C, and D are appropriately dimensioned constant matrices. The form of the *time-varying* state model is

$$\dot{x}(t) = A(t)x(t) + B(t)u(t)$$
$$y(t) = C(t)x(t) + D(t)u(t) \tag{1.27}$$

where $A(\cdot)$, $B(\cdot)$, $C(\cdot)$, and $D(\cdot)$ may now be time-dependent matrices. An example of a linear varying circuit is given in Figure 1.8.

The discrete-time time-varying state model is given by the system

$$x(k+1) = A(k)x(k) + B(k)u(k)$$
$$y(k) = C(k)x(k) + D(k)u(k) \tag{1.28}$$

where k is shorthand for $k\Delta t$. In these equations, the value of the state at the $(k+1)$st time instant (i.e., $x(k+1)$) depends on the past values of the state and input via the time-dependent matrices $A(k)$ and $B(k)$, and similarly for the output sequence $y(k)$. Note that k could just as well designate location or some other ordered discrete set of variables; time, however, is the typical underlying variable.

Finally, the form of the time-invariant discrete-time state model is simply equation 1.28 with the various matrices set to constants, i.e.,

$$x(k + 1) = Ax(k) + Bu(k)$$
$$y(k) = Cx(k) + Du(k) \tag{1.29}$$

Chapter 9 fully develops and explains the discrete-time state model.

CONCLUDING REMARKS

This chapter has introduced a number of basic system concepts and notations. The notions of system, state, linearity, time invariance, causality, etc., all introduced here play important roles in the chapters ahead as the concept of the state model, its theory, and numerical implementation are developed.

Chapter 2 provides some motivation for using the state model as opposed to high order scalar differential equation models. Some of its advantages such as easy handling of initial conditions are pointed out. Chapter 3 formally presents the linear, time-varying state model and illustrates the construction of a state model. In chapter 4, the relationship between differential equations and state models is explored and problems associated with determining and setting up initial conditions motivate the ideas of controllability and observability as glimpsed in chapter 5. Chapters 6 through 8 develop the numerical solution to the state equations and succeeding chapters explore the analytic solution and the notions of controllability, observability, and stability in the context of the analytic solution. Numerical methods pertinent to these studies are presented at various points in the text.

PROBLEMS

1. For a linear system characterized by an operator L, show that the output $y(\cdot) \equiv 0$ if the input $u(\cdot) \equiv 0$, i.e., the identically zero input function produces an identically zero output function.

2. (Causality and linearity) Using the result of problem 1 and definition 1.4, prove that a linear system is causal if and only if, whenever an input $u(t) = 0$ for $t \leqslant T$, the resultant output is zero for $t \leqslant T$. This is called zero-input zero-output causality.

3. Consider the block diagram of Figure P1.3. Suppose the delay block has an output which is identical to the input delayed by a time units.

(i) The system is

    (a) causal   **(d)** distributed

    (b) noncausal   **(e)** time-invariant

    (c) lumped   **(f)** time-varying

(ii) A state model for this system is:

    (a) $\dot{x}(t) = ax(t) + u(t)$
         $y(t) = x(t) + u(t)$

    (b) $\dot{x}(t) = -ax(t) + u(t)$
         $y(t) = x(t) + u(t)$

    (c) $\dot{x}(t) = -x(t + a) + u(t)$
         $y(t) = x(t) + u(t)$

    (d) $\dot{x}(t) = -x(t - a) + u(t)$
         $y(t) = -x(t - a) + u(t)$

    (e) $\dot{x}(t) = -x(t - a) + u(t)$
         $y(t) = x(t) + u(t)$

    (f) none of the above

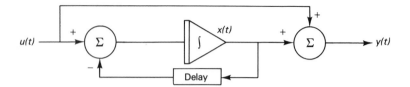

Figure P1.3   Block diagram for problem 3.

4. Given that V_i is the input and V_o the output, describe the circuit shown in Figure P1.4. (That is, is it linear or nonlinear, etc...?)

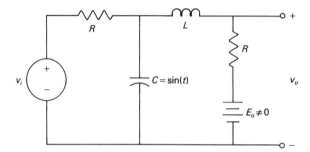

Figure P1.4   Circuit for problem 4.

5. Consider a system represented by an operator $W:U \rightarrow Y$ such that for any u_1 and u_2 for which $u_1 u_2 \equiv 0$, $W(u_1 + u_2) = Wu_1 + Wu_2$. A system satisfying this property is said to be *weakly additive* [13]. Show that for a weakly additive system, zero-input zero-output causality (see problem 2) is equivalent to definition 1.4.

6. (Linearity of ordinary differential equations) Assuming that all initial conditions are zero, show, without using frequency domain techniques, that equation 1.12 is linear. (*Hint*: Let y_1 be the response to u_1 and y_2 the response to u_2, and note that a response is valid if and only if it satisfies the differential equation.)

7. (Linearity of the operation of differentiation) If y is the response to the input u for the system described by equation 1.12, derive the response to the derivative of u, i.e., to $\dot{u}$. Do not use frequency domain techniques and assume that all initial conditions are zero. What problems occur if the initial conditions are not zero?

8. (Time invariance of ordinary differential equations) If $y(t)$ is the response to $u(t)$ for the system described by equation 1.12, then, given a real number T, what is the response to $u(t - T)$? Assume that all initial conditions are zero and do not use frequency domain techniques. Is such a result valid if the initial conditions are not zero? Explain.

9. Determine whether or not the following systems are causal. Explain.
   (a) $\dot{x}(t + T) = f(x(t), u(t), t)$
   $\quad\quad y(t) = g(x(t), u(t), t)$
   (b) $\dot{x}(t) = f(x(t + T), u(t), t)$
   $\quad\quad y(t) = g(x(t), u(t), t)$

10. Suppose a physical system is represented by the following system of equations, where $u(t)$ is the input and $y(t)$ is the output:

$$\begin{bmatrix} \dot{x}_1(t) \\ \dot{x}_2(t) \end{bmatrix} = \begin{bmatrix} x_1(t-1) + |u(t)| \\ x_1(t) \end{bmatrix}$$

$$y(t) = x_1(t + 1) + x_2(t) - u(t)$$

Circle *all* appropriate descriptions of the system:

| Linear | Time invariant | Causal | Lumped | Continuous time |
| Not linear | Not time-invariant | Noncausal | Distributed | Discrete time |

11. If $y(t) = 2e^{-t}\cos(3t + 1)$ is the response of a relaxed time invariant linear system to the impulsive input $u(t) = 0.5\delta(t)$, what is the response to $u(t) = \delta(t + 1)$?

12. The response of a linear system (which is initially relaxed) to a step input (designated $1^+(t)$) is $y(t) = (1 - e^{-t})1^+(t)$. What is the response to the input $u(t) = 2r(t)$, where $r(t)$ is the ramp function, i.e., $r(t) = t$ for $t \geqslant 0$ and 0 elsewhere?

13. The response of a linear time-invariant causal system to the input signal $u_1(t)$ *is* $y_1(t) = \exp(-t)1^+(t)$. The response to a second admissible input signal $u_2(t)$ *is* $y_2(t) = \cos(2t)1^+(t)$. Compute the response to the signal

$$u_3(t) = 2u_1(t) + \frac{du_2}{dt}(t + 1)$$

14. A causal time-invariant system has step response $(1 - e^{-t})1^+(t)$. Compute the response to the input signal $u(t) = 1^+(t - 1)1^+(2 - t)$ for $t \leqslant 2$.

15. The transfer function of a linear, time-invariant, causal, lumped, single-input single output (SISO) system is

$$H(s) = \frac{s^2 + 1}{s^2 + 2s - 1}$$

Construct the associated differential equation in the output $y(t)$ and input $u(t)$.

16. The response of a linear, time-invariant, causal system to the pulse $u_1(t) = 1^+(t)1^+(2 - t)$ is

$$y(t) = [1 - e^{-t}]1^+(t)1^+(2 - t) + e^{-(t-2)}1^+(t - 2).$$

Find the response of the system to the input $u_2(t) = (t - 2)1^+(t - 2)1^+(4 - t)$ for $t \leqslant 4$.

17. Suppose a linear time-invariant system is connected in series with a linear time-varying system.

(a) Is the concatenated system still linear?

(b) If the order were reversed, would the new system have the same responses as the old? If your answer is "no," give an example.

(*Hint on b*: Consider a system composed of a differentiator (the output is the derivative of the input) followed by a system whose output is the multiplication (scaling) of the input by a time dependent differentiable function $\alpha(t)$.)

REFERENCES

1. N. H. McClamroch, *State Models of Dynamic Systems: A Case Study Approach* (New York: Springer-Verlag, 1981).

2. Bernard Friedland, *Control Systems Design: An Introduction to State Space Methods* (New York: McGraw Hill, 1986).

3. David G. Luenberger, *Introduction to Dynamical Systems: Theory, Models, and Applications* (New York: Wiley, 1979).

4. Clare D. McGillem and George R. Cooper, *Continuous and Discrete Signal and System Analysis* (New York: Holt, Rinehart and Winston, 1974).

5. R. DeCarlo and R. Saeks, *Interconnected Dynamical Systems* (New York: Marcel Dekker, 1982).

6. Thomas Kailath, *Linear Systems* (Englewood Cliffs, NJ: Prentice Hall, 1980).

7. R. Saeks, "Causality in Hilbert Space," *SIAM Review*, Vol. 12, No. 3, July 1970, pp. 357–383.

8. A. Feintuch and R. Saeks, *System Theory: A Hilbert Space Approach* (New York: Academic Press, 1982).

9. R. M. Desantis, R. Saeks, and L. Tung, "Basic Optimal Estimation and Control Problems in Hilbert Space," *Math System Theory*, Vol. 12, 1978, pp. 175–203.

10. R. W. Newcomb, *Concepts of Linear Systems and Controls* (Belmont, CA: Brooks/Cole, 1968).

11. A. Oppenheim and A. Willsky, *Signals and Systems* (Englewood Cliffs, NJ: Prentice Hall, 1983).

12. L. A. Zadeh and C. A. Desoer, *Linear System Theory: The State Space Approach* (New York: McGraw Hill, 1963).

13. R. DeCarlo and R. Saeks, "Representation of Weakly Additive Operators," *Proceedings of the American Mathematical Society*, Vol. 59, No. 1, August 1976, pp. 55–61.

14. R. A. Gabel and R. A. Roberts, *Signals and Linear Systems* (New York: Wiley, 1974).

2

STATE MODELS

Motivation and Overview

INTRODUCTION

This chapter briefly introduces the concept of a state model. A comparison of a second-order scalar differential equation and a first-order matrix differential equation model of a series RLC circuit serve to motivate the use of the state model for analysis purposes. The exposition provides few details, just enough to whet the appetite, hopefully. More complete explanations and fuller discussions follow in later chapters.

Three specific objectives underlie the presentation of a state model in this chapter. The first is to describe two mathematical models of a series RLC circuit, and the second is to show that one of these, the state model, is a superior analytical tool. The third and more subtle goal is to demonstrate that each of the two models induces a different perception of the series RLC circuit.

To reinforce the importance of this third goal, consider the oblique view of a room as sketched in Figure 2.1. How does the drawing's perspective influence "what you see" and "what you imagine the room to be like"? In point of fact, the way the room is drawn induces a perspective in the viewer's mind, and this perspective shades or biases the information conveyed by the sketch.

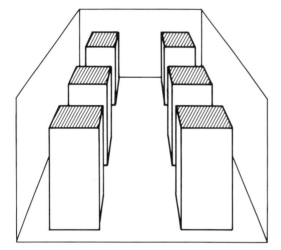

Figure 2.1   Side view of a room.

Figure 2.2 presents a top view of the same room. What perspective does this top view induce in your imagination? How do the top and side views differ? Which view do you prefer and why? Does one provide more information than the other? Certainly, such questions as these indicate that an observer will perceive a room in a way dependent upon his or her vantage point. This fact is most important and very subtle.

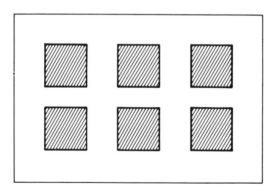

Figure 2.2   Top view of room depicted in Figure 2.1.

In a manner similar to the way the room is presented, a mathematical model of a physical system induces a particular perspective also. One can liken it to the way that the focus, aperture, exposure, and angle of a camera influence the quality of a snapshot. In fact the vantage point set up by a mathematical model is far more consequential than the actual use of the model in analysis and design. A major part of the discussion in this chapter deals with the perspectives induced by the scalar and matrix differential equation models of the series *RLC* circuit. Both of these models contain information adequate for an engineering interpretation of the physical behavior of the circuit; however, the matrix model provides a more complete description of the physics of the circuit and displays this information differently. In fact, the solution of the matrix differential equation is conceptually simpler than the solution of the scalar differential equation.

BASIC DIFFERENTIAL EQUATIONS — A REVIEW

The form of a *first-order scalar homogeneous differential equation* is

$$\dot{x}(t) - \lambda x(t) = 0, \; x(0) = x_0 \tag{2.1}$$

where $x(0) = x_0$ is the given initial condition, λ is a parameter arising from physical considerations, and $\dot{x}(t)$ is the time derivative of $x(\cdot)$ evaluated at time t. The solution of this equation takes the special form

$$x(t) = e^{\lambda t} x_0 = \left[1 + \lambda t + \frac{(\lambda t)^2}{2!} + \frac{(\lambda t)^3}{3!} + \cdots \right] x_0 \tag{2.2}$$

where the series in brackets is the Taylor series of $e^{\lambda t}$.

This differential equation is rather simple with a rather simple solution. It generalizes to an equation where λ is a matrix and $x(t)$ is a vector of interesting variables, for example, capacitor voltages and inductor currents in circuit theory. If λ becomes a matrix A, then the solution has the same form as the scalar case; i.e., $x(t) = e^{At} x_0$, where

$$e^{At} = \exp(At) = I + At + \frac{(At)^2}{2!} + \frac{(At)^3}{3!} + \cdots \tag{2.3}$$

with I the identity matrix. This form is conceptually elegant in that the scalar case becomes a special form of the matrix case.

The *homogeneous second-order scalar differential equation* has the form

$$\frac{d^2 i}{dt^2}(t) + \frac{R}{L} \frac{di}{dt}(t) + \frac{1}{LC} i(t) = 0 \tag{2.4}$$

To uniquely specify a solution to this differential equation, both $i(0)$ and its derivative $\dot{i}(0)$ are needed; otherwise the solution remains indeterminate. For most cases the solution has the form

$$i(t) = \alpha_1 e^{\lambda_1 t} + \alpha_2 e^{\lambda_2 t} \tag{2.5}$$

where α_1 and α_2 are constants that depend on the initial conditions and λ_1 and λ_2 are constants that depend on the physical parameters R, L, and C.

To intuitively derive this solution, let us postulate a solution of the form $i(t) = \alpha e^{\lambda t}$, motivated by our experience with the first-order case. Plugging this into equation 2.4 produces

$$\alpha e^{\lambda t} \left[\lambda^2 + \frac{R}{L} \lambda + \frac{1}{LC} \right] = 0 \tag{2.6}$$

Equation 2.6 suggests two possible solutions. The first is $\alpha e^{\lambda t} = 0$, which is uninteresting since it means that somehow our physical system in not functioning. The second possibility,

$$\lambda^2 + \frac{R}{L} \lambda + \frac{1}{LC} = 0 \tag{2.7}$$

is more interesting. The quadratic nature of this equation implies that there are two possible values the variable λ can assume. Solving the equation with the quadratic formula yields

$$\lambda_1 = -\frac{R}{2L} + \sqrt{(\frac{R}{2L})^2 - (\frac{1}{LC})}$$

$$\lambda_2 = -\frac{R}{2L} - \sqrt{(\frac{R}{2L})^2 - (\frac{1}{LC})}$$

(2.8)

Hence, provided that $\lambda_1 \neq \lambda_2$, two independent solutions of equation 2.4 of the form $\alpha_i e^{\lambda_i t}$ always exist [1]. Moreover, since equation 2.4 is a linear equation, the sum of any two solutions is always a solution. Consequently, the solution form given in equation 2.5 is valid whenever $\lambda_1 \neq \lambda_2$.

Let us consider more closely the solution of the two types of differential equations just introduced. Clearly, the solution of the second-order differential equation is less intuitive and more difficult to understand than the solution of the first-order scalar equation. This is because the second-order equation permits more varied forms of interaction as indicated by its mathematical solution: the sum of two independent first-order solutions, i.e., the sum of two exponentials. Also, the second-order differential equation submerges the effect of the initial conditions into the constants α_1 and α_2, whereas the effect of the initial conditions is explicit in the first-order solution.

The more varied type of behavior of the solution of the second-order differential equation is a microcosm upon which more complicated behaviors (higher order behaviors) build. In fact, for all practical purposes, it is possible to view more complicated behaviors as weighted sums of second-order behaviors. However, a first-order matrix differential equation offers the best of these worlds and more.

THE SERIES *RLC* CIRCUIT

With the preceding background in differential equations under our belts, we can profitably consider the structure and physics of the series *RLC* circuit. Figure 2.3 shows such a circuit. Three elements, a resistor with resistance R, a capacitor with capacitance C, and an inductor with inductance L, are connected in this circuit. Associated with each element is a voltage and a current with the indicated polari-

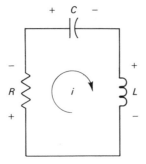

Figure 2.3   Schematic of a series *RLC* circuit.

ties and directions. The voltage-current relationship for each element is typically called the component dynamics of that element. Ohms law characterizes the voltage-current relationship for the resistor:

$$v_R = R\, i_R \tag{2.9}$$

For the capacitor the dynamical equation is

$$v_C(t) = \frac{1}{C} \int_0^t i_C(\tau)\, d\tau + v_C(0) \tag{2.10}$$

and for the inductor

$$v_L(t) = L \frac{di_L}{dt}(t) \tag{2.11}$$

In addition to these element equations, there are two conservation laws governing the interaction of the voltages and currents in the circuit. The first of these is Kirchhoff's voltage law, or, simply, KVL. Conceptually, it is analogous to the classical notions of conservation of energy and conservation of momentum. The KVL states that the sum of the voltages around a closed loop is zero,

$$v_R + v_C + v_L = 0 \tag{2.12}$$

Similarly, there is a Kirchhoff's current law, or, simply, KCL. This conservation law requires that the sum of the currents entering a node be zero:

$$i_R = i_C = i_L \tag{2.13}$$

A SECOND ORDER SCALAR DIFFERENTIAL EQUATION MODEL

Substitutuing equations 2.9, 2.10, 2.11, and 2.13 into equation 2.12 yields

$$L \frac{di_L}{dt} + R\, i_L + \frac{1}{C} \int_0^t i_L(\tau)\, d\tau + v_C(0) = 0. \tag{2.14}$$

This is an integro-differential equation since it involves both derivatives and integrals of the unknown function i_L. Taking the derivative and dividing by L produces the second-order scalar differential equation model of the series RLC circuit:

$$\frac{d^2 i_L}{dt^2} + \frac{R}{L} \frac{di_L}{dt} + \frac{1}{LC} i_L = 0 \tag{2.15}$$

The initial conditions of the differential equation are $i_L(0) = i_0$ and $\dot{i}_L(0) = f(i_0, v_0)$, where $v_0 = v_C(0)$ and the function $f(\cdot, \cdot)$ indicates that $\dot{i}_L(0)$ is a function of the initial inductor current and the initial capacitor voltage. These initial conditions permit us to specify a unique solution to the differential equation.

Recall that for most cases the solution of this differential equation is given by equation 2.5, i.e.,

$$i_L(t) = \alpha_1 e^{\lambda_1 t} + \alpha_2 e^{\lambda_2 t} \tag{2.16}$$

where we assume that $\lambda_1 \neq \lambda_2$ and that these constants are given by equation 2.8.

Now the constants α_1 and α_2 depend on the initial conditions. More specifically,

$$i_0 = i_L(0) = \left[\alpha_1 e^{\lambda_1 t} + \alpha_2 e^{\lambda_2 t} \right]_{t=0} \tag{2.17}$$

which implies that

$$i_0 = \alpha_1 + \alpha_2 \tag{2.18}$$

Similarly,

$$v_0 = v_C(0) = - v_R(0) - v_L(0) = - Ri_0 - L \frac{di_L}{dt}(0) \tag{2.19}$$

which, after appropriate substitutions and simplifications, produces

$$v_0 = - (R + L\lambda_1)\,\alpha_1 - (R + L\lambda_2)\,\alpha_2 \tag{2.20}$$

Finding α_1 and α_2, then requires solving equations 2.18 and 2.20 simultaneously, which is equivalent to solving the matrix equation

$$\begin{bmatrix} i_0 \\ v_0 \end{bmatrix} = \begin{bmatrix} 1 & 1 \\ -(R+L\lambda_1) & -(R+L\lambda_2) \end{bmatrix} \begin{bmatrix} \alpha_1 \\ \alpha_2 \end{bmatrix} \tag{2.21}$$

by Gaussian elimination (see chapter 6) or some other technique. The resulting solution is

$$\begin{aligned} \alpha_1 &= - \frac{R+L\lambda_2}{L(\lambda_1-\lambda_2)}\, i_0 - \frac{1}{L(\lambda_1-\lambda_2)}\, v_0 \\[2mm] \alpha_2 &= \frac{R+L\lambda_1}{L(\lambda_1-\lambda_2)}\, i_0 + \frac{1}{L(\lambda_1-\lambda_2)}\, v_0 \end{aligned} \tag{2.22}$$

Some remarks are now in order. The preceding derivation is rather tedious and time-consuming. Moreover, the large amount of manipulation tends to hide the physical interactions of the circuit elements. Certainly, the mathematical solution in conjunction with some physical reasoning about the circuit will provide an adequate picture of the behavior of the circuit. The point, however, is that the second-order scalar differential equation model does not adequately reflect this behavior. Hence, we may conclude that the derivation and result are unintuitive and not "cost-effective." The latter is true because the capacitor voltage v_C is not given explicitly; further manipulations are needed to obtain an expression for it.

THE FIRST-ORDER MATRIX EQUATION MODEL

We now turn to the first-order vector-matrix equation model of the series RLC circuit. Motivation for the use of matrices in analysis and design stems from the need to handle large numbers of equations in an organized, efficient manner. In our case only two variables are of prime importance: the capacitor voltage and the inductor current. Knowledge of these two parameters allows one to compute all other voltages and currents in the circuit. What, then, is the equation form we are searching for? We begin by defining $x(t)$ and its derivative $\dot{x}(t)$ as the vectors

$$x(t) = \begin{bmatrix} v_C(t) \\ i_L(t) \end{bmatrix} \text{ and } \dot{x}(t) = \begin{bmatrix} \dot{v}_C(t) \\ \dot{i}_L(t) \end{bmatrix} \tag{2.23}$$

The first-order matrix equation model will have the form

$$\dot{x}(t) - Ax(t) = 0, \quad x(0) = x_0. \tag{2.24}$$

where A is a 2×2 matrix and zero is the zero vector. More detailedly, the model will have the form

$$\begin{bmatrix} \dot{v}_C(t) \\ \dot{i}_L(t) \end{bmatrix} - \begin{bmatrix} a_{11} & a_{12} \\ a_{21} & a_{22} \end{bmatrix} \begin{bmatrix} v_C(t) \\ i_L(t) \end{bmatrix} = \begin{bmatrix} 0 \\ 0 \end{bmatrix} \tag{2.25}$$

with initial condition vector

$$x_0 = \begin{bmatrix} v_C(0) \\ i_L(0) \end{bmatrix} = \begin{bmatrix} v_0 \\ i_0 \end{bmatrix} \tag{2.26}$$

A model of this form is called a *homogeneous state model*, homogeneous because there are no forcing functions or inputs. The quantity $x(t)$ is known as a *state vector*, and $v_C(t)$ and $i_L(t)$ are called *state variables*.

Physical processes involving the motion of particles and objects (in this case voltages and currents in a circuit) generally have a behavior described by a differential equation. The variables of the equation are intrinsic to the physical process. They refer to physical quantities we can measure: things that we can understand and feel, so to speak. Generally, the word *state* refers to a *set of physical properties* and the word *state variable* to a *specific parameter* which is *part of the state*. So in the case under consideration, $x(t)$ is the state vector and the state variables are the capacitor voltage and the inductor current. All other voltages and currents in the circuit are determined by these state variables.

The task is to find constants a_{11}, a_{12}, a_{21}, and a_{22} for which the first-order matrix equation 2.25 holds given the initial condition vector 2.26. Note that this requires finding expressions for both the derivative of the capacitor voltage and the derivative of the inductor current in terms of both the capacitor voltage and the inductor current.

To construct the necessary matrix A of equation 2.24, observe that

$$\dot{v}_C(t) = \frac{1}{C} i_C(t) = \frac{1}{C} i_L(t) \tag{2.27}$$

In addition,

$$\dot{i}_L(t) = \frac{1}{L} v_L(t) = -\frac{1}{L} v_C(t) - \frac{R}{L} i_L(t) \tag{2.28}$$

Hence,

$$\begin{bmatrix} \dot{v}_C(t) \\ \dot{i}_L(t) \end{bmatrix} = \begin{bmatrix} 0 & \dfrac{1}{C} \\ -\dfrac{1}{L} & -\dfrac{R}{L} \end{bmatrix} \begin{bmatrix} v_C(t) \\ i_L(t) \end{bmatrix} \tag{2.29}$$

This is the first-order matrix differential equation model of the series RLC circuit having no sources. The crucial question is, Why did we seek such a model? To answer this question, recall that throughout the development we have hinted at various philosophical and physically motivated reasons for constructing a matrix equation model. However, the tools of analysis and design are mathematical tools, and the form of the solution of this first-order matrix equation is an exact replica of the solution of the first-order scalar equation. Indeed, looking back, we see that the scalar form is simply a special case of the matrix form, so the generalization is very intuitive. More importantly, the mathematics is consistent; in particular

$$x(t) = e^{At} x_0 = [I + At + \frac{(At)^2}{2!} + \frac{(At)^3}{3!} + \cdots] x_0 \qquad (2.30)$$

To see that this is in fact the solution to equation 2.24 consider that

$$\dot{x}(t) = [A + A^2 t + A \frac{(At)^2}{2!} + \cdots] x_0 \qquad (2.31)$$

which is obtained by termwise differentiation of the series in equation 2.30. Clearly, then,

$$\dot{x}(t) = A e^{At} x_0 \qquad (2.32)$$

so that $\dot{x}(t) - Ax(t) = 0$, showing that the solution form 2.30 satisfies equation 2.24.

   The actual solution of equation 2.29 (the matrix model of the series RLC circuit) is

$$\begin{bmatrix} v_C(t) \\ i_L(t) \end{bmatrix} = \frac{1}{L(\lambda_1 - \lambda_2)} \begin{bmatrix} R + L\lambda_1 & -(R + L\lambda_2) \\ -1 & 1 \end{bmatrix} \qquad (2.33)$$
$$\times \begin{bmatrix} e^{\lambda_1 t} & 0 \\ 0 & e^{\lambda_2 t} \end{bmatrix} \begin{bmatrix} 1 & R + L\lambda_2 \\ 1 & R + L\lambda_1 \end{bmatrix} \begin{bmatrix} v_0 \\ i_0 \end{bmatrix}$$

where λ_1 and λ_2 are as before and arise naturally in the matrix context according to the following discussion. Consider

$$\det(\lambda I - A) = \begin{vmatrix} \begin{bmatrix} \lambda & 0 \\ 0 & \lambda \end{bmatrix} - \begin{bmatrix} 0 & \frac{1}{C} \\ -\frac{1}{L} & -\frac{R}{L} \end{bmatrix} \end{vmatrix} = \lambda^2 + \frac{R}{L}\lambda + \frac{1}{LC} \qquad (2.34)$$

This equation is known as the *characteristic equation* of both the matrix A and the circuit. The zeros of this polynomial are λ_1 and λ_2 and are given by equation 2.8. The quantities λ_1 and λ_2 are called the *natural frequencies* of the RLC circuit and the *eigenvalues* of the matrix A. They identify the physical behavior of the circuit or physical system just as a fingerprint identifies a person. The theory behind these computations follows in a later chapter.

DISCUSSION AND SUMMARY

As indicated throughout the discussion of the first-order matrix equation model, the initial condition vector, $[v_0, i_0]^t$ is explicit in the matrix model itself and the solution vector is an explicit function of this initial condition vector. This is in direct contrast to the second-order scalar model.

Recall also that in computing the solution to the scalar model, a large amount of algebraic manipulation was required to incorporate the initial condition information into the solution form. No such algebra is necessary in the first-order matrix model. However, computing the matrix exponential solution for the more general matrix models is not always clear cut, although the matrix formulation does provide values for all state variables.

Notice also that the matrix solution provides simultaneous expressions for both $v_C(t)$ and $i_L(t)$, in contrast to the solution of the scalar model. Hence, more information is displayed in the matrix model since it more completely reflects the physical behavior of the circuit. In point of fact, the underlying perspective of the matrix model is far more encompassing: it opens a door to a new world of analysis and design. Of course, there is always a tradeoff. In this case we have traded conceptual simplicity and a more complete description of the physical behavior of the circuit for some difficulty in computing the matrix e^{At}. The good news is that such a matrix formulation is amenable to efficient numerical techniques [3], making the matrix model a superior analytical tool.

PROBLEMS

1. Construct a scalar differential equation model for a current source in parallel with a parallel *RLC* circuit. What variable is germane to the model?

2. Construct a first-order matrix model for the circuit of problem 1. (*Hint*: the input I must be accounted for.)

3. Consider the scalar differential equation

$$2 \frac{d^2 i}{dt^2}(t) + 3\frac{di}{dt} - 9\,i(t) = 0$$

   (a) Find the natural frequencies of the equation.
   (b) Find two independent solutions of the equation.
   (c) Verify that the solutions in part *b* satisfy the equation.

4. (a) Show that the matrix

$$\Phi(t) = \begin{bmatrix} 1 & -2 \\ 1 & -1 \end{bmatrix} \begin{bmatrix} e^t & 0 \\ 0 & e^{-t} \end{bmatrix} \begin{bmatrix} -1 & 2 \\ -1 & 1 \end{bmatrix}$$

satisfies the matrix differential equation

$$\dot{M}(t) = A M(t), M(0) = I$$

where

$$A = \begin{bmatrix} -3 & 4 \\ -2 & 3 \end{bmatrix}$$

(b) For the matrix A of part a, show that a solution to $\dot{x}(t) = A x(t), x(0) = [1 \ 1]'$, is given by

$$x(t) = \Phi(t) \begin{bmatrix} 1 \\ 1 \end{bmatrix}$$

5. If

$$\begin{bmatrix} \dot{x}_1(t) \\ \dot{x}_2(t) \end{bmatrix} = \begin{bmatrix} -3 & 4 \\ -2 & 3 \end{bmatrix} \begin{bmatrix} x_1(t) \\ x_2(t) \end{bmatrix}$$

find the natural frequencies of the system represented by the equation.

6. Use the series definition of the matrix exponential to obtain a closed-form expression for $y(t)$ in the state model given by

$$\dot{x}(t) = \begin{bmatrix} 1 & 0 \\ 0 & -1 \end{bmatrix} x(t), \quad x(0) = \begin{bmatrix} 1 \\ 1 \end{bmatrix}$$

$$y(t) = [1 \ -1] \ x(t)$$

7. Using the Taylor series definition of the matrix exponential, find a closed-form expression for $\exp[At]$ when

$$A = \begin{bmatrix} 1 & 1 \\ 0 & 1 \end{bmatrix}$$

8. Generalize the result of problem 7 to find a closed-form expression $\exp(At)$ for the $n \times n$ matrix

$$A = \begin{bmatrix} \lambda & 1 & 0 & & & 0 & 0 \\ & & & \cdots & & & \\ 0 & \lambda & 1 & & & 0 & 0 \\ & & & \cdot & & & \\ & & & & \cdot & & \\ & & & & & \cdot & \\ 0 & 0 & 0 & \cdots & \cdots & \lambda & 1 \\ 0 & 0 & 0 & & & 0 & \lambda \end{bmatrix}$$

(*Hints*: Consider the 3 × 3 case first; consult [2].)

9. Show that if

$$A = \begin{bmatrix} A_1 & 0 \\ 0 & A_2 \end{bmatrix}$$

where A_1 is $n_1 \times n_1$ and A_2 is $n_2 \times n_2$, then

$$\exp[At] = \begin{bmatrix} \exp[A_1 t] & 0 \\ 0 & \exp[A_2 t] \end{bmatrix}$$

10. Construct an example which shows that if $A_1 A_2 \neq A_2 A_1$, then $\exp[A_1 t]\exp[A_2 t]$ $\neq \exp[(A_1 + A_2)t]$.

11. Prove that $\exp[A_1 t]\exp[A_2 t] = \exp[(A_1 + A_2)t]$ if and only if $A_1 A_2 = A_2 A_1$.

12. As a hint at the numerical difficulties associated with computing the matrix exponential, perform the following experiment. First compute

$$\exp \begin{bmatrix} 1 & 0 \\ 0 & -6 \end{bmatrix}$$

(i) "exactly"

(ii) approximately, by summing the first four terms in the Taylor series, and

(iii) approximately, by using the fact that $\exp(A) = [\exp(-A)]^{-1}$: i.e., sum the first four terms in the Taylor series of

$$\exp \begin{bmatrix} -1 & 0 \\ 0 & 6 \end{bmatrix}$$

and compute the inverse, which will give the desired approximation. Discuss your results.

REFERENCES

1. W. Hayt and J. Kemmerly, *Engineering Circuit Analysis* (New York: McGraw-Hill, 1978).

2. C. A. Desoer, *Notes for a Second Course on Linear Systems* (New York: Van Nostrand, 1970).

3. G. E. Forsythe, Michael A. Malcolm, and Cleve E. Moler, *Computer Methods for Mathematical Computation* (Englewood Cliffs, NJ: Prentice Hall, 1977).

3

LINEAR STATE MODELS FOR LUMPED SYSTEMS

THE LINEAR TIME-VARYING STATE MODEL

Intrinsic to a state model are three types of variables: input variables, state variables, and output variables, all generally vectors. The state model identifies the dynamics and interaction of these variables. Equation 1.9 first introduced the structure of a linear time-varying state model for lumped systems. Figure 3.1 mirrors this equation as a block diagram.

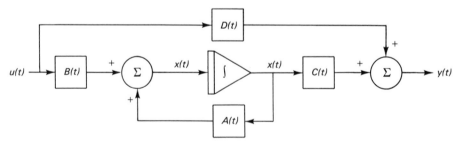

Figure 3.1 Block diagram of a linear, time-varying state model for lumped systems.

In the figure the input variables and the output variables characterize the interface between the physical system and the external world. The input $u(\cdot)$ reflects the excitations delivered to the physical system whereas the output $y(\cdot)$ reflects the signal vector returned to the external world. The block containing the matrix $B(\cdot)$ is an input filter or window; it defines the "filtered" interaction between the excitations of the external world and the state, or internal, variables of the system. Typically, state variables correspond to "meaningful" physical variables or possibly linear combinations of such variables. The matrix $A(\cdot)$ is a feedback matrix which defines the internal interaction among the states, whereas $C(\cdot)$ acts as an output filter or buffer between the measurable output signals $y(\cdot)$ and the internal state variables $x(\cdot)$. Finally, the block containing $D(\cdot)$ represents the feed-forward gains or straight-through connections. This intuitive description motivates the following mathematical definition:

Definition 3.1   A state model for a lumped linear system is a set $\{A(\cdot), B(\cdot), C(\cdot), D(\cdot)\}$ of four matrices which defines a first-order, degree-n vector differential equation

$$\begin{aligned} \dot{x}(t) &= A(t)x(t) + B(t)u(t) \\ y(t) &= C(t)x(t) + D(t)u(t) \end{aligned} \quad x(t_0) = x_0 \quad\quad (3.1)$$

where

(i) $x(t)\epsilon\mathbf{R}^n$ for each time t is the *state vector* (vector of state variables)

(ii) $u(t)\epsilon\mathbf{R}^m$ for each t is the system input vector

(iii) $y(t)\epsilon\mathbf{R}^r$ for each t is the system output vector

(iv) $A(\cdot), B(\cdot), C(\cdot)$, and $D(\cdot)$ are $n \times n$, $n \times m$, $r \times n$, and $r \times m$ matrices whose entries are piecewise-continuous real-valued functions of time

(v) $x(t_0)$ is the initial state vector or initial condition for equation 3.1.

The restriction to real-valued functions is convenient but unnecessary; all of the results (to be derived subsequently) extend directly to complex-valued functions.

Time-invariant state models are of course special cases of time-varying state models. They have the same form as equation 3.1 with the time dependence of A, B, C, and D deleted. Such a form was in fact introduced in equation 1.26. Much of this text will focus on the time-invariant model and its special properties.

CLASSIFICATION OF RESPONSES

Several types of solutions, or more commonly, system responses are pertinent: the zero-input (system) response, the zero-input state response, the zero-state (system) response, and the zero-state state response. The *zero-input (system) response* is the response $y(\cdot)$ given an initial condition $x(t_0)$ with the input $u(\cdot)$ set to zero. This is similar [1] to the *homogeneous solution* of the system of differential equations defined by equation 3.1. Similarly the *zero-input state response* $x(\cdot)$, results when $u(\cdot) \equiv 0$ given $x(t_o) = x_o$. The *zero-state system response* to an input $u(\cdot)$ is the function $y(\cdot)$ which results when $x(t_0) = \theta$, the zero vector. Finally the *zero-state state response* is $x(\cdot)$, when $x(t_o) = \theta$. The *complete response* is the sum of the zero-input response and the zero-state response. In this context, the solutions $x(\cdot)$ and $y(\cdot)$ of equation 3.1 for a given initial condition and input are the state trajectory and output trajectory, respectively. These trajectories describe the evolution with time of the internal and external system variables. As discussed in chapter 1, for each t_o, $x(t_o)$ sums up the totality of the interaction of the system with the external world for all t less than t_o. That is, future values of the state vector depend only on $x(t_o)$ and the input $u(t)$ for $t \geqslant t_o$. This remarkable property becomes more transparent after describing existence and uniqueness conditions regarding equation 3.1 in chapter 10.

CIRCUIT-RELATED EXAMPLES

To add depth to the preceding discussion, let us consider several illustrations of the construction of state models for simple circuits.

EXAMPLE 3.3

Write a set of state equations for the circuit of Figure 3.2 using v_R as the output and I as the input.

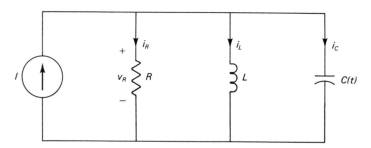

Figure 3.2   Schematic of parallel *RC* with time-varying capacitor.

Fabricating a state model for this circuit requires first choosing appropriate state variables. The model arises out of the relationship between the derivatives of the chosen state variables, the state variables themselves, and the external excitations. Let us choose v_C and i_L as state variables.

The dynamics of the time-varying capacitor has the current-voltage relationship

$$i_C(t) = \frac{d}{dt} [C(t)v_C(t)] = \dot{C}(t)v_C(t) + C(t)\dot{v}_C(t) .$$

Rearranging terms produces

$$\dot{v}_C(t) = \frac{1}{C(t)} [i_C(t) - \dot{C}(t) v_C(t)]$$

Here, the derivative of the capacitor voltage is dependent on the capacitor current and the capacitor voltage. The capacitor current is not a state variable; hence, it is necessary to find it in terms of I and the two designated state variables. To accomplish this, observe that the KCL and KVL for the circuit imply that

$$i_C = I - i_R - i_L = I - \frac{v_R}{R} - i_L = I - \frac{v_C}{R} - i_L$$

Hence,

$$\dot{v}_C(t) = \left[-\frac{1}{C(t)} \right] i_L(t) - \left[\frac{\frac{1}{R} + \dot{C}(t)}{C(t)} \right] v_C(t) + \left[\frac{1}{C(t)} \right] I \qquad (3.2)$$

which is an expression for the derivative of $v_C(t)$ in terms of the state variables $i_L(t)$ and $v_C(t)$, and the input I. An application of the KVL to the inductor dynamics produces

$$i_L(t) = \frac{1}{L} v_L(t) = \frac{1}{L} v_C(t) \qquad (3.3)$$

Writing equations 3.2 and 3.3 in matrix form yields

$$\begin{bmatrix} \dot{i}_L(t) \\ \dot{v}_C(t) \end{bmatrix} = \begin{bmatrix} 0 & \frac{1}{L} \\ -\frac{1}{C(t)} & -\frac{\frac{1}{R} + \dot{C}(t)}{C(t)} \end{bmatrix} \begin{bmatrix} i_L(t) \\ v_C(t) \end{bmatrix} + \begin{bmatrix} 0 \\ \frac{1}{C(t)} \end{bmatrix} I$$

$$v_R(t) = [0 \quad 1] \begin{bmatrix} i_L(t) \\ v_C(t) \end{bmatrix} + [0] I$$

The procedure used to construct the state model in example 3.1 was ad hoc. More complicated circuits demand a more systematic approach. The theory of graphs [2,3] provides the tools for achieving such an approach. The common graph-theoretic concepts are well known so the discussion that follows is brief and by no means complete.

A *branch* in a circuit graph represents a single circuit element. A *node* denotes a junction of two or more branches. Connecting branches together at nodes produces *loops*, which are closed connected paths having no self-intersections—i.e., each node in a loop has exactly two loop branches incident upon it. A *tree* is a set of branches without loops for which there exists a path from each node to every other node. A tree is a *connected subgraph*, any branch not contained in the tree is called a *link*, and the set of links defines the *cotree*. Given a connected graph, a set of branches forms a *cutset* if (*i*) the deletion of the branches (but not their end points) produces a disconnected graph, and (*ii*) reconnecting any one branch yields a connected graph. For a rigorous and expanded discussion of these terms see [2,3].

In order to write a set of state equations for a circuit, we must assume that the circuit has no loops of only independent voltage sources and no cutsets of only independent current sources. The presence of such loops or cutsets would in general violate the KVL and KCL, respectively. We draw a graph for the circuit and establish a tree. The tree should contain all the independent voltage sources and as many capacitors as possible, whereas the cotree should contain as many inductors as possible as well as the independent current sources. Such a tree is called a *normal*, or *proper*, tree. The independent voltage sources and independent current sources are the model inputs. We designate the capacitor voltages in the tree and the inductor currents in the cotree as the necessary state variables. A state model results by expressing the derivative of each of the designated state variables as a linear combination of the state variables and inputs. Putting these equations in matrix form and writing the desired output equation produces a state model for the circuit.

In order to express the derivative of each state variable as a linear combination of all the state variables, it is important to properly label each branch in the graph with appropriate voltage polarities and current directions. Mesh analysis and nodal analysis [3] provide a very effective and systematic approach to the task. Even these techniques, however, require some ingenuity in eliminating unwanted variables (voltages and/or currents which are not state variables) so as to produce a valid state model. Fortunately, numerical algorithms exist that lessen the amount of ingenuity required. These algorithms use the various incidence matrices and the individual element dynamics to construct a large sparse set of equations that describes the circuit topology and the physics of the circuit elements. By then applying reduction techniques to this set or a reordered set of equations, one arrives at a state model [4].

Figure 3.3 shows a tree (solid lines) and cotree (dashed lines) for the circuit of Figure 3.2. In example 3.2, the need for the graph theoretic approach is acute.

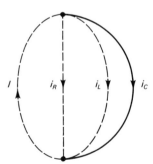

Figure 3.3 Graph of parallel *RLC* circuit with designated tree and cotree.

EXAMPLE 3.2.

Using graph theoretic techniques, construct a state model for the circuit in Figure 3.4 with v_{R_2} as the output.

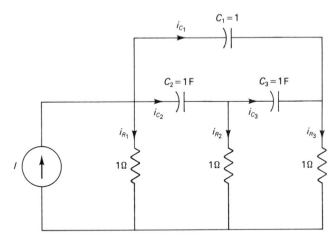

Figure 3.4 Circuit containing cutset of capacitors and one independent source.

Step 1. Draw a tree for the circuit. The tree chosen here is shown in Figure 3.5 below.

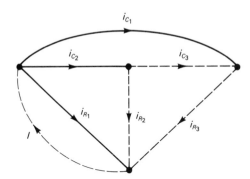

Figure 3.5 Graph with designated tree for circuit of Figure 3.4.

Step 2. Choose state variables. In accordance with the requirements of the algorithm previously mentioned, one chooses v_{C_1} and v_{C_2} as state variables as per the chosen tree. If one were to have picked all three capacitor voltages as state variables, then an invalid state model would have resulted because a state model is in fact a differential equation whose initial conditions are in general freely assignable. An arbitrary choice of capacitor voltages at, say, $t = 0$ would violate the KVL for the capacitor loop. Hence, permitting such a choice is mathematically inconsistent and physically impossible.

Step 3. Within the loop of capacitors, determine the relationship between the derivative of the state variables and the current through capacitor 3. To do so, observe that $v_{C_1} - v_{C_2} - v_{C_3} = 0$. Then, given that $C_1 = C_2 = C_3 = 1$, differentiating produces

$$\dot{v}_{C_1} - \dot{v}_{C_2} - i_{C_3} = 0 \qquad (3.4)$$

Step 4. Write equations for the derivatives of v_{C_1} and v_{C_2} in terms of the link currents. For this task, it is a simple matter to show that

$$\dot{v}_{C_1} + i_{C_3} - i_{R_3} = 0 \qquad (3.5a)$$

$$\dot{v}_{C_2} - i_{C_3} - i_{R_2} = 0 \qquad (3.5b)$$

Step 5. Write the cotree resistor currents in terms of the branch voltages. Ultimately, we will write all equations as a single matrix equation wherein the necessary manipulations to construct the state model become transparent. As such, the only variables to appear on the right side of an equals sign are state variables and inputs.

Observe that since $R_2 = 1$, and $R_3 = 1$,

$$i_{R_2} - v_{R_1} = -v_{C_2} \qquad (3.6a)$$

and

$$i_{R_3} - v_{R_1} = -v_{C_1} \qquad (3.6b)$$

Step 6. Write the resistor branch voltage v_{R_1}, in terms of the cotree current. Here, observe that $i_{R_1} = I - i_{R_2} - i_{R_3}$. Thus $R_1 = 1$,

$$v_{R_1} + i_{R_2} + i_{R_3} = I \qquad (3.7)$$

Step 7. Writing equations 3.4 through 3.7 in matrix form produces

$$
\begin{bmatrix}
1 & -1 & -1 & 0 & 0 & 0 \\
1 & 0 & 1 & 0 & -1 & 0 \\
0 & 1 & -1 & -1 & 0 & 0 \\
0 & 0 & 0 & 1 & 0 & -1 \\
0 & 0 & 0 & 0 & 1 & -1 \\
0 & 0 & 0 & 1 & 1 & 1
\end{bmatrix}
\begin{bmatrix}
\dot{v}_{C_1} \\
\dot{v}_{C_2} \\
i_{C_3} \\
i_{R_2} \\
i_{R_3} \\
v_{R_1}
\end{bmatrix}
=
\begin{bmatrix}
0 & 0 & 0 \\
0 & 0 & 0 \\
0 & 0 & 0 \\
0 & -1 & 0 \\
-1 & 0 & 0 \\
0 & 0 & 1
\end{bmatrix}
\begin{bmatrix}
v_{C_1} \\
v_{C_2} \\
I
\end{bmatrix}
\qquad (3.8)
$$

Step 8. Either invert the matrix on the left-hand side of equation 3.8 and solve to express each variable on the left-hand side as a linear combination of state variables and inputs or execute row reduction techniques on both sides of the equation to obtain a 2×2 identity in the upper left-hand corner with zeros to the right. In either case, the resulting state model is

$$
\begin{bmatrix}
\dot{v}_{C_1} \\
\dot{v}_{C_2}
\end{bmatrix}
=
\begin{bmatrix}
-\frac{1}{3} & 0 \\
0 & -\frac{1}{3}
\end{bmatrix}
\begin{bmatrix}
v_{C_1} \\
v_{C_2}
\end{bmatrix}
+
\begin{bmatrix}
\frac{1}{3} \\
\frac{1}{3}
\end{bmatrix}
I
$$

$$v_{R_2} = [\tfrac{1}{3} \quad -\tfrac{2}{3}] \begin{bmatrix} v_{C_1} \\ v_{C_2} \end{bmatrix} + [\tfrac{1}{3}] I \qquad (3.9)$$

Examples 3.1 and 3.2 illustrate the construction of a linear state model for simple lumped circuits. Again, for more complicated circuits, graph-theoretic approaches become necessary. An excellent exposition appears in [4].

Although the foregoing development was cast in circuit-theoretic terms, the concept of a state model finds application in the whole gamut of physical systems. For an excellent discussion on constructing state models for a wide range of physical processes, see [5].

MECHANICAL EXAMPLES

A wide variety of physical processes (other than circuits) have state variable representations. These include paint shakers, automatic door openers often found in supermarkets, DC motor position control devices, flow control for city park fountains, gyroscopes, inverted pendulums on moving carts, satellite dynamics, power systems, and turbine engines. As an illustration of such a mechanical system, consider a simple moving mass constrained by frictional and spring forces.

EXAMPLE 3.3.

This example [6] illustrates the construction of a state model describing the displacement x of the simple mass-spring-dashpot system of Figure 3.6. Two critical concepts underlie the construction of the state model: (*i*) Newton's second law, in the form "the sum of the forces acting on the mass M is zero"; and (*ii*) specification of the forces F, F_1, and F_2 in terms of the position x, the velocity $\dot{x}$, and the acceleration $\ddot{x}$ of the mass. Note the similarity between the conservation-of-force law and the KCL, which requires that the total current entering a node be zero. Summing the forces on M with due consideration to direction implies that

$$F = F_1 + F_2 \qquad (3.10)$$

Since acceleration is the second derivative of position, the force F must satisfy

$$F = m\ddot{x} - f_0 \qquad (3.11)$$

where $m\ddot{x}$ is the force the mass exerts in opposition to the spring and to frictional forces, and f_0 is an applied force.

Now, the force F_1 exerted by the spring equals some nonlinear function of the position of the mass, i.e.,

$$F_1 = f_1(x) \qquad (3.12)$$

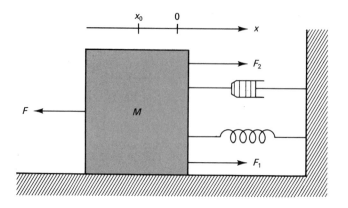

Figure 3.6 Mass-spring system having spring force F_1 and frictional force F_2.

The frictional force generated by the dashpot, however, is proportional to the velocity of the mass:

$$F_2 = f_2(\dot{x}) \qquad (3.13)$$

Substituting equations 3.11 through 3.13 into equation 3.10 produces the nonlinear differential equation

$$m\ddot{x} = f_1(x) + f_2(\dot{x}) + f_0 \qquad (3.14)$$

For small displacements, the spring and frictional forces are described by linear equations: $F_1 = f_1(x) = -k_1 x$ for some spring constant k_1 and $F_2 = f_2(\dot{x}) = -b_1\dot{x}$ for some damping coefficient b_1. Thus for small displacements, 3.14 becomes

$$m\ddot{x} = -b_1\dot{x} - k_1 x + f_0 \qquad (3.15)$$

A state model can be constructed by noting that the derivative of position, $x \triangleq x_1$, is velocity, $\dot{x} \triangleq x_2$, and acceleration is $\ddot{x}$. Hence, one chooses position and velocity as state variables to produce the state model

$$\begin{bmatrix} \dot{x}_1 \\ \dot{x}_2 \end{bmatrix} = \begin{bmatrix} 0 & 1 \\ -\dfrac{k_1}{m} & -\dfrac{b_1}{m} \end{bmatrix} \begin{bmatrix} x_1 \\ x_2 \end{bmatrix} + \begin{bmatrix} 0 \\ \dfrac{1}{m} \end{bmatrix} f_0$$

$$(3.16)$$

$$y = \begin{bmatrix} 1 & 0 \end{bmatrix} \begin{bmatrix} x_1 \\ x_2 \end{bmatrix}$$

where we have taken the system output to be the position of the mass.

The next example involving masses, springs, and dashpots is more practical and interesting than the previous.

EXAMPLE 3.4.

Consider Figure 3.7 which depicts a simplified one-wheel version of an automobile suspension system. Here, the springs and dashpot (shock absorber) are assumed to operate in their linear regions.

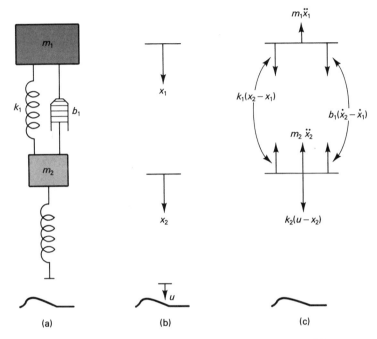

Figure 3.7   (a) Simplified one-wheel automobile suspension system; (b) Displacement diagram;   (c) Force diagram.

In Figure 3.7(a), m_1 is taken to be one-fourth of the mass of the car; k_1 is the spring constant of the usual coil spring between the wheel and the car frame; b_1 is the damping coefficient of the shock absorber; m_2 is the mass of the wheel assumed to be concentrated at a point; and k_2 is a spring constant associated with the "springiness" of the tire. In Figure 3.7(b), x_1 represents the displacement of the car frame, x_2 the displacement of the center of the wheel, and u an input representing the variation in the surface of the tire from an equilibrium road surface. The tire is assumed to be in contact with the road surface at all times.

The first step in constructing the desired state model requires specification of a differential equation relating the motions of the masses m_1 and m_2. This in turn requires summing the forces acting on m_1 and m_2 just as one sums the currents entering various nodes in a circuit to produce a circuit model.

From Figure 3.7(c), summing the forces on m_1 leads to

$$m_1\ddot{x}_1 = k_1(x_2 - x_1) + b_1(\dot{x}_2 - \dot{x}_1) \tag{3.17}$$

and summing the forces on m_2 leads to

$$m_2 \ddot{x}_2 = -k_1(x_2 - x_1) - b_1(\dot{x}_2 - \dot{x}_1) + k_2(u - x_2) \qquad (3.18)$$

Since these are two second-order differential equations, a state model representation requires a minimum of four state variables. The obvious choices include $x_1, x_2, x_3 \triangleq \dot{x}_1$, and $x_4 \triangleq \dot{x}_2$. The resulting state model takes the form

$$\begin{bmatrix} \dot{x}_1 \\ \dot{x}_2 \\ \dot{x}_3 \\ \dot{x}_4 \end{bmatrix} = \begin{bmatrix} 0 & 0 & 1 & 0 \\ 0 & 0 & 0 & 1 \\ -\dfrac{k_1}{m_1} & \dfrac{k_1}{m_1} & -\dfrac{b_1}{m_1} & \dfrac{b_1}{m_1} \\ \dfrac{k_1}{m_2} & -\dfrac{(k_1+k_2)}{m_2} & \dfrac{b_1}{m_2} & -\dfrac{b_1}{m_2} \end{bmatrix} \begin{bmatrix} x_1 \\ x_2 \\ x_3 \\ x_4 \end{bmatrix} + \begin{bmatrix} 0 \\ 0 \\ 0 \\ \dfrac{k_2}{m_2} \end{bmatrix} u$$

$$y = [1\ 0\ 0\ 0] \begin{bmatrix} x_1 \\ x_2 \\ x_3 \\ x_4 \end{bmatrix}$$

where the output has been taken as the position of the car frame.

Other state models grow out of analyses similar to that presented in the preceding example. The problems at the end of the chapter, as well as of subsequent chapters, ask for the development of various state models for a variety of physical systems.

CONCLUDING REMARKS

In this chapter we have formally introduced and rigorized the notion of the linear lumped state model, considered various types of responses in that model, and constructed the state models for some simple circuits and mechanical systems. By constrast, the notion of a state model for a distributed system is a more complicated mathematical concept. In the lumped case, the state of the system $x(\cdot)$, evaluated at a particular time t (i.e., $x(t)$) is a vector in $\mathbf{R}^n$. For a linear system containing a delay of T seconds, the state of the system at time t is $x(t) \epsilon \mathbf{R}^n$ adjoined to the function segment $x(\cdot)$ defined over the interval $[t - T, t]$. Mathematically, this is a delicate problem and will not be considered further.

Although not directly considered in this chapter, most physical processes have nonlinear state models—i.e., the differential and algebraic equations describing the processes may be shaped into a form equivalent to equations 1.23 and 1.25. Unfortunately, these models seldom have known analytic solutions, unlike the linear time-varying state model. The development of a solution to the linear time-varying state model follows in chapters 10 through 13. Suffice it to say that the construction of the desired solution to a nonlinear state model forces one to resort to numerical simulation of the model. Chapter 8 covers the details of numerical simulation for the nonlinear state model as well as for linear state models.

PROBLEMS

1. With V and I as the circuit inputs and V_R the circuit output, write a set of state equations for the circuit shown in Figure P3.1. (*Hint*: Use a mesh current approach.)

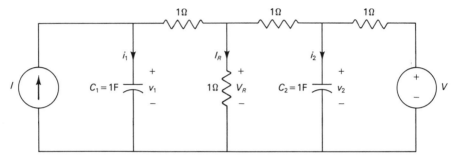

Figure P3.1   Circuit for problem 1.

2. Write a set of state equations for the circuit of Figure P3.2, assuming that I is the input and V_{R_1} the output.

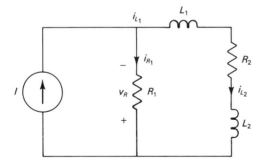

Figure P3.2   Circuit for problem 2.

3. Consider the circuit shown in Figure P3.3. Choose i_1 and i_2 as state variables. Assuming that I is the input and v_R the output, construct the associated state model. How

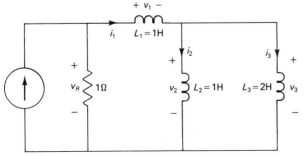

Figure P3.3   Circuit for problem 3.

does the model change if v_3 is also an output? What are the natural frequencies of the circuit? (*Hint*: Review chapter 2.)

4. Consider the system block diagram shown in Fig. P3.4, in which $j = \sqrt{-1}$. Given that $x_1(0) = x_2(0) = 1$, compute the zero-input response.

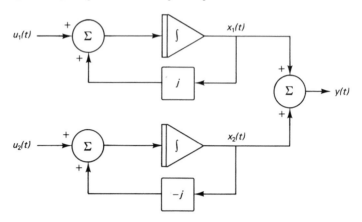

Figure P3.4   Circuit for problem 4.

5. What is the relationship between the zero-input and zero-state responses of a system and the transient and steady-state responses of the system?

6. Write a state model for the circuit of Figure P3.6, in which the triangular block is an infinite gain operational amplifier.

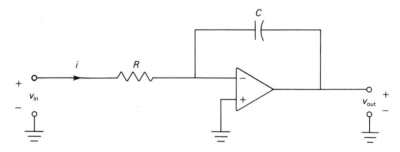

Figure P3.6   Circuit for problem 6.

7. Consider the state model

$$\dot{x}(t) = A(t)x(t) + B(t)u(t) , \; x(t_o) = x_o$$
$$y(t) = C(t)x(t) + D(t)u(t)$$

(a) Show that the zero-input state response is linear in the initial condition, x_o;

(b) Based on your answer to (a), what can you say about the zero-input response?

(c) Show that the zero-state response is linear in the input.

(d) Show that the zero-state response is time-invariant in the input when A, B, C, and D are constant matrices.

(*Hint*: The technique of proof for all of the preceding is essentially the same. The following fact may prove useful: if two functions, say $x^1(t)$ and $x^2(t)$, satisfy the differential equation defined by the linear state model, and if $x^1(t') = x^2(t')$ for some time t', then $x^1(t) = x^2(t)$ for all t).

8. Draw a tree for the circuit shown in Figure P3.8. Based on the tree, define a set of admissible state variables. Now construct a state model.

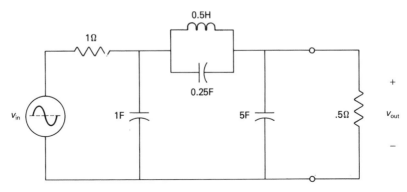

Figure P3.8   Circuit for problem 8.

9. A linear time-invariant state model has a zero-input state response $x(t)$ which results from the initial condition $x(0) = [1 \quad 1]'$. What is the zero-input state response to the initial condition $\hat{x}(1) = [1 \quad 1]'$?

10. Consider the differential equation

$$\dot{x}(t) = f(x(t), u(t))$$
$$y(t) = g(x(t), u(t)) \tag{1}$$

Suppose $f(\cdot, \cdot)$ and $g(\cdot, \cdot)$ have continuous partial derivatives of all orders. Let $x_0(\cdot)$ and $u_0(\cdot)$ be the nominal state trajectory and the nominal input, respectively. Recall that the Taylor series for $f(\cdot, \cdot)$ about $x_0(\cdot)$ and $u_0(\cdot)$ has the form

$$f(x, u) = f(x_0, u_0) + \left.\frac{\partial f}{\partial x}\right|_{x_0, u_0} [\Delta x] + \left.\frac{\partial f}{\partial u}\right|_{x_0, u_0} [\Delta u] + \textit{higher order terms}$$

For small perturbations Δx and Δu about the nominal values, $x(t) = x_0(t) + \Delta x(t)$ and $u(t) \approx u_0(t) + \Delta u(t)$.

(a) Derive an approximating linear state model for equation 1 about the nominal state trajectory $x_0(t)$ and the nominal input $u_0(t)$. Justify your steps.

(b) The equations of an inverted pendulum are

$$\dot{x}_1(t) = x_2(t)$$
$$\dot{x}_2(t) = \frac{g}{\ell}\sin(x_1(t)) + u(t)$$

What are the linearized state equations about the equilibrium solution $x_1(\cdot) = x_2(\cdot) = u(\cdot) \equiv 0$?

11. As discussed in [5], a mathematical model for a magnetic microphone is given by

$$L\frac{dI}{dt} + (R_c + R)I + B\frac{dZ}{dt} = 0$$

$$M\frac{d^2Z}{dt^2} + u\frac{dZ}{dt} + KZ - BI = F$$

where I is the current through the coil and Z is the displacement of the diaphragm element. The input is the force F on the diaphragm, and the output is the voltage drop RI.

The parameter values are as follows: $R_c = 9.0\,\Omega$, $R = 1.0\,\Omega$, $L = 10^{-3}\,\text{H}$, $M = 0.01\,\text{kg}$, $\mu = 0.1\,(\text{N-sec})/\text{m}$, $K = 0.5\,\text{N/m}$, and $B = 0.3\,(V\text{-sec})/\text{m}$. Develop a state model for the microphone.

12. Also discussed in [5], the vertical ascent of a rocket from the earth has the following dynamics. Let m denote the constant mass of the rocket, r the distance of the rocket from the center of the earth, T the thrust of the rocket, and R the radius of the earth. Assume that the only forces on the rocket are an inverse square gravitational force plus the thrust; the vertical motion of the rocket is described by

$$m\frac{d^2r}{dt^2} = -mg\left[\frac{R}{r}\right]^2 + T$$

Suppose that $g = 32.2\,\text{ft/sec}^2$, and $R = 4{,}000.0$ miles, and consider the thrust as the input variable. Choose state variables and write state equations for the vertical motion of the rocket.

13. For the state model developed in problem 12, use the results of problem 10 to find linearized state models at (a) $r = 1.5R$ and (b) $r = 2R$.

14. Consider the circuit shown in Figure P3.14. Choose i_{L_1} and i_{L_2} as state variables and construct a state model for the circuit. Let $i_{L_1} - i_{L_2}$ be the desired output; note the indicated current directions.

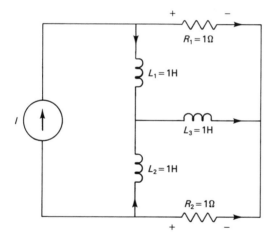

Figure P3.14   Circuit for problem 14.

15. In chapter 2, we found that the solution to $\dot{x}(t) = Ax(t)$ for constant A was $x(t) = \exp[At]x(0)$. Under certain conditions, the solution to $\dot{x}(t) = A(t)x(t)$ is $x(t) = \exp[M(t)]x(0)$, where

$$M(t) = \int_0^t A(q)\,dq$$

As a first step in understanding the subtleties involved in the proof of this statement:
(a) What is the form of the derivative of $\exp[M(t)]$? Justify your answer.

(b) Suppose that

$$M(t) = \begin{bmatrix} 0 & t & t - t^2/2 \\ 0 & 0 & t^3/3 \\ 0 & 0 & 0 \end{bmatrix}$$

Show that the derivative of $\exp[M(t)]$ at $t = 1$ is

$$\begin{bmatrix} 0 & 1 & 0.6667 \\ 0 & 0 & 1 \\ 0 & 0 & 0 \end{bmatrix}$$

REFERENCES

1. E. A. Coddington and N. Levinson, *Theory of Ordinary Differential Equations* (New York: McGraw-Hill, 1955), chapters 1 and 2.

2. W. Mayeda, *Graph Theory* (New York: Wiley, 1972).

3. S. P. Chan, *Introductory Topological Analysis of Electrical Networks* (New York: Holt, Rinehart and Winston, 1968).

4. L. O. Chua and P. M. Lin, *Computer Aided Analysis of Electronic Circuits; Algorithms and Computational Techniques* (Englewood Cliffs, NJ: Prentice Hall, 1975).

5. N. H. McClamrock, *State Models of Dynamic Systems: A Case Study Approach* (New York: Springer-Verlag, 1980).

6. R. W. Newcomb, *Concepts of Linear Systems and Control* (Belmont, CA: Cole, 1968).

4

STATE MODELS FROM ORDINARY DIFFERENTIAL EQUATIONS – I

INTRODUCTION

Chapter 2 furnished an impetus for the state model concept by contrasting a state model of an *RLC* circuit with an equivalent second order scalar differential equation model. To expand that development, this chapter links nth order time-invariant scalar differential equation models with nth degree or more usually nth order canonical state models. Block diagram schematics of the resulting *canonical* state models depict possible analog simulation diagrams, the first step in setting up simulations on an analog or hybrid computer.

Ordinary differential equation models of a circuit or system reflect differential relationships between a system input and a system output. For example, if u denotes a scalar input and y a scalar output, and if $m \leqslant n$, then an nth order scalar ordinary differential equation model has the form

$$y^{(n)} + a_1 y^{(n-1)} + \cdots + a_{n-1}\dot{y} + a_n y = b_m u + b_{m-1}\dot{u} + \cdots + b_0 u^{(m)} \quad (4.1)$$

where the parenthetical superscripts denote the order of the derivative. For the present all, initial conditions are assumed to be zero. Such models arise naturally in most scientific and engineering disciplines.

Equation 4.1 depicts an *input-output* type of model often called an *external* system description, because it hides the internal structure of the system. For example, the capacitor voltage of the series *RLC* circuit does not appear in the second-order scalar differential equation model of chapter 2. By contrast, in examples 3.1 and 3.2 the schematics identified the internal structure of the circuits, and the chosen state variables corresponded to physically meaningful internal quantities. The key issue, then, is can one choose state variables and construct meaningful state models for systems represented by external models, and if so, how?

An input-output differential equation model provides no clue as to what physically meaningful internal quantities are appropriate state variables. But does this imply that the choice is arbitrary? Not at all, for the form of equation 4.1 suggests making each choice a linear combination of the derivatives of the input and output variables. Thus, since $y^{(n)}$ is the highest appearing derivative, there is a need for n state variables.

Different choices of state variables produce different (but of course equivalent) state models for the same circuit. Consequently, a state model for a given system or circuit is not unique. To illustrate, consider the state model of the series *RLC* circuit example (see equation 2.29). If $R = C = L = 1$, then

$$\begin{bmatrix} \dot{v}_C \\ \dot{i}_L \end{bmatrix} = \begin{bmatrix} 0 & 1 \\ -1 & -1 \end{bmatrix} \begin{bmatrix} v_C \\ i_L \end{bmatrix} \tag{4.2}$$

Choosing the scattering variables [1], $(v_C + i_L)$ and $(v_C - i_L)$, as state variables, the following equivalent state model results:

$$\begin{bmatrix} \dot{(v_C + i_L)} \\ \dot{(v_C - i_L)} \end{bmatrix} = \begin{bmatrix} -0.5 & -0.5 \\ 1.5 & -0.5 \end{bmatrix} \begin{bmatrix} (v_C + i_L) \\ (v_C - i_L) \end{bmatrix} \tag{4.3}$$

Of course, scattering variables are but one possible choice: any two independent linear combinations of v_C and i_L would suffice to yield a valid state model. Relationships called *nonsingular state transformations* link all such choices into an equivalence class [2,3]. For example, the state variables of equations 4.2 and 4.3 satisfy

$$\begin{bmatrix} v_c + i_L \\ v_C - i_L \end{bmatrix} = \begin{bmatrix} 1 & 1 \\ 1 & -1 \end{bmatrix} \begin{bmatrix} v_C \\ i_L \end{bmatrix} \tag{4.4}$$

which is a nonsingular state transformation. A later section of the chapter develops this notion further.

Within the realm (the equivalence class) of all possible choices of state variables for a state model of equation 4.1, a special set of choices leads to the so-called *canonical state models*. The most common of these are the observable, the observability, the controllable, and the controllability canonical forms. Canonical models reflect a more ordered and structured approach to state model formulation. They are generally more amenable to system analysis and control design than noncanonical models.

CANONICAL FORMS

Rather than presenting a general development of the n-dimensional canonical forms necessary to represent 4.1, a series of concrete examples with $n = 3$, $m = 3$, and $b_0 = 0$ better serves the instructional intent. Comments on the appropriate generalizations follow the examples, with the actual extensions relegated to the problems.

EXAMPLE 4.1.

In this example the goal is to construct and define the observable canonical form of a time-invariant state model for the external system description

$$\ddot{y} + a_1\ddot{y} + a_2\dot{y} + a_3 y = b_3 u + b_2 \dot{u} + b_1 \ddot{u} \tag{4.5}$$

and then depict the structure of the model as an analog computer simulation diagram. Again, we assume for the present that all initial conditions are zero.

As a notational convenience, let $D = \dfrac{d}{dt}$ be the derivative operator. The analogous inverse operator D^{-1}, represents an integral operator. Then with this notation, equation 4.5 becomes

$$D^3 y + a_1 D^2 y + a_2 D y + a_3 y = b_3 u + b_2 D u + b_1 D^2 u \tag{4.6}$$

(For all practical purposes these D-operators correspond to frequency domain multiplication by the Laplace transform variable.) Combining terms in like powers of D on the right-hand side and then multiplying on the left by D^{-3} produces

$$y = D^{-1}\left[b_1 u - a_1 y\right] + D^{-2}\left[b_2 u - a_2 y\right] + D^{-3}\left[b_3 u - a_3 y\right] \tag{4.7}$$

Defining the state variable $x_3 = y$ and then differentiating equation 4.7 by multiplying on the left by D yields

$$\dot{x}_3 = D x_3 = D y \tag{4.8}$$

$$= \left[b_1 u - a_1 y\right] + D^{-1}\left[b_2 u - a_2 y\right] + D^{-2}\left[b_3 u - a_3 y\right]$$

After substituting x_3 for y, we implicitly define x_2 via the equation

$$\dot{x}_3 \triangleq \left[b_1 u - a_1 x_3\right] + x_2 \tag{4.9}$$

Hence,

$$x_2 = D^{-1}\left[b_2 u - a_2 x_3\right] + D^{-2}\left[b_3 u - a_3 x_3\right] \tag{4.10}$$

which implies that

$$\dot{x}_2 = \left[b_2 u - a_2 x_3\right] + D^{-1}\left[b_3 u - a_3 x_3\right] \tag{4.11}$$

Defining x_1 in a manner similar to x_2 yields

$$\dot{x}_2 = \left[b_2 u - a_2 x_3 \right] + x_1 \tag{4.12}$$

and consequently,

$$\dot{x}_1 = \left[b_3 u - a_3 x_3 \right] \tag{4.13}$$

Recalling that $x_3 = y$ and combining equations 4.9, 4.12, and 4.13 into a matrix description produces the *observable canonical form* of a state model for equaiton 4.5:

$$\begin{bmatrix} \dot{x}_1 \\ \dot{x}_2 \\ \dot{x}_3 \end{bmatrix} = \begin{bmatrix} 0 & 0 & -a_3 \\ 1 & 0 & -a_2 \\ 0 & 1 & -a_1 \end{bmatrix} \begin{bmatrix} x_1 \\ x_2 \\ x_3 \end{bmatrix} + \begin{bmatrix} b_3 \\ b_2 \\ b_1 \end{bmatrix} u$$

$$y = \begin{bmatrix} 0 & 0 & 1 \end{bmatrix} \begin{bmatrix} x_1 \\ x_2 \\ x_3 \end{bmatrix} \tag{4.14}$$

The analog simulation diagram associated with equation 4.14 appears in Figure 4.1. Patching this schematic onto an analog computer (with appropriate scaling) sets up an analog simulation of the system. However, before executing such a simulation, two questions arise: (*i*) What are the appropriate initial values for the state variables x_i as determined from initial conditions on y and u and their derivatives (if they are nonzero), and (*ii*) Is there a mechanism for setting up these conditions? The next chapter sketches solutions to these questions.

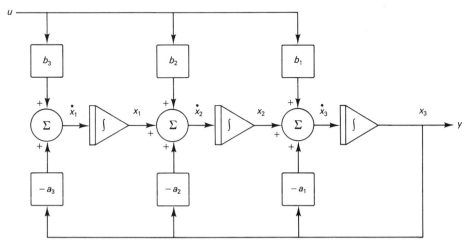

Figure 4.1   Block diagram of observable canonical form of equation 4.5.

This state model (equation 4.14) and its associated block diagram given in Figure 4.1, have a dual counterpart known as the *controllable* or *standard canonical form.* (Some authors also use the term, *phase variable form.*) Intuitively, the observable and controllable forms are dual [4] to each other because in a sense their block diagrams are inverses of each other. To see this, reverse the flow (direction of all arrows and integrations) in Figure 4.1, interchange the positions of u and y and x_i and $\dot{x}_i$, make all summers connection points and all connection points summers, and for consistency reflect the resulting image about a vertical axis. This inversion process produces Figure 4.2, which is the block diagram representation of the controllable canonical form, dual to the observable form. The corresponding state model formulation is

$$\begin{bmatrix} \dot{x}_1 \\ \dot{x}_2 \\ \dot{x}_3 \end{bmatrix} = \begin{bmatrix} 0 & 1 & 0 \\ 0 & 0 & 1 \\ -a_3 & -a_2 & -a_1 \end{bmatrix} \begin{bmatrix} x_1 \\ x_2 \\ x_3 \end{bmatrix} + \begin{bmatrix} 0 \\ 0 \\ 1 \end{bmatrix} u \qquad (4.15)$$

$$y = \begin{bmatrix} b_3 & b_2 & b_1 \end{bmatrix} \begin{bmatrix} x_1 \\ x_2 \\ x_3 \end{bmatrix}$$

Note that the state variables x_1, x_2, and x_3, in equation 4.15 correspond to *different* entities from those in equation 4.14. What is the nonsingular state transformation relating the two sets of state variables?

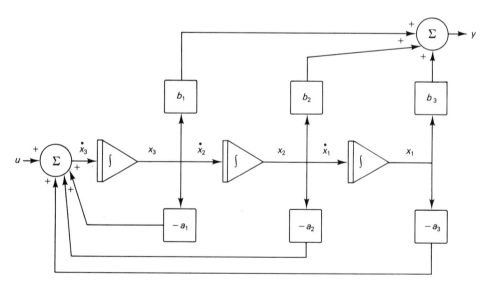

Figure 4.2   Controllable canonical form of equation 4.5.

Observe that the "A-matrix" of the two forms are transposes of each other and that the "B-matrix" of the observable form is the transpose of the "C-matrix" of the controllable form; and conversely. This is a heuristic interpretation of duality, to be fleshed out formally in a later chapter on observability and controllability.

The preceding ad hoc derivation of the controllable canonical form does not in fact verify that equation 4.15 represents the differential equation 4.5. To do that, two procedures are possible: (*i*) take the model given by equation 4.15 or the diagram of Figure 4.2 and derive equation 4.5, or (ii) derive the model equation 4.15 directly from equation 4.5. For our purposes, the latter route is more instructive and is presented in the following example.

EXAMPLE 4.2.

In this example we develop the controllable or standard canonical form [4,5,6,7] of a state model representation of equation 4.5. Constructing this form requires some cleverness. The first step is to develop the canonical state model for the auxiliary system

$$D^3 \hat{y} + a_1 D^2 \hat{y} + a_2 D \hat{y} + a_3 \hat{y} = u \qquad (4.16)$$

where the coefficients a_1, a_2, and a_3 coincide with those in equation 4.5 and the response $\hat{y}$ results from the input u. Then, assuming that all initial conditions are zero, the response of the system description given by equation 4.5 is the superposition of the responses of the auxiliary system given by equation 4.16, to the inputs $b_3 u$, $b_2 D u$, and $b_1 D^2 u$. In particular, $y = b_3 \hat{y} + b_2 D \hat{y} + b_1 D^2 \hat{y}$.

Keeping this view in mind, the next step is to construct the canonical state model for equation 4.16. In terms of the differential operator D, equation 4.16 has the equivalent representation.

$$\hat{y} = D^{-3} \left[u - a_3 \hat{y} \right] - a_2 D^{-2} \hat{y} - a_1 D^{-1} \hat{y}$$

As such define the state variable $x_1 = \hat{y}$, which implies that

$$\dot{x}_1 = D \hat{y} = D^{-2} \left[u - a_3 x_1 \right] - a_2 D^{-1} x_1 - a_1 x_1 \qquad (4.17)$$

Now define $x_2 = \dot{x}_1$ which implies that

$$\dot{x}_2 = D^{-1} \left[u - a_3 x_1 \right] - a_2 x_1 - a_1 x_2 \qquad (4.18)$$

Similarly, define $x_3 = \dot{x}_2$ which from equation 4.18, yields

$$\dot{x}_3 = u - a_3 x_1 - a_2 x_2 - a_1 x_3 \qquad (4.19)$$

From the discussion at the beginning of the example and the preceding choice of state variables,

$$y = b_3 x_1 + b_2 D x_1 + b_1 D^2 x_1 = b_3 x_1 + b_2 x_2 + b_1 x_3 \qquad (4.20)$$

Hence, equations 4.20 and 4.19, together with the equivalences $x_2 = \dot{x}_1$ and $x_3 = \dot{x}_2$, produce the controllable canonical state model given by equation 4.15.

In addition to the observable and controllable dual forms, there is another set of dual canonical forms called *observability* and *controllability canonical forms* [4,6]. The observability canonical state representation has the form

$$\begin{bmatrix} \dot{x}_1 \\ \dot{x}_2 \\ \dot{x}_3 \end{bmatrix} = \begin{bmatrix} 0 & 1 & 0 \\ 0 & 0 & 1 \\ -a_3 & -a_2 & -a_1 \end{bmatrix} \begin{bmatrix} x_1 \\ x_2 \\ x_3 \end{bmatrix} + \begin{bmatrix} \beta_1 \\ \beta_2 \\ \beta_3 \end{bmatrix} u \qquad (4.21)$$

$$y = \begin{bmatrix} 1 & 0 & 0 \end{bmatrix} \begin{bmatrix} x_1 \\ x_2 \\ x_3 \end{bmatrix}$$

the quantities β_1, β_2, and β_3 are called the *Markov parameters* [4] of the system and are given by $\beta_1 = b_1$, $\beta_2 = b_2 - a_1 b_1$, and $\beta_3 = a_1^2 b_1 - a_1 b_2 - a_2 b_1 + b_3$. The analog simulation diagram corresponding to equation 4.21 appears in Figure 4.3.

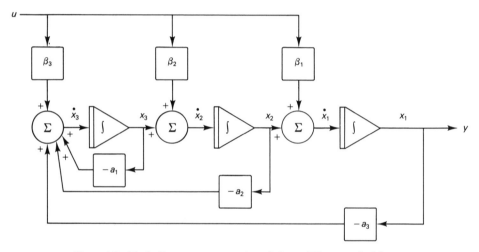

Figure 4.3   Block diagram representation of observability canonical form.

Rather than dive into a tangled derivation of this canonical form, the text below will sketch the salient points. By parroting the fine points of the two previous derivations, the reader should be able to plug the gaps.

To sketch the derivation of equation 4.21, rewrite equation 4.5 as the integral equation

$$y = D^{-1} \Big[b_1 u - a_1 y \Big] + D^{-2} \Big[b_2 u - a_2 y \Big] + D^{-3} \Big[b_3 u - a_3 y \Big]$$

Identifying y with the state variable x_1 (i.e., define $x_1 = y$), implicitly define x_2 via

$$\dot{x}_1 = Dy = b_1 u + x_2 \qquad (4.22)$$

Identifying x_2 with the appropriate part of Dy, permits us to implicitly define x_3 as

$$\dot{x}_2 = \Big[b_2 - a_1 b_1 \Big] u + x_3 \qquad (4.23)$$

in which case

$$\dot{x}_3 = \Big[a_1^2 b_1 - a_1 b_2 - a_2 b_1 + b_3 \Big] u - a_1 x_3 - a_2 x_2 - a_3 x_1 \qquad (4.24)$$

Letting $y = x_1$, equations 4.22 through 4.24 produce the observability canonical state model given by equation 4.21. By duality, the controllability canonical form of equation 4.5 has the state model representation

$$
\begin{bmatrix} \dot{x}_1 \\ \dot{x}_2 \\ \dot{x}_3 \end{bmatrix} = \begin{bmatrix} 0 & 0 & -a_3 \\ 1 & 0 & -a_2 \\ 0 & 1 & -a_1 \end{bmatrix} \begin{bmatrix} x_1 \\ x_2 \\ x_3 \end{bmatrix} + \begin{bmatrix} 1 \\ 0 \\ 0 \end{bmatrix} u
\tag{4.25}
$$

$$
y = \begin{bmatrix} \beta_1 & \beta_2 & \beta_3 \end{bmatrix} \begin{bmatrix} x_1 \\ x_2 \\ x_3 \end{bmatrix}
$$

Equations 4.14, 4.15, 4.21, and 4.25 define four important canonical state models for equation 4.1 when $n = 3$ and $m = 2$. The derivations of these equations extend naturally to the general n-dimensional form of equation 4.1 in the obvious manner. For example, the general controllable canonical form for $m = n$ and $b_0 = 0$ is

$$
\begin{bmatrix} \dot{x}_1 \\ \dot{x}_2 \\ \cdot \\ \cdot \\ \cdot \\ \dot{x}_{n-1} \\ \dot{x}_n \end{bmatrix} = \left\{ \begin{array}{ccccc} 0 & 1 & 0 & & 0 \\ 0 & 0 & 1 & & 0 \\ & & \cdot & & \\ & & & \cdot & \\ & & & & \cdot \\ 0 & 0 & 0 & \cdots & 1 \\ -a_n & -a_{n-1} & -a_{n-2} & & -a_1 \end{array} \right\} \begin{bmatrix} x_1 \\ x_2 \\ \cdot \\ \cdot \\ \cdot \\ x_{n-1} \\ x_n \end{bmatrix} + \begin{bmatrix} 0 \\ 0 \\ \cdot \\ \cdot \\ \cdot \\ 0 \\ 1 \end{bmatrix} u
\tag{4.26}
$$

$$
y = [b_n \ b_{n-1} \cdots b_2 \ b_1] \begin{bmatrix} x_1 \\ x_2 \\ \cdot \\ \cdot \\ \cdot \\ x_{n-1} \\ x_n \end{bmatrix}
$$

CONCLUDING REMARKS

Several intents underlie the presentation of the foregoing forms. First, the previous examples forge a kinship between an analog simulation diagram and the mathematical state model, linking the image of a block diagram with a set of equations. Second, canonical state models provide an organized way of representing a high order scalar time-invariant differential equation as opposed to an arbitrary representation. Hence, identification theory, realization theory, and general system analysis become more straightforward and more transparent. Third, delineation of the forms demonstrates the nonuniqueness of the state model representation. Indeed, any linear time-invariant state model

$$\dot{x} = Ax + Bu$$
$$y = Cx + Du \tag{4.27}$$

has a zero-state equivalent state model representation

$$\dot{z} = [TAT^{-1}]z + [TB]u$$
$$y = [CT^{-1}]z + Du \tag{4.28}$$

under the nonsingular state transformation $z = Tx$ where T is $n \times n$ nonsingular, and as set out in chapter 3, A is $n \times n$, B is $n \times m$, C is $r \times n$, and D is $r \times m$. The term *zero-state equivalent* means that if $x(t_0) = z(t_0) = \theta$, and if the inputs to equations 4.27 and 4.28 coincide, then the responses coincide. The reader should note however, that although the controllable and observable canonical forms of a particular differential equation are always zero-state equivalent, there will not always exist a nonsingular state transformation linking the two forms. Finally, certain controller design procedures [7] build around the canonical forms. To illustrate, consider the controllable canonical form

$$\begin{bmatrix} \dot{x}_1 \\ \dot{x}_2 \\ \dot{x}_3 \end{bmatrix} = \begin{bmatrix} 0 & 1 & 0 \\ 0 & 0 & 1 \\ -a_3 & -a_2 & -a_1 \end{bmatrix} \begin{bmatrix} x_1 \\ x_2 \\ x_3 \end{bmatrix} + \begin{bmatrix} 0 \\ 0 \\ 1 \end{bmatrix} u$$

given in equation 4.15. This has the usual symbolic form $\dot{x} = Ax + Bu$. As indicated in equations 2.34 and 2.8, systems so modeled have natural modes of oscillation (eigenvalues, natural frequencies, or natural vibrations) given by the roots of the polynomial

$$\pi_A(\lambda) \triangleq \det\left[\lambda I - A\right] = \det \begin{bmatrix} \lambda & -1 & 0 \\ 0 & \lambda & -1 \\ a_3 & a_2 & \lambda + a_1 \end{bmatrix} \tag{4.29}$$

$$= a_3 + a_2\lambda + a_1\lambda^2 + \lambda^3$$

The subscript on $\pi_A(\lambda)$ indicates the *characteristic polynomial* [5,7,8,9] of the matrix A. Clearly, the polynomial is computable by inspection, which is not the case for arbitrary state model representations.

Many times the natural frequencies of a system or plant are too oscillatory and more damping is desired. To improve system damping (say, to mild disturbances), an engineer might design a "state feedback controller" to move the given natural frequencies to ones with a higher degree of damping. Equivalently, he or she would change the coefficients of the characteristic polynomial via the state feedback controller.

Symbolically, a state feedback control takes the form

$$u \leftarrow Fx + u$$

where F is a constant $m \times n$ feedback matrix—i.e., the given external input u is replaced by the sum of a feedback term, Fx, and an external input u. The resulting "compensated" state model dynamics are

$$\dot{x} = (A + BF)x + Bu \qquad (4.30)$$

Figure 4.4 shows a block diagram of this relationship.

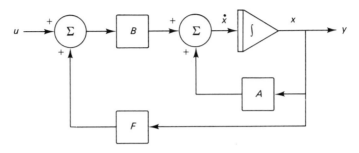

Figure 4.4   Block diagram of feedback compensation of equation 4.30.

Relative to the controllable canonical form given in equation 4.15 having three states and one input, $F = \begin{bmatrix} f_3 & f_2 & f_1 \end{bmatrix}$; hence,

$$A + BF = \begin{bmatrix} 0 & 1 & 0 \\ 0 & 0 & 1 \\ f_3 - a_3 & f_2 - a_2 & f_1 - a_1 \end{bmatrix}$$

implying that

$$\pi_{A+BF}(\lambda) = (a_3 - f_3) + (a_2 - f_2)\lambda + (a_1 - f_1)\lambda^2 + \lambda^3 \qquad (4.31)$$

Therefore, changing a given $\pi_A(\lambda)$ to a desired $\pi_{A+BF}(\lambda)$ is straightforward for systems in the controllable canonical form. When a single input system is in an arbitrary form, the engineer will often transform it to its controllable form, design an appropriate compensator, and transform the resulting design back to the given state coordinates [7]. Figure 4.5 illustrates the flow of one kind of control design procedure. Dealing with control design is perhaps one of the more striking reasons for introducing canonical state models.

A final question is, Are there canonical forms for multi-input systems? Indeed there are, and they have fascinating side effects which are beyond the scope of this text. The interested reader should consult [10].

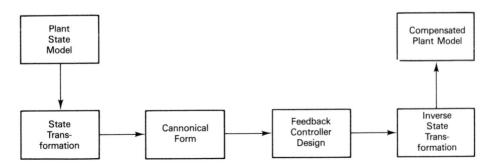

Figure 4.5   Flowchart for one kind of compensator design procedure.

PROBLEMS

1. Assuming all initial conditions are zero, show, for equation 4.1, that if y is the response to u, then $D^p y$ is the response to $D^p u$.

2. Construct a detailed derivation of the observability canonical form.

3. Derive the controllability canonical form from equation 4.5, and draw the associated block diagram.

4. Using the state transformation $z = Tx$, derive in detail the state model representation of equation 4.28 from the representation given in equation 4.27.

5. Let x denote the vector of state variables associated with the controllable canonical form. Let z denote that of the observable canonical form. Find the state transformation T for which $z = Tx$ in regard to equations 4.14 and 4.15. Under what conditions is the transformation nonsingular?

6. Compute the observable and controllable canonical state models for the differential equation $2\ddot{y} - 4\dot{y} + y = 2u + 4\dot{u} + 2\ddot{u}$. What is the characteristic polynomial of the associated A-matrices? What are the eigenvalues or natural frequencies of the system?

7. Suppose the state model representation of a plant is in the controllable canonical form given by equation 4.15. Find the state feedback $F = [f_3\, f_2\, f_1]$ which will give the plant the new characteristic polynomial

$$\pi_{A+BF}(\lambda) = \lambda^3 - \left[-\hat{a}_3 - \hat{a}_2\lambda - \hat{a}_1\lambda^2 \right]$$

In other words, what is the form of f_i?

8. A plant has the differential equation model $D^3 y + 3D^2 y + 3Dy + y = u$. Find the state feedback such that the compensated system has a characteristic polynomial with zeros at $-0.5, -1 \pm 2j$.

9. Construct the observable and controllable canonical forms and the associated analog simulation diagram for the differential equation

$$a_4 D^3 y + a_3 D^2 y + a_2 Dy + a_1 y = b_4 u + b_3 Du + b_2 D^2 u + b_1 D^3 u$$

10. Repeat problem 9 for the observability and controllability canonical forms.

11. The characteristic equation (or polynomial) of the 2×2 matrix

$$A = \begin{bmatrix} -1 & 1 \\ -1 & -1 \end{bmatrix}$$

is $\pi_A(\lambda) = \lambda^2 + 2\lambda + 2$. Let I be the identity matrix. Show by direct calculation that $\pi_A(A) = A^2 + 2A + 2I = 0$. This illustrates the well-known Caley-Hamilton theorem: For any square matrix A, $\pi_A(A) = 0$. The theorem is useful throughout much of the rest of the text.

12. Let $T(t)$ be $n \times n$ and nonsingular for all t. Given the state model

$$\dot{x}(t) = A(t)x(t) + B(t)u(t)$$
$$y(t) = C(t)x(t) + D(t)u(t)$$

and the state transformation $z(t) = T(t)x(t)$, find an equivalent state model in terms of the new state variables $z(t)$.

13. Investigate the possibility of putting the time-varying scalar differential equation

$$\dddot{y}(t) + a_1(t)\ddot{y}(t) + a_2(t)\dot{y}(t) + a_3(t)y(t) = u(t)$$

into the controllable and observable canonical forms. What assumptions did you make? (*Hint*: Derive the controllable form first and then the observable form, using a time-varying state transformation.)

14. Consider $\dot{x} = Ax + Bu$ where A is $n \times n$ and B is $n \times 1$. If $\det[Q] \neq 0$, where $Q = [B \,|\, AB \,|\, \cdots \,|\, A^{n-1}B]$, then the system can be converted into the controllable canonical form as follows:

(a) Compute Q^{-1} and designate the last row of Q^{-1} by the row vector v^t, where t indicates the transpose of the column vector v.

(b) Compute the matrices V and V^{-1}, where

$$
V = \begin{bmatrix} v^t \\ v^t A \\ . \\ . \\ . \\ v^t A^{n-1} \end{bmatrix}
$$

The equivalent system $\dot{z} = VAV^{-1}z + VBu$ is then in controllable canonical form. Now, by designing controllers in the z-coordinate system, one can invert the state transformation to reflect a designed feedback controller in the x-coordinate system [7]. Accordingly, given the system

$$
\dot{x} = \begin{bmatrix} 1 & 1 \\ 0 & -1 \end{bmatrix} x + \begin{bmatrix} 0 \\ 1 \end{bmatrix} u
$$

$$
y = [1 \;\; 1] x
$$

(i) find a feedback controller to assign the eigenvalues $\{-1, -2\}$ to the system. (*Hint*: Be sure to state the answer in the x-coordinate frame.)

(ii) Find the differential equation relating y and its derivatives to u and its derivatives.

(iii) Find the transfer function of the system.

(iv) Suppose $u(t) = t \, 1^+(t)$ and the response $y(t)$ is to have a rise time of 4 seconds and an overshoot of 20%. Find the necessary pole locations for such a response, and then find a state feedback which will assign these poles.

15. Look up the definitions and forms of a *parallel* and a *cascade* realization [5,6,11,12] of the differential equation 4.1. Compare these with the forms in this chapter.

16. If equation 4.26 duly represents the general controllable canonical form, then, by duality, what is the general observable canonical form?

17. Appropriately extend the derivations of the controllable and observable canonical forms to account for the general form of equation 4.1.

18. Recall that multiplication by s in the frequency domain corresponds to differentiation in the time domain. In view of this fact, consider

$$
y(s) = \frac{s^2 + 2s + 1}{s^2 + 3s + 2} u(s)
$$

where $y(s)$ and $u(s)$ are the Laplace transforms of the functions $y(t)$ and $u(t)$, respectively.

(a) Write a differential equation characterizing the relationship between $y(t)$ and $u(t)$,

(b) Put the differential equation found in (a) into the controllable canonical state model form,

(c) What is the characteristic polynomial of the A-matrix?

(d) What are the system's natural frequencies or eigenvalues?

19. Consider the state model

$$
\begin{bmatrix} \dot{x}_1 \\ \dot{x}_2 \end{bmatrix} = \begin{bmatrix} 1 & -1 \\ 1 & 1 \end{bmatrix} \begin{bmatrix} x_1 \\ x_2 \end{bmatrix} + \begin{bmatrix} 1 \\ 0 \end{bmatrix} u
$$

$$
y = \begin{bmatrix} 1 & -1 \end{bmatrix} \begin{bmatrix} x_1 \\ x_2 \end{bmatrix}
$$

Find a differential equation which is equivalent to this state model.

20. Let $\dot{x} = Ax + Bu$, where

$$
A = \begin{bmatrix} a_1 & a_2 & a_3 \\ 1 & 0 & 0 \\ 0 & 1 & 0 \end{bmatrix}, \quad B = \begin{bmatrix} 1 \\ 0 \\ 0 \end{bmatrix}
$$

(a) Find the characteristic polynomial $\pi(\lambda)$ of A.

(b) If $a_1 = 2$, $a_2 = -2$, and $a_3 = 3$, find the state feedback $F = [f_1 \; f_2 \; f_3]$ such that the spectrum of $A + BF$ is $\{0, -1, +1\}$.

21. Let $\dot{x} = Ax + Bu$, where

$$
A = \begin{bmatrix} 0 & 1 & 0 \\ 0 & 0 & 1 \\ 1 & 2 & 3 \end{bmatrix}, B = \begin{bmatrix} 0 \\ 0 \\ 1 \end{bmatrix}
$$

Find the state feedback such that the eigenvalues of $A + BF$ are -1 and $-1 \pm j$.

22. Find the controllable and observable canonical state model forms for problem 13 of chapter 3.

23. Can the differential equations of problem 11 of chapter 3 be put into the controllable state model form? Explain and/or derive.

24. The differential equation [13] for a field-controlled dc servomotor takes the form

$$
T_f T_m \frac{d^3\theta}{dt^3}(t) + (T_f + T_m)\frac{d^2\theta}{dt^2}(t) + \frac{d\theta}{dt}(t) = \frac{K_T}{R_f B}e_f(t)
$$

where $\theta(t)$ is the angular position of the rotor, T_f is the field time constant, T_m is the motor time constant, K_T is the torque constant for the motor, B is the damping constant, R_f is the effective resistance of the motor field circuit, and $e_f(t)$ is the field excitation voltage. Put the differential equation into a state model controllable canonical form.

25. The scalar differential equation

$$
\dddot{y} + a_1\ddot{y} + a_2\dot{y} + a_3 y = u
$$

has response $y(t) = e^{-t}\sin(t)1^+(t)$ when $u(t) = \sin(t)1^+(t)$, assuming all initial conditions are zero. Again, if all initial conditions are zero, find the solution to

$$
\dddot{y} + a_1\ddot{y} + a_2\dot{y} + a_3 y = 2u + \dot{u}
$$

26. Let $H(s) = n(s)/d(s)$ be a proper rational transfer function; i.e., $n(s) = b_0 s^n + b_1 s^{n-1} + \dots + b_{n-1}s + b_n$ and $d(s) = s^n + a_1 s^{n-1} + \dots + a_{n-1}s + a_n$, for real scalars $b_0, b_1, \dots, b_n, a_1, \dots, a_n$. Show constructively that every such transfer function can be realized by an nth-order state model. Assume that $n(s)$ and $d(s)$ have no common factors, i.e., that they are relatively prime polynomials. If they are relatively prime, the state model is said to be minimal. What if they are not relatively prime? Explain.

REFERENCES

1. R. Saeks, *Generalized Networks* (New York: Holt, Rinehart and Winston, 1971).

2. F. Gantmacher, *Theory of Matrices,* Vol. 1, (New York: Chelsea, 1960).

3. G. Birkhoff and S. Maclane, *A Survey of Modern Algebra,* 4th ed. (New York: Macmillan, 1977).

4. Thomas Kailath, *Linear Systems* (Englewood Cliffs NJ: Prentice Hall, 1980).

5. C. A. Desoer, *Notes for a Second Course on Linear Systems* (New York: Van Nostrand, 1970).

6. David G. Luenberger, *Introduction to Dynamical Systems: Theory, Models, and Applications* (New York: Wiley, 1979).

7. M. W. Wonham, *Linear Multivariable Control: A Geometric Approach* (New York: Springer-Verlag, 1979).

8. G. Strang, *Linear Algebra and Its Applications* (New York: Academic Press, 1980).

9. K. Hoffman, and R. Kunze, *Linear Algebra* (Englewood Cliffs, NJ: Prentice Hall, 1971).

10. L. Zadeh and C. Desoer, *Linear System Theory: The State Space Approach* (New York: McGraw-Hill, 1963).

11. A. V. Oppenheim and A. S. Willsky, *Signals and Systems* (Englewood Cliffs, NJ: Prentice Hall, 1983).

12. David G. Luenberger, "Canonical Forms for Linear Multivariable Systems," *IEEE Transactions on Automatic Control,* June 1967, pp. 290–293.

13. S. M. Shinners, *Modern Control System Theory and Application* (Reading, MA: Addison-Wesley, 1972).

5

STATE MODELS
FROM ORDINARY
DIFFERENTIAL EQUATIONS — II

The Problem of Initial Conditions

INTRODUCTION

The last chapter described four canonical state models from an nth order scalar differential equation. This entailed defining artificial intermediate variables called state variables. Each of the four canonical models had state variables which characterized a different view of the internal response of the system. Yet the input-output characteristics of each model were the same. In turn, for each form, this requires appropriately determined initial conditions $x(0)$, dependent on the initial conditions $y(0), y^{(1)}(0), \ldots, y^{(n-1)}(0)$ of the original differential equation and the input $u(t)$ and its derivatives, all evaluated at $t = 0$. Clearly, if the initial condition $x(0)$ for the canonical model was chosen improperly, the response of the state model representation and the response of the original differential equation model would be very different.

However, we know that each state model has an infinite number of equivalent representations obtained through a nonsingular state transformation $Tz = x$. Thus, although the initial condition problem is motivated by the canonical state model realization of an nth order differential equation, the problem is present for every state model realization of a differential equation. Accordingly, we will assume no

special structure on any of the matrices A, B, C, and D of the realization.
This discussion underlies the following problem.

Problem 1: *Given the nth order scalar differential equation*

$$y^{(n)} + a_1 y^{(n-1)} + \ldots + a_n y = b_n u + b_{n-1} u^{(1)} + \ldots + b_1 u^{(n-1)} + b_0 u^{(n)} \qquad (5.1)$$

having a state model representation

$$\begin{aligned}
\dot{x}(t) &= Ax(t) + Bu(t) \\
y(t) &= Cx(t) + Du(t)
\end{aligned} \qquad (5.2)$$

where A is $n \times n$, B is $n \times 1$, C is $1 \times n$ and D is 1×1, determine the initial state vector $x(0) = [x_1(0), x_2(0), \ldots, x_n(0)]^t$ from the given initial conditions $y(0), y^{(1)}(0), \ldots, y^{(n-1)}(0)$ and the input values $u(0), u^{(1)}(0), \ldots, u^{(n-1)}(0)$.

Computing the appropriate $x(0)$ which solves this problem allows an analog simulation, based on the diagrams of the previous chapter, to proceed smoothly. However, what if the integrators in a patched-up analog simulation board are not accessible. Can one then find an input excitation $u(\cdot)$ to set up $x(0)$ given some known $x(0^-)$? For simplicity, suppose $u(\cdot)$ is impulsive—i.e., $u(\cdot)$ is a linear combination of impulses and derivatives of impulses. Then we can state this problem as follows.

Problem 2: *Given $\dot{x}(t) = Ax(t) + Bu(t)$, a known $x(0^-)$, and a desired $x(0^+)$, find an impulsive input $u(t) = \xi_0 \delta(t) + \xi_1 \delta^{(1)}(t) + \ldots + \xi_{n-1} \delta^{(n-1)}(t)$ to instantaneously change $x(0^-)$ to $x(0^+)$.*

The goal is to formulate each of the preceding two problems as a linear matrix equation. In the case of Problem 1 the solution will be $x(0^+)$, and for Problem 2 it will be the vector $[\xi_0, \xi_1, \ldots, \xi_{n-1}]^t$. These problems, their formulation as a linear matrix equation, and their solution foreshadow our more realistic and comprehensive investigations of observability and controllability.

Note that the use of impulses is for convenience; it is an ill-advised if not unrealistic engineering practice.

PROBLEM 1: DETERMINING INITIAL CONDITIONS

Let us begin by casting problem 1 in a more general context: determine the state vector $x(t)$ from "measurements" $y(t), y^{(1)}(t), \ldots, y^{(n-1)}(t)$ and the input $u(t)$ and its derivatives for the system given by equations 5.2 when $D = [0]$, i.e., the zero matrix. (The case of the nonzero D-matrix is presented in the problems at the end of the chapter.) The objective is to derive a linear matrix equation whose solution is $x(t)$. Since the available data are clearly $y(t)$ and its derivatives and $u(t)$ and its derivatives, the logical course of action is to sequentially differentiate $y(t) = Cx(t)$ to produce

$$y(t) = Cx(t)$$

$$\dot{y}(t) = C\dot{x}(t) = CAx(t) + CBu(t)$$

$$\ddot{y}(t) = C\ddot{x}(t) = CA\dot{x}(t) + CB\dot{u}(t)$$
$$= CA^2 x(t) + CABu(t) + CB\dot{u}(t)$$
$$\vdots$$
$$y^{(n-1)}(t) = CA^{n-1}x(t) + CA^{n-2}Bu(t) + ... + CBu^{(n-2)}(t)$$

(5.3)

The set of equations terminates at $y^{(n-1)}(t)$ because there are only n initial conditions. Analytically, this follows from the Caley-Hamilton Theorem [1], which states that $\pi_A(A) = [0]$; in particular, $\pi_A(A) = [0]$ implies that

$$A^n = -a_n I - a_{n-1}A - ... - a_1 A^{n-1}$$

In other words, the nth power of A is a linear combination of lower powers of A. Hence, for any p, A^p can be expressed as a linear combination of the zeroth through $(n-1)$st powers of A.

Again, taking derivatives of the "measured", possibly noisy output variables lacks engineering soundness, i.e., realism. However, our purpose here is pedagogical. In chapter 21, we develop a dynamic observer which estimates the state vector using only integrators.

In matrix form, equation 5.3 becomes

$$
\begin{bmatrix} y(t) \\ \dot{y}(t) \\ \ddot{y}(t) \\ \cdot \\ \cdot \\ \cdot \\ y^{(n-1)}(t) \end{bmatrix}
=
\begin{bmatrix} C \\ CA \\ CA^2 \\ \cdot \\ \cdot \\ \cdot \\ CA^{n-1} \end{bmatrix}
x(t) +
\begin{bmatrix} 0 & 0 & 0 & & 0 \\ CB & 0 & 0 & & 0 \\ CAB & CB & 0 & ... & 0 \\ \cdot & & & & \cdot \\ \cdot & & & & \cdot \\ \cdot & & & & \cdot \\ CA^{n-2}B & ... & CB & & 0 \end{bmatrix}
\begin{bmatrix} u(t) \\ \dot{u}(t) \\ \ddot{u}(t) \\ \cdot \\ \cdot \\ \cdot \\ u^{(n-1)}(t) \end{bmatrix}
$$

(5.4)

Designate this equation by the more compact notation

$$Y(t) = R x(t) + T U(t)$$

(5.5)

where each of the variables $Y(t)$, R, $x(t)$, T, and $U(t)$ is identified with its obvious counterpart in equation 5.4. To compute $x(0) = [x_1(0), \cdots, x_n(0)]^t$, evaluate equation 5.5 at $t = 0$ and solve

$$[Y(0) - TU(0)] = R x(0)$$

(5.6)

for $x(0)$. Since A is $n \times n$ and C is $1 \times n$ and R is $n \times n$, equation 5.6 has a unique solution if and only if R is nonsingular, i.e., $\det[R] \neq 0$. If R is nonsingular, then R^{-1} exists and

$$x(0) = R^{-1}[Y(0) - TU(0)]$$

(5.7)

Note that a (not necessarily unique) solution may exist even when R^{-1} does not exist. This occurs whenever the vector $[Y(0) - TU(0)]$ can be expressed as a linear combination of the columns of R— i.e., whenever $[Y(0) - TU(0)]$ lies in the *column space* or, more frequently, *image space* of the matrix R. In such cases many initial states will produce the same response for a fixed input. Hence, the initial state of the system cannot be uniquely reconstructed by input-output measurements (more on this a little later).

The solution for $x(0)$ in equation 5.7 appears in terms of R^{-1}. This is a convenient form for the purposes of instructional exposition. With regard to numerical computations however, it is much better to use a sophisticated Gaussian elimination procedure which might be able to utilize the techniques of *equilibration, partial pivoting*, and *iterative improvement* [2]. Such an algorithm computes the solution to an equation of the form of equation 5.6 without explicitly constructing R^{-1}. More on the whys and wherefores of such numerical topics is presented in the next chapter. Suffice it to say here that, because computers represent numbers by an approximate binary or hexadecimal representation, round-off errors are present, and algorithms which fail to take cognizance of this fact may execute floating point operations which magnify the effect of such errors to produce erroneous answers [3, 4].

EXAMPLE 5.1.

Consider the observable canonical form of

$$\dddot{y} + a_1\ddot{y} + a_2\dot{y} + a_3 y = b_3 u + b_2 \dot{u} + b_1 \ddot{u} \tag{5.8}$$

which is

$$\begin{bmatrix} \dot{x}_1 \\ \dot{x}_2 \\ \dot{x}_3 \end{bmatrix} = \begin{bmatrix} 0 & 0 & -a_3 \\ 1 & 0 & -a_2 \\ 0 & 1 & -a_1 \end{bmatrix} \begin{bmatrix} x_1 \\ x_2 \\ x_3 \end{bmatrix} + \begin{bmatrix} b_3 \\ b_2 \\ b_1 \end{bmatrix} u$$

$$y = \begin{bmatrix} 0 & 0 & 1 \end{bmatrix} \begin{bmatrix} x_1 \\ x_2 \\ x_3 \end{bmatrix} \tag{5.9}$$

This pair of equations has the form $\dot{x} = Ax + Bu$ and $y = Cx$, which implies that

$$R = \begin{bmatrix} C \\ CA \\ CA^2 \end{bmatrix} = \begin{bmatrix} 0 & 0 & 1 \\ 0 & 1 & -a_1 \\ 1 & -a_1 & a_1^2 - a_2 \end{bmatrix}$$

Since $\det[R] = -1$ for any choice of a_1 or a_2, R is nonsingular with inverse

$$R^{-1} = \begin{bmatrix} a_2 & a_1 & 1 \\ a_1 & 1 & 0 \\ 1 & 0 & 0 \end{bmatrix}$$

Noting that

$$T = \begin{bmatrix} 0 & 0 & 0 \\ CB & 0 & 0 \\ CAB & CB & 0 \end{bmatrix} = \begin{bmatrix} 0 & 0 & 0 \\ b_1 & 0 & 0 \\ b_2 - a_1 b_1 & b_1 & 0 \end{bmatrix}$$

it follows that

$$\begin{bmatrix} x_1(0) \\ x_2(0) \\ x_3(0) \end{bmatrix} = \begin{bmatrix} a_2 & a_1 & 1 \\ a_1 & 1 & 0 \\ 1 & 0 & 0 \end{bmatrix} \begin{bmatrix} \begin{bmatrix} y(0) \\ \dot{y}(0) \\ \ddot{y}(0) \end{bmatrix} - \begin{bmatrix} 0 \\ b_1 u(0) \\ (b_2 - a_1 b_1) u(0) + b_1 \dot{u}(0) \end{bmatrix} \end{bmatrix} \quad (5.10)$$

PROBLEM 2: SETTING UP INITIAL CONDITIONS

Concomitant with the problem of determining initial conditions for a state model at a particular time is the problem of setting them up by exciting the input with, say, impulses. This changes the initial state to a desired state in zero time. Convenience, mathematical simplicity, and some physical insight motivate the choice of impulsive inputs. For example, prior to pulling the trigger on a rifle, the bullet in the chamber is stationary. The trigger action then sets off a contained mini-explosion. The resulting expansion of gases more or less "instantaneously" changes the velocity of the bullet from zero to some initial velocity v_0. An impulse function represents the explosion as a mathematical ideal: the "instantaneous" change in velocity of the bullet corresponds to an "instantaneous" change in initial condition. The following example heuristically demonstrates the mathematical nature of this choice.

EXAMPLE 5.2.

Consider the scalar state model of

$$\dot{x}(t) = \alpha x(t) + \beta u(t), \quad x(0^-) = x_0 \quad (5.11)$$

which depicts some physical process. Suppose $t > 0^-$, $\beta > 0$, and $u(t) = \xi \delta(t)$ is the impulsive input. Then from differential equation theory,

$$x(t) = e^{\alpha t} x_0 + \int_{0^-}^{t} e^{\alpha(t-\tau)} \beta u(\tau) d\tau$$

Using $u(\tau) = \xi \delta(\tau)$ and the sifting property [5] of the impulse function,

$$x(t) = e^{\alpha t} x_0 + \beta \xi e^{\alpha t} \int_{0^-}^{t} e^{-\alpha \tau} \delta(\tau) d\tau = e^{\alpha t} x_0 + \beta \xi e^{\alpha t}$$

Setting $t = 0^+$ and $x(0^+) = x(0^-) + \beta \xi$ and solving for ξ yields

$$\xi = \beta^{-1}[x(0^+) - x(0^-)]$$

Consequently, $u(t) = \xi\delta(t)$ will instantaneously change a given $x(0^-)$ to a desired $x(0^+)$.

Our present goal is to extend the foregoing scenario to the state model $\dot{x} = Ax + Bu$, where A is $n \times n$ and B is $n \times 1$. To do this requires several pieces of information. First, the solution of the state dynamics, $\dot{x} = Ax + Bu$, takes the form

$$x(t) = e^{At}x_0 + \int_{0^-}^{t} e^{A(t-\tau)}Bu(\tau)d\tau \qquad (5.12)$$

(A derivation of this formula is presented in chapter 13.) Second, for impulsive inputs, the integral term satisfies the equality

$$\int_{0^-}^{t} e^{A(t-\tau)}B\delta^{(i)}(\tau)d\tau = (-1)^i e^{At}\left[\frac{d^i}{d\tau^i}e^{-A\tau}B\right]_{\tau=0} = e^{At}A^iB \qquad (5.13)$$

where $\delta^{(i)}(\tau)$ denotes the ith derivative of the impulse function $\delta(\tau)$. This formula comes about via integration by parts.

Theorem 5.1. Let

$$\dot{x}(t) = Ax(t) + Bu(t), \quad x(0^-) = x_0$$

where A is $n \times n$ and B is $n \times 1$. Suppose

$$u(t) = \xi_0\delta(t) + \xi_1\delta^{(1)}(t) + \dots + \xi_{n-1}\delta^{(n-1)}(t) \qquad (5.14)$$

Then

$$x(0^+) - x(0^-) = [B \,|\, AB \,|\dots|\, A^{n-1}B] \begin{bmatrix} \xi_0 \\ \xi_1 \\ \cdot \\ \cdot \\ \xi_{n-1} \end{bmatrix} \qquad (5.15)$$

The proof of this theorem proceeds as follows. Substitute the form of the impulsive input, equation 15.14, into the solution of the state dynamics, equation 5.12. Now apply the sifting property of equation 5.13 and put the result in matrix form.

Corollary 5.1. Let $Q = [B \,|\, AB \,|\dots|\, A^{n-1}B]$. Then there exists a unique input $u(t)$ of the form

$$u(t) = \xi_0\delta(t) + \xi_1\delta^{(1)}(t) + \dots + \xi_{n-1}\delta^{(n-1)}(t)$$

which will instantaneously change $x(0^-)$ to $x(0^+)$ if and only if $\det[Q] = \det[B \,|\, AB \,|\dots|\, A^{n-1}B] \neq 0$. Furthermore, $[\xi_0, \xi_1, \dots, \xi_{n-1}]^t = Q^{-1}[x(0^+) - x(0^-)]$.

As with the solution of equation 5.6, a solution of equation 5.15 will exist whenever $[x(0^+) - x(0^-)]$ lies in the image space of the matrix

$Q = [B \ AB \ ... \ A^{n-1}B]$— i.e., whenever $[x(0^+) - x(0^-)]$ can be expressed as a linear combination of the columns of Q. If B is $n \times m$ with $m > 1$, the solution is seldom unique. But such nonuniqueness is desirable, because it allows the "control design engineer" the opportunity to pick an "optimal" input from a family of acceptable ones.

Corollary 5.2. There exists a solution to equation 5.15 whenever $[x(0^+) - x(0^-)]$ lies in the image of Q.

The actual computation of a solution in such cases uses the Moore-Penrose pseudo inverse [2]. Chapter 19 addresses the said computation.

EXAMPLE 5.3.

Consider the controllable canonical form of the differential equation

$$\dddot{y} + a_1\ddot{y} + a_2\dot{y} + a_3y = b_3u + b_2\dot{u} + b_1\ddot{u}$$

which appears in equation 4.15. The associated Q-matrix is

$$Q = \begin{bmatrix} B & | & AB & | & A^2B \end{bmatrix} = \begin{bmatrix} 0 & 0 & 1 \\ 0 & 1 & -a_1 \\ 1 & -a_1 & a_1^2 - a_2 \end{bmatrix}$$

Thus, Q is nonsingular for any arbitrary choice of a_1 and a_2. Hence,

$$\begin{bmatrix} \xi_0 \\ \xi_1 \\ \xi_2 \end{bmatrix} = Q^{-1}\begin{bmatrix} x(0^+) - x(0^-) \end{bmatrix} = \begin{bmatrix} a_2 & a_1 & 1 \\ a_1 & 1 & 0 \\ 1 & 0 & 0 \end{bmatrix}\begin{bmatrix} x_1(0^+) - x_1(0^-) \\ x_2(0^+) - x_2(0^-) \\ x_3(0^+) - x_3(0^-) \end{bmatrix} \quad (5.16)$$

The necessary input as per equation 5.14 then becomes

$$u(t) = \xi_0 \delta(t) + \xi_1 \delta^{(1)}(t) + \xi_2\delta^{(2)}(t)$$

with ξ_0, ξ_1, and ξ_2 as computed in equation 5.16.

GENERALIZATION OF PROBLEM 1

Problems 1 and 2 sprung from the initial condition problem of a state model representation of a scalar differential equation. Conceptually, the essence of problem 1 is the determination of the internal status of a system from input-output observations. In particular, the fundamental question is, Can one reconstruct the state $x(t)$ of the system model at a particular time given (*i*) the system state model, (*ii*) complete knowledge of the system input $u(t)$, and (*iii*) exact measurements of the system response $y(t)$? The flip side of this coin is the control question of problem 2. Here the objective is to drive the internal system "status quo" to a more desirable one. Specifically, given the system state model, can one construct an input $u(t)$ which will drive the state vector from one point to another?

Generalized Problem 1: *Given the state model*

$$\dot{x}(t) = Ax(t) + Bu(t)$$
$$y(t) = Cx(t) + Du(t)$$

with A $n \times n$, B $n \times m$, C $r \times n$, and D $r \times m$, and given exact input-output measurements $(u(t), y(t))$ over the time interval, $[t_0, t_1]$, compute the state vector $x(t)$ for any given $t \in [t_0, t_1]$.

The word "exact" in the statement of the problem means that, at least theoretically, it is possible to compute exactly the first $n - 1$ derivatives of both $u(t)$ and $y(t)$. The equation which solves this problem has a derivation which replicates that of problem 1. The resulting equation again has the structure of equation 5.5, i.e.,

$$R\, x(t) = [Y(t) - TU(t)] \tag{5.17}$$

where $x(t)$, $Y(t)$, $U(t)$, and R are the same as in equations 5.5 and 5.4 and T differs only in that there are D-matrices along the diagonal. Also, the matrix R is $rn \times n$ and T is now $rn \times nm$.

To deepen our understanding of this generalized problem, we consider the various aspects of the solution for a rectangular R-matrix having more rows than columns.

Question 1: *When does there exist such a solution?*

Answer: *A solution exists whenever $[Y(t) - TU(t)]$ lies in the column space of R — i.e., whenever the equations are consistent.*

Theoretically, consistent equations should always exist if (*i*) the model accurately mirrors the physical process under study, (*ii*) the input-output measurements $(u(t), y(t))$ are not noisy, and (*iii*) all derivatives are accurately computed. Practically speaking, however, state models, as well as all other models, only mirror an idealized or dominant behavior of a physical process. All measurements contain noise. Numerical round-off error and the progressive algorithmic error induced in the calculation of the derivatives of noisy signals make "accurate computation" unlikely. Hence, in a realistic setting the equations are typically inconsistent. On the other hand, frequently they are "almost consistent" in the mathematical sense that, given the data $[Y(t) - TU(t)]$, there often exists a vector v (depending on t) whose Euclidean norm

$$||v||_2 < \epsilon ||Y(t) - TU(t)||_2$$

for some reasonably small ϵ such that $[Y(t) - TU(t)] + v$ lies in the column space of R. That is to say, equation 5.17 becomes consistent with the addition of the perturbation vector v to the right-hand side. Therefore, investigating the consistent case sheds much light on the more realistic inconsistent one. In fact, the actual solution of the equations is virtually identical to the theoretical one. Accordingly, what follows centers on the consistent case with comments addressing the subtleties of the inconsistent one.

Question 2: *If there exists a solution to equation 5.17, is it ever unique?*

Answer: *Yes, if and only if rank[R] = n.*

Some explanation is in order. First, rank[R] denotes the rank of the matrix R [1,6]. Rank[R] equals the number of linearly independent columns of R. Since R is a constant matrix, the number of linearly independent rows and columns coincide, i.e., row rank equals column rank. Hence, the maximum rank of the $rn \times n$ matrix R is n.

It is convenient to view each column of R as a vector in $\mathbf{R}^{rn}$ and the transpose of each row as a vector in $\mathbf{R}^n$. Rank[R] is the largest set of linearly independent column (row) vectors. Recall that a set of vectors $\{v_1,...,v_q\}$ is linearly independent [1,6] if and only if the equation

$$\alpha_1 v_1 + \alpha_2 v_2 + ... + \alpha_q v_q = \theta \tag{5.18}$$

implies that $\alpha_1 = \alpha_2 = ... = \alpha_q = 0$ — i.e., all scalars α_i equal zero.

These definitions set up the basic theory of matrix rank. Chapter 19 discusses the difficult problem of numerically computing the rank of a matrix in the context of the *singular value decomposition* (SVD) [7] of a matrix. Suffice it to say here that the SVD of a matrix provides a means for effective computation of rank even if the matrices C and A (which determine R) are "noisy."

Our concern in this chapter is whether or not R is *full rank*, i.e., whether or not rank[R]=n. Checking whether det[R'R] $\neq$ 0 provides a simple solution. However, such a method is numerically dangerous and misleading. Consider that

$$\det \begin{bmatrix} \epsilon & 0 \\ 0 & \epsilon \end{bmatrix} = \epsilon^2$$

could approximate zero for small ϵ. Yet, the matrix unequivocally has rank two. In particular, if $fl(\epsilon^2) = 0$ (the floating point equivalent of ϵ^2 in the computer memory), the numerical rank would not be two.

Question 3: *If the solution to equation 5.17 is unique, how does one compute it for a rectangular matrix R?*

Answer: *Since R has at least as many rows as columns, the unique solution is given by*

$$x = R^{-L}[Y - TU] \tag{5.19}$$

where R^{-L} is any left inverse of R.

Briefly, if R is $rn \times n$ and rank[R]=n, then any matrix R^{-L} for which $R^{-L}R = I$ is a *left inverse* of R. A particular left inverse known as the *Moore-Penrose pseudo (left) inverse* is

$$R^{-L} = (R'R)^{-1}R' \tag{5.20}$$

Clearly, $R^{-L}R = (R'R)^{-1}R'R = I$. Although often useful, equation 5.20 is a numerically ill-conditioned method for computing R^{-L} in that computation of $(R^t R)^{-1}$ is sensitive to floating-point round-off. Chapter 19 considers this general problem in detail.

The left inverse is not unique unless $r = 1$, in which case $R^{-L} = R^{-1}$, which exists since rank$[R] = n$ by assumption. The solution to equation 5.17, however, is unique regardless of the choice of R^{-L}. The nonuniqueness of R^{-L} follows because equation 5.17 represents a set of nr equations in n unknowns. By picking different sets of n linearly independent equations, one obtains different (square) matrix equations whose solutions are identical. The computation of each solution, however, requires the inversion of a different matrix. Hence, the existence of many left inverses which give rise to the same answer should come as no surprise.

With regard to possible inconsistent equations, suppose the observations are noisy and the entries of the state model matrices are nominal — i.e., actual values could deviate from nominal by as much as 10 or 15 percent. Then in all probability, equation 5.17 would be inconsistent and no solution would exist. In such cases, one computes a *least squares* solution. The least squares solution of equation 5.17 is the state vector x for which the square of the usual Euclidean norm

$$|| Rx - [Y - TU] ||_2^2 = (Rx - [Y - TU])^t (Rx - [Y - TU]) \quad (5.21)$$

is a minimum. Since equation 5.21 represents a quadratic nonnegative function, it has a minimum at that x, causing the derivative to be zero. Taking the derivative, rearranging, and setting the result equal to zero produces

$$2(R^t Rx - R^t[Y - TU]) = \theta$$

which is zero if and only if

$$R^t Rx = R^t[Y - TU]$$

Since rank$[R] = n$, $R^t R$ is invertible. Solving for x implies that

$$x = (R^t R)^{-1} R^t[Y - TU]$$

is the least squares solution. Hence, selecting $R^{-L} = (R^t R)^{-1} R^t$ (equation 5.20) produces the least squares solution according to equation 5.19 [7]. The resulting x is closest to satisfying equation 5.17 in the sense of Euclidean distance. Historically, Gauss first introduced this well-used method of least squares in 1795 [8].

A final question remains: What if rank$[R] < n$? In that case the solution of equation 5.17 is not unique even if it exists. A fuller discussion of this question and its consequences is undertaken in chapter 21, which tackles the system observability question. The following theorem summarizes the development so far.

Theorem 5.2. Suppose rank$[R] = n$. Then for any $t \in [t_0, t_1]$, a unique solution to generalized problem 1 exists if and only if $[Y(t) - TU(t)]$ lies in the column space of R. Moreover, when it exists, the solution is given by

$$x = (R^t R)^{-1} R^t[Y - TU]$$

EXAMPLE 5.4.

Let a system have the representation

$$\dot{x} = \begin{bmatrix} 1 & 0 & 0 \\ 0 & -1 & 0 \\ 0 & 0 & 2 \end{bmatrix} x + \begin{bmatrix} 1 & 0 \\ 0 & 0 \\ 0 & 1 \end{bmatrix} u$$

$$y = \begin{bmatrix} 1 & 0 & 1 \\ 0 & 1 & 0 \end{bmatrix} x$$

Suppose the known input is $u(t) = [1^+(t) \ \ 1^+(t)]^t$ and the observed response is

$$y(t) = \begin{bmatrix} 2e^t + 1.5e^{2t} - 1.5 \\ e^{-t} \end{bmatrix} 1^+(t)$$

Compute $x(1)$.

The solution requires solving $Rx(1) = [Y(1) - TU(1)]$. The following steps provide details.

Step 1. Compute $Y(1)$.

$$Y(1) = \begin{bmatrix} y(1) \\ \dot{y}(1) \\ \ddot{y}(1) \end{bmatrix} = \begin{bmatrix} 2e + 1.5e^2 - 1.5 \\ e^{-1} \\ \hline 2e + 3e^2 \\ -e^{-1} \\ \hline 2e + 6e^2 \\ e^{-1} \end{bmatrix}$$

Step 2. Compute $U(1)$.

$$U(1) = [1 \ 1 \ | \ 0 \ 0 \ | \ 0 \ 0]^t$$

Step 3. Compute R.

$$R = \begin{bmatrix} C \\ CA \\ CA^2 \end{bmatrix} = \begin{bmatrix} 1 & 0 & 1 \\ 0 & 1 & 0 \\ 1 & 0 & 2 \\ 0 & -1 & 0 \\ 1 & 0 & 4 \\ 0 & 1 & 0 \end{bmatrix}$$

Step 4. Compute T.

$$
T = \begin{bmatrix} 0 & 0 & 0 \\ CB & 0 & 0 \\ CAB & CB & 0 \end{bmatrix} = \left[\begin{array}{cc|cc|cc} 0 & 0 & 0 & 0 & 0 & 0 \\ 0 & 0 & 0 & 0 & 0 & 0 \\ \hline 1 & 1 & 0 & 0 & 0 & 0 \\ 0 & 0 & 0 & 0 & 0 & 0 \\ \hline 1 & 2 & 1 & 1 & 0 & 0 \\ 0 & 0 & 0 & 0 & 0 & 0 \end{array}\right]
$$

Step 5. Compute $Rx(1) = Y(1) - TU(1)$.

$$
\begin{bmatrix} 1 & 0 & 1 \\ 0 & 1 & 0 \\ 1 & 0 & 2 \\ 0 & -1 & 0 \\ 1 & 0 & 4 \\ 0 & 1 & 0 \end{bmatrix} x(1) = \left[\begin{array}{c} 2e + 1.5e^2 - 1.5 \\ e^{-1} \\ \hline 2e + 3e^2 - 2 \\ -e^{-1} \\ \hline 2e + 6e^2 - 3 \\ e^{-1} \end{array}\right]
$$

A solution exists because $(2e - 1) \times$ [column 1 of R] plus $e^{-1} \times$ [column 2 of R] plus $(1.5\ e^2 - 0.5) \times$ [column 3 of R] equals $[Y(1) - TU(1)]$; i.e., $[Y(1) - TU(1)]$ lies in the column space of R. Note that determining this is equivalent to actually computing the solution. Also, all measurements are exact.

Step 6. Computing the solution requires constructing a left inverse R^{-L} of R, satisfying $R^{-L}R = I$. Recall that R^{-L} is not unique but that the solution is. The Moore-Penrose pseudo inverse computed according to $R_1^{-L} = (R^tR)^{-1}R^t$ is

$$
R_1^{-L} = \begin{bmatrix} 1 & 0 & 0.5 & 0 & -0.5 & 0 \\ 0 & 0.3333 & 0 & -0.3333 & 0 & 0.3333 \\ -0.2857 & 0 & -0.07143 & 0 & 0.3571 & 0 \end{bmatrix}
$$

Two other left inverses, computed by inspection, are

$$
R_2^{-L} = \begin{bmatrix} 2 & 0 & -1 & 0 & 0 & 0 \\ 0 & 1 & 0 & 0 & 0 & 0 \\ -1 & 0 & 1 & 0 & 0 & 0 \end{bmatrix} \tag{5.22}
$$

and

$$
R_3^{-L} = \begin{bmatrix} 0 & 0 & 2 & 0 & -1 & 0 \\ 0 & 0 & 0 & 0 & 0 & 1 \\ -1 & 0 & 1 & 0 & 0 & 0 \end{bmatrix} \tag{5.23}
$$

Since the data are exact and the equations consistent, all three left inverses produce the same answer:

$$x(1) = R_i^{-L}[Y(1) - TU(1)] = \begin{bmatrix} 2e - 1 \\ e^{-1} \\ 1.5e^2 - 0.5 \end{bmatrix}$$

As a final caution, the reader should be aware that taking derivatives of the output y is an unrealistic engineering practice, done here for theoretical and pedagogical reasons.

GENERALIZATION OF PROBLEM 2

The generalization of problem 2 requires instantaneously changing the system state using impulsive inputs. The richness of the generalization lies in the multi-input structure of the $n \times m$ B-matrix.

Generalized Problem 2: *Given* $\dot{x}(t) = Ax(t) + Bu(t), x(0^-) = x_0,$ *where A is $n \times n$ and B is $n \times m$, find an impulsive vector input which will drive $x(0^-)$ to $x(0^+)$.*

The theorem which describes the solution procedure effectively duplicates theorem 5.1.

Theorem 5.3.   Let $\dot{x}(t) = Ax(t) + Bu(t),$ $x(0^-) = x_0,$ where A is $n \times n$ and B is $n \times m.$ Suppose

$$u(t) = \xi_0 \delta(t) + \xi_1 \delta^{(1)}(t) + \ldots + \xi_{n-1} \delta^{(n-1)}(t)$$

where $\xi_i \epsilon \mathbf{R}^m$ and $\delta^{(i)}(t)$ represents the ith derivative of the δ-function. Let $Q = [B | AB | \ldots | A^{n-1}B]$. Then

$$[x(0^+) - x(0^-)] = Q \begin{bmatrix} \xi_0 \\ \xi_1 \\ \vdots \\ \xi_{n-1} \end{bmatrix} \qquad (5.24)$$

where Q is an $n \times nm$ matrix.

As might be expected, the proof of this theorem duplicates the proof of theorem 5.1. The interesting aspects of the solution to equation 5.24 arise out of the fact that Q has in general more columns than rows.

Question 1: *When does a solution exist?*

Answer: *A solution exists whenever $[x(0^+) - x(0^-)]$ lies in the column space of Q.*

Sometimes $[x(0^+) - x(0^-)]$ will not lie in the column space of Q. However, is it possible for there always to exist a solution regardless of the value of $[x(0^-) - x(0^+)]$?

Question 2: *For arbitrary* $x(0^+)$ *and* $x(0^-)$, *when will there always exist a solution?*

Answer: *There will always exist a solution regardless of the values of* $x(0^+)$ *and* $x(0^-)$ *whenever* $rank[Q]=n$.

This condition regarding rank arises because $[x(0^+) - x(0^-)]$ is an arbitrary vector in $\mathbf{R}^n$. If $rank[Q]=n$, it is always possible to pick a set of n linearly independent columns of Q which form a basis for $\mathbf{R}^n$. If $rank[Q]<n$, there are vectors in $\mathbf{R}^n$ which cannot be expressed as a linear combination of its columns.

Question 3: *If* $rank[Q]=n$, *is the solution ever unique?*

Answer: *Yes, whenever* Q *is* $n \times n$ — *i.e., whenever* $m=1$ —*in which case* Q^{-1} *exists; otherwise the solution is not unique.*

Question 4: *How does one compute the solution when* $m > 1$ *and* $rank[Q]=n$?

Answer: *If* $rank[Q]=n$ *and* $m > 1$, *the solution is given by*

$$\begin{bmatrix} \xi_0 \\ \xi_1 \\ \vdots \\ \xi_{n-1} \end{bmatrix} = Q^{-R}[x(0^+) - x(0^-)] \tag{5.25}$$

where Q^{-R} *is any right inverse of* Q.

The matrix Q has a right inverse only when $rank[Q]=n$. A *right inverse of* Q is any matrix Q^{-R} for which $QQ^{-R} = I$. The use of the right inverse is somewhat different from the use of the left and the usual inverse of a matrix. One does not solve equation 5.24 by multiplying on the left by Q^{-R} to obtain equation 5.25; rather, the claim is simply that equation 5.25 is a solution to equation 5.24 in that it satisfies equation 5.24, i.e.,

$$Q\begin{bmatrix} \xi_0 \\ \xi_1 \\ \vdots \\ \xi_{n-1} \end{bmatrix} = QQ^{-R}[x(0^+) - x(0^-)] = [x(0^+) - x(0^-)]$$

The *Moore-Penrose pseudo (right) inverse* of Q is a particular right inverse of Q that is given by the formula

$$Q^{-R} = Q^t(QQ^t)^{-1} \tag{5.26}$$

As with the left inverse of a matrix, the right inverse is not unique whenever $m > 1$. In this case, however, neither is the solution. Each right inverse produces a different solution to the problem. The right inverse of equation 5.26 produces a solution vector $\text{col}[\xi_0, \xi_1, ..., \xi_{n-1}]$ whose Euclidean norm is smaller than that of any other solution. Physically speaking, this is the *minimum-energy solution*. Again, the numerical computation of Q^{-R} should be carried out not via equation 5.26 but rather through the SVD.

Question 5: *If* $rank(Q) = n$ *and* $m > 1$, *is it possible to characterize all the possible solutions to equation 5.24?*

Answer: *Yes, every solution to equation 5.24 has the form*

$$\begin{bmatrix} \xi_0 \\ \xi_1 \\ \vdots \\ \xi_{n-1} \end{bmatrix} = Q^{-R}[x(0^+) - x(0^-)] + V\alpha \qquad (5.27)$$

where the matrix $V = [v_1,...,v_p]$, *in which* v_i *is the ith column of V, and where* α *is a vector in* $\mathbf{R}^p$ *and the columns of V, viz.,* $v_1, ..., v_p$, *are a basis for the null space of Q. The null space of Q, denoted N[Q], is simply the set of all vectors v for which* $Qv = \theta$; *in the usual mathematical notation,*

$$N[Q] = \{v \mid Qv = \theta\}$$

We will assume that the dimension of $N[Q]$ — i.e., the number of required basis vectors for $N[Q]$ — is p.

Clearly, equation 5.26 makes sense. For consider multiplying both sides of equation 5.27 by Q:

$$Q\begin{bmatrix} \xi_0 \\ \xi_1 \\ \vdots \\ \xi_{n-1} \end{bmatrix} = QQ^{-R}[x(0^+) - x(0^-)] + QV\alpha$$

By definition, $QQ^{-R} = I$, the identity matrix, and $Qv = \theta$, since each column of V lies in N[Q]. Hence, for any α, equation 5.27 produces a valid solution of equation 5.24.

EXAMPLE 5.5.

For the purpose of illustrating the above ideas, consider the state dynamics

$$\dot{x}(t) = \begin{bmatrix} 1 & 0 & 0 \\ 0 & 0 & 0 \\ 0 & 0 & -1 \end{bmatrix} x(t) + \begin{bmatrix} 1 & 0 \\ 0 & 1 \\ 0 & 1 \end{bmatrix} u(t) \quad x(0^-) = \begin{bmatrix} 1 \\ -1 \\ 1 \end{bmatrix}$$

where for each instant of time $x(t) \in \mathbf{R}^3$ and $u(t) \in \mathbf{R}^2$. Suppose the desired initial state is $x(0^+) = [0\ 0\ 0]^t$. Then the problem is precisely generalized problem 2. The solution entails constructing an impulsive input $u(t) = $

$\xi_0 \delta(t) + \xi_1 \delta^{(1)}(t) + \xi_2 \delta^{(2)}(t)$, where $\xi_i \in \mathbf{R}^2$ for $i = 0,1,2$. According to equation 5.25,

$$\begin{bmatrix} \xi_0 \\ \xi_1 \\ \xi_2 \end{bmatrix} = Q^{-R} \begin{bmatrix} -1 \\ 1 \\ -1 \end{bmatrix}$$

where

$$Q = \begin{bmatrix} 1 & 0 & 1 & 0 & 1 & 0 \\ 0 & 1 & 0 & 0 & 0 & 0 \\ 0 & 1 & 0 & -1 & 0 & 1 \end{bmatrix}$$

The Moore-Penrose pseudo right inverse for Q is

$$Q_1^{-R} = Q^t[Q\,Q^t]^{-1} = \begin{bmatrix} 0.3333 & 0 & 0 \\ 0 & 1 & 0 \\ 0.3333 & 0 & 0 \\ 0 & 0.5 & -0.5 \\ 0.3333 & 0 & 0 \\ 0 & -0.5 & 0.5 \end{bmatrix}$$

This leads to the solution

$$[\xi_0^t \ \xi_1^t \ \xi_2^t]^t = [-0.3333 \ 1 \ -0.3333 \ 1 \ -0.3333 \ -1]^t \qquad (5.28)$$

which has Euclidean norm 1.825742. For clarity, the resulting input takes the form

$$u(t) = \begin{bmatrix} -0.3333 \\ 1 \end{bmatrix} \delta(t) + \begin{bmatrix} -0.3333 \\ 1 \end{bmatrix} \delta^{(1)}(t) + \begin{bmatrix} -0.3333 \\ -1 \end{bmatrix} \delta^{(2)}(t)$$

A second right inverse, computed by inspection, is

$$Q_2^{-R} = \begin{bmatrix} 1 & 0 & 0 \\ 0 & 1 & 0 \\ 0 & 0 & 0 \\ 0 & 1 & 0 \\ 0 & 0 & 0 \\ 0 & 0 & 1 \end{bmatrix}$$

This leads to a solution

$$u(t) = \begin{bmatrix} -1 \\ 1 \end{bmatrix} \delta(t) + \begin{bmatrix} 0 \\ 1 \end{bmatrix} \delta^{(1)}(t) + \begin{bmatrix} 0 \\ -1 \end{bmatrix} \delta^{(2)}(t) \qquad (5.29)$$

The Euclidean norm of $[-1 \ 1 \ 0 \ 1 \ 0 \ -1]^t$ is 2. Thus, the solution has a norm which exceeds the "minimum norm" solution given by equation 5.28.

Every solution which will drive $x(0^-) = [1 \ -1 \ 1]^t$ to θ must satisfy equation 5.27, which we write as

$$
\begin{bmatrix} \xi_0 \\ \xi_1 \\ \xi_2 \end{bmatrix} = \begin{bmatrix} -0.3333 \\ 1 \\ -0.3333 \\ 1 \\ -0.3333 \\ -1 \end{bmatrix} + \begin{bmatrix} 0 & 0 & 1 \\ 0 & 0 & 0 \\ 0 & 1 & -1 \\ 1 & 0 & 0 \\ 0 & -1 & 0 \\ 1 & 0 & 0 \end{bmatrix} \begin{bmatrix} \alpha_1 \\ \alpha_2 \\ \alpha_3 \end{bmatrix}
$$

where the 6×3 matrix is V mentioned in question 5, whose columns are a basis for the null space of Q. These basis vectors were computed by inspection. Observe that if $[\alpha_1 \ \alpha_2 \ \alpha_3]^t = [0 \ -0.3333 \ -0.6666]$, we obtain the same solution as equation 5.29.

Thus far, our attention has focused on the case where rank$[Q] = n$. The case where rank$[Q] < n$ is deferred to chapter 20. The only remaining aspect is the problem of numerical computation. Again, the computation of right inverses and bases for the null space of a matrix falls out of the singular value decomposition (SVD) covered in chapter 19.

CONCLUDING REMARKS

Given the initial conditions and input for a scalar differential equation model of a system, our development has pointed out how to determine the initial condition vector of the derived state model so that the responses agree. The development then detailed how to set up the initial condition vector using impulsive inputs. By viewing these questions in a more general context, the development tackled simplified versions of the problems of state reconstruction and state adjustment using impulsive inputs, both for multi-input multi-output state models.

Under certain rank conditions, solutions to these problems were sketched. For Generalized Problem 1, assuming rank$[R] = n$, the left inverse of R, denoted R^{-L}, provided the desired answer. Similarly, the solution to Generalized Problem 2 required that rank$[Q] = n$. The right inverse of Q, denoted Q^{-R}, became the focal point where a different but usable solution was computed for each Q^{-R}. The null space of Q linked all the solutions — i.e., the difference vector between any two solution vectors must be an element of $N[Q]$.

Solutions to generalized problems 1 and 2 were not given when rank$[R] < n$ or rank$[Q] < n$. These questions and the completion of the associated investigations occur in the chapters on controllability, observability, and their numerical aspects. In fact in those chapters the R-matrix becomes known as the observability matrix since its properties determine what states are reconstructable and hence observable. The Q-matrix becomes the controllability matrix since its properties determine what states can be driven to zero and vice versa.

When the R-matrix has rank less than n, then a certain subspace of states cannot be detected at the output. Intuitively the system A-matrix interaction with the output matrix C (look at the structure of R) filters out certain state information. In particular every state in the null space of R becomes filtered out. The null space of R becomes the unobservable subspace. If rank$[R]=n$, there is no null space. Hence when rank$[R]<n$, only certain states are reconstructable. Similarly when rank$[Q]<n$, only certain states within the column space of Q can be transferred to each other in finite time. The column space of Q becomes known as the controllable subspace.

The many ideas of this chapter are meant as an "introduction to" and a "preparation for" the many physically motivated system theoretic concepts to come. Expectation is the watchword.

PROBLEMS

1. Define, from whatever source is available, the terms Gaussian elimination, equilibration, partial pivoting, and iterative improvement.

2. In the context of the observable canonical form, compute the state vector at $t = 0$ for

$$\dddot{y} + \ddot{y} + \dot{y} + y = u + \dot{u} + \ddot{u}$$

where $\ddot{y} = y(0) = 1$, $\dot{y}(0) = -1$, and $u(t) = sin(2\pi t)1^+(t+1)$

3. In the context of the observability canonical form, compute the state vector at $t = 0$ for

$$\dddot{y} + \ddot{y} + \dot{y} + y = u + \dot{u} + \ddot{u}$$

where $\ddot{y}(0) = y(0) = 1$, $\dot{y}(0) = -1$, and $u(t) = sin(2\pi t)1^+(t+1)$

4. In the context of the controllable canonical form, compute the state vector at $t = 0$ for

$$\dddot{y} + \ddot{y} + \dot{y} + y = u + \dot{u} + \ddot{u}$$

where $\ddot{y}(0) = y(0) = 1$, $\dot{y}(0) = -1$, and $u(t) = sin(2\pi t)1^+(t+1)$

5. Find an impulsive input which will set up the initial conditions computed in problem 4, with regard to the controllable canonical form, at $t = 0^+$, given $x(0^-) = \theta$.

6. Using the results of problem 2, given $x(0^-) = \theta$, find an impulsive input which will set up the computed initial conditions at $x(0^+)$ with regard to the observable canonical form.

7. For the differential equation

$$2y^{(4)} + 4y^{(3)} + 2y^{(2)} + 4y = 6u + 4u^{(1)} + 2u^{(3)}$$

with initial conditions $y^{(3)}(0) = y^{(2)}(0) = y^{(1)}(0) = y(0) = 1$, assuming the input and its derivatives are zero at $t = 0^-$, find the appropriate initial condition vector $x(0)$ and an appropriate input to set up these initial conditions. (If necessary, use a computer to invert the matrix.)

8. Consider the differential equation

$$\ddot{y} - \dot{y} - y = 2\ddot{u}$$

where $u(t) = (1 - e^{-t})1^+(t)$ and $y(0) = \dot{y}(0) = \ddot{y}(0) = 1$.

(a) Put the equation in observable canonical form by *inspection* if possible.

(b) Determine the necessary initial state vector $x(0)$.

(c) Construct an impulsive input which will set up these initial conditions.

9. Consider the differential equation

$$\ddot{y} + 2\dot{y} - 2y = u - \dot{u}$$

where $\dot{y}(0) = y(0) = 1$ and $u(t) = \sin(t)\,1^+(t)$.
(a) Put the equation in observable canonical form.
(b) Find the appropriate initial state vector $x(0)$.
(c) Find an impulsive input which will set up this initial condition at 0^+ assuming that $x(0^-) = \theta$.

10. Consider the differential equation

$$\ddot{y} + 2\dot{y} - 2y = u - \dot{u} + \ddot{u}$$

where $\dot{y}(0) = y(0) = 1$ and $u(t) = \sin(t)1^+(t)$.
(a) Put the equation in controllable canonical form.
(b) Find the appropriate initial state vector $x(0)$.
(c) Find an impulsive input which will set up this initial condition at 0^+ assuming that $x(0^-) = \theta$.

11. Given that $\dot{x} = Ax + Bu$, where

$$A = \begin{bmatrix} a_1 & a_2 & a_3 \\ 1 & 0 & 0 \\ 0 & 1 & 0 \end{bmatrix}, B = \begin{bmatrix} 1 \\ 0 \\ 0 \end{bmatrix}$$

(a) Find the characteristic polynomial $\pi(\lambda)$ of A.
(b) If $a_1 = 2$, $a_2 = -2$, and $a_3 = 3$, find the state feedback $F = [f_1\ f_2\ f_3]$ such that the spectrum of $A + BF$ is $\{0, -1, +1\}$.

12. For the system shown in Figure P5.12,
(a) Write the state equation.
(b) Is the system realization controllable, i.e., does rank$[Q] = n$? Is it observable, i.e., does rank$[R] = n$?
(c) What is the transfer function from $u(\cdot)$ to $y(\cdot)$?

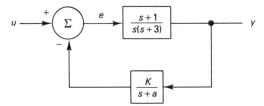

Figure P5.12 Block diagram for problem 12.

13. Consider the scalar differential equation

$$\dot{x}(t) = \lambda x(t) + \beta u(t)$$

Find two different nonimpulsive inputs which will drive the initial state $x(t_0)$ to $x(t_1)$. (*Hint*: Select an input form such as $\alpha 1^+(t)$, where α is a scalar to be determined in terms of λ, β, $x(t_0)$, and $x(t_1)$. Are your results valid if $\beta = 0$? Note that if $\beta = 0$, the system is said to be *uncontrollable*—i.e., there does not exist an input which will drive the system from an arbitrary initial state $x(t_0)$ to an arbitrary final state $x(t_1)$.)

14. Consult a standard linear algebra text such as [1] or [9] for the purpose of understanding the notions of the rank of a rectangular constant matrix, left and/or right inverse of a rectangular matrix, null space of a rectangular matrix, and image space of a rectangular matrix. What do the rows and/or columns of the matrix have in common with these notions when viewed as vectors in a vector space? Investigate the application of these ideas to the solution of a linear system of equations. Other helpful references might be [2], [10], and [7].

15. Suppose the observed response $y(t)$ in example 5.4 were measured as

$$y(t) = \begin{bmatrix} 2.1e^t + 1.4e^{2t} - 1.4 \\ 0.98e^{-t} - 0.02 \end{bmatrix} 1^+(t)$$

   due to noise and modeling errors.
   (a) Is equation 5.17 consistent at $t = 1$?
   (b) Compute the least squares solution.
   (c) Compute the solutions using equations 5.22 and 5.23 as left inverses of R.
   (d) Compare the resulting solutions by computing the Euclidean norm of equation 5.21 in each case.

16. Consider the usual linear time-invariant state model. Let

$$A = \begin{bmatrix} 0 & 1 \\ 0 & 0 \end{bmatrix} ; \ B = \begin{bmatrix} 1 & 1 \\ 0 & 1 \end{bmatrix} ; \ C = [1 \ \ -1]; \ D = [0]$$

   Let $u(t) = (-6 \ 6)^T t \ 1^+(t)$, and suppose the measured output $y(t) = 3 - 3t^2 + t^3$. Find $x(1)$.

17. Consider the time-varying state model

$$\dot{x}(t) = Ax(t) + Bu(t)$$

$$y(t) = C(t)x(t) + Du(t)$$

   where A, B, and D are constant matrices of dimensions 2×2, 2×1, and 1×1, respectively, and $C(t)$ is a time-varying matrix of dimension 1×2.

   (a) Suppose $y(t)$ and $u(t)$ are known for all time but no initial state vector is known. Derive a set of two linear algebraic equations whose solution at any time t will yield the unknown state vector $x(t)$. Put the equations in matrix form.

   (b) Suppose now that

$$A = \begin{bmatrix} 1 & 1 \\ 1 & 1 \end{bmatrix} ; \ B = \begin{bmatrix} 1 \\ 2 \end{bmatrix} ; \ C = [t - 1 \ \ 1]; \ D = [1]$$

   Suppose further that $u(t) = \sin(t)1^+(t + 1)$ and $y(t) = (t \exp(t) + 1)$ $1^+(t + 1)$. Find $x(0)$.

18. Verify equation 5.13.

19. Given a single-input, single-output state model (equation 5.2), develop an algorithm to construct the differential equation of equation 5.1.
   Hint:
   (*i*) Append the equation for $y^{(n)}$ to equation 5.4.
   (*ii*) Write this equation in the form of equation 5.5 as

$$\hat{Y}(t) = \hat{R}x(t) + \hat{T}\hat{U}(t)$$

   (*iii*) Form the matrix $[\hat{R} \,|\, I \,|\, \hat{T}]$.
   (*iv*) Upper triangularize $\hat{R}$ and thereby $[\hat{R} \,|\, I \,|\, \hat{T}]$ so that at least the bottom row of the reduced $\hat{R}$ is zero.
   (*v*) See [11] for further details.

20. Generalize the procedure of problem 19 to multi-input, multi-output systems.

21. Consider the system

$$\dot{x}(t) = Ax(t) + Bu(t)$$
$$y(t) = C(t)x(t) + Du(t)$$

where

$$A = \begin{bmatrix} -1 & 0 \\ -1 & 0 \end{bmatrix} \quad B = \begin{bmatrix} 1 \\ 1 \end{bmatrix} \quad C(t) = \begin{bmatrix} e^t & 0 \\ 0 & e^t \end{bmatrix} \quad D = \begin{bmatrix} 1 \\ 0 \end{bmatrix}$$

It is known that when $u(t) = e^{-t}1^+(t+1)$, the response to a particular $x(0)$ is

$$y(t) = \begin{bmatrix} 1 + t + e^{-t} \\ 1 + t \end{bmatrix} 1^+(t+1)$$

(a) Using only $y(t)$ and $\dot{y}(t)$ and the known input and its derivatives, derive a set of equations in A, B, $C(t)$, and D whose solution will yield the state $x(t)$.

(b) Substitute into the equations developed in (a), and compute the initial state $x(0)$ using the appropriate left or right inverse.

(c) Suppose that Vader's Vassels were jamming your observation equipment on Cirus 3 so that instead of the initial state computed in part (b), your calculations produced the initial state $x(0^-) = (2\ 2)^t$. For the given system, find an impulsive input which will drive this initial state to $x(0^+) = (0\ 0)^t$ as follows:

  (i) What is the form of the impulsive input?

  (ii) Either derive or write by inspection the form of the equations to be solved.

  (iii) Substitute the appropriate matrices and find a solution.

(e) Characterize the set of all possible solutions to part (c).

(f) From your answer to part (d), can you find (at least guess at) the minimum energy solution to (c). Justify your answer at least intuitively.

REFERENCES

1. K. Hoffman and R. Kunze, *Linear Algebra* (Englewood Cliffs, NJ: Prentice-Hall, 1971).

2. G. E. Forsythe, M. A. Malcolm, and C. B. Moler, *Computer Methods for Mathematical Computations* (Englewood Cliffs, NJ: Prentice-Hall, 1977).

3. J. H. Wilkenson, *Rounding Errors in Algebraic Processes* (Englewood Cliffs, NJ: Prentice-Hall, 1963).

4. W. Miller, and C. Wrathall, *Software for Roundoff Analysis of Matrix Algorithms* (New York: Academic Press, 1980).

5. Clare D. McGillem and George R. Cooper, *Continuous and Discrete Signal and System Analysis* (New York: Holt, Rinehart and Winston, 1974).

6. G. Birkhoff and S. Maclane, *A Survey of Modern Algebra,* 4th ed. (New York: Macmillan, 1977).

7. V. Klema, and A. Laub, "The Singular Value Decomposition: Its Computation and Some Applications," *IEEE Transactions on Automatic Control,* Vol. AC-25, No. 2, April 1980, pp.164–176.

8. H. W. Sorenson, "Least-squares Estimation from Gauss to Kalman," *IEEE Spectrum,* July 1970, pp.63–68.

9. A. H. Zemanian, *Distribution Theory and Transform Analysis* (New York: McGraw-Hill, 1965).

10. C. A. Desoer, *Notes for a Second Course on Linear Systems* (New York: Van Nostrand, 1970).

11. C. C. Blackwell, "On Obtaining the Coefficients of the Output Transfer Function from a State Space Model and an Output Model of a Linear Constant Coefficient System," *IEEE Transactions on Automatic Control,* Vol. AC-29, No. 12, December 1984, pp.1122–1124.

6

INTRODUCTION TO NUMERICAL COMPUTATION AND GAUSSIAN ELIMINATION

INTRODUCTION

This chapter centers on the numerical solution of the matrix equation $Mx = b$ by the method of Gaussian elimination. The persistent need to solve linear systems of equations in this form motivates this important discussion. In past chapters many equations had the form $Mx = b$, for example chapter 2, equation 2.21, which defined the constants in an analytic solution to a differential equation. When finding the state model matrices describing a circuit (chapter 3, equation 3.8), once again one solved an equation of the form $Mx = b$. Chapter 4 explained the notion of a time-invariant state transformation which required finding the inverse matrix, T^{-1}. Numerical calculation of T^{-1} uses Gaussian elimination. Equations of the form $Mx = b$ also cropped up in chapter 5: equation 5.6, equations 5.15, 5.17, etc. The next two chapters covering the Newton-Raphson solution of nonlinear algebraic equations and the numerical simulation of the nonlinear state model will directly utilize Gaussian elimination.

These needs make it important to provide some basic understanding of the hows and whys of a numerical implementation of Gaussian elimination. Before entering upon the discussion, however, it is important to realize that algorithms which fail to account for the inherent finite word length and the finite range of numbers in a digital computer typically fail to give good performance [1]. For example, in the floating-point arithmetic of a digital computer, the associative law of addition, $a + [b + c] = [a + b] + c$, does not hold [2]. Thus, the usual approach to computing the eigenvalues of a matrix — e.g., given a 5×5 square matrix, (i) compute its characteristic polynomial and (ii) find the roots by some root-finding scheme — is a numerical disaster. Even the quadratic formula for finding the roots of a second-order polynomial must be carefully implemented [3]. A sound, reliable numerical approach is to use the QR-algorithm [1,3]; this algorithm, however, is not in any reasonable sense adaptable to hand calculation.

Again, the difficulties rest with the representation of numbers in a digital computer and the finite range of numbers represented. For example, the eight-bit binary representation of the base 10 decimal 0.4 is

$$(0.4)_{10} = 0/2^0 + 0/2^1 + 1/2^2 + 1/2^3 + \ldots$$

$$\approx (0.0110011)_2 = (0.398438\ldots)_{10} \tag{6.1}$$

The base 2 number, indicated by the subscript 2, is the computer's floating-point representation, and the difference between 0.4 and 0.398438... is the well-recognized round-off error. Technically, *round-off is the error in the difference between the true value of a base-10 number and the base-10 equivalent of its binary representation (base-2) in the computer, which has a fixed number of bits.* The often observed computer error messages of *overflow* and *underflow* sport the problem of the finite range of number representations.

An example of how round-off could affect the solution of an initial condition determination as discussed in chapter 5 is the following.

EXAMPLE 6.1

Consider the infinite precision-matrix equation

$$\begin{bmatrix} 1 & 0.1 & 0.01 \\ 0.1 & 1 & 0.01 \\ 0.1 & 0.01 & 1 \end{bmatrix} \begin{bmatrix} x_1(0) \\ x_2(0) \\ x_3(0) \end{bmatrix} = \begin{bmatrix} 300 \\ 1110 \\ 10020 \end{bmatrix} \tag{6.2}$$

where the desired initial conditions $x_i(0)$ all have units of millivolts. Suppose for pedagogical purposes that the smallest nonzero floating-point number representation (base 10) in the computer is 0.1 (one-tenth). Then the floating-point representation of equation 6.2 becomes

$$\begin{bmatrix} 1 & 0.1 & 0 \\ 0.1 & 1 & 0 \\ 0.1 & 0 & 1 \end{bmatrix} \begin{bmatrix} \hat{x}_1(0) \\ \hat{x}_2(0) \\ \hat{x}_3(0) \end{bmatrix} = \begin{bmatrix} 300 \\ 1110 \\ 10020 \end{bmatrix} \tag{6.3}$$

where the carets indicate the approximate nature of the solution variables. Solving equations 6.2 and 6.3 for exact solutions then yields

$$\begin{bmatrix} \hat{x}_1(0) \\ \hat{x}_2(0) \\ \hat{x}_3(0) \end{bmatrix} = \begin{bmatrix} 190.91 \\ 1090.91 \\ 10000.91 \end{bmatrix} \neq \begin{bmatrix} 100 \\ 1000 \\ 10000 \end{bmatrix} = \begin{bmatrix} x_1(0) \\ x_2(0) \\ x_3(0) \end{bmatrix}$$

This contrived example demonstrates the unwanted upshot of the inaccurate representation of numbers in the computer, assuming exact algorithmic calculations. Exact calculations can rarely be performed, and errors in algorithms will often magnify and even introduce further round-off. This brings up a second aspect of the problem of numerical computations: the stability of the algorithm. The following hypothetical example depicts the potential for catastrophe arising out of stability-related errors.

EXAMPLE 6.2

Consider the calculation of $\exp(-6)$ via the Taylor series. By definition,

$$\exp(-6) = 1 + (-6) + \frac{(-6)^2}{2!} + \frac{(-6)^3}{3!} + \frac{(-6)^4}{4!} + \dots \quad (6.4)$$

$$= 1 - 6 + \frac{36}{2!} - \frac{216}{3!} + \frac{1296}{4!} - \frac{7776}{5!} + \dots$$

Now, how would a computer evaluate this series? Briefly, it would add successive terms until additional terms make no difference in the floating-point representation of the sum. More specifically, let S_k represent the sum of the first k terms in the Taylor series of equation 6.4. The computer would decide it has computed the correct number for $\exp(-6)$ when $fl\{S_k\} = fl\{S_{k+1}\}$, where fl designates the floating-point representation of the operand in braces. For the sake of argument, suppose the computer evaluates only the first eight terms in the series. Then, from equation 6.4, this "unstable" approach produces the absurd answer

$$\exp(-6) \approx 17.1143$$

The problem, of course, is that of adding and subtracting relatively large numbers (whose possible round-off is larger than the desired answer) to obtain a very small number.

Can one more accurately compute $\exp(-6)$ without all this adding and subtracting? Yes, but not directly. First observe that $\exp(-6) = [\exp(6)]^{-1}$ and that computing $\exp(6)$ does not require any subtractions. Taking the first eight terms in a Taylor series of $\exp(6)$ produces

$$\exp(6) \approx 1 + 6 + \frac{36}{2!} + \frac{216}{3!} + \dots + = 341.8$$

Taking the reciprocal of this number yields $\exp(-6) \approx 2.926 \times 10^{-3}$, which much better approximates the correct answer of 2.479×10^{-3}. The conclusion to all this calculation is that reasonable caution must be exercised in developing a numerical algorithm for solving $Mx = b$.

The stumbling blocks presented in examples 6.1 and 6.2 pinpoint two hazards to numerical computation. With these potential hazards in mind, the following statement poses the objective of this chapter.

Objective: Assuming that the n × n matrix M is nonsingular, find a numerically efficient and stable algorithm for solving the equation Mx = b for x.

Two questions immediately come to mind:

Question 1: What is meant by "numerically efficient"?

Answer: In this context, "numerically efficient" means allocating minimal storage and affording a minimal number of operations (essentially multiplications).

Question 2: What is meant by "numerically stable"?

Answer: "Numerically stable" means that the algorithm proceeds with minimal magnification and minimal new introduction of round-off error.

One technique which meets these criteria is a form of Gaussian elimination [3–7] with pivoting. The value of Gaussian elimination, and indeed the justification for the viewpoint set out here, is best seen by contrasting it with two other methods of solution, *Cramer's rule* and the *inversion method*.

Cramer's rule (entrenched in undergraduate courses such as basic circuit theory [8]) is a technique for solving for each entry of $x = [x_1,...,x_n]^t$ in $Mx = b$ as a ratio of determinants, specifically,

$$x_i = \frac{\det[M_i]}{\det[M]} \tag{6.5}$$

where M_i is the matrix M with the ith column replaced by the vector b. Computation of x via Cramer's rule requires computation of $n + 1$ determinants. By a direct approach, each determinant will require the sum of $(n - 1)!$ products containing an n-fold product each. Thus, by direct application, Cramer's rule requires $(n + 1) \times (n) \times (n-1)!$ multiplications. (How many years would it take to solve a 20×20 system of equations by a computer which executes a multiplication every microsecond?) On the other hand, it is possible to compute each determinant with about $n^3/3$ operations. However, this is still too costly in number of multiplications, as well as being numerically sensitive because of the large number of additions and subtractions.

The other usual approach to solving $Mx = b$ is the inversion method. Here, one explicitly computes M^{-1} and finds $x = M^{-1}b$. One criticism of this method is that construction of M^{-1} essentially requires solution of $Mx = b$ n times. In particular, let η_i be the n-vector with zeros everywhere except in the ith position, which is unity, and let $M^{-1} = [q_1,...,q_n]$ have columns q_i. Then, computation of M^{-1} requires solving $Mq_i = \eta_i$ for $i = 1,...,n$. Futher, one must execute the matrix-vector multiplication $M^{-1}b$, which potentially introduces more round-off. So, unless there is some more compelling reason for computing M^{-1}, inefficiency rules out this method for solving $Mx = b$.

BASIC GAUSSIAN ELIMINATION

In presenting Gaussian elimination, we begin with the solution of $Mx = b$ when $M = L$, a lower triangular matrix, and progress through an upper triangular case and finally a general case. The goal is to cover the rudiments of Gaussian elimination without delving into clever numerical tricks such as pivoting. Accordingly, special strategies and discussions of numerical accuracy and efficiency are topics of the next two sections.

Throughout this section the underlying assumption is that M is $n \times n$ and nonsingular. A lower triangular matrix L has all its entries equal to zero above the diagonal. Since the determinant of a lower triangular matrix is the product of its diagonal entries, L is nonsingular if and only if all its diagonal entries are nonzero. The lower triangularity of L makes the solution of $Mx = Lx = b$ straightforward. The following example illustrates the simplicity of the solution for this case and motivates a recursive solution scheme.

EXAMPLE 6.3

Consider the matrix equation

$$\begin{bmatrix} \ell_{11} & 0 & 0 \\ \ell_{21} & \ell_{22} & 0 \\ \ell_{31} & \ell_{32} & \ell_{33} \end{bmatrix} \begin{bmatrix} x_1 \\ x_2 \\ x_3 \end{bmatrix} = \begin{bmatrix} b_1 \\ b_2 \\ b_3 \end{bmatrix} \tag{6.6}$$

Step 1. Solve for x_1. Find x_1, x_2, and x_3. We have $\ell_{11}x_1 = b_1$. Therefore,

$$x_1 = \frac{b_1}{\ell_{11}} \tag{6.7}$$

Step 2. Solve for x_2. Since we know x_1, we have, from equation 6.6, $\ell_{21}x_1 + \ell_{22}x_2 = b_2$. Thus,

$$x_2 = \frac{b_2 - \ell_{21}x_1}{\ell_{22}} \tag{6.8}$$

Step 3. Solve for x_3. Using the same maneuvers as in steps 1 and 2, we obtain

$$x_3 = \frac{b_3 - \ell_{31}x_1 - \ell_{32}x_2}{\ell_{33}} \tag{6.9}$$

where x_1 is given by equation 6.7 and x_2 by equation 6.8.

This "forward" moving recursion (building on previously computed x_i's at each step) is known as *forward substitution*. The technique not only is recursive but avoids explicitly computing the inverse of L. Generally, this method is preferable to directly computing L^{-1} and subsequently multiplying by $L^{-1}b$. However, one circumstance in which directly computing L^{-1} becomes desirable is when, for a given L, one must solve the equation $Lx = b_i$ for a series of vectors b_i. Here, the additional overhead in computing the inverse pays off, especially if the number of b_i's is large. (See problem 6 for a forward-substitution technique to compute L^{-1}.)

Theorem 6.1. Given the linear equation $Lx = b$, where $L = [\ell_{ij}]$ is $n \times n$ and nonsingular, $b = [b_1, \ldots, b_n]^t$ is in $\mathbf{R}^n$, and the unknown vector $x = [x_1, \ldots, x_n]^t$ is in $\mathbf{R}^n$, the recursive formula for computing the solution x is

$$x_k = \frac{b_k - \sum_{j=1}^{k-1} \ell_{kj} x_j}{\ell_{kk}} \tag{6.10}$$

for $k = 1, 2, \ldots, n$.

The proof of equation 6.10 is the obvious straightforward generalization of the technique used in example 6.3. The structure of the formula immediately suggests a similar recursion for upper triangular matrices $U = [u_{ij}]$, i.e., matrices with zeros below the diagonal ($u_{ij} = 0$ for $i > j$). Analogous to a lower triangular matrix, an upper triangular matrix is nonsingular if and only if all its diagonal entries are nonzero.

EXAMPLE 6.4

Consider the linear system of equations $Ux = b$, where U is nonsingular and

$$\begin{bmatrix} u_{11} & u_{12} & u_{13} \\ 0 & u_{22} & u_{23} \\ 0 & 0 & u_{33} \end{bmatrix} \begin{bmatrix} x_1 \\ x_2 \\ x_3 \end{bmatrix} = \begin{bmatrix} b_1 \\ b_2 \\ b_3 \end{bmatrix} \tag{6.11}$$

A quick glance indicates that x_3 is easiest to solve for.

Step 1. Compute x_3. Clearly,

$$x_3 = \frac{b_3}{u_{33}} \tag{6.12}$$

Step 2. Compute x_2. Similarly, $u_{22}x_2 + u_{23}x_3 = b_2$, so that

$$x_2 = \frac{b_2 - u_{23}x_3}{u_{22}} \tag{6.13}$$

Step 3. Compute x_1. Finally,

$$x_1 = \frac{b_1 - u_{12}x_2 - u_{13}x_3}{u_{11}} \tag{6.14}$$

where x_2 and x_3 are given by equation 6.13 and 6.12, respectively.

This backwards recursion method is known as *back substitution*. Note that the technique avoids explicitly computing U^{-1}. The following theorem (whose proof is left as an exercise) presents the general formula.

Theorem 6.2. Given a nonsingular $n \times n$ upper triangular matrix U and a vector $b = [b_1,...,b_n]^t$ in $\mathbf{R}^n$, the recursion formula for computing the solution x in $Ux = b$ is

$$x_k = \frac{b_k - \displaystyle\sum_{j=k+1}^{n} u_{kj}x_j}{u_{kk}} \qquad (6.15)$$

for $k = n, n-1,...,1$.

The two special cases of theorems 6.1 and 6.2 do not directly confront the real problem of solving $Mx = b$. Imagine, however, that M has the factorization $M = LU$, where L is lower triangular and U is upper triangular. Then

$$Mx = LUx = L(Ux) = b \qquad (6.16)$$

and the straightforward course of action is (*i*) solve $Ly = b$ for vector y by forward substitution, and then (*ii*) solve $Ux = y$ by back substitution for the desired vector x. This factorization idea unifies theorems 6.1 and 6.2 in the following theorem.

Theorem 6.3. Suppose that the $n \times n$ nonsingular matrix M admits a factorization $M = LU$. Then the solution to $Mx = LUx = b$ is accomplished by solving

(*i*) $Ly = b$ for y via

$$y_k = \frac{b_k - \displaystyle\sum_{j=1}^{k-1} \ell_{kj}y_j}{\ell_{kk}} \qquad (6.17)$$

for $k = 1,2,...,n$, and

(*ii*) $Ux = y$ for x according to

$$x_k = \frac{y_k - \displaystyle\sum_{j=k+1}^{n} u_{kj}x_j}{u_{kk}} \qquad (6.18)$$

for $k = n, n-1,...,1$.

If an LU-factorization of M is known, all we need do is automatically execute equations 6.17 and 6.18. However, seldom is an LU factorization obvious. Two questions arise:

(*i*) *When does M have an LU factorization?*
and more importantly,

(*ii*) *How does one compute the factorization when it exists?*
The following theorem, describing the *Crout algorithm* [4], partially answers the question of "when" and fully accounts for "how."

Theorem 6.4. (*Crout Algorithm*). Let M be as before with all principal minors nonzero. Then M has a unique factorization $M = LU$ with L a lower triangular and U an upper triangular matrix constrained to have 1's on its diagonal.

Further, the entries of $L = [\ell_{ij}]$ and $U = [u_{ij}]$ are computed row by row according to the respective formulas

$$\ell_{ik} = m_{ik} - \sum_{j=1}^{k-1} \ell_{ij} u_{jk} \qquad \text{for} \quad i \geqslant k \qquad (6.19)$$

and

$$u_{ki} = \frac{m_{ki} - \sum_{j=1}^{k-1} \ell_{kj} u_{ji}}{\ell_{kk}} \qquad \text{for} \quad i \geqslant k \qquad (6.20)$$

For detailed proofs of this theorem, see [9] and [10, pp. 35–40]. Also in [10] are extensions to block matrices; these extensions are attractive for parallel processing techniques.

Some questions and answers are as follows. What are the principal minors of a matrix? These are the principle determinants [11] illustrated in the next example. Are the preceding formulas correct, and how are they implemented? A formal proof would answer both questions. However, in the interests of pedagogy, let us consider a simple 3×3 example. In addition to illustrating how equations 6.19 and 6.20 work, the example will indicate an efficient means of storing the L and U factors in the space originally allocated to M.

EXAMPLE 6.5

Consider a 3×3 matrix

$$M = \begin{bmatrix} m_{11} & m_{12} & m_{13} \\ m_{21} & m_{22} & m_{23} \\ m_{31} & m_{32} & m_{33} \end{bmatrix}$$

Step 1. What are the principal minors of M? Quite simply, there are three:

(*i*) m_{11},

(*ii*) $\det \begin{bmatrix} m_{11} & m_{12} \\ m_{21} & m_{22} \end{bmatrix} = m_{11}m_{22} - m_{21}m_{12}$,

(*iii*) $\det[M]$.

Our presumption, of course, is that M is nonsingular, so $\det[M] \neq 0$, and in addition we assume $m_{11} \neq 0$ and $m_{11}m_{22} - m_{21}m_{12} \neq 0$. Ordinarily one would not check the principal minors of M. Why? Because if a principal minor is zero, the recursive procedure of equations 6.19 and 6.20 will break down at some step. At this point one can permute the rows of M and proceed to compute the LU factorization of a permuted M matrix. This technique is called *pivoting* and will be discussed in detail in the section "Advanced Gaussian Elimination." Further, computing determinants is numerically ill-conditioned, from both an operations count and the vantage point of round-off error.

Step 2.   Now suppose M has the LU factorization

$$\begin{bmatrix} m_{11} & m_{12} & m_{13} \\ m_{21} & m_{22} & m_{23} \\ m_{31} & m_{32} & m_{33} \end{bmatrix} = \begin{bmatrix} \ell_{11} & 0 & 0 \\ \ell_{21} & \ell_{22} & 0 \\ \ell_{31} & \ell_{32} & \ell_{33} \end{bmatrix} \begin{bmatrix} 1 & u_{12} & u_{13} \\ 0 & 1 & u_{23} \\ 0 & 0 & 1 \end{bmatrix} \qquad (6.21)$$

where U is required to have 1's along its diagonal. Consider the first row of L times the matrix U. The simple relationships between the first rows of M, L, and U become

(*i*) $m_{11} = \ell_{11}$, which one stores in the location of m_{11}.

(*ii*) $m_{12} = \ell_{11} u_{12}$, so that $u_{12} = m_{12}/\ell_{11}$, which is stored in the location of m_{12}.

(*iii*) $m_{13} = \ell_{11} u_{13}$, so that $u_{13} = m_{13}/\ell_{11}$, which is stored in the location of m_{13}.

Step 3.   Now consider the second row of L times the matrix U. Again, the simple relationships become

(*i*) $m_{21} = \ell_{21}$, which one stores in the m_{21} location.

(*ii*) $m_{22} = \ell_{21} u_{12} + \ell_{22}$, which implies that $\ell_{22} = m_{22} - \ell_{21} u_{12}$, which is stored in the location of m_{22}.

(*iii*) $m_{23} = \ell_{21} u_{13} + \ell_{22} u_{23}$, which implies that $u_{23} = [m_{23} - \ell_{21} u_{13}]/\ell_{22}$, which is stored in the location of m_{23}.

Exercise.   Continue this example to establish the last row of L.

At this point, the rudiments of basic elements of Gaussian elimination have been delineated. Figure 6.1 illustrates the structure of the procedure so far. Only several more points remain to end the development. First it is advantageous to place these "rudiments" in their proper numerical perspective: What modifications are necessary to make the approach numerically sound? The next section undertakes this task. Out of this discussion springs the need for two significant modifications to the present algorithm, scaling and pivoting. These techniques

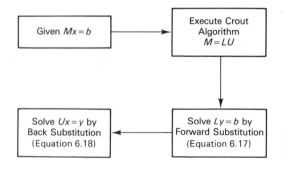

Given $Mx = b$ → Execute Crout Algorithm $M = LU$

Solve $Ux = y$ by Back Substitution (Equation 6.18) ← Solve $Ly = b$ by Forward Substitution (Equation 6.17)

Figure 6.1   Flow chart on basic Gaussian elimination

bring the numerical conditioning of the algorithm to a satisfactory level. Also pivoting enables one to relax the assumption of nonzero principal minors; only the nonsingularity of M is necessary.

ADVANTAGES, CAUTIONS, AND POINTERS

As indicated in example 6.5, the method is efficient in terms of storage. In particular, assuming U has a unity diagonal, the LU decomposition of M can be stored in the same space allocated to M. Since $LU = M$, M is not really destroyed, but simply written differently. Also, in terms of number of operations, the method proves frugal: for an $n \times n$ matrix M, the procedure requires around $n^3/3$ multiplications whereas the inversion method needs about n^3 and Cramer's rule many, many more.

The algorithm, however, needs fine tuning. As mentioned earlier, it is neither stable nor capable of producing accurate answers for all types of problems in the face of floating-point computations. One technique for improving the ease of computation of a particular problem is scaling. Consider the equation

$$\begin{bmatrix} 10^2 & 0 \\ 10^3 & 2.12 \times 10^8 \end{bmatrix} \begin{bmatrix} x_1 \\ x_2 \end{bmatrix} = \begin{bmatrix} 10 \\ 0 \end{bmatrix} \tag{6.22}$$

The entry 2.12×10^8 dominates the other entries. However, by "magnitude scaling" the variable x_2 as $\hat{x}_2 = x_2/10^6$, a set of equations with less lopsided entries comes about, viz.,

$$\begin{bmatrix} 10^2 & 0 \\ 10^3 & 212 \end{bmatrix} \begin{bmatrix} x_1 \\ \hat{x}_2 \end{bmatrix} = \begin{bmatrix} 10 \\ 0 \end{bmatrix} \tag{6.23}$$

This sort of scaling amounts to column scaling the M-matrix; the technique is related to the magnitude and frequency scaling common to circuit analysis and synthesis [12,13].

A further equalization of the entries occurs with the technique of row scaling. Here one tries to obtain "reasonably" close magnitudes for the various rows of the M-matrix through left multiplication of a diagonal scaling matrix. For example multiply both sides of equation 6.23 by

$$\begin{bmatrix} 0.01 & 0 \\ 0 & 0.001 \end{bmatrix}$$

produces the modified equation

$$\begin{bmatrix} 1 & 0 \\ 1 & 0.212 \end{bmatrix} \begin{bmatrix} x_1 \\ x_2 \end{bmatrix} = \begin{bmatrix} 0.1 \\ 0 \end{bmatrix} \tag{6.24}$$

All entries of the new M-matrix of equation 6.24 approximate each other in magnitude. Matrices in which the maximum elements in each row and column have about the same magnitude are said to be *equilibrated*. Thus, the M-matrix is equilibrated. A properly equilibrated matrix leads to improved numerical accuracy.

A word of caution, however, is in order: a given matrix can have several equilibrated forms, each giving a drastically different numerical result. For example, the matrix

$$M = \begin{bmatrix} 1 & 1 & 1 \\ 10^9 & -1 & 1 \\ 10^9 & 1 & 0 \end{bmatrix}$$

has the two equilibrated forms [3]

$$M_1 = \begin{bmatrix} 10^{-9} & 1 & 1 \\ 1 & -1 & 1 \\ 1 & 1 & 0 \end{bmatrix} \quad \text{and} \quad M_2 = \begin{bmatrix} 1 & 1 & 1 \\ 1 & -10^{-9} & 10^{-9} \\ 1 & 10^{-9} & 0 \end{bmatrix}$$

The choice of algorithm is then crucial: the matrix M_2 leads to instability for any computer with less than nine-decimal-place accuracy. Unfortunately, a foolproof equilibration algorithm does not exist [3,4].

Despite these cautions, scaling does serve a useful purpose. Support for scaling is derived from analogies with circuit analysis and synthesis. For example, in filter design, the usual approach takes "specifications" for the filter's passband, cutoffs, and rejection bands and transforms these specifications to those for a canonical ("equilibrated") low-pass filter with cutoff "1." After designing the canonical low-pass filter, the approach calls for an inverse transformation. Appropriately scaled element values [12,13] realize the filter.

A second extension of the basic Gaussian elimination algorithm is *pivoting*. Intuitively, (partial) pivoting amounts to the intelligent permutation of the rows of M to improve numerical accuracy. The following examples (cast in the context of the basic scheme) demonstrate the need for and use of pivoting by (*i*) indicating how floating-point round-off leads to a faulty answer and (*ii*) showing how a simple interchange of rows eliminates errors in the solution.

EXAMPLE 6.6

Let us solve $Mx = b$ using the basic Gaussian elimination algorithm with floating-point arithmetic truncated to three-decimal-place accuracy. Here *truncated* means chopped, not rounded up or down. The blind application of the algorithm leads to an erroneous answer. For consider the linear equation $Mx = b$ defined by

$$\begin{bmatrix} 7 \times 10^{-3} & 3 & 2 \\ 0 & 5 & 7 \\ 6 & 9 & 8 \end{bmatrix} \begin{bmatrix} x_1 \\ x_2 \\ x_3 \end{bmatrix} = \begin{bmatrix} -1 \\ 2 \\ -1 \end{bmatrix} \tag{6.25}$$

whose solution is easily seen to be $[x_1, x_2, x_3] = [0, -1, 1]$

Step 1. We compute the LU factorization of M using truncated arithmetic:

$$M = \begin{bmatrix} 7 \times 10^{-3} & 0 & 0 \\ 0 & 5 & 0 \\ 6 & -2.56 \times 10^3 & 1.87 \times 10^3 \end{bmatrix} \begin{bmatrix} 1 & 4.28 \times 10^2 & 2.85 \times 10^2 \\ 0 & 1 & 1.4 \\ 0 & 0 & 1 \end{bmatrix}$$

To illuminate the calculation of these numbers observe that

$$\ell_{33} = 8 - fl\{6 \times 2.85 \times 10^2\} + fl\{2.56 \times 1.4 \times 10^3\}$$
$$= fl\{8 - 1.71 \times 10^3 + 3.58 \times 10^3\} = 1.87 \times 10^3$$

Step 2. We solve the equation $Ly = b$ for y using forward substitution:

$$\begin{bmatrix} 7 \times 10^{-3} & 0 & 2 \\ 0 & 5 & 7 \\ 6 & -2.56 \times 10^3 & 1.87 \times 10^3 \end{bmatrix} \begin{bmatrix} y_1 \\ y_2 \\ y_3 \end{bmatrix} = \begin{bmatrix} -1 \\ 2 \\ -1 \end{bmatrix}$$

Again, using floating-point calculations with truncated three-place accuracy yields

$$[y_1, y_2, y_3] = [-1.42 \times 10^2, 0.4, 1]$$

Step 3. Solving for $[x_1, x_2, x_3]$ via

$$\begin{bmatrix} 1 & 4.28 \times 10^2 & 2.85 \times 10^2 \\ 0 & 1 & 1.4 \\ 0 & 0 & 1 \end{bmatrix} \begin{bmatrix} x_1 \\ x_2 \\ x_3 \end{bmatrix} = \begin{bmatrix} -1.42 \times 10^2 \\ 0.4 \\ 1 \end{bmatrix}$$

produces

$$[x_1, x_2, x_3] = [1, -1, 1] \neq [0, -1, 1] \tag{6.26}$$

an erroneous answer.

The error in example 6.6 arises from the influence of the small term, 7×10^{-3}, in the LU factorization. The next example shows how a more accurate answer results by a simple row interchange called a *pivot*.

EXAMPLE 6.7

Multiply both sides of equation 6.25 by a so-called permutation matrix P which interchanges rows 1 and 3 of M and b. The equation then becomes $PMx = Pb$, or

$$\begin{bmatrix} 6 & 9 & 8 \\ 0 & 5 & 7 \\ 7 \times 10^{-3} & 3 & 2 \end{bmatrix} \begin{bmatrix} x_3 \\ x_2 \\ x_1 \end{bmatrix} = \begin{bmatrix} -1 \\ 2 \\ -1 \end{bmatrix} \tag{6.27}$$

Step 1. Executing an *LU* factorization of *M* yields

$$M = LU = \begin{bmatrix} 6 & 0 & 0 \\ 0 & 5 & 0 \\ 7 \times 10^{-3} & 2.99 & -2.18 \end{bmatrix} \begin{bmatrix} 1 & 1.5 & 1.33 \\ 0 & 1 & 1.4 \\ 0 & 0 & 1 \end{bmatrix}$$

Step 2. Solving $Ly = b$ by forward substitution, we obtain

$$[y_1, y_2, y_3] = [-0.166, 0.4, 1.0]$$

Step 3. Solving $Ux = y$ by back substitution produces

$$[x_1, x_2, x_3] = [0.004, -1, 1]$$

which far better approximates the correct answer of $[0, -1, 1]$ than $[1, -1, 1]$ of equation 6.26.

Regarding these examples, one might speculate small ℓ_{ii} entries (relative to other entries) lead to floating-point inaccuracies whereas ℓ_{ii} of comparable magnitudes avoid such inaccuracies. A detailed discussion of pivoting takes place in the next chapter. Sometimes pivoting will fail to alleviate the numerical inaccuracies of some problems. Such problems are said to be ill-conditioned. The next section briefly compares well-conditioned versus ill-conditioned problems.

ADVANCED GAUSSIAN ELIMINATION

Error magnification occurs whenever the matrix entry ℓ_{11} has a small magnitude relative to the magnitude of other entries in the first row of *M*. Such error magnification continues and expands whenever other ℓ_{ii} are also small. Example 6.7 showed how to avoid this error magnification through a simple row interchange. Essentially, the method is to permute the rows of *M* so that the entries ℓ_{ii} maintain reasonable magnitudes relative to other row entries. This row interchange procedure is called *partial pivoting*. The more general procedure of *complete pivoting* allows the interchange of both rows and columns. The pivoting technique controls the error magnification effects of the Gaussian elimination algorithm. Of course, it does not in general alleviate numerical problems resulting from sensitivity to round-off or parameter uncertainties.

EXAMPLE 6.8

Consider the matrix equation $Mx = b$ defined by

$$\begin{bmatrix} 1 & -2 & 1 \\ 9 & 10 & 10 \\ 10 & 11.1 & 0 \end{bmatrix} \begin{bmatrix} x_1 \\ x_2 \\ x_3 \end{bmatrix} = \begin{bmatrix} -3 \\ 0 \\ 11.1 \end{bmatrix} \qquad (6.28)$$

Step 1. (Pivot as necessary). The entry with the largest absolute magnitude in column 1 is 10. Therefore, we pivot by interchanging rows 3 and 1.

Specifically, multiply M and b by

$$P_1 = \begin{bmatrix} 0 & 0 & 1 \\ 0 & 1 & 0 \\ 1 & 0 & 0 \end{bmatrix}$$

This results in the equation $P_1 M x = P_1 b$, or

$$\begin{bmatrix} 10 & 11.1 & 0 \\ 9 & 10 & 10 \\ 1 & -2 & 1 \end{bmatrix} \begin{bmatrix} x_1 \\ x_2 \\ x_3 \end{bmatrix} = \begin{bmatrix} 11.1 \\ 0 \\ -3 \end{bmatrix} \qquad (6.29)$$

Step 2. *(Compute row 1 of L and U).* Row 1 of L becomes

$$[\ell_{11}, 0, 0] = [10, 0, 0]$$

Using the formulas of the Crout algorithm (equation 6.20),

$$[1, u_{12}, u_{13}] = [1, m_{12}/\ell_{11}, m_{13}/\ell_{11}] = [1, 1.11, 0]$$

Step 3. *(Compute row 2 of L and U).* At this point there does not appear to be a need to pivot. Using equation 6.19, the second row of L becomes

$$[\ell_{21}, \ell_{22}, 0] = [m_{21}, m_{22} - \ell_{21} u_{12}, 0] = [9, 0.01, 0] \qquad (6.30)$$

Note that ℓ_{22} is very small relative to the other entries in the row and column; therefore, pivoting is necessary and the rows must be recalculated.

Step 4. We recalculate row 2 of L and U after interchanging rows 2 and 3 in $P_1 M$ of equation 6.29 or, equivalently, of M and b via multiplication by the pivot matrix

$$P_2 = \begin{bmatrix} 1 & 0 & 0 \\ 0 & 0 & 1 \\ 0 & 1 & 0 \end{bmatrix}$$

The new b-vector becomes $b = [11.1, -3, 0]^t$, and row 1 of both L and U remains invariant with respect to the new permutation. The procedure now continues with a second go-around on the calculation of row 2 of L and U. Again, using equation 6.14,

$$[\ell_{21}, \ell_{22}, 0] = [m_{21}, m_{22} - \ell_{21} u_{12}, 0] = [1, -3.11, 0]$$

which is much more balanced than equation 6.30. Using 6.20

$$[0, 1, u_{23}] = [0, 1, (m_{23} - scrl_{21} u_{13})/\ell_{22}] = [0, 1, -0.32154]$$

Step 5. Compute row 3 of L and U. Performing the necessary computations yields

$$[\ell_{31}, \ell_{32}, \ell_{33}] = [9, 0.01, 10.00322]$$

and of course the third row of U is $[0, 0, 1]$.

Step 6. The preceding steps boil down to the equation

$$PMx = P_2P_1Mx = LUx = P_2P_1b = Pb$$

where $P = P_2P_1$ is a *permutation matrix*. In particular,

$$\begin{bmatrix} 0 & 0 & 1 \\ 1 & 0 & 0 \\ 0 & 1 & 0 \end{bmatrix} \begin{bmatrix} 1 & -2 & 1 \\ 9 & 10 & 10 \\ 10 & 11.1 & 0 \end{bmatrix} \begin{bmatrix} x_1 \\ x_2 \\ x_3 \end{bmatrix}$$

$$= \begin{bmatrix} 10 & 0 & 0 \\ 1 & -3.11 & 0 \\ 9 & 0.01 & 10.00322 \end{bmatrix} \begin{bmatrix} 1 & 1.11 & 0 \\ 0 & 1 & -0.32154 \\ 0 & 0 & 1 \end{bmatrix} \begin{bmatrix} x_1 \\ x_2 \\ x_3 \end{bmatrix}$$

$$= \begin{bmatrix} 0 & 0 & 1 \\ 1 & 0 & 0 \\ 0 & 1 & 0 \end{bmatrix} \begin{bmatrix} -3 \\ 0 \\ 11.1 \end{bmatrix} = \begin{bmatrix} 11.1 \\ -3 \\ 0 \end{bmatrix} \qquad (6.31)$$

Step 7. Solving for y by forward substitution produces

$$[y_1, y_2, y_3] = [1.11, 1.32154, -1]$$

and solving for x by back substitution results in

$$[x_1, x_2, x_3] = [0, 1, -1]$$

the correct answer.

The next step of the development formalizes these loosely sketched notions of pivoting into a theorem.

Definition 6.1. A matrix P is a *permutation matrix* if each entry of P is either zero or one and each row and each column contains only a single one.

The P matrix of equation 6.31 has the form

$$P = P_2P_1 = \begin{bmatrix} 1 & 0 & 0 \\ 0 & 0 & 1 \\ 0 & 1 & 0 \end{bmatrix} \begin{bmatrix} 0 & 0 & 1 \\ 0 & 1 & 0 \\ 1 & 0 & 0 \end{bmatrix} = \begin{bmatrix} 0 & 0 & 1 \\ 1 & 0 & 0 \\ 0 & 1 & 0 \end{bmatrix}$$

and is thus a permutation matrix.

Exercise: Prove that the product of permutation matrices is a permutation matrix.

Theorem 6.5. Let M be a nonsingular square matrix. Then there exists an upper triangular matrix U with 1's along the diagonal, a lower triangular matrix L, and a permutation matrix P such that

$$PM = LU \qquad (6.32)$$

A proof of this theorem and the following corollaries can be found in [4].

Corollary 6.5a. Let the principal minors of M be nonzero. Then P can be chosen as the identity matrix, and the LU factorization of theorem 6.5 is unique.

Corollary 6.5b.   If M is singular, then equation 6.32 holds with the proviso that U is singular with one or more zeros on the diagonal.

As an illustration of corollary 6.5b, the matrix

$$M = \begin{bmatrix} 1 & 1 & 1 & 1 \\ 2 & 3 & 1 & 1 \\ 1 & 1 & 1 & 1 \\ 1 & 2 & 2 & 2 \end{bmatrix}$$

is clearly singular (rows 1 and 3 coincide), yet it has the factorization $PM = LU$ given by

$$\begin{bmatrix} 1 & 0 & 0 & 0 \\ 0 & 1 & 0 & 0 \\ 0 & 0 & 0 & 1 \\ 0 & 0 & 1 & 0 \end{bmatrix} \begin{bmatrix} 1 & 1 & 1 & 1 \\ 2 & 3 & 1 & 1 \\ 1 & 1 & 1 & 1 \\ 1 & 2 & 2 & 2 \end{bmatrix} = \begin{bmatrix} 1 & 0 & 0 & 0 \\ 2 & 1 & 0 & 0 \\ 1 & 1 & 2 & 0 \\ 1 & 0 & 0 & -1 \end{bmatrix} \begin{bmatrix} 1 & 1 & 1 & 1 \\ 0 & 1 & -1 & -1 \\ 0 & 0 & 1 & 1 \\ 0 & 0 & 0 & 0 \end{bmatrix}$$

Exercise: Construct the preceding factorization.

Theorem 6.5 culminates the main thrust of the development of Gaussian elimination. Tying up the remaining loose ends, note that as a general rule, after executing a partial pivot, say, at the kth row, the new kk element satisfies

$$|m_{kk}| = \max_i |m_{ik}|$$

for $i = k, k+1, ..., n$. In complete pivoting (where row and column interchanges are permissible), one chooses the new m_{kk} element to satisfy

$$|m_{kk}| = \max_{i,j} |m_{ij}|$$

for $i = k, k+1, ..., n$ and $j = k, k+1, ..., n$. This procedure, however, generates too much computational overhead. In fact, complete pivoting costs about as much to implement as a regular Gaussian elimination, whereas partial pivoting costs are negligible. Thus, although the round-off error magnification for complete pivoting is two to four times smaller than that of partial pivoting, cost effectiveness precludes its use. When the availability of techniques for identifying and avoiding to some extent excessive error magnification is coupled with the experimentally observed fact that the probability of "trouble" with partial pivoting is very small, it becomes clear that Gaussian elimination is a reliable, stable accurate approach to the solution of $Mx = b$.

ITERATIVE IMPROVEMENT AND ILL-CONDITIONED PROBLEMS

A technique known as *iterative improvement* exists for reducing the error induced by a Gaussian elimination procedure. Let $\hat{x}$ denote the solution to $Mx = b$ produced by a Gaussian elimination procedure, and define $\hat{b} = M\hat{x}$. Then

$$r = b - \hat{b} = b - M\hat{x} = Mx - M\hat{x} = M(x - \hat{x}) = -Me$$

where the solution error $e = \hat{x} - x$. Using double the precision of (or at least a higher precision than) the Gaussian elimination procedure used to compute x, solve

$$Me = -r$$

for e and set $\hat{x}_{new} = \hat{x} - e$. One then iteratively executes the procedure according to

$$[\hat{x}]^{j+1} = [\hat{x}]^j - [e]^j$$

After several iterations the improvement in $\hat{x}$ usually ceases.

Despite our efforts, some problems resist all attempts at computing an accurate floating-point solution. These problems are termed *ill-conditioned*. Their ill-conditionedness arises from the structure of M or, qualitatively speaking, from the relative dependence on the linear equations which specify M. For example, suppose

$$M = \begin{bmatrix} 1 & 1 \\ 1 & 1+e \end{bmatrix}$$

for small e. Each row of M determines an equation in two unknowns and hence represents information about the system. Now, rows 1 and 2 are "almost" the same. For all practical purposes, they contain the same information. On the other hand, if

$$M = \begin{bmatrix} 1 & -1 \\ 1 & 1 \end{bmatrix}$$

then the rows and columns of M are orthogonal vectors and the information in M is "maximally" independent. The relative dependence or independence of the rows of M is associated with the condition of M. One can quantitatively define that condition via the singular value decomposition (SVD) [3] of M, to be discussed in chapter 19.

The reader should note that reliable implementation of the techniques communicated in this chapter are available in the libraries IMSL [14], EISPACK [15], and LINPACK [16]. In fact, one can choose various options depending on the type of problem. Of course, other equally reliable implementations exist. For small, well-conditioned systems of equations almost all the different schemes work well, and some scientific calculators now even have provision for solving such problems as a standard option.

As a final remark, the approach taken in this chapter is amenable to sparse matrix techniques when dealing with the solution of a large sparse set of equations. For a sampling of these ideas, see [17,18].

PROBLEMS

1. For the base-10 numbers 0.7 and 1.7,

  (i)   Compute the eight-bit and 16-bit binary representations,

(ii) Compute the associated round-off.

2. Using only five terms in a Taylor series expansion, compute a reasonable approximation for exp(A), where

$$A = \begin{bmatrix} 2 & 1 \\ 1 & -8 \end{bmatrix}$$

(*Hint:* exp(A) = [exp($-A$)]$^{-1}$.)

3. Using Cramer's rule (equation 6.5), solve

$$\begin{bmatrix} 1 & 2 & 3 \\ -3 & 1 & 0 \\ -2 & 0 & 4 \end{bmatrix} \begin{bmatrix} x_1 \\ x_2 \\ x_3 \end{bmatrix} = \begin{bmatrix} -2 \\ 0 \\ 3 \end{bmatrix}$$

4. Represent the equation

$$\begin{bmatrix} 0.5 & 0.3 \\ 0.3 & 0.25 \end{bmatrix} \begin{bmatrix} x \\ y \end{bmatrix} = \begin{bmatrix} 2 \\ 1.55 \end{bmatrix}$$

as the base-10 equivalent of a four-bit binary floating-point representation. Solve the resulting equation and compare with the correct answer.

5. Solve $Lx = b$ by forward substitution when

$$L = \begin{bmatrix} 1 & 0 & 0 & 0 \\ 2 & 3 & 0 & 0 \\ 4 & 5 & -5 & 0 \\ -4 & -3 & -2 & -1 \end{bmatrix} \quad \text{and} \quad b = \begin{bmatrix} 1 \\ -1 \\ -11 \\ -3 \end{bmatrix}$$

6. Compute L^{-1} for the L-matrix of problem 5 using forward substitution. (*Hint:* $LL^{-1} = I$. Let x_1 designate the first column of L^{-1}. Then $Lx_1 = \eta_1 = [1\ 0\ 0\ 0]^t$.)

7. (i) Verify the forward substitution formula of equation 6.10.

 (ii) Verify the back substitution formula of equation 6.15.

8. Solve $Ux = b$ by back substitution when

$$U = \begin{bmatrix} -4 & -3 & -2 & -1 \\ 0 & -5 & 5 & 4 \\ 0 & 0 & 3 & 2 \\ 0 & 0 & 0 & 1 \end{bmatrix} \quad \text{and} \quad b = \begin{bmatrix} -1 \\ -1 \\ 5 \\ 1 \end{bmatrix}$$

9. Develop a technique for computing U^{-1} using back substitution.

10. Write a computer program which inputs L and b and calculates the solution x of $Lx = b$ by forward substitution.

11. Repeat problem 10 for the equation $Ux = b$ via the back substitution formula.

12. Let

$$M = \begin{bmatrix} 1 & 1 & 1 \\ 1 & 0 & 1 \\ 1 & 1 & 0 \end{bmatrix}$$

(i) Compute the principal minors of M. Observe that M has zeros on its diagonal yet has nonzero principal minors.

(ii) Use the Crout algorithm to factor M into LU.

(iii) Solve the equation $Mx = b$ for x using forward and back substitution when $b = [0\ 1\ 1]^t$.

13. Solve the equation $Mx = b$ by first executing the Crout algorithm to decompose M into LU and then solving by forward and back substitution when

$$M = \begin{bmatrix} 1 & 0 & 0 & 1 & 0 \\ 0 & -1 & 2 & 0 & 0 \\ 1 & 0 & -1 & 0 & 0 \\ 0 & -2 & 0 & 1 & 0 \\ 1 & 0 & 0 & 0 & 1 \end{bmatrix}$$

14. Write a computer program which first executes a Crout algorithm on a nonsingular matrix M and then solves the equation $Mx = b$ by forward and back substitution. The program should input M and b and output x. It should also use the storage scheme indicated by example 6.5. Check your program using the method of problem 13 or with programs developed in [14,19].

15. Consider the equation $Mx = b$ as applied to

$$\begin{bmatrix} 0 & 1 & 1 \\ 1 & 0 & 1 \\ 10 & 20 & 0 \end{bmatrix} \begin{bmatrix} x_1 \\ x_2 \\ x_3 \end{bmatrix} = \begin{bmatrix} 1 \\ 1 \\ 15 \end{bmatrix}$$

(i) Row scale M and b to obtain M' and b',

(ii) Find a permutation matrix P and lower and upper triangular matrices L and U such that $PM' = LU$ has a numerically sound structure,

(iii) Solve the equation in part (ii) by forward and back substitution.

16. Modify the program of problem 14 to incorporate partial pivoting. Check your program on problem 15.

17. Consider the equation $Mx = b$ as applied to

$$\begin{bmatrix} 0 & 1 & -2 & 0 \\ 4 & 0 & 3 & 2 \\ -2 & -1 & 2 & 2 \\ -2 & -7 & 0 & 4 \end{bmatrix} \begin{bmatrix} x_1 \\ x_2 \\ x_3 \\ x_4 \end{bmatrix} = \begin{bmatrix} 3 \\ 0 \\ -3 \\ -6 \end{bmatrix}$$

(i) Find matrices $P, L,$ and U such that $PM = LU$ has a numerically sound structure,

(ii) Solve the resulting equation for $x = [x_1, x_2, x_3, x_4]^t$.

18. (i) Execute an LU factorization of the tridiagonal (band) matrix

$$M = \begin{bmatrix} 1 & 2 & 0 & 0 & 0 \\ 0.1 & 1 & 2 & 0 & 0 \\ 0 & 0.1 & 1 & 2 & 0 \\ 0 & 0 & 0 & 1 & 2 \\ 0 & 0 & 0 & 0.1 & 1 \end{bmatrix}$$

(ii) Comment on the structure of L and U. (See [20] and [4].)

(iii) Develop an algorithm for reducing this type of matrix to upper triangular form. Assume that no pivoting is needed.

REFERENCES

1. Alan J. Laub, "Numerical Linear Algebra Aspects of Control Design Computations," *IEEE Transactions on Automatic Control*, Vol. AC-30, No. 2, pp. 97–108.

2. G. Dahlquist and A. Bjorck, *Numerical Methods* (Englewood Cliffs, NJ: Prentice Hall, 1974).

3. George Forsythe, et al., *Computer Methods for Mathematical Computations* (Englewood Cliffs, NJ: Prentice Hall, 1981).

4. John R. Rice, *Matrix Computations and Mathematical Software* (New York: McGraw-Hill, 1981).

5. George Forsythe and Cleve Moler, *Computer Solutions of Linear Algebraic Systems* (Englewood Cliffs, NJ: Prentice Hall, 1967).

6. Stephen M. Pizer, *Numerical Computing and Mathematical Computations* (Englewood Cliffs, NJ: Prentice Hall, 1977).

7. Gilbert Strang, *Linear Algebra and its Applications* (New York: Academic Press, 1980).

8. W. Hayt and J. Kemmerly, *Engineering Circuit Analysis* (New York: McGraw-Hill, 1981).

9. G. H. Golub and C. F. Van Loan, *Matrix Computations* (Baltimore: Johns Hopkins University Press, 1983).

10. F. R. Gantmacher, *Matrix Theory,* Vol. 1 (New York: Chelsea, 1960).

11. G. Birkhoff and S. Maclane, *A Survey of Modern Algebra,* 4th ed. (New York: Macmillan, 1977).

12. Gobind Daryanani, *Principles of Active Network Synthesis and Design* (New York: Wiley, 1976).

13. Gabor C. Temes and Jack W. LaPatra, *Introduction to Circuit Synthesis and Design* (New York: McGraw-Hill, 1977).

14. IMSL Library, IMSL, Inc., Houston, Texas.

15. B. T. Smith et al., *Matrix Eigensystem Routines–EISPACK Guide,* 2nd ed., Lecture Notes in Computer Science, Vol. 6 (New York: Springer-Verlag, 1976). SIAM, 1979).

16. J. Dongarra et al., *LINPACK User's Guide* (Philadelphia: SIAM, 1979).

17. Iain S. Duff, "A Survey of Sparse Matrix Research," *Proceedings of IEEE,* Vol. 65, No. 4, April 1977, pp. 500–535.

18. Donald J. Rose and Ralph A. Willoughby, eds., *Sparse Matrices and Their Applications,* (Proceedings of a Symposium on Sparse Matrices and their Applications, Sept. 9–10, 1971), IBM Thomas J. Watson Research Center, Yorktown Heights, NY (New York: Plenum Press, 1972).

19. J. J. Dongarra, C. B. Moler, J. R. Bunch, and G. W. Stewart, *Linpack User's Guide* (Philadelphia: SIAM, 1979).

20. Melvin J. Maron, *Numerical Analysis: A Practical Approach* (New York: Macmillan, 1982).

7

NEWTON-RAPHSON TECHNIQUES

INTRODUCTION

Consider the following problem: for some (possibly vector-valued) function of a vector, say, $F(x)$, find a point or points x^* such that $F(x^*) = \theta$, where θ is the zero vector. The usual scalar quadratic function $F(x) = ax^2 + bx + c$ is a special well-known case. Given some system of nonlinear equations described by some continuous and differentiable function $F(\cdot): \mathbf{R}^n \rightarrow \mathbf{R}^n$, where $x = [x_1, \ldots, x_n]^t$ and

$$F(x) = \begin{bmatrix} F_1(x) \\ F_2(x) \\ \vdots \\ F_n(x) \end{bmatrix} \tag{7.1}$$

it is unlikely that one can find a specific solution x^* to satisfy $F(x^*) = \theta$ in a finite number of steps. Indeed, in view of the use of finite word-length arithmetic, finding the exact solution x^* becomes even more unlikely. The upshot is that one has to resort to either approximation methods or a numerically implementable algorithm. In that regard, this chapter will devise and investigate the Newton-Raphson algorithm.

The Newton-Raphson algorithm is an iterative process which successively computes approximations x^k, to the solution x^* of equation 7.1. Such an iteration process is described by some function $\Phi(\cdot)$ where

$$x^{k+1} = \Phi(x^k)$$

The search for $F(x^*) = \theta$ then becomes equivalent to a search for a *fixed point* of the function $\Phi(\cdot)$ — i.e., $F(x^*) = \theta$ becomes equivalent to $\Phi(x^*) = x^*$, where x^* is said to be a fixed point of $\Phi(\cdot)$. The iteration process, $x^{k+1} = \Phi(x^k)$, ceases for example whenever $||x^{k+1} - x^k||_2 < \epsilon$ for some tolerance $\epsilon > 0$.

DERIVATION OF THE NEWTON-RAPHSON ITERATION ALGORITHM

Let $F(\cdot) : \mathbf{R}^m \to \mathbf{R}^n$ be continuously differentiable at least in a neighborhood of x^* where it is assumed that $F(x^*) = \theta$. Then the function $F(\cdot)$ is (Frechet) differentiable at the point $x^* \in R^n$ if an $n \times n$ matrix M exists for which

$$\lim_{x \to x^*} \frac{||F(x) - F(x^*) - M(x - x^*)||}{||x - x^*||} = 0$$

$F(\cdot)$ is differentiable in a neighborhood of x^* if it is differentiable at every point in the neighborhood. If it is so differentiable for any x in a sufficiently small neighborhood of x^* (i.e., $||x^* - x||_2 < \delta$ for some small $\delta > 0$), then the following multidimensional Taylor series holds:

$$\theta = F(x^*) = F(x) + \frac{\partial F}{\partial x}\big|_x(x^* - x) + \textit{higher order terms} \qquad (7.2)$$

If δ is chosen sufficiently small ($||x^* - x||_2 < \delta$), the higher order terms become negligible and the following first-order linear approximation results:

$$\theta \approx F(x) + \frac{\partial F}{\partial x}\big|_x(x^* - x) \qquad (7.3)$$

The derivative $\dfrac{\partial F}{\partial x}$ of F is called the *Jacobian matrix,* denoted $J_F(\cdot)$; and is given by

$$J_F(x) = \frac{\partial F}{\partial x}\big|_x = \begin{bmatrix} \dfrac{\partial F_1}{\partial x_1} & \dfrac{\partial F_1}{\partial x_2} & \cdots & \dfrac{\partial F_1}{\partial x_n} \\[2mm] \dfrac{\partial F_2}{\partial x_1} & \dfrac{\partial F_2}{\partial x_2} & \cdots & \dfrac{\partial F_2}{\partial x_n} \\[2mm] & & \vdots & \\[2mm] \dfrac{\partial F_n}{\partial x_1} & \dfrac{\partial F_n}{\partial x_2} & \cdots & \dfrac{\partial F_n}{\partial x_n} \end{bmatrix} \qquad (7.4)$$

By identifying x^{k+1} (the newest estimate of x^*) with x^* in equation 7.3 and x^k with x, we obtain the implicitly defined *Newton-Raphson iteration process:*

$$\theta = F(x^k) + J_F(x^k)(x^{k+1} - x^k) \qquad (7.5)$$

With regard to the iteration process $\Phi(x^k) = x^{k+1}$ defined earlier, if the inverse of $J_F(x^k)$ exists at each x^k, then equation 7.5 becomes

$$x^{k+1} = x^k - J_F^{-1}(x^k)F(x^k) \tag{7.6}$$

where the right hand side is $\Phi(\cdot)$.

EXAMPLE 7.1

Let us find the voltage and current of the nonlinear resistor for the circuit shown in Figure 7.1. The Kirchhoff voltage and current laws imply that $i = I - v$. Thus, to find v and i it is necessary to solve

$$F(v,i) = \begin{bmatrix} F_1(v,i) \\ F_2(v,i) \end{bmatrix} = \begin{bmatrix} v^3 - v - i \\ v + i - I \end{bmatrix} = \begin{bmatrix} 0 \\ 0 \end{bmatrix}$$

The Jacobian of $F(v,i)$ and its inverse take the form

$$J_F(v,i) = \begin{bmatrix} 3v^2 - 1 & -1 \\ 1 & 1 \end{bmatrix}, \quad [J_F(v,i)]^{-1} = \frac{1}{3v^2} \begin{bmatrix} 1 & 1 \\ -1 & 3v^2 - 1 \end{bmatrix}$$

The iteration formula 7.6 then becomes

$$\begin{bmatrix} v^{k+1} \\ i^{k+1} \end{bmatrix} = \begin{bmatrix} v^k \\ i^k \end{bmatrix} - \frac{1}{3(v^k)^2} \begin{bmatrix} 1 & 1 \\ -1 & 3(v^k)^2 - 1 \end{bmatrix} \begin{bmatrix} (v^k)^3 - v^k - i^k \\ v^k + i^k - I \end{bmatrix}$$

Using the initial guess $(v^0, i^0) = (2,1)$ produces iterates $(v^1, i^1) = (1.42, -0.42)$ and $(v^2, i^2) = (1.11, -0.11)$. The actual answer is $(v*, i*) = (1,0)$.

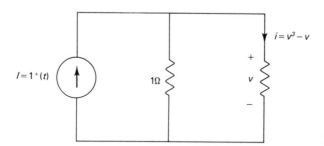

$i = 1^+(t)$ 1Ω v $i = v^3 - v$

Figure 7.1   Nonlinear resistive circuit.

Observe that the Newton-Raphson iteration has various possibilities for a stopping criterion. For example, one could use

$$\left\| F(x^k) \right\|_2 = \left\| F(v^k, i^k) \right\|_2 < \epsilon$$

for some positive ϵ close to 0. Thus, in the preceding example, $\| F(1.0019, -0.0019) \|_2 = 0.0057$, a small number. Or one could look at the norm of the difference between successive iterates, which in the foregoing case is

$$\left\| x^{k+1} - x^k \right\|_2 = \left\| \begin{matrix} v^{k+1} - v^k \\ i^{k+1} - i^k \end{matrix} \right\|_2$$

When this is sufficiently small, the process has "converged." In general, a weighted sum of the two possibilities is appropriate, i.e.,

$$\mathcal{E}_1 \left\| F(x^k) \right\|_2 + \mathcal{E}_2 \left\| x^k - x^{k-1} \right\|_2 < \epsilon \tag{7.7}$$

for some positive ϵ near zero and appropriate nonnegative scalar weights $\mathcal{E}_1$ and $\mathcal{E}_2$. One usually chooses $\mathcal{E}_1$ and $\mathcal{E}_2$ to be less than or equal to 1. Equation 7.7 represents a general stopping criterion for the Newton-Raphson scheme.

Note that one seldom solves for x^{k+1} via equation 7.6, where the Jacobian matrix is explicitly inverted. Instead, one would solve

$$J_F(x^k)(x^{k+1} - x^k) = -F(x^k) \tag{7.8}$$

for $x^{k+1} - x^k$ using Gaussian elimination. If, however, the Jacobian matrix is constant, it is advantageous to use the Crout algorithm to obtain the factorization $J_F = L_F U_F$ and then construct $J_F^{-1} = U_F^{-1} L_F^{-1}$. Since the inverses would be computed only once, one could update the approximations to x^* through equation 7.6 very conveniently. This technique has further advantages when J_F is sparse.

Another instance when one might compute the inverse of $J_F(\cdot)$ explicitly in some form such as $U^{-1}L^{-1}$, is when only a few entries of J_F change from $J_F(x^k)$ to $J_F(x^{k+1})$. In particular, suppose that $J_F(x^{k+1})$ is a low-rank perturbation on $J_F(x^k)$ — i.e., low rank relative to the dimension of J_F. Such a relationship might have the form $J_F(x^{k+1}) = J_F(x^k) + rc^t$, where r and c are vectors — for example,

$$\begin{bmatrix} 1 & 0 & -1 \\ 1 & 8 & 0 \\ -1 & 1 & 1 \end{bmatrix} = \begin{bmatrix} 1 & 0 & -1 \\ 1 & 2 & 0 \\ -1 & 1 & 1 \end{bmatrix} + \begin{bmatrix} 0 \\ 2 \\ 0 \end{bmatrix} \begin{bmatrix} 0 & 3 & 0 \end{bmatrix}$$

Householder's formula [1] then says that

$$J_F^{-1}(x^{k+1}) = [I - J_F^{-1}(x^k)r(I + c^t J_F^{-1}(x^k)r)^{-1}c^t]J_F^{-1}(x^k) \tag{7.9}$$

where $(I + c^t J_F^{-1}(x^k)r)^{-1}$ is the inverse of a 1×1 matrix (consistent with the rank-1 perturbation rc^t). Thus,

$$J_F^{-1}(x^{k+1}) = \begin{bmatrix} 1 & -6 & 0 \\ 0 & 1 & 0 \\ 0 & -6 & 1 \end{bmatrix} \begin{bmatrix} -2 & 1 & -2 \\ 1 & 0 & 1 \\ -3 & 1 & -2 \end{bmatrix} = \begin{bmatrix} -8 & 1 & -8 \\ 1 & 0 & 1 \\ -9 & 1 & -8 \end{bmatrix} \tag{7.10}$$

where

$$J_F^{-1}(x^k) = \begin{bmatrix} -2 & 1 & -2 \\ 1 & 0 & 1 \\ -3 & 1 & -2 \end{bmatrix}$$

Hence, one would explicitly compute, say, $J_F^{-1}(x^k)$ and then use Householder's formula to update successive inverses, $J_F^{-1}(x^{k+1})$. The problems at the end of the chapter develop a more general version of Householder's formula.

EXAMPLE 7.2

Consider the nonlinear set of equations

$$F(x) = \begin{bmatrix} F_1(x) \\ F_2(x) \end{bmatrix} = \begin{bmatrix} x_1 - 0.25x_2 + \dfrac{\pi}{2} \\ \sin(x_1) + x_2 - 1 \end{bmatrix} = \begin{bmatrix} 0 \\ 0 \end{bmatrix}$$

which arise from a digital simulation of the nonlinear state model of a pendulum. The Jacobian of $F(x)$ takes the form

$$J_F(x) = \begin{bmatrix} 1 & -0.25 \\ \cos(x_1) & 1 \end{bmatrix} = \tag{7.11}$$

$$\begin{bmatrix} 1 & 0 \\ \cos(x_1) & 1 + 0.25\cos(x_1) \end{bmatrix} \begin{bmatrix} 1 & -0.25 \\ 0 & 1 \end{bmatrix} = L(x_1)U$$

Since $\det[J_F(x)] = 1 + 0.25\cos(x_1) \neq 0$ for all x_1, $J_F(x)^{-1}$ exists for all $x \in \mathbf{R}^2$. Hence, an execution of the Newton-Raphson scheme is possible and would specifically require solving

$$L(x_1^k)U[x^{k+1} - x^k] = -F(x^k)$$

at each iteration, where $x^k = [x_1^k \ x_2^k]^t$. Since U is constant and since $L(x_1^k)$ has the form

$$L(x_1^k) = I + rc^t = \begin{bmatrix} 1 & 0 \\ 0 & 1 \end{bmatrix} + \begin{bmatrix} 0 \\ \cos(x_1^k) \end{bmatrix} \begin{bmatrix} 1 & 0.25 \end{bmatrix} \tag{7.12}$$

it is convenient to explicitly compute U^{-1} as

$$U^{-1} = \begin{bmatrix} 1 & 0.25 \\ 0 & 1 \end{bmatrix}$$

and to successively update L^{-1} according to Householder's formula:

$$L^{-1}(x_1^k) = \begin{bmatrix} 1 & 0 \\ 0 & 1 \end{bmatrix} - \frac{\cos(x_1^k)}{1 + 0.25\cos(x_1^k)} \begin{bmatrix} 0 & 0 \\ 1 & 0.25 \end{bmatrix} \tag{7.13}$$

This is a somewhat different application of Householder's formula than that described just prior to the example. Here we are viewing $L(x_1^k)$ as a rank-1 perturbation on the identity matrix, as indicated in equation 7.12. In equation 7.8 one sets $J_F(x^k) = I$ to obtain equation 7.13. Thus, in this special case, a numerical implementation of the Newton-Raphson iteration formula would take the form

$$\begin{bmatrix} x_1^{k+1} \\ x_2^{k+1} \end{bmatrix} = \begin{bmatrix} x_1^k \\ x_2^k \end{bmatrix} - U^{-1}L^{-1}(x_1^k) \begin{bmatrix} x_1^k - 0.25x_2^k + \dfrac{\pi}{2} \\ \sin(x_1^k) + x_2 - 1 \end{bmatrix} \tag{7.14}$$

Assuming an initial guess of $[x_1^0 \ x_2^0] = [0 \ 0]$, one obtains Table 7.1, which summarizes the iterations:

TABLE 7.1 ITERATIONS FOR EQUATION 7.14

Iteration	1	2	3
x_1^k	$-0.672676 \; \dfrac{\pi}{2}$	$-0.699028 \; \dfrac{\pi}{2}$	$-.699135 \; \dfrac{\pi}{2}$
x_2^k	2.056637	1.891064	1.890389
$\|F(x^k)\|_2$	0.185931	0.752×10^{-3}	0.416×10^{-7}

Notice how $\|F(x^k)\|_2 \to 0$, which indicates convergence to the correct solution.

A MODIFIED NEWTON-RAPHSON ALGORITHM

One problem with the Newton-Raphson algorithm is finding an initial guess x^0 such that the iteration process converges. Often the iterations will diverge for a poorly chosen initial guess.

EXAMPLE 7.3

Let $F:\mathbf{R} \to \mathbf{R}$ be given by $F(x) = \tan^{-1}x$. The solution to $F(x) = 0$ is of course $x^* = 0$. The Newton-Raphson iteration scheme is

$$x^{k+1} = x^k - (1 + (x^k)^2)\tan^{-1}x^k \tag{7.15}$$

which diverges for any x^0 satisfying

$$\tan^{-1}|x^0| > \frac{2|x^0|}{1 + (x^0)^2} \tag{7.16}$$

To illustrate, observe that $x^0 = 1.5$ is one such x^0. Table 7.2 summarizes the resultant iterates:

TABLE 7.2 ITERATIONS FOR EQUATION 7.15

Iteration	0	1	2	3
x^k	1.5	-1.69408	2.321127	-5.114088

Clearly, the absolute value of the iterates diverges.

It is possible to modify the Newton-Raphson scheme so as to obtain a globally convergent set of iterates. To see how this is done, consider the modified Newton-Raphson iteration formula

$$x^{k+1} = x^k - \lambda^k d^k \tag{7.17}$$

where

(i) $d^k = J_F^{-1}(x^k)F(x^k)$ defines the Newton "search" direction, and

(ii) λ^k is a positive scalar to be chosen according to the instructions that follow.

Observe that if $\lambda^k = 1$, equation 7.17 becomes the usual Newton-Raphson iteration scheme. The question then becomes: *How does one choose λ^k at each k so that the resultant iteration sequence is globally convergent?* A scheme proposed by Navid and Wilson [2] is based on the equivalence $F(x) = \theta$ if and only if $\|F(x)\|_2^2 = F^t(x)F(x) = 0$. Specifically $F(x^k - \lambda^k d^k) \approx \theta$ if and only if $\|F(x^k - \lambda^k d^k)\|^2 \approx 0$. The idea then is to choose λ^k so that

$$\left\|F(x^{k+1})\right\|_2^2 = \left\|F(x^k - \lambda^k d^k)\right\|_2^2 < \left\|F(x^k)\right\|_2^2 \qquad (7.18)$$

Indeed, it would be optimal to choose λ^k so that $\|F(x^k - \lambda^k d^k)\|_2^2$ is minimized along the search direction d^k.

The scheme proposed by Navid and Wilson does not yield an optimal choice for λ^k at each k. Under proper conditions, however it does yield a globally convergent sequence. Accordingly, we have the following *modified Newton-Raphson algorithm:*

1. Choose $\lambda^k = 1$,
2. Compute $\|F(x^k)\|_2^2$,
3. Compute $\|F(x^k - \lambda^k d^k)\|_2^2$,
4. If $\|F(x^k)\|_2^2 \leqslant \|F(x^k - \lambda^k d^k)\|_2^2$, set $\lambda^k_{new} = 0.5 \lambda^k_{old}$ and return to step 3,
5. If $\|F(x^k)\|_2^2 > \|F(x^k - \lambda^k d^k)\|_2^2$, proceed to the next Newton iteration.

EXAMPLE 7.4

Consider the two-variable nonlinear equation [3]

$$F(x,y) = \begin{bmatrix} F_1(x,y) \\ F_2(x,y) \end{bmatrix} = \begin{bmatrix} xy - 2x - 2y + 3 \\ xy - 2y + 1 \end{bmatrix} = \begin{bmatrix} 0 \\ 0 \end{bmatrix} \qquad (7.19)$$

For this equation, the modified Newton-Raphson iteration formula takes the form

$$\begin{bmatrix} x^{k+1} \\ y^{k+1} \end{bmatrix} = \begin{bmatrix} x^k \\ y^k \end{bmatrix} - \lambda^k \begin{bmatrix} y^k - 2 & x^k - 2 \\ y^k & x^k - 2 \end{bmatrix}^{-1} \begin{bmatrix} x^k y^k - 2x^k - 2y^k + 3 \\ x^k y^k - 2y^k + 1 \end{bmatrix} \qquad (7.20)$$

Suppose an initial guess is $[x^0, y^0] = [2.1, 3]$, so that $\|F(x^0, y^0)\|_2^2 = 2.5$. Then the modified Newton-Raphson scheme of equation 7.20 reduces to

$$\begin{bmatrix} x^1 \\ y^1 \end{bmatrix} = \begin{bmatrix} 2.1 \\ 3 \end{bmatrix} - \lambda^0 \begin{bmatrix} 1.1 \\ -20 \end{bmatrix} \qquad (7.21)$$

where $[1.1 - 20]^t$ has become the search direction and the objective is to choose λ^0 to ensure that $\|F(x^1, y^1)\|_2 < \|F(x^0, y^0)\|_2$. Table 7.3 summarizes the results for various λ^0's chosen according to the modified Newton-Raphson algorithm; recall that $\|F(x^0, y^0)\|_2^2 = 2.5$.

TABLE 7.3 IMPLEMENTATION OF MODIFIED NEWTON-RAPHSON

λ^0	x^1	y^1	$\|F(x^1,y^1)\|_2^2$		
1.00	1.00	23.0	968.000	>	2.5
0.50	1.55	13.0	58.00	>	2.5
0.25	1.825	8.0	4.363	>	2.5
0.125	1.9625	5.5	1.910	<	2.5

Hence, one would choose $\lambda^0 = 0.125$ according to this method.

In this particular case it is also possible to choose the optimal λ^0 as follows. Observe that

$$F(x^1,y^1) = F(2.1-\lambda 1.1, 3+\lambda 20) = \begin{bmatrix} -22\lambda^2 + 0.9\lambda - 0.9 \\ -22\lambda^2 - 1.3\lambda + 1.3 \end{bmatrix}$$

where we have deleted the superscript zero for convenience. Then the squared norm of F takes the form

$$\left\|F(x^1,y^1)\right\|_2^2 = (-22\lambda^2 + 0.9\lambda - 0.9)^2 + (-22\lambda^2 - 1.3\lambda + 1.3)^2$$
$$= 968\lambda^4 + 17.6\lambda^3 - 15.1\lambda^2 - 5\lambda + 2.5$$

The minimum of this function occurs at one of the points where the derivative is zero. In particular, $\dfrac{d\|F(x^1,y^1)\|_2^2}{d\lambda} = 3872\lambda^3 + 52.8\lambda^2 - 30.2\lambda - 5 = 0$ has a single real root at $\lambda = 0.127307$. Thus, the optimal choice for λ^0 is 0.127307.

The modified Newton-Raphson algorithm builds on the following global convergence theorem [2], included here for completeness.

Theorem 7.1. Let $F: D \subset \mathbf{R}^n \to \mathbf{R}^n$ be continuously differentiable and have a nonsingular derivative $F'(x) = J_F(x)$ on the open set D. Assume that there exists an $x^* \in D$ for which $F(x^*) = \theta$. Then, given any $x^0 \in D$, there exists a sequence $\{\lambda^k\}$ such that the iterates

$$x^{k+1} = x^k - \lambda^k J_F(x^k)^{-1} F(x^k), \quad k = 0,1,\dots \tag{7.22}$$

converge to the point x^*. Moreover, there exists an m, depending on x^0, such that for $k \geqslant m$ we may take $\lambda^k = 1$ and hence achieve a faster rate of convergence to x^*.

The general problem and associated theory of the convergence of the Newton-Raphson scheme tends to be a complex issue. Suffice it to say that if the initial guess is sufficiently close to the solution point, the algorithm will converge at least quadratically. If the initial guess is too far from the sought-after solution point, one may implement a modified Newton-Raphson scheme with a slower rate of convergence. The reader is directed to [4] for further details.

PROBLEMS

1. Use the modified Newton-Raphson scheme to obtain a convergent sequence for example 7.3.

2. This problem develops an alternative view of the Newton-Raphson iteration scheme.

Definition: A stable limiting point x of a (possibly nonlinear) differential equation $\dot{x}(t) = \Phi(x,t)$ is a *finite limiting value* of some solution trajectory; i.e., assuming the limit exists, $x = \lim_{t \to \infty} x(t)$ for some solution trajectory $x(t)$.

(a) Prove the following theorem: If $F(\cdot): \mathbf{R}^n \to \mathbf{R}^n$ has a continuous nonsingular derivative, then the stable limiting points x of the differential equation

$$\dot{x}(t) = -J_F^{-1}[x(t)]F(x(t)), \quad x(0) = x_0 \tag{1}$$

satisfy $F(x) = 0$, where $J_F[x(t)] = \partial F/\partial x|_{x=x(t)}$.
(*Hint:* Use the chain rule to obtain an expression for dF/dt and then modify equation 1 accordingly.)

(b) Approximate $\dot{x}(t)$ by the forward Euler formula with unit step size, i.e., $\dot{x}(t) \approx x^{k+1} - x^k$. What is the resulting form of equation 1?

(c) Look closely at the proof of the theorem in part (a). What can one say about the convergence of the Newton-Raphson scheme?

(d) Investigate the use of other approximations to $\dot{x}(t)$ in constructing other iteration formulas.

3. Prove Householder's formula: Suppose $D = D_0 + REC$, where D_0 is $n \times n$, R is $n \times r$, E is $r \times r$, C is $r \times n$, R, E, and C are full rank, and D_0^{-1} exists and is known. Then

$$D^{-1} = [I - D_0^{-1}RE[I + CD_0^{-1}RE]^{-1}C]D_0^{-1}$$

(*Hint:* Consider the product DD^{-1}.)
Remark: The power of this formula rests in the observation that computing the $n \times n$ inverse D^{-1} only requires computing the inverse of an $r \times r$ matrix $[I + CD_0^{-1}RE]^{-1}$, given that D_0^{-1} is known.

4. (Householder's formula and LU decompositions) [5]. Suppose $D_0 = L_0U_0$ has a known inverse $D_0^{-1} = U_0^{-1}L_0^{-1}$ and that for some parameter vector α, $D(\alpha) = D_0 + R(\alpha)C(\alpha)$ for appropriate matrices $R(\alpha)$ and $C(\alpha)$.

(a) Show that

$$D^{-1}(\alpha) = [I - U_0^{-1}L_0^{-1}R(\alpha)[I + C(\alpha)U_0^{-1}L_0^{-1}R(\alpha)]^{-1}C(\alpha)]U_0^{-1}L_0^{-1}$$

(b) What are the advantages of this approach over a Householder implementation of L or U or both?

5. Using Householder's formula, compute the inverse of the following matrix for $\alpha = 0,1,2,$ and 3:

$$\begin{bmatrix} 2 & 0 & 1 & 0 \\ \alpha & 1 & 1 & -\alpha \\ -1 & -1 & 0 & 1 \\ 0 & 1 & -1 & 2 \end{bmatrix}$$

6. Find v and i for the nonlinear resistive circuit sketched in Figure P7.6.

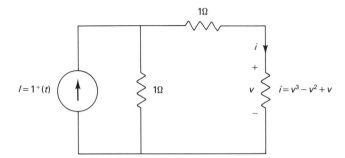

Figure P7.6 Circuit for problem 6.

7. Solve for the capacitor voltage and current and the nonlinear resistor voltage and current at time $t = 1$ for the circuit shown in Figure P7.7. Assume that $v_c(0) = 0$ and $v_c(1) = v_c(0) + 0.5(i_c(0) + i_c(1))$. (*Hint*: Do you have four equations in four unknowns?)

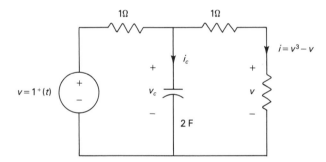

Figure P7.7 Circuit for problem 7.

8. Solve the system of equations

$$x(1 - 0.5y) - 2.1 = 0$$
$$y(1.05 - 0.05x) - 0.84 = 0$$

for x and y using the Newton-Raphson iteration.

9. The input current i and the output voltage v of a nonlinear circuit at a particular time satisfy the following two relationships:

        (*i*) $\ln(v) - i + 0.5 = 0$
      (*ii*) $v^2 - iv - 0.6857 = 0$

(a) Write these equations in vector form, i.e., $F(v,i) = \theta$, where θ is the zero vector in $\mathbf{R}^2$.

(b) What is the Newton-Raphson formula for the solution of these equations? Define all terms.

(c) Compute the Jacobian of $F(v,i)$.

(d) Compute $[J_F(v,i)]^{-1}$ at the initial guess $(v,i) = (2,2)$.

(e) Perform four iterations of the Newton-Raphson formula to obtain an approximate value for the solution.

REFERENCES

1. A. S. Householder, *Principles of Numerical Analysis* (New York: McGraw-Hill, 1953), p. 79.

2. N. Navid and A. Wilson, "A Theory and an Algorithm for Analog Fault Diagnosis," *IEEE Transactions on Circuits and Systems,* Vol. CAS-26, No. 7, July 1979, pp. 440–457.

3. L. Rapisarda, "Multifrequency Fault Diagnosis for Linearized Circuits," Ph.D. Dissertation, School of Electrical Engineering, Purdue University, May 1983.

4. J. Stoer and R. Bulirsh, *Introduction to Numerical Analysis* (New York: Springer-Verlag, 1980).

5. J. M. Bennet, "Triangular Factors of Modified Matrices," *Numerische Mathematik,* Vol. 7, (1965), pp. 217–221.

8

DIGITAL SIMULATION
OF STATE MODELS

INTRODUCTION

Chapters 1–5 developed a number of fundamental perspectives about state models, implicit in some of which were equations of the form $Mx = b$, whose solution was taken up in chapter 6. Chapter 7 presented the Newton-Raphson technique for the solution of nonlinear algebraic equations, which also crop up in the usual course on the analysis of systems. By contrast, the silence surrounding the numerical solution of differential equations (the state model represents, of course, a system of differential equations) was everywhere. This chapter speaks to that silence by describing some techniques used in the digital simulation of state models, many of which directly utilize the Newton-Raphson algorithm. The discussion here concentrates on basics and not on the sophisticated algorithms found in [1], which are topics for advanced courses in numerical methods. The aim is to familiarize the reader with some fundamentals and some simple, often useful simulation schemes.

The first question we ask is, *What is digital simulation of a state model?* To answer this question, recall that state models are (*i*) approximations to the time-dependent dynamics of a physical process and (*ii*) a special class of differential equations which, for the nonlinear case and many time-varying linear cases, do not have analytic solutions. Simulation, therefore, is the mimicking of the response behavior of a physical process to a set of input stimuli and initial conditions by numerically generating a solution to the state model differential equations representing the process. This numerical generation of a response signal entails another approximation of the continuous-time differential equations by a discrete-time system or system of (usually nonlinear) difference equations. The various schemes for digital simulation, such as Euler's technique, Simpson's method, and the trapezoidal method are simply different ways of generating the discrete-time equivalent of the continuous-time state model [2].

A second question is, *What is the importance of simulation?* The answer has many facets. One important role of simulation in analysis is the verification of the accuracy of the modeling dynamics. Specifically, simulation studies can determine the degree to which the model approximates a known physical process and hence verify its accuracy. Since many physical processes under analysis have certain known behavioral responses, one verifies the model by seeing whether its responses coincide with the observed responses of the process. These types of studies often suggest modifications in the model to better approximate the process. Once a model reasonably represents a process, simulation serves as a cost-effective experimental laboratory for evaluating, investigating, and predicting the behavior of the process. Running experiments on such real-world systems as rocket launches, drug use on humans, and jet and car engines would be too expensive and could be unethical without reasonable certainty as to the result. Hence, simulation serves as an essential aid in understanding the nature of the physical processes, ensuring safe performance of many physical systems such as cars, aircraft, TVs, manufacturing systems, and robots.

One of the most widespread uses of simulation is in the design, testing, and evaluation of feedback controllers—i.e., the actual development of the control algorithm, the evaluation of the design, and the optimization of parameters, as well as the evaluation of performance in the face of system disturbances.

All these reasons, coupled with the fact that most real-world systems have models that defy analytic solution, give simulation an honored place in the application of state variable theory.

PROBLEM SETTING

The general nonlinear time-varying state model, repeated as equation 8.1, underlies the simulation studies of this chapter:

$$\dot{x}(t) = f(x(t),u(t),t), \quad x(t_0) = x_0 \tag{8.1a}$$

$$y(t) = g(x(t),u(t),t) \tag{8.1b}$$

The linear state model, of course, is a special case, and hence the subsequent development applies. As before, throughout this chapter the existence of a unique solution is presupposed. Chapter 10 picks up the existence and uniqueness question.

Since $g(\cdot,\cdot,\cdot)$ explicitly defines $y(t)$ in terms of $x(t)$, $u(t)$, and t, the crux of a simulation rests with the numerical solution of equation 8.1a for the state vector $x(t)$ over some interval $t_0 \leqslant t \leqslant t_n$ given the initial state x_0 and the input signal $u(\cdot)$ over $[t_0,t_n]$. Computing values of $x(t)$ at every t is of course impossible.

Practically speaking, one partitions $[t_0, t_n]$ as $\{t_0, t_1, t_2, ..., t_n\}$ and computes approximate values, denoted x^j, of the actual solution $x(t_j)$. Then interpolating polynomials, splines, or other functions are fitted to the known points $\{(t_0, x^0), (t_1, x^1), (t_2, x^2), ..., (t_j, x^j), ..., (t_n, x^n)\}$ [2–5,6,7]. The resultant interpolating function provides all the desired values of $x(t)$ over $t_0 \leqslant t \leqslant t_n$. Figure 8.1 illustrates the use of *constant polynomial interpolation* to approximate a fictitious $x(t)$ over the interval $t_0 \leqslant t \leqslant t_n$. Figure 8.2 uses linear interpolating polynomials to present a somewhat different, apparently more accurate, picture.

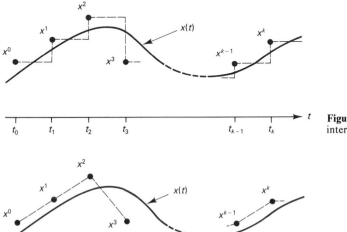

Figure 8.1 Constant polynomial interpolation of a fictitious $x(t)$.

Figure 8.2 Linear polynomial interpolation of a fictitious $x(t)$.

Finally, the question looming in the background surfaces: *How does one go about computing the estimates, x^j of $x(t_j)$ over $t_0 \leqslant t \leqslant t_n$?* To this end, one chooses a subinterval $[t_{k-m}, t_k]$ partitioned as $t_{k-m} < t_{k-m+1} < ... < t_{k-1} < t_k$ over which $x^j \approx x(t_j)$ is known for $j = k-m, ..., k-1$. With these m known estimates, one aims to obtain the estimate x^k using either of two possible approaches.

The first approach, termed the *derivative approximation*, generates an interpolation of $\dot{x}(t_k)$ as a linear combination of the points $\{x^k, x^{k-1}, ..., x^{k-m}\}$. Denote this interpolation by $\text{APPROX}(x^k, x^{k-1}, ..., x^{k-m})$. The estimate x^k is obtained by solving

$$\text{APPROX}(x^k, x^{k-1}, ..., x^{k-m}) - f(x^k, u(t_k), t_k) = 0$$

by a Newton-Raphson or some other algorithm. The approach deals directly with the differential equation.

The second approach is termed the *integral approximation* because it deals with the integral equation

$$x(t_k) = x(t_{k-m}) + \int_{t_{k-m}}^{t_k} f(x(q), u(q), q) \, dq \qquad (8.2)$$

The equivalence to equation 8.1a over $[t_{k-m}, t_k]$ is clear: differentiating equation 8.2 yields equation 8.1a and integrating equation 8.1a results in equation 8.2. By expressing the integral on the right side of equation 8.2 as a linear combination of the points $f^j = f(x^j, u(t_j), t_j)$ denoted by the interpolating function $\text{APPROX}(f^k, f^{k-1}, \ldots, f^{k-m})$, one generates the algebraic equation

$$\text{APPROX}(f^k, f^{k-1}, \ldots f^{k-m}) - x^k + x^{k-m} = 0 \qquad (8.3)$$

which is solvable for x^k by Newton-Raphson iteration.

INTEGRAL APPROXIMATION: EULER AND TRAPEZOIDAL SCHEMES

Before discussing any formal theoretical development of the preceding approaches, we consider how constant and linear polynomial interpolation lead to the well-known Euler and trapezoidal rules for solving equation 8.2. First, what strategy does one employ when assembling $\text{APPROX}(f^k, f^{k-1}, \ldots f^{k-m})$ of equation 8.3? Recall that this function estimates a definite integral

$$\int_{t_{k-m}}^{t_k} f(x(q), u(q), q) \, dq \qquad (8.4)$$

which measures the area beneath the curve $f(x(q), u(q), q)$ for $t_{k-m} \leqslant q \leqslant t_k$. Implicitly, the idea requires measuring this area as a linear combination of the values $\{f^k, f^{k-1}, \ldots f^{k-m}\}$, where again $f^j = f(x^j, u(t_j), t_j)$. Suppose the points $\{f^k, f^{k-1}, \ldots f^{k-m}\}$ are interpolated by constant polynomials between successive points, as depicted in Figure 8.3. Here one builds the area beneath the curve piece by piece, and hence,

$$\int_{t_{k-m}}^{t_{k-m+1}} f[x(q), u(q), q] dq$$

$$\approx (t_{k-m+1} - t_{k-m}) f^{k-m} + (t_{k-m+2} - t_{k-m+1}) f^{k-m+1}$$

Notice that the area under each piece is given by the formula

$$\text{AREA} = (t_j - t_{j-1}) f^{j-1}$$

and thus,

$$x^j = x^{j-1} + (t_j - t_{j-1}) f^{j-1} \qquad (8.5)$$

which is known as the *forward Euler formula* since f^{j-1} and x^{j-1} are both known. Hence, x^j is an explicit function of prior known points. Unfortunately, this formula often leads to algorithmic instabilities [4,7]. However, a little twist in the computation of the area leads to the so-called *backward Euler formula,* which has much better simulation characteristics. Instead of computing area in the forward direction, we use a backward-directed approach, i.e.,

$$\text{AREA} = (t_{j+1} - t_j) f^{j+1}$$

in which case

$$\int_{t_{k-m}}^{t_{k-m+2}} f[x(q),u(q),q]dq$$

$$\approx (t_{k-m+1} - t_{k-m})f^{k-m+1} + (t_{k-m+2} - t_{k-m+1})f^{k-m+2}$$

The backward Euler formula becomes

$$x^j = x^{j-1} + (t_j - t_{j-1})f^j = x^{j-1} + (t_j - t_{j-1})f(x^j,u(t_j),t_j) \qquad (8.6)$$

For general functions $f(\cdot,\cdot,\cdot)$, computing x^j requires solving the nonlinear equation

$$x^j - x^{j-1} - (t_j - t_{j-1})f(x^j,u(t_j),t_j) = 0 \qquad (8.7)$$

via Newton-Raphson iteration. In contrast to the forward Euler formula, the backward Euler formula implicitly defines x^j since f^j depends on x^j. Thus, the backward Euler formula is called an *implicit scheme* and the forward Euler formula an *explicit scheme*.

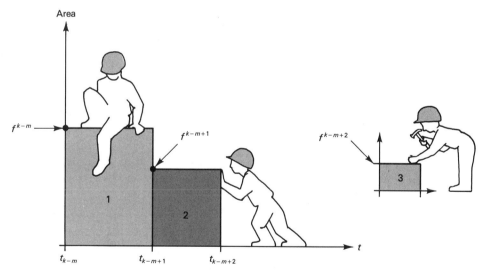

Figure 8.3 Building area by constant polynomial interpolation.

EXAMPLE 8.1

As a simple illustration, suppose one needs to simulate the *RC* circuit of Figure 8.4 over the time interval [0,3] when the initial capacitor voltage is

$v_c(0) = 3$. A little arithmetic produces the homogeneous state dynamical equation

$$\dot{v}_c(t) = -v_c(t) = f[v_c(t)], \quad v_c(0) = 3 \qquad (8.8)$$

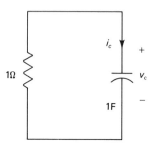

Figure 8.4 Parallel RC circuit for illustrating numerical simulation.

Of course, numerical simulation here has little practical purpose other than a pedagogical one since the analytic solution to equation 8.8 takes the form

$$v_c(t) = 3\,e^{-t}\,1^+(t) \qquad (8.9)$$

However, for the sake of illustration, using the (explicit) forward and (implicit) backward Euler formulas, the goal is to numerically compute estimates of $v_c(t)$ at $t = 0,1,2,$ and 3. First observe that

$$v_c(k) = v_c(k-1) + \int_{k-1}^{k} f[v_c(q)]\,dq$$

where the function $f[v_c(q)]$ is assumed constant over the intervals [0,1], [1,2], and [2,3]. The forward Euler formula implies that

$$v_c^k = v_c^{k-1} + [k - (k-1)]f^{k-1} = v_c^{k-1} + (-v_c^{k-1})$$

in which case, since $v_c^0 = 3$, it follows that $v_c^1 = v_c^2 = v_c^3 = 0$.

On the other hand, the backward Euler formula requires that

$$v_c^k = v_c^{k-1} + [k - (k-1)]f^k = v_c^{k-1} + (-v_c^k) \qquad (8.10)$$

Since $f[v_c(t)] = -v_c(t)$ is linear in v_c, solution via Newton-Raphson technique is unnecessary. Thus,

$$v_c^k = 0.5\,v_c^{k-1}$$

which leads to $v_c^0 = 3$, $v_c^1 = 1.5$, $v_c^2 = 0.75$, and $v_c^3 = 0.375$. A graph superimposing this result, that of the forward Euler formula and that of the analytic solution appears in Figure 8.5. Of the two Euler approaches, the backward one clearly demonstrates superiority.

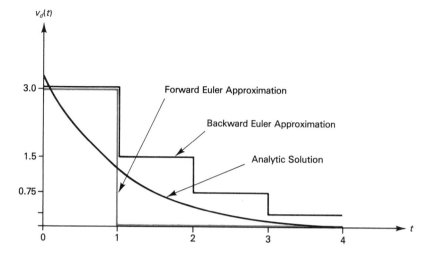

Figure 8.5   Forward and backward Euler approximation to analytic solution.

Let us now return to consideration of equation 8.3 and use linear polynomial interpolation of $f(\cdot,\cdot,\cdot)$ instead of constant interpolation. In this regard, consider Figure 8.6 with the task of approximating

$$\int_{t_{k-m}}^{t_{k-m+2}} f(x(q),u(q),q)dq \approx \text{AREA 1} + \text{AREA 2}$$

AREA 1 is easily calculated to be

$$\text{AREA 1} = 0.5 \, (t_{k-m+1} - t_{k-m}) \, [f^{k-m+1} + f^{k-m}]$$

with a similar expression for AREA 2. Thus,

$$\int_{t_{k-m}}^{t_{k-m+2}} f[x(q),u(q),q]dq \approx 0.5 \, \Delta t \, [f^{k-m} + f^{k-m+1}]$$

$$+ 0.5 \, \Delta t \, [f^{k-m+1} + f^{k-m+2}]$$

where $\Delta t = t_{k-m+1} - t_{k-m} = t_{k-m+2} - t_{k-m+1}$. This leads to the so called *trapezoidal rule*

$$x^j = x^{j-1} + 0.5 \, \Delta t \, [f^{j-1} + f^j] \qquad (8.11)$$

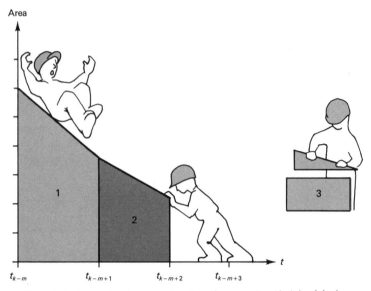

Figure 8.6 Piecewise linear construction of area under $f(x(q),u(q),q)$.

EXAMPLE 8.2

To illustrate the trapezoidal rule, consider again Figure 8.4 and its linear state model, equation 8.8. Applying the trapezoidal rule to that equation produces

$$v_c^j = v_c^{j-1} + 0.5 \, \Delta t \, [f^{j-1} + f^j] = v_c^{j-1} + 0.5 \, [-v_c^{j-1} - v_c^j] \qquad (8.12)$$

which implies

$$v_c^j = 0.3333 \ldots v_c^{j-1}$$

Therefore since $v_c^0 = 3$, it follows that $v_c^1 = 1 \approx v_c(1)$, $v_c^2 = 0.333\ldots$ $\approx v_c(2)$, $v_c^3 = 0.111 \ldots \approx v_c(3)$, etc. Making use of the assumption of linear interpolation, one obtains the plot of Figure 8.7, which compares favorably with the analytic solution.

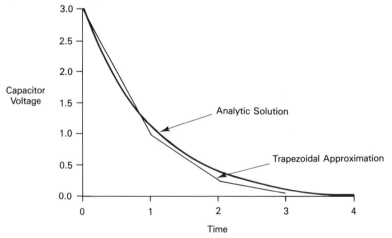

Figure 8.7 Trapezoidal simulation results of example 8.2.

Note that solution of both the backward Euler formula and the trapezoidal rule required some manipulation of the equation to obtain the desired estimates. If the state model of equation 8.8 contained two state variables (i.e., if the circuit contained two capacitors), then solution of equations 8.10 and 8.12 would require a matrix inversion. In general, for nonlinear systems, a solution requires a Newton-Raphson technique.

Even though the implicit trapezoidal rule appears more difficult to implement than an explicit formula, its accuracy seems superior. Generally speaking, implicit formulas have good accuracy despite some difficulty in implementation. By contrast, explicit formulas have a straightforward implementation with poorer accuracy for a given step size. So called *predictor-corrector routines* offer a viable compromise between the two positions. These routines use an explicit formula in a predictor step to obtain an initial estimate of $x(t_j)$ denoted by $[x_j]^0$. One then sequentially applies an implicit formula as a corrector to obtain a series of improved estimates denoted by $[x_j]^i, i = 1, 2, 3,..., p$. The number of iterations typically borders on 3 or 4 since the idea is to compromise the amount of computation between the explicit and implicit implementations. Too many iterations would consume more work, without improved accuracy, than a direct implementation of an implicit scheme.

EXAMPLE 8.3

In this example, we return to the *RC* circuit of the previous two examples with the aim of constructing a solution using a so-called forward Euler predictor and a trapezoidal corrector at each step. The predictor is simply a straightforward application of the Euler formula. Recall that the trapezoidal rule satisfies $x^j = x^{j-1} + 0.5 \, \Delta t \, [f^{j-1} + f^j]$ which leads to the corrector formula

$$[x^j]^i = x^{j-1} + 0.5 \, [f^{j-1} + [f^j]^i] \qquad (8.13)$$

where $[x^j]^i$ designates the *i*th estimate of $x(t_j)$ and $[f^j]^i = f([x^j]^i, u(t_j), t_j)$.

Using this approach to estimate $x(1)$, we then execute the forward Euler formula in a predictor step to produce $[x^1]^0 = 0$ as per example 8.1. Utilizing equation 8.13, one obtains

$$[x^1]^1 = 3 + 0.5[-3 - 0] = 1.5$$
$$[x^1]^2 = 3 + 0.5[-3 - 1.5] = 0.75$$

and

$$[x^1]^3 = 3 + 0.5[-3 - 0.75] = 1.13$$

Let us now define $x^1 = 1.13$ as our best estimate; too many corrector steps make the scheme computationally expensive.

Next, using an Euler predictor to obtain $[x^2]^0$ yields

$$[x^2]^0 = 1.13 + 1(-1.13) = 0.0$$

Executing the trapezoidal corrector results in $[x^2]^1 = 0.57, [x^2]^2 = 0.28$, and $[x^2]^3 = 0.42$. Accordingly, let us take $[x^2]^3$ as our best estimate of $x(2)$ — i.e., $x^2 = 0.42$.

Linearly interpolating these points produces a graph similar to Figure 8.7.

The preceding cases all illustrate scalar solutions. Multidimensional cases are analogous. For example, evaluation of the integral in the hypothetical second-order equation 8.14 in the next section requires computing the area underneath a curve such as that in Figure 8.8.

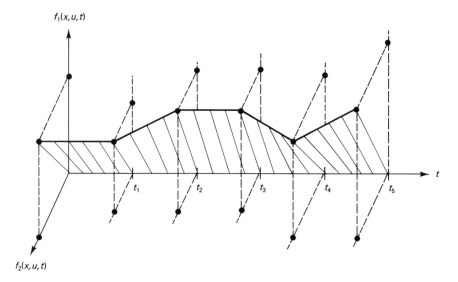

Figure 8.8   Integral approximation of a two dimensional fictitious $f(\cdot,\cdot,\cdot)$.

INTEGRAL APPROXIMATION: THEORETICAL DEVELOPMENT

To initiate the theoretical development of integral approximation, recall this chapter's central aim of sequentially constructing estimates $x^j \approx x(t_j)$ over $[t_0, t_n]$ at the points $\{t_0, t_1, \ldots, t_{k-m}, \ldots, t_k, \ldots, t_n\}$— in effect, constructing $x^1, x^2, x^3, \ldots$ This is accomplished by considering a subinterval $[t_{k-m}, t_k]$ (with m often dependent on k) over which $x^j, j = k-m, \ldots, k-1$, are known and x^k is to be computed, where, again,

$$x^k \approx x(t_k) = x(t_{k-m}) + \int_{t_{k-m}}^{t_k} f(x(q), u(q), q)\, dq \qquad (8.14)$$

To accomplish this approximate the integral by an AREA, i.e.,

$$x^k = x^{k-m} + \text{AREA}(f^k(x^k), f^{k-1}, \ldots, f^{k-m}) \qquad (8.15)$$

where the function AREA($\cdot$) is a weighted linear combination of $\{f^k(x^k), f^{k-1},...f^{k-m}\}$:

$$\text{AREA}(f^k(x^k), f^{k-1},...f^{k-m}) = \sum_{j=0}^{m} \xi_j f^{k-j} \qquad (8.16)$$

The ξ_j's are the weights in the linear combination and distinguish the various formulas. Calculation of these weights typically depends on some form of polynomial interpolation. The theoretical development presented here essentially communicates how to determine such weights, assuming *Lagrange polynomial interpolation* [2,7].

The first step in computing $\{\xi_j \mid j = 0,...,m\}$ entails fitting a polynomial $p(t)$ through the points

$$(t_{k-m}, f^{k-m}),...,(t_{k-1}, f^{k-1}), (t_k, f^k)$$

where (t_k, f^k) is the only unknown. For Lagrange interpolation,

$$p(t) = \sum_{j=k-m}^{k} \frac{(t-t_{k-m})...(t-t_{j-1})(t-t_{j+1})...(t-t_k)}{(t_j-t_{k-m})...(t_j-t_{j-1})(t_j-t_{j+1})...(t_j-t_k)} f^j \qquad (8.17)$$

$$= \sum_{j=k-m}^{k} \zeta_j(t) f^j$$

Note that $p(t)$ is an explicit polynomial function in t with the f^j viewed as constants. Hence, from equations 8.14, 8.15, and 8.17,

$$x^k = x^{k-m} + \int_{t_{k-m}}^{t_k} p(q)\, dq = x^{k-m} + \int_{t_{k-m}}^{t_k} \sum_{j=k-m}^{k} \zeta_j(q) f^j\, dq \qquad (8.18)$$

Only the known polynomial function $\zeta_j(q)$ depends on the variable of integration. Also, since equation 8.18 is a definite integral, the integrals of the $\zeta_j(q)$ over $[t_{k-m}, t_k]$ are explicit. Therefore,

$$x^k = x^{k-m} + \sum_{j=k-m}^{k} \left[\int_{t_{k-m}}^{t_k} \zeta_j(q)\, dq \right] f^j = x^{k-m} + \sum_{j=k-m}^{k} \xi_j f^j \qquad (8.19)$$

where

$$\xi_j = \int_{t_{k-m}}^{t_k} \zeta_j(q)\, dq \qquad (8.20)$$

These equations determine a basis on which many of the common integration formulas, such as the trapezoidal rule and Simpson's rule (see shortly), arise. These formulas also suggest some clear classifications into explicit, implicit and predictor-corrector types of schemes. To see this, first rewrite equation 8.19 as

$$x^k = x^{k-m} + \left[\sum_{j=k-m}^{k-1} \xi_j f^j \right] + \xi_k f(x^k, u(t_k), t_k) \qquad (8.21)$$

The category of *explicit formulas* presupposes that $\xi_k = 0$, and hence, explicit formulas have the general structure

$$x^k = x^{k-m} + \sum_{j=k-m}^{k-1} \xi_j f^j \tag{8.22}$$

where the newest estimate of the state depends only on past known values. Such formulas are clearly easy to evaluate but often produce estimates with undesirable inaccuracies.

The category of *implicit formulas* presupposes that $\xi_k \neq 0$ in equation 8.21 making x^k the solution of the generally nonlinear equation

$$\xi_k f(x^k, u(t_k), t_k) - x^k + \beta = 0 \tag{8.23}$$

where β is a known vector and is given by

$$\beta = x^{k-m} + \sum_{j=k-m}^{k-1} \xi_j f^j$$

Newton-Raphson iteration is the preferred method of solution of equation 8.23. Implicit schemes typically have very good accuracy at the trade-off of high implementation costs. A common, very simple implicit formula is *Simpson's rule,* which has the form

$$x^k = x^{k-2} + \frac{t_k - t_{k-1}}{3} [f^{k-2} + 4f^{k-1} + f^k]$$

Finally, we have the compromise category of *predictor-corrector formulas,* which offer straightforward implementation and accuracy approaching that of the implicit formulas. The execution of an explicit predictor step constitutes the first phase of such an algorithm. The predictor step yields an initial estimate according to the formula

$$[x^k]^0 = x^{k-\hat{m}} + \sum_{j=k-\hat{m}}^{k-1} \hat{\xi}_j f^j \tag{8.24}$$

where $\hat{m} \leqslant m$. (Often, $\hat{m} = m$.) This step precedes the iterative application of the corrector steps, executed according to the formula

$$[x^k]^{j+1} = x^{k-m} + \sum_{i=k-m}^{k-1} \xi_i f^i + \xi_k [f^k]^j \tag{8.25}$$

where

$$[f^k]^j = f([x^k]^j, u(t_k), t_k) \tag{8.26}$$

Clearly, the iterative nature of the formula allows easy implementation, while its implicit aspect leads to accuracy approaching that of the equivalent implicit formula.

A flowchart for the predictor-corrector scheme appears in Figure 8.9. The so-called Adams method, a good fourth order predictor-corrector formula, has the form

$$x^k = x^{k-1} + \frac{h}{24}\left[55f^{k-1} - 59f^{k-2} + 37f^{k-3} - 9f^{k-4}\right] \qquad (8.27)$$

and that of the corrector is

$$[x^k]^{i+1} = x^{k-1} + \frac{h}{24}\left[9[f^k]^i + 19f^{k-1} - 5f^{k-2} + f^{k-3}\right] \qquad (8.28)$$

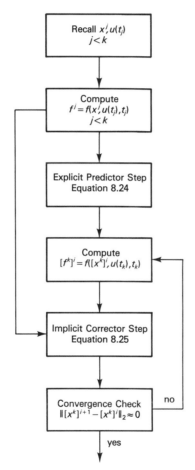

Figure 8.9 Flow chart for predictor-corrector scheme.

INTEGRAL APPROXIMATION: APPLICATION

This section illustrates the basic simulation techniques to the case of a state model for a nonlinear pendulum. Specifically, the goal is to compute numerical estimates for the state variables at $t = 0.5$ and $t = 1.0$. Accordingly, suppose that a certain pendulum has the two-dimensional nonlinear state model

$$\dot{x}_1 = x_2, \qquad\qquad x_1(0) = -\frac{\pi}{2}$$

$$\dot{x}_2 = -4 \sin [x_1], \quad x_2(0) = 0$$

Integrating both sides of this differential equation yields the integral equation

$$\begin{bmatrix} x_1(t) \\ x_2(t) \end{bmatrix} = \begin{bmatrix} x_1(0) \\ x_2(0) \end{bmatrix} + \int_0^t \begin{bmatrix} x_2(q) \\ -4 \sin [x_1(q)] \end{bmatrix} dq \qquad (8.29)$$

Three numerical evaluations of this integral follow: an explicit Euler method, an implicit trapezoidal method, and a predictor-corrector method.

Explicit Euler Method

An Euler approximation to equation 8.29 for $t = 0.5$ produces

$$\begin{bmatrix} x_1(0.5) \\ x_2(0.5) \end{bmatrix} \approx \begin{bmatrix} x_1(0) \\ x_2(0) \end{bmatrix} + 0.5 \begin{bmatrix} x_2(0) \\ -4 \sin [x_1(0)] \end{bmatrix} = \begin{bmatrix} -\dfrac{\pi}{2} \\ 2 \end{bmatrix} \qquad (8.30)$$

Similarly, for $t = 1.0$, we have

$$\begin{bmatrix} x_1(1) \\ x_2(1) \end{bmatrix} \approx \begin{bmatrix} -\dfrac{\pi}{2} \\ 2 \end{bmatrix} + 0.5 \begin{bmatrix} 2 \\ 4 \end{bmatrix} = \begin{bmatrix} -0.363\dfrac{\pi}{2} \\ 4 \end{bmatrix} = \begin{bmatrix} x_1^1 \\ x_2^1 \end{bmatrix}$$

Implicit Trapezoidal Method

The approximation to equation 8.29 according to the trapezoidal technique is

$$\begin{bmatrix} x_1^k \\ x_2^k \end{bmatrix} = \begin{bmatrix} x_1^{k-1} \\ x_2^{k-1} \end{bmatrix} + 0.25 \left[\begin{bmatrix} x_2^{k-1} \\ -4 \sin (x_1^{k-1}) \end{bmatrix} + \begin{bmatrix} x_2^k \\ -4 \sin (x_1^k) \end{bmatrix} \right] \qquad (8.31)$$

This produces the nonlinear vector equation

$$\begin{bmatrix} x_1^k - 0.25\, x_2^k - x_1^{k-1} - 0.25\, x_2^{k-1} \\ \sin(x_1^k) + x_2^k - x_2^{k-1} + \sin(x_1^{k-1}) \end{bmatrix} \triangleq \begin{bmatrix} F_1(x_1^k, x_2^k) \\ F_2(x_1^k, x_2^k) \end{bmatrix} = \begin{bmatrix} 0 \\ 0 \end{bmatrix}$$

Solution of this equation requires some kind of nonlinear equation-solving technique such as the Newton-Raphson scheme. Here the associated Jacobian is

$$J_F(x_1, x_2) = \begin{vmatrix} \dfrac{\partial F_1}{\partial x_1} & \dfrac{\partial F_1}{\partial x_2} \\ \dfrac{\partial F_2}{\partial x_1} & \dfrac{\partial F_2}{\partial x_2} \end{vmatrix} = \begin{bmatrix} 1 & -0.25 \\ \cos[x_1] & 1 \end{bmatrix}$$

and the resulting inverse is

$$J_F^{-1}(x_1, x_2) = [1 + 0.25 \cos(x_1)]^{-1} \begin{bmatrix} 1 & 0.25 \\ -\cos(x_1) & 1 \end{bmatrix}$$

Now recall the Newton-Raphson iteration formula

$$[x^k]^{i+1} = [x^k]^i - J_F^{-1}([x^k]^i)F([x^k]^i)$$

where $x^k = [x_1^k, x_2^k]^t$ and $F(x) = [F_1(x_1,x_2), F_2(x_1,x_2)]^t$. Accordingly, for $t = 0.5$ or $k = 1$ and $[x^0]^1 = [-\frac{\pi}{2}, 0]^t$, equation 8.31 becomes

$$\begin{bmatrix} x_1^1 - 0.25x_2^1 + \dfrac{\pi}{2} \\ \sin(x_1^1) + x_2^1 - 1 \end{bmatrix} = \begin{bmatrix} F_1(x_1,x_2) \\ F_2(x_1,x_2) \end{bmatrix} = \begin{bmatrix} 0 \\ 0 \end{bmatrix}$$

The corresponding best estimates are computed as

$$\begin{bmatrix} x_1^1 \\ x_2^1 \end{bmatrix}^2 = \begin{bmatrix} -\dfrac{\pi}{2} \\ 0 \end{bmatrix} - \begin{bmatrix} 1 & 0.25 \\ 0 & 1 \end{bmatrix}\begin{bmatrix} 0 \\ -2 \end{bmatrix} = \begin{bmatrix} -0.682\ \dfrac{\pi}{2} \\ 2 \end{bmatrix}$$

and

$$\begin{bmatrix} x_1^1 \\ x_2^1 \end{bmatrix}^3 = \begin{bmatrix} -0.682\ \dfrac{\pi}{2} \\ 2 \end{bmatrix} - 0.893\begin{bmatrix} 1 & 0.25 \\ -0.479 & 1 \end{bmatrix}\begin{bmatrix} 0 \\ 0.122 \end{bmatrix} = \begin{bmatrix} -0.699\ \dfrac{\pi}{2} \\ 1.891 \end{bmatrix}$$

and finally,

$$\begin{bmatrix} x^1 \\ x^2 \end{bmatrix}^4 = \begin{bmatrix} -0.699\ \dfrac{\pi}{2} \\ 1.890 \end{bmatrix}$$

Using this last value as the best estimate at $t = 0.5$ for the initial condition at $t = 1.0$ or $k = 2$ produces

$$\begin{bmatrix} x_1^2 - 0.25x_2^2 + 0.398\ \dfrac{\pi}{2} \\ \sin(x_1^2) + x_2^2 - 2.780 \end{bmatrix} = \begin{bmatrix} F_1(x_1^2, x_2^2) \\ F_2(x_1^2, x_2^2) \end{bmatrix} = \begin{bmatrix} 0 \\ 0 \end{bmatrix}$$

Executing the Newton-Raphson iteration scheme produces Table 8.1.

TABLE 8.1 VALUES FOR IMPLICIT TRAPEZOIDAL METHODS

	Iteration		
	1	2	3
x_1^2	$0.0955\ \dfrac{\pi}{2}$	$0.0345\ \dfrac{\pi}{2}$	$0.0355\ \dfrac{\pi}{2}$
x_2^2	3.102	2.724	2.724

A Predictor-Corrector Approach

To compute an estimate of the state vector at $t = 0.5(k = 1)$, we will use an Euler predictor and a trapezoidal corrector. From equation 8.30, the Euler predictor yields the initial uncorrected estimate as

$$\begin{bmatrix} x_1(0.5) \\ x_2(0.5) \end{bmatrix}^1 = \begin{bmatrix} -\dfrac{\pi}{2} \\ 2 \end{bmatrix}$$

The trapezoidal corrector takes the form

$$\begin{bmatrix} x_1^1 \\ x_2^1 \end{bmatrix}^{j+1} = \begin{bmatrix} x_1(0) \\ x_2(0) \end{bmatrix} + 0.25 \left[\begin{bmatrix} x_2(0) \\ -4\sin[x_1(0)] \end{bmatrix} + \begin{bmatrix} [x_2^1]^j \\ -4\sin([x_1^1]^j) \end{bmatrix} \right]$$

Hence, executing the necessary computations produces Table 8.2.

TABLE 8.2 TRAPEZOIDAL CORRECTOR VALUES AT $t = 0.5$

	Iteration			
	1	2	3	4
x_1^1	$-0.682\,\dfrac{\pi}{2}$	$-0.682\,\dfrac{\pi}{2}$	$-0.701\,\dfrac{\pi}{2}$	$-0.701\,\dfrac{\pi}{2}$
x_2^1	2	1.878	1.878	1.892

To round out the development, we will use an Euler predictor and a Simpson's rule corrector to compute the state vector at $t = 1$. Using the value of the state at the fourth iteration given in Table 8.2, an Euler predictor gives the initial estimate as

$$\begin{bmatrix} x_1^2 \\ x_2^2 \end{bmatrix}^1 = \begin{bmatrix} -0.099\,\dfrac{\pi}{2} \\ 3.676 \end{bmatrix}$$

The first iteration of the Simpson's rule corrector is as follows:

$$\begin{bmatrix} x_1^2 \\ x_2^2 \end{bmatrix}^2 = \begin{bmatrix} -\dfrac{\pi}{2} \\ 0 \end{bmatrix} + \frac{1}{6} \left[\begin{bmatrix} 0 \\ 4 \end{bmatrix} + 4 \begin{bmatrix} 1.892 \\ 3.567 \end{bmatrix} + \begin{bmatrix} 3.676 \\ 0.6197 \end{bmatrix} \right] = \begin{bmatrix} 0.1929 \\ 3.1482 \end{bmatrix} \dfrac{\pi}{2}$$

Continuing the process for several iterations produces Table 8.3.

TABLE 8.3 SIMPSON'S RULE CORRECTOR VALUES AT $t = 2$

	Iteration			
	1	2	3	4
x_1^2	$0.1929\,\dfrac{\pi}{2}$	$0.1369\,\dfrac{\pi}{2}$	$0.1049\,\dfrac{\pi}{2}$	$0.1109\,\dfrac{\pi}{2}$
x_2^2	3.148	2.846	2.903	2.936

A comparison of the results of the three methods follows in Table 8.4.

TABLE 8.4 COMPARISON OF RESULTS

	Explicit Euler Method	Implicit Trapezoidal	Predictor Corrector	Actual Value
x_1^1	$-\dfrac{\pi}{2}$	$-0.699\dfrac{\pi}{2}$	$-0.701\dfrac{\pi}{2}$	$-0.684\dfrac{\pi}{2}$
x_2^1	2	1.890	1.892	1.951
x_1^2	$-0.363\dfrac{\pi}{2}$	$0.0354\dfrac{\pi}{2}$	$0.1109\dfrac{\pi}{2}$	$0.131\dfrac{\pi}{2}$
x_2^2	4	2.724	2.936	2.798

Clearly, these methods do not produce the same numerical estimate. This lack of agreement arises mainly from the large step size. A significantly smaller step size, say, 0.05, would result in much closer agreement.

THE DERIVATIVE APPROXIMATION

Consider again equation 8.1a, rewritten here as

$$\dot{x}(t) = f(x, u, t), \quad x(t_o) = x_o \tag{8.32}$$

This section focuses on the numerical solution of equation 8.32 via a polynomial approximation to the derivative $\dot{x}(t)$ of the state trajectory. The task is to find a solution over some interval $[t_o, t_n]$ by finding a polynomial $p(t)$ such that $\dot{p}(t) \approx \dot{x}(t)$ over various subintervals of $[t_o, t_n]$ and then execute predetermined floating-point calculations. For a pictorial view, see Figure 8.10.

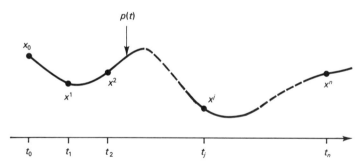

Figure 8.10   Pictorial representation of approximation to $x(t)$.

Here again, x^j designates the numerical approximation to the correct value of the state trajectory at t_j — i.e., $x(t_j)$. To "numerically solve" means to intelligently compute approximate values x^k for $x(t_j)$, $j = 1,....,n$, over the interval $[t_o, t_n]$. As before, the focus is on the subinterval $[t_{k-m}, t_k]$. By fitting an interpolating polynomial or spline function through the $(m+1)$ points (t_j, x^j), $j = 0,1,...,m$, one obtains a means for approximating $x(t)$ continuously over $[t_{k-m}, t_k]$.

To compute the approximation for $\dot{x}(t)$ evaluated at $t = t_k$ — i.e., $\dot{x}(t_k)$ — suppose x^j is known for $j = k - m,...,k-1$. Then given $\{(t_j, x^j) \mid j = k-m,...,k\}$, it is possible to fit an interpolating polynomial $p(t)$ through these points, differentiate, and evaluate at $t = t_k$ to obtain the estimate of the derivative $\dot{x}(t_k)$, which, in general, depends on x^k. The *Lagrange interpolation formula* [2,7] facilitates the task. With regard to x^j, we have

$$p(t) = \sum_{j=k-m}^{k} \frac{(t-t_{k-m})\cdots(t-t_{j-1})(t-t_{j+1})\cdots(t-t_k)}{(t_j-t_{k-m})\cdots(t_j-t_{j-1})(t_j-t_{j+1})\cdots(t_j-t_k)} \, x^j \quad (8.33)$$

Hence, if $p(t)$ is properly computed, $\dot{p}(t_k) \approx \dot{x}(t_k)$.

EXAMPLE 8.4

We will find the Lagrange interpolating polynomial and its derivative for the pair of points (t_{k-1}, x^{k-1}) and (t_k, x^k). From equation 8.33,

$$p(t) = \frac{t-t_k}{t_{k-1}-t_k} x^{k-1} + \frac{t-t_{k-1}}{t_k-t_{k-1}} x^k = \frac{(t-t_{k-1})x^k - (t-t_k)x^{k-1}}{t_k - t_{k-1}}$$

Differentiating with respect to t produces

$$\dot{p}(t) = \frac{x^k - x^{k-1}}{t_k - t_{k-1}}$$

In particular,

$$\dot{p}(t_k) = \frac{x^k - x^{k-1}}{t_k - t_{k-1}}$$

which is the Backward Euler approximation to $\dot{x}(t_k)$.

A third-order approximation to $\dot{x}(t_k)$, assuming equally spaced time instants of width h is

$$\dot{x}(t_k) \approx \frac{11}{6h} x^k - \frac{3}{h} x^{k-1} + \frac{3}{2h} x^{k-2} - \frac{1}{3h} x^{k-3} \quad (8.34)$$

For other approximations, see [6,7].

The preceding discussions lead naturally to some general observations:

1. The Lagrange interpolation formula has the more compact form

$$p(t) = \sum_{j=k-m}^{k} \zeta_j(t) \, x^j \quad (8.35)$$

for the interpolation of $x(t)$.

2. Differentiating equation 8.35 produces

$$\dot{p}(t) = \sum_{j=k-m}^{k} \dot{\zeta}_j(t) \, x^j \quad (8.36)$$

3. Defining $\dot{\zeta}_j(t_k) = d_{k-j}$ (known quantities), one fabricates

$$\dot{x}(t_k) \approx \dot{p}(t_k) = \sum_{j=0}^{m} d_j x^{k-j} = d_o x^k + d_1 x^{k-1} + \cdots + d_m x^{k-m} \quad (8.37)$$

This approximation is capable of establishing a method for solving $\dot{x}(t) = f(x,u,t)$. Again, consider the subinterval $[t_{k-m},t_k]$ of $[t_o,t_n]$ over which one must compute estimates x^k of $x(t_k)$ given that $x^{k-m},...,x^{k-1}$ are known a priori. Since $\dot{x}(t_k) = f(x(t_k),a(t_k),t_k)$, the numerical problem reduces to solving

$$\sum_{j=0}^{m} d_j x^{k-j} = f(x^k, u(t_k), t_k) \tag{8.38}$$

Rewriting produces the equivalent form,

$$f(x^k, u(t_k), t_k) - d_o x^k - \sum_{j=1}^{m} d_j x^{k-j} = \theta \tag{8.39}$$

This equation is nonlinear and algebraic in x^k. The Newton-Raphson scheme again provides the means of solution. In particular, one solves

$$J_F\left[[x^k]^i\right]\left[[x^k]^{i+1} - [x^k]^i\right] = -F\left[[x^k]^i\right] \tag{8.40}$$

where $[x^k]^i$ is the ith estimate of x^k from the Newton-Raphson algorithm

$$F(x^k) = f(x^k, a(t_k), t_k) - d_o x^k - \sum_{j=1}^{m} d_j x^{k-j} \tag{8.41}$$

and where

$$J_F(x^k) = \frac{\partial f}{\partial x}(x^k, a(t_k), t_k) - d_o I \tag{8.42}$$

When $f(x,u,t)$ has the form of the linear state model, equation 8.39 itself becomes linear allowing one to dispense with the Newton-Raphson iteration formula of 8.40. In particular, suppose

$$\dot{x}(t) = f(x,u,t) = A(t) x(t) + B(t) u(t) \tag{8.43}$$

Then equation 8.40 becomes

$$A(t_k)x^k - d_o x^k + B(t_k)u(t_k) - \sum_{j=1}^{m} d_j x^{k-j} = \theta$$

and rearranging produces

$$\left[A(t_k) - d_o I\right] x^k = \sum_{j=1}^{m} d_j x^{k-j} - B(t_k) u(t_k) \tag{8.44}$$

which may be solved directly for x^k via Gaussian elimination.

EXAMPLE 8.5

Consider a linearized pendulum state model

$$\begin{bmatrix} \dot{x}_1 \\ \dot{x}_2 \end{bmatrix} = \begin{bmatrix} 0 & 1 \\ -1 & 0 \end{bmatrix} \begin{bmatrix} x_1 \\ x_2 \end{bmatrix} + \begin{bmatrix} 0 \\ 1 \end{bmatrix} 1^+(t), \quad x(0) = \begin{bmatrix} \frac{5\pi}{6} \\ 0 \end{bmatrix}$$

Using the backward Euler approximation to $\dot{x}(1)$ yields

$$\begin{bmatrix} x_1^1 + \dfrac{5\pi}{6} \\ x_2^1 - 0 \end{bmatrix} = \begin{bmatrix} 0 & 1 \\ -1 & 0 \end{bmatrix} \begin{bmatrix} x_1^1 \\ x_2^1 \end{bmatrix} + \begin{bmatrix} 0 \\ 1 \end{bmatrix}$$

which, according to equation 8.43, becomes

$$\begin{bmatrix} -1 & 1 \\ -1 & -1 \end{bmatrix} \begin{bmatrix} x_1^1 \\ x_2^1 \end{bmatrix} = \begin{bmatrix} \dfrac{5\pi}{6} \\ -1 \end{bmatrix}$$

Consequently,

$$\begin{bmatrix} x_1^1 \\ x_2^1 \end{bmatrix} = \begin{bmatrix} 0.309(\dfrac{-5\pi}{6}) \\ 1.809 \end{bmatrix}$$

CONCLUDING REMARKS

The various numerical integration formulas represent discrete-time systems whose solutions approximate the solution of the continuous-time differential (or integral) equation of interest. This suggests that some consideration of the quality of the approximation is both necessary and useful. In fact, the literature calls the difference between the solution of the approximate discrete-time system and the true analytic solution of the given continuous-time differential equation the *discretization error*. Discretization error is then subcategorized into two parts: *local* discretization error and *global* discretization error.

Local discretization error marks the induced error after one step of a particular algorithm. Suppose $\dot{x} = f(x,u,t)$ with initial condition $x(t_{n-1}) = x^{n-1}$. Here the initial condition depends on the estimated value x^{n-1} of the state, not the true value of the state at t_{n-1}. The difference $x^n - x(t_n)$ defines the local discretization error, where $x(t_n)$ depicts the true solution of the differential equation given $x(t_{n-1}) \approx x^{n-1}$. The difference represents the error accumulated in one step, assuming infinite-precision arithmetic.

Global discretization error marks the error between the estimated solution x^n and the true solution $x(t_n)$ over the entire interval $[t_0,t_n]$ given the initial condition $x(t_0)$. Mathematically, the global discretization error is

$$e_n = x^n - x(t_n)$$

Of course, the magnitude of the global discretization error depends on the particular algorithm used in the simulation.

If the local error can be kept small, the global error is often small also. For differential equations which are said to be "stable" and very well behaved, the sum of the local discretization errors places an upper bound on the global discretization error. However, for so-called unstable linear differential equations (whose solutions tend to infinity as t approaches infinity), the global error exceeds the sum of the local errors.

These types of distinctions suggest a general framework for analyzing how well the numerical algorithms approximate the analytic solution. Intuitively, one expects that as the (assumed uniform) step size h tends to zero, the solution of the discrete-time equivalent model would tend to the true analytic solution of the continuous-time differential equation. Proper analysis of such convergence notions requires a reasonable background of knowledge regarding the existence and uniqueness of differential equations, discrete-time state models and difference equations, and the stability of such systems of equations. These are dealt with in subsequent chapters; meanwhile the reader is directed to [1–7] for further information.

The last type of error intrinsic to numerical simulation is round-off error. Round-off error depends on both the machine epsilon of the computer and the way a particular algorithm has been programmed. It is possible to link this error with the step size h. It turns out that an expression for the total error, i.e., the sum of the round-off error and the global discretization error (also linked to h) can be used to choose an optimal value for h [2]. This optimal h depends on the particular computer, program, and algorithm in operation. Again, adequate development of these ideas is beyond our present scope.

PROBLEMS

1. (a) Consider the scalar nonlinear differential equation

$$\dot{x}(t) = e^{-t}x^3(t) \quad x(0) = 2$$

Use an Euler predictor and two iterations of a trapezoidal corrector to compute an approximate value for $x(0.01)$.

(b) Now use an Euler predictor and two iterations of a Simpson's rule corrector to obtain an approximate value for $x(0.02)$.

(c) Obtain an analytical expression for $x(t)$ by solving the differential equation using separation of variables. Compare with (a) and (b).

2. Consider the differential equation

$$\begin{bmatrix} \dot{x}_1 \\ \dot{x}_2 \end{bmatrix} = \begin{bmatrix} x_1 x_2 \\ x_2(x_1 - 1) \end{bmatrix}, \quad \begin{bmatrix} x_1(0) \\ x_2(0) \end{bmatrix} = \begin{bmatrix} 2 \\ 1 \end{bmatrix}$$

(a) Write this differential equation as an integral equation in vector form.

(b) Find $x(0.1)$ using an explicit Euler approach.

(c) Find $x(0.1)$ using the implicit trapezoidal approach.

3. After appropriate scaling, the state model for the vertical ascent of a rocket is

$$\begin{bmatrix} \dot{x}_1(t) \\ \dot{x}_2(t) \end{bmatrix} = \begin{bmatrix} x_2(t) \\ -\begin{bmatrix} \dfrac{1}{x_1(t)} \end{bmatrix}^2 \end{bmatrix} + \begin{bmatrix} 0 \\ 1 \end{bmatrix} u(t), \quad \begin{bmatrix} x_1(0) \\ x_2(0) \end{bmatrix} = \begin{bmatrix} 1 \\ 0 \end{bmatrix} \tag{1}$$

given that the rocket thrust $u(t) = 2 \times 1^+(t)$.

(a) Write the preceding differential equation as an integral equation of the form

$$x(t) = x(0) + \int_0^t f(x(q), u(q), q)\, dq$$

(b) What is the function $f(\cdot,\cdot,\cdot)$?

(c) Compute an estimate of $x(0.3)$ using an Euler predictor and one iteration of a trapezoidal corrector.

(d) For the result of part (c), use an Euler predictor and one iteration of a trapezoidal corrector to obtain an estimate of $x(0.6)$.

(e) For the result of part (c), use an Euler predictor and one iteration of a Simpson's rule corrector to obtain an estimate for $x(0.6)$. Simpson's rule takes the form

$$x(t_k) = x(t_{k-2}) + \frac{\Delta t}{3} [f^{k-2} + 4 f^{k-1} + f^k]$$

(f) Use an implicit trapezoidal rule to obtain estimates of $x(0.3)$ and $x(0.6)$. In using Newton-Raphson iteration to solve the implicit equations, two iterations will suffice. You may use a computer to solve these problems provided you write your own programs. If so, turn in your program listing and output properly labeled.

(g) Tabularize the results of parts (c), (d), (e), and (f), and comment appropriately.

4. Consider Example 3.1. Let $R = L = 1$ and $C(t) = 1 + e^{-t}$. Assume $x(0) = [1 \ -1]^t$ and $I = 1^+(t)$. Find numerical approximations to $x(0.5)$ and $x(1)$ using

(a) the Euler method.

(b) the implicit trapezoidal method.

(c) an Euler predictor and two iterations of a trapezoidal corrector. Tabulate the results and make a comparative discussion.

(*Suggestion*: Write a program.)

5. An appropriately normalized scalar differential equation describing the vertical motion of a rocket [8] whose mass changes with time is

$$\frac{d^2h}{dt^2} + \left(\frac{1}{1+h}\right)^2 = -\frac{1}{m}\frac{dm}{dt}$$

where h is the altitude above the earth's surface, dm/dt is the exhaust mass flow rate, and the rocket mass (including fuel) is $m(t) = 1 + K (dm/dt)$ for $t \geqslant 0, K \neq 0$.

(a) Write a nonlinear state model choosing altitude and velocity as state variables and $u(t) = -(dm/dt)(t)$ as the input.

(b) Write the result of (a) as an integral equation.

(c) Suppose $h(0) = h^{(1)}(0) = (0)$ and $u(t) = 2e^{-t}1^+(t)$. A stable *implicit* integration scheme is the *backward Euler* formula,

$$x^k = x^{k-1} + \Delta t f^k$$

Write an equation whose solution yields an approximation to $x(0.1)$ using the backward Euler formula. *Do not solve.*

(d) Write your answer to (c) in the form $F(x^1) = \theta$.

(e) What is the general Newton-Raphson iteration formula for solving $F(x^1) = \theta$? Explain how one could numerically solve the equation (briefly), compute $J_F(x^1)$, and solve it.

6. The normalized equations for a rocket in rectilinear motion [8] are

$$\begin{bmatrix} \dot{x}_1(t) \\ \dot{x}_2(t) \\ \dot{x}_3(t) \end{bmatrix} = \begin{bmatrix} x_2(t) \\ 0 \\ 0 \end{bmatrix} + \begin{bmatrix} 0 \\ 1/x_3(t) \\ -1 \end{bmatrix} u(t), \qquad x(0) = \begin{bmatrix} 2 \\ 1 \\ 2 \end{bmatrix}$$

where $u(t) = (1 - e^{-t}) 1^+(t) 1^+(1 - t)$.

(a) Write these equations as an integral equation.
(b) Write down the forward Euler formula and the implicit trapezoidal formula for the integral form of the equation $\dot{x}(t) = f(x(t), u(t))$.
(c) Use the forward Euler formula to compute estimates for $x(0.2)$ and $x(0.4)$ (step size $= 0.2$).
(d) Use the forward Euler formula and one iteration of a trapezoidal corrector to compute estimates of $x(0.2)$ and $x(0.4)$ (step size $= 0.2$).
(e) Use Euler predictors and Simpson's rule correctors to obtain an estimate of $x(0.6)$.
(f) Use the formulas

$$x^k = x^{k-1} + \frac{h}{24}\left[9f^k + 19f^{k-1} - 5f^{k-2} + f^{k-3}\right]$$

and

$$x^k = x^{k-1} + \frac{h}{24}\left[55f^{k-1} - 59f^{k-2} + 37f^{k-3} - 9f^{k-4}\right]$$

in a predictor-corrector mode to obtain estimates of $x(0.8)$, $x(1.0)$, $x(1.2)$, and $x(1.4)$.

7. Consider the nonlinear differential equation

$$\begin{bmatrix} \dot{x}_1 \\ \dot{x}_2 \end{bmatrix} = \begin{bmatrix} -10^{-3}x_1x_2 \\ 10^{-3}x_2(x_1-80) \end{bmatrix} + \begin{bmatrix} -1 \\ 0 \end{bmatrix} u, \quad \begin{bmatrix} x_1(0) \\ x_2(0) \end{bmatrix} = \begin{bmatrix} 990 \\ 10 \end{bmatrix}$$

where $u(t) = 90 \, 1^+(t)$. Loosely speaking, this equation represents the spread of a disease in a closed community [8].

(a) The equation has the form $\dot{x} = f(x,t,u)$. In this context, what is the function $f(x,t,u)$?
(b) What is F^k and $[f^k]^i$ in terms of the given equation?
(c) Use the Euler method to compute an approximation for the state $x(1)$ (step size $= 1$).
(d) Use the Euler method on the results of part (c) to compute $x(2)$.
(e) Correct the result obtained in part (c) by using one iteration of a trapezoidal corrector.
(f) In terms of the given differential equation, write down the form of the *implicit* trapezoidal equations for the calculation of $[x_1^k \; x_2^k]^t$. Then write them in the form $F(x) = \theta$, where θ is the zero vector.
(g) What is the Jacobian matrix of the equation $F(x) = 0$ which was computed in part (f).
(h) Use Newton-Raphson iteration to solve part (f) for estimates of $x(1)$ and $x(2)$.

8. Consider the nonlinear differential equation

$$\begin{bmatrix} \dot{x}_1 \\ \dot{x}_2 \end{bmatrix} = \begin{bmatrix} x_2 \\ 1 - \cos(x_1) \end{bmatrix} + \begin{bmatrix} 0 \\ 1 \end{bmatrix} \cos(2t) \, 1^+(t), \quad x(0) = \begin{bmatrix} 0 \\ 1 \end{bmatrix}$$

Accurately solve this system of equations over the interval $[0,2]$.

REFERENCES

1. IMSL Library, IMSL Inc., Houston, Texas.

2. George Forsythe and Cleve Moler, *Computer Solutions of Linear Algebraic Systems* (Englewood Cliffs, NJ: Prentice Hall, 1967).

3. G. Dahlquist and A. Bjorck, *Numerical Methods* (Englewood Cliffs, NJ: Prentice Hall, 1974).

4. George Forsythe, et al., *Computer Methods for Mathematical Computations* (Englewood Cliffs, NJ: Prentice Hall, 1981).

5. Stephen M. Pizer, *Numerical Computing and Mathematical Computations* (Englewood Cliffs, NJ: Prentice Hall, 1977).

6. L. O. Chua and P. M. Lin, *Computer Aided Analysis of Electronic Circuits: Algorithms and Computational Techniques* (Englewood Cliffs, NJ: Prentice Hall, 1975).

7. J. Stoer and R. Bulirsch, *Introduction to Numerical Analysis* (New York: Springer-Verlag, 1980).

8. N. H. McClamroch, *State Models of Dynamic Systems: A Case Study Approach* (New York: Springer-Verlag, 1980).

9

LINEAR DISCRETE-TIME STATE MODELS: BASICS AND PARALLELS WITH THE CONTINUOUS TIME CASE

INTRODUCTION

Equation 1.28 of Chapter 1 displayed the time invariant discrete-time state model structure without any elaboration. The simulation development of the previous chapter provides one aspect of the motivational history missing from Chapter 1. A simulation scheme represents a conversion of the continuous time state model to an approximate discrete time model. This chapter introduces some basic information about the nature of the discrete-time state model. To motivate the branching of continuous-time systems into discrete-time systems, recall the forward Euler approximation

$$\dot{x}(t_k) \approx \frac{x(t_{k+1}) - x(t_k)}{t_{k+1} - t_k} \tag{9.1}$$

where $x(t_k)$ can be thought of as the sampled value of the continuous-time signal $x(t)$ [1]. Substituting this into the continuous-time linear time-varying state model (equation 3.1) produces

$$\frac{x(t_{k+1}) - x(t_k)}{t_{k+1} - t_k} \approx A(t_k) x(t_k) + B(t_k) u(t_k) \tag{9.2}$$

141

Let $h = t_{k+1} - t_k$ a fixed step size for all k. Then, simplifying equation 9.2 yields

$$x(t_{k+1}) \approx [1 + hA(t_k)] x(t_k) + h B(t_k) u(t_k) \tag{9.3}$$

Next, designate $x(t_k)$ by $x(k)$ and define a new A-matrix by $A(k) = [I + hA(t_k)]$ and a new B-matrix by $B(k) = h B(t_k)$. Finally, after tacking on the usual output equation to equation 9.3, one obtains the linear time-variant (nonstationary) discrete-time state model

$$x(k+1) = A(k)x(k) + B(k)u(k)$$

$$y(k) = C(k)x(k) + D(k)u(k) \tag{9.4}$$

where $y(k)$ and $u(k)$ stand for $y(t_k)$ and $u(t_k)$, respectively. As we shall shortly see, many parallels exist between the discrete-time state model and its continuous-time counterpart.

BASIC VOCABULARY AND EXAMPLES

Discrete-time signals are defined at discrete-time instants, as depicted in Figure 9.1. The input signal shown is a sequence of points denoted $\{u(k)\}_{k=0}^{\infty}$, where k begins

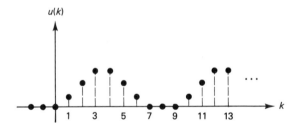

Figure 9.1. Illustration of a discrete-time signal.

at zero merely for convenience. Although k runs through the integers, $u(k)$ could correspond to $u(t_k)$, where t_k is some real number related to k only in that $t_{k-1} < t_k < t_{k+1}$.

Two basic discrete-time signals are the unit step sequence

$$1^+(k) = \begin{cases} 1, & k \geqslant 0 \\ 0, & k < 0 \end{cases} \tag{9.5}$$

and the unit impulse sequence

$$\delta(k) = \begin{cases} 1, & k = 0 \\ 0, & \text{otherwise} \end{cases} \tag{9.6}$$

Intuitively, one can think of a discrete-time system as a processor of discrete-time input sequences (like the unit step or impulse sequence) to produce discrete-time output sequences. For example, the difference equation (the discrete-time analog of a differential equation)

$$y(k+1) + y(k) = 2u(k), \quad y(0) = 0$$

represents a time-invariant discrete-time linear system. For $\{u(k)\}_0^\infty = \{\delta(k)\}_0^\infty = \{1,0,0,0,...\}$, it produces the output sequence $\{y(k)\}_0^\infty = \{0,2,-2,2,-2,2,...\}$. This output sequence is the discrete-time impulse response.

The difference equation

$$y(k) = [u(k)]^2$$

represents a nonlinear time-invariant discrete-time system. Pictorially, Figure 9.2 sketches the concept. Figure 9.3 depicts a linear discrete-time system where

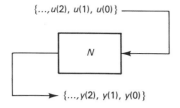

Figure 9.2. The discrete-time system as a processor of sequences.

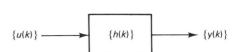

Figure 9.3 Linear, time-invariant, discrete-time system with impulse response sequence $\{h(k)\}$.

$\{h(k)\}_{k=0}^\infty$ denotes the discrete-time impulse response, which relates the input sequence $\{u(k)\}_{k=0}^\infty$ to the output sequence $\{y(k)\}_{k=0}^\infty$ by the formula

$$y(k) = \sum_{j=0}^\infty h(k-j)\, u(j) \tag{9.7}$$

The definitions of linearity, time invariance, causality, etc., for continuous-time systems all carry over to their discrete-time counterparts with the appropriate conceptual modification. For example, a discrete-time system L is *linear* in the input if, for input sequences $\{u_1(k)\}_{k=0}^\infty$ and $\{u_2(k)\}_{k=0}^\infty$ and scalar ξ,

$$L[u_1(k) + \xi u_2(k)] = Lu_1(k) + \xi Lu_2(k) \tag{9.8}$$

whenever the system is initially relaxed, i.e., all initial conditions are zero. (For convenience, equation 9.8 drops the sequence notation $\{\cdot\}$.) Similarly, a (possibly nonlinear) system N is *time-invariant* if, whenever $y(k) = Nu(k)$, then for any integer K, $y(k-K) = Nu(k-K)$, where, again, we have dispensed with the sequence notation.

In addition to these various characterizations, the different kinds of responses are duplicated for discrete-time systems. In particular, the notions of zero-input state-response and system-response and zero-state state-response and system-response have meaning as sequences.

Despite the natural link between discrete- and continuous-time systems, discrete-time systems have an identity of their own, as illustrated by the following example.

EXAMPLE 9.1

Consider Figure 9.4, which shows the block diagram of a *binary communication channel* [2]. Suppose the channel specifications are given by the following:

(*i*) $w(k+2) = u(k+2) + u(k+1) + u(k)$.

(*ii*) $y(k+2) = z(k+2) + y(k+1) + y(k)$.

(*iii*) $v(k)$ is some channel input noise process.

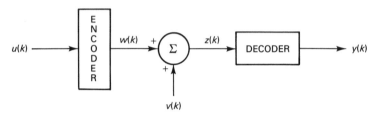

Figure 9.4 Binary communication channel.

A binary communication channel requires that all sequences be combinations of 1's and/or 0's and that arithmetic operations be modulo 2. Hence, $1 + 1 = 1 - 1 = 0$. Given these Galois operations, the communication channel defined has a scalar second-order difference equation model

$$y(k+2) + y(k+1) + y(k) = \qquad\qquad (9.9)$$

$$u(k) + u(k+1) + u(k+2) + v(k+2)$$

Equation 9.9 is second order since the highest "difference" to appear is $k+2$. Also, the equation is clearly linear in both the input sequence $\{u(k)\}_0^\infty$ and the "input" noise process sequence $\{v(k)\}_0^\infty$.

In the context of differential equations, we defined the derivative operator as $D^j y(t) = y^{(j)}(t)$. In the discrete-time context, we define the delay operator to be $z^{-j} y(k) = y(k-j)$, where z^{-j} means a delay of j units.

The goal is to construct a discrete-time state model for the binary communication channel by converting the difference equation model of 9.9 into a discrete-time linear time-invariant state model. Our approach is to parrot the techniques of the continuous-time case. Using the delay operator z^{-j} and its inverse z^j, equation 9.9 has the equivalent form

$$y(k) = u(k) + v(k) + z^{-1}[u(k) + y(k)] + z^{-2}[u(k) + y(k)] \quad (9.10)$$

The approach proceeds as follows.

Step 1. We define the discrete-time state variable

$$x_1(k) = z^{-1}[u(k) + y(k)] + z^{-2}[u(k) + y(k)]$$

in which case equation 9.10 becomes

$$y(k) = u(k) + v(k) + x_1(k) \qquad\qquad (9.11)$$

Step 2.

$$x_1(k+1) = z\,x_1(k) = u(k) + y(k) + z^{-1}[u(k) + y(k)]$$
$$= v(k) + x_1(k) + z^{-1}[u(k) + y(k)]$$

since, from equation 9.11, $y(k) = v(k) + x_1(k) + u(k)$. This is necessary because $y(k)$ is not a state variable.

Step 3. We define $x_2(k) = z^{-1}[u(k) + y(k)]$, which implies that

$$x_2(k+1) = z\,x_2(k) = u(k) + y(k) = x_1(k) + v(k)$$

Step 4. Combining the results of these equations, one arrives at the discrete-time state model

$$\begin{bmatrix} x_1(k+1) \\ x_2(k+1) \end{bmatrix} = \begin{bmatrix} 1 & 1 \\ 1 & 0 \end{bmatrix} \begin{bmatrix} x_1(k) \\ x_2(k) \end{bmatrix} + \begin{bmatrix} 0 & 1 \\ 0 & 1 \end{bmatrix} \begin{bmatrix} u(k) \\ v(k) \end{bmatrix} \qquad (9.12)$$

$$y(k) = [1 \quad 0] \begin{bmatrix} x_1(k) \\ x_2(k) \end{bmatrix} + [1 \quad 1] \begin{bmatrix} u(k) \\ v(k) \end{bmatrix}$$

Observe that if $x_1(0) = x_2(0) = 0$ and the noise process $v(k) = 0$ for all k, then $y(k) = u(k)$ for all k, as expected; i.e., there is perfect communication.

ANALOGIES WITH CONTINUOUS-TIME SYSTEMS

The preceding example, in conjunction with the canonical forms developed in chapter 4, suggests some interesting general results. First, the arbitrary constant-coefficient (time-invariant or stationary) linear nth-order difference equation

$$y(k+n) + a_1 y(k+n-1) + \ldots + a_n y(k) \qquad (9.13)$$
$$= b_n u(k) + b_{n-1} u(k+1) + \ldots + b_0 u(k+n)$$

has discrete-time canonical state model form analogous to those described in chapter 4. Derivations of discrete-time canonical forms parallel those of their continuous-time counterparts upon interpreting the derivative operator D^j as the inverse delay operator z^j. Furthermore, if one is given n initial conditions $\{y(k), y(k+1), \ldots, y(k+n-1)\}$ and knowledge of the input sequence $\{u(k), u(k+1), \ldots, u(k+n-1)\}$, then one can determine the initial state vector $x(k)$ via an equation whose structure is identical to that derived in chapter 5. In particular, given the time-invariant linear discrete-time state model

$$x(k+1) = Ax(k) + Bu(k)$$
$$y(k) = Cx(k) + Du(k) \qquad (9.14)$$

where A is $n \times n$, B is $n \times m$, C is $r \times n$ and D is $r \times m$, then, to reconstruct the state $x(k)$ from input-output measurements, one considers the set of equations

(*i*) $y(k) = Cx(k) + Du(k)$

$(ii)\quad y(k+1) = CAx(k) + CBu(k) + Du(k+1)$ $\qquad$ (9.15)

$(iii)\quad y(k+2) = CA^2x(k) + CABu(k) + CBu(k+1) + Du(k+2)$

$$\vdots$$

In direct correspondence with the continuous-time case, one obtains

$$\begin{bmatrix} y(k) \\ y(k+1) \\ \vdots \\ y(k+n-1) \end{bmatrix} = Rx(k) + T \begin{bmatrix} u(k) \\ u(k+1) \\ \vdots \\ u(k+n-1) \end{bmatrix} \qquad (9.16)$$

where

$$R = \begin{bmatrix} C \\ CA \\ \vdots \\ CA^{n-1} \end{bmatrix} \qquad (9.17)$$

and

$$T = \begin{bmatrix} D & O & \cdots & O & O \\ CB & D & \cdots & O & O \\ \vdots & & & & \vdots \\ CA^{n-2}B & CA^{n-3}B & \cdots & CB & D \end{bmatrix}$$

Another result is that generalized problem 2 of chapter 5 has a discrete-time version: find an input sequence $\{u(n+k-1), u(n+k-2), ..., u(k+1), u(k)\}$ which will drive $x(k)$ to a desired $x(n+k)$. The solution to this problem proceeds by considering the set of n equations

$(i)\quad x(k+1) = Ax(k) + Bu(k)$

$(ii)\quad x(k+2) = A^2x(k) + ABu(k) + Bu(k+1)$ $\qquad$ (9.18)

$(iii)\quad x(k+3) = A^3x(k) + A^2Bu(k) + ABu(k+1) + Bu(k+2)$

$$\vdots$$

This set of equations has the matrix form

$$[x(n+k) - A^n x(k)] = Q \begin{bmatrix} u(k+n-1) \\ u(k+n-2) \\ \vdots \\ u(k) \end{bmatrix} \qquad (9.19)$$

where again, $Q = [B\ AB\ ...\ A^{n-1}\ B]$.

Equations 9.16 and 9.19 both have structures which duplicate their continuous-time counterparts in chapter 5. Hence, the discussions of that chapter with regard to existence, uniqueness, and calculation of the solution apply in the discrete-time context. The following example and the problems at the end of the chapter help to clarify the links.

EXAMPLE 9.2

Consider the nonstationary (time-varying) discrete-time state dynamical equation

$$x(k+1) = A(k)x(k) + B(k)u(k), \quad x(0) \text{ given}$$

(a) It is desired to drive $x(0)$ to $x(3)$ for a given $x(0)$ and $x(3)$. Derive a vector-matrix equation whose solution will yield a suitable input sequence $\{u(2), u(1), u(0)\}$.

(b) Suppose

$$A(k) = \begin{bmatrix} k & 1 \\ 0 & k \end{bmatrix}, \quad B(k) = \begin{bmatrix} 0 \\ k \end{bmatrix}$$

and suppose $x(0) = [2\ 2]^t$ and $x(3) = [0\ 0]^t$. Find the minimum-energy input sequence $\{u(2), u(1), u(0)\}$ which will drive $x(0)$ to $x(3)$.

(c) Characterize the set of all input sequences which will drive $x(0)$ to $x(3)$ for the problem of part (b).

Solution to part (a): It is straightforward to show that

$$x(1) = A(0)x(0) + B(0)u(0)$$

$$x(2) = A(1)A(0)x(0) + A(1)B(0)u(0) + B(1)u(1)$$

and

$$x(3) = A(2)A(1)A(0)x(0) + A(2)A(1)B(0)u(0)$$
$$+ A(2)B(1)u(1) + B(2)u(2)$$

In matrix form,

$$[x(3) - A(2)A(1)A(0)x(0)] \tag{9.20}$$

$$= [B(2) \mid A(2)B(1) \mid A(2)A(1)B(0)] \begin{bmatrix} u(2) \\ u(1) \\ u(0) \end{bmatrix}$$

This is the matrix equation required by part (a).

Solution to part (b): Here, observe that $B(0) = [0\ 0]^t$ and that

$$A(2)A(1)A(0) = \begin{bmatrix} 0 & 2 \\ 0 & 0 \end{bmatrix}$$

Therefore, equation 9.20 becomes

$$\begin{bmatrix} -4 \\ 0 \end{bmatrix} = \begin{bmatrix} 0 & 1 & 0 \\ 2 & 2 & 0 \end{bmatrix} \begin{bmatrix} u(2) \\ u(1) \\ u(0) \end{bmatrix}$$

Solving for the minimum-energy solution via the Moore-Penrose pseudo right inverse produces

$$\begin{bmatrix} u(2) \\ u(1) \\ u(0) \end{bmatrix} = \begin{bmatrix} 4 \\ -4 \\ 0 \end{bmatrix} = \begin{bmatrix} -1 & 0.5 \\ 1 & 0 \\ 0 & 0 \end{bmatrix} \begin{bmatrix} -4 \\ 0 \end{bmatrix}$$

Solution to part (c): As per the solution of the generalized problem 2 of chapter 5, the null space of

$$\begin{bmatrix} 0 & 1 & 0 \\ 2 & 2 & 0 \end{bmatrix}$$

characterizes the difference in the possible solutions. By inspection, it becomes clear that every solution has the form

$$\begin{bmatrix} u(2) \\ u(1) \\ u(0) \end{bmatrix} = \begin{bmatrix} 4 \\ -4 \\ 0 \end{bmatrix} + \alpha \begin{bmatrix} 0 \\ 0 \\ 1 \end{bmatrix}$$

for arbitrary α.

COMMENTS ON THE SOLUTION OF THE TIME-INVARIANT DISCRETE-TIME STATE MODEL

In order to fashion some intuition for the behavior of a discrete-time state model response, consider the scalar state dynamical equation

$$x(k+1) = \lambda x(k) + u(k), \quad x(0) = x_0 \tag{9.21}$$

Let $u(k) = \xi 1^+(k)$. The terms in the state response sequence are

$x(0) = x_0$

$x(1) = \lambda x_0 + \xi$

$x(2) = \lambda x(1) + \xi = \lambda^2 x_0 + (\lambda + 1)\xi$

$x(3) = \lambda x(2) + \xi = \lambda^3 x_0 + (\lambda^2 + \lambda + 1)\xi$

$$\vdots$$

and, in general,

$$x(k) = \lambda^k x_0 + \sum_{i=0}^{k-1} \lambda^i \xi \tag{9.22}$$

Now, like integrals, summations have closed-form expressions. The idea is to find a simple analytic expression which characterizes the summation in equation 9.22. There are two cases: $\lambda = 1$ and $\lambda \neq 1$. If $\lambda = 1$, then

$$x(k) = x_0 + k\xi \tag{9.23}$$

and if $\lambda \neq 1$, then it is a simple matter to show that

$$x(k) = \lambda^k x_0 + \left[\frac{1 - \lambda^k}{1 - \lambda}\right]\xi \tag{9.24}$$

Equation 9.24 specifies the terms in the state response sequence as functions of the system parameter λ and the initial condition x_0. The part of the solution contributed by $\lambda^k x_0$ is the *zero-input state-response sequence*, and the part contributed by

$$\left[\frac{1 - \lambda^k}{1 - \lambda}\right]\xi$$

is the *zero-state state-response sequence*.

A prime question that arises out of this example is, In general, when does there exist a solution and when is it unique? *For a discrete-time system of the general form* $x(k+1) = f(x(k),u(k),k)$ *with* $x(0) = x_0$ *and* $f(\cdot,\cdot,\cdot)$ *a single-valued function, there will always exist one and only one solution for each initial condition vector.* The problem of existence and uniqueness is much more complex for continuous-time systems, as the next chapter illustrates. With this knowledge tucked away, let us consider the zero-input state response of the time-invariant state homogeneous state dynamical equation

$$x(k+1) = Ax(k), \quad x(0) = x_0 \tag{9.25}$$

After writing the value of each term in the sequence, $\{x(0), x(1), x(2), ...\}$, in terms of A and x_0, one observes that

$$x(k) = A^k x_0 \tag{9.26}$$

Evaluating A^k for matrices of any size at all is a difficult matter. However, in many circumstances A has the factorization $A = TDT^{-1}$, where D is a diagonal matrix (see problem 15). When this is the case,

$$x(k) = TD^k T^{-1} x_0 \tag{9.27}$$

where, if $D = \text{diag}[\lambda_1,...,\lambda_n]$, then $D^k = \text{diag}[\lambda_1^k,...,\lambda_n^k]$, a much easier expression to calculate.

Again, by considering the individual terms $x(0)$, $x(1)$, $x(2)$, $x(3)$, etc., in the state response sequence, it is straightforward to show that the response to

$$x(k+1) = Ax(k) + Bu(k), \quad x(0) = x_0 \tag{9.28}$$

is given by the convolution sum

$$x(k) = A^k x_0 + \sum_{j=0}^{k-1} A^{k-1-j}Bu(j) \tag{9.29}$$

This formula is derived in detail in chapter 16. Note that if A has the factorization $A = TDT^{-1}$, then equation 9.29 simplifies to

$$x(k) = TD^k T^{-1} x_0 + T\sum_{j=0}^{k-1} D^{k-1-j}(T^{-1}B)u(j) \tag{9.30}$$

which again is far easier to evaluate. The actual conditions on when A has such a factorization are discussed in chapter 15, which also considers methods for calculating the various matrices.

An example illustrates the use of equation 9.30.

EXAMPLE 9.3

Using equation 9.30, construct the state response for the discrete-time state model

$$x(k+1) = \begin{bmatrix} 1 & -0.5 \\ 0.5 & 0 \end{bmatrix} x(k) + \begin{bmatrix} 2 \\ -2 \end{bmatrix} u(k)$$

where $u(k) = 1^+(k)$ and $x(0) = [2 \ -2]^t$.

The trick, of course, is to properly factor the A-matrix. In this case,

$$A = TDT^{-1} = \begin{bmatrix} 1 & 1 \\ 1 & -1 \end{bmatrix} \begin{bmatrix} 0.5 & 1 \\ 0 & 0.5 \end{bmatrix} \begin{bmatrix} 0.5 & 0.5 \\ 0.5 & -0.5 \end{bmatrix}$$

Observe that although D is not diagonal,

$$D^k = \begin{bmatrix} (0.5)^k & k(0.5)^{k-1} \\ 0 & (0.5)^k \end{bmatrix}$$

Since $u(k) = 1^+(k)$ is constant, equation 9.30 requires computing a sum of D's raised to appropriate powers. Specifically,

$$\sum_{j=0}^{k-1} D^j = \begin{bmatrix} \dfrac{1-(0.5)^k}{1-0.5} & \dfrac{1-k(0.5)^{k-1}+(k-1)(0.5)^k}{(1-0.5)^2} \\ 0 & \dfrac{1-(0.5)^k}{1-0.5} \end{bmatrix}$$

where, for $\lambda = 0.5$,

$$\sum_{j=0}^{k-1} \lambda^j = \frac{1-\lambda^k}{1-\lambda}$$

and

$$\sum_{j=0}^{k-1} j\lambda^{j-1} = \frac{d}{d\lambda}\sum_{j=0}^{k-1}\lambda^j = \frac{d}{d\lambda}\left[\frac{1-\lambda^k}{1-\lambda}\right]$$

Evaluating the terms in equation 9.30 produces

$$x(k) = \begin{bmatrix} 1 & 1 \\ 1 & -1 \end{bmatrix} \begin{bmatrix} 8-6k(0.5)^{k-1}+8(k-1)(0.5)^k \\ 4-2(0.5)^k] \end{bmatrix}$$

for $k \geqslant 0$.

PROBLEMS

1. Give precise definitions of the following concepts for a discrete-time system and/or state model.
- **(a)** Causality
- **(b)** zero-input state-response sequence
- **(c)** zero-state state response sequence
- **(d)** zero-input response sequence
- **(e)** zero-state response sequence

2. The equation of motion of an ideal pendulum is

$$\ddot{\theta}(t) = -\sin[\theta(t)] + u(t)$$

where θ is the angle of the pendulum bob with respect to the vertical and $u(t)$ is some control input.
- **(a)** Compute the linearized state model observable canonical form of this differential equation, assuming small changes in θ.
- **(b)** Using an Euler approximation on your answer in (a), choose a suitable sampling period T, and construct a discrete-time state model.
- **(c)** Accurately simulate (a) and compare with a simulation of (b) over the interval [0,2], assuming $\theta(0) = \pi/6$ and $\dot{\theta}(0) = 0$.

3. Consider the discrete-time state model

$$x(k+1) = Ax(k) + Bu(k)$$

$$y(k) = Cx(k)$$

- **(a)** Derive an expression for the zero-input state response.
- **(b)** Derive an expression for the state response when $x(0) = \theta$.
- **(c)** What is the complete response of the system?

4. Consider the homogeneous form of the discrete-time time-invariant state model

$$x(k+1) = Ax(k)$$

for $k = 0, 1, 2, \ldots$, and A an $n \times n$ matrix. Hypothesize a solution of the form $x(k) = \lambda^k x(0)$ for some fixed vector $x(0)$ and some constant λ. What conditions must λ and $x(0)$ satisfy for the hypothesis to be valid? (*Hint*: Look up the notions of eigenvalue and eigenvector.)

5. Consider the continuous-time differential equation

$$\dot{x}(t) = Ax(t) \tag{1}$$

and a corresponding discrete-time system

$$x(k+1) = Fx(k) \tag{2}$$

where k denotes the time interval $k(\Delta t)$.
- **(a)** Using the approximation

$$\dot{x}(t) \approx \frac{x(t + \Delta t) - x(t)}{\Delta t}$$

  find F in terms of A.
- **(b)** What conditions on A guarantee that the solutions of equations (1) and (2) as per the approximation derived in (a) are the same?

6. Put the following difference equation into the controllability canonical form:

$$y(k+3) + 2y(k+2) + y(k) = u(k+1) - u(k) - u(k+3)$$

7. The *impulse response* of a discrete-time system is the response due to the input $\delta(kT) = 1$ when $k = 0$ and $\delta(kT) = 0$ when $k \neq 0$, assuming all initial conditions are zero. Suppose a certain discrete-time system has a z-domain transfer function matrix

$$h(z) = \frac{z^2 + z}{z^2 + \frac{5}{6}z + \frac{1}{6}}$$

(a) Find the impulse response sequence $h(kT)$.

(b) Look up the definition of Markov parameters. What are the Markov parameters for this system?

8. Consider the difference equation

$$y(k+3) + a_1 y(k+2) + a_2 y(k+1) + a_3 y(k)$$
$$= b_3 u(k) + b_2 u(k+1) + b_1 u(k+2) + b_0 u(k+3)$$

(a) Put the equation into the controllable canonical form of the discrete-time time-invariant state model.

(b) Put the equation into the observable canonical form.

9. Put the difference equation

$$0.5y(k+3) + 2y(k+2) - 2y(k) = u(k) - u(k+2) + 2u(k+3)$$

(a) into the controllable canonical form.

(b) into the observable canonical form.

10. Consider the discrete-time state model

$$x(k+1) = Ax(k) + Bu(k)$$
$$y(k) = Cx(k) + Du(k)$$

where A is $n \times n$, B is $n \times m$, C is $r \times n$, and D is $r \times m$. Suppose $u(k)$ and $y(k)$ are known for all $k \geqslant 0$.

(a) Derive a set of equations whose solution would yield the initial state $x(0)$. In other words, given input and output information, derive a set of equations whose solution allows one to determine the initial condition $x(0)$.

(b) Discuss the solution of the set of equations derived in (a).

11. The discrete-time difference equation

$$y(k+3) + y(k+2) + y(k) = u(k+2) + u(k+1) - u(k)$$

gives rise to a state model which has the controllability canonical form

$$\begin{bmatrix} x_1(k+1) \\ x_2(k+1) \\ x_3(k+1) \end{bmatrix} = \begin{bmatrix} 0 & 0 & -1 \\ 1 & 0 & 0 \\ 0 & 1 & -1 \end{bmatrix} \begin{bmatrix} x_1(k) \\ x_2(k) \\ x_3(k) \end{bmatrix} + \begin{bmatrix} 1 \\ 0 \\ 0 \end{bmatrix} u(k)$$

$$y(k) = [1 \quad 0 \quad -1] \begin{bmatrix} x_1(k) \\ x_2(k) \\ x_3(k) \end{bmatrix}$$

(a) Given that

$$\begin{bmatrix} y(0) \\ y(1) \\ y(2) \end{bmatrix} = \begin{bmatrix} 1 \\ 0 \\ 1 \end{bmatrix}$$

and

$$\begin{bmatrix} u(0) \\ u(1) \\ u(2) \end{bmatrix} = \begin{bmatrix} 1 \\ 2 \\ 9 \end{bmatrix}$$

find the initial state $x(0)$.

(b) Given that $x(0) = [1 \ 1 \ 1]^t$, find the input sequence $\{u(0), u(1), u(2)\}$, which will make $x(3) = 0$.

12. Consider the discrete-time system

$$x(k+1) = Ax(k) + Bu(k)$$

where A is $n \times n$ and B is $n \times m$,

(a) Derive an expression for $x(n)$ in terms of $x(0)$ and the input sequence $[u(0), u(1), ..., u(n-1)]$.

(b) Given an arbitrary $x(0)$ and an arbitrary desired $x(n)$, under what conditions can the equation of part (a) always be solved for some input sequence $[u(0), ..., u(n-1)]$ which will drive $x(0)$ to $x(n)$.

(c) Given that

$$A = \begin{bmatrix} 1 & -1 \\ 0 & 1 \end{bmatrix}$$

and

$$B = \begin{bmatrix} 2 \\ 1 \end{bmatrix}$$

find an input sequence $[u(0), u(1)]$ which will drive $x(0) = (1 \ 1)^t$ to $x(2) = (-2 \ -0.5)^t$.

13. Consider the difference equation

$$y(k+2) + 2y(k+1) - 2y(k) = u(k) - u(k+1)$$

where $y(1) = y(0) = 1$ and $u(k) = \sin(k)1^+(k)$.

(a) Put the equation into observable canonical form.

(b) Find the appropriate initial state $x(0)$.

(c) Find an input sequence which will set up this initial condition at $x(2)$.

14. Consider the difference equation

$$y(k+2) + 2y(k+1) - 2y(k) = u(k) - u(k+1) + u(k+2)$$

where $y(1) = y(0) = 1$ and $u(k) = \sin(k)1^+(k)$.

(a) Put the equation into controllable canonical form.

(b) Find the appropriate $x(0)$ consistent with the initial conditions on the difference equation.

(c) Find an input sequence which will set up this initial condition at $x(2)$.

(d) For the model of part (a), compute a state feedback which will assign the poles $0.5 \pm j$ to the system.

(e) Discuss the asymptotic behavior (as $k \to \infty$) of the zero-input state response sequence.

15. Construct an inductive proof which shows that if $A = TDT^{-1}$, where A is $n \times n$ and $D = \text{diag}[\lambda_1, ..., \lambda_n]$, then $A^k = TD^k T^{-1}$.

16. Consider the discrete-time state dynamical equation

$$x(k+1) = Ax(k) + Bu(k)$$

where $u(k) = 1^+(k)$ and

$$A = \begin{bmatrix} 1 & 1 \\ 0 & 1 \end{bmatrix} \begin{bmatrix} \lambda_1 & 0 \\ 0 & \lambda_2 \end{bmatrix} \begin{bmatrix} 1 & -1 \\ 0 & 1 \end{bmatrix}, \quad B = \begin{bmatrix} 1 \\ 0 \end{bmatrix}$$

If $x(0) = [1 \ 1]'$, find a closed-form expression for the state response sequence.

17. Consider the nonstationary discrete-time state model

$$x(k+1) = A(k)x(k) + B(k)u(k)$$
$$y(k) = C(k)x(k) + D(k)u(k)$$

(a) Construct a set of n equations whose solution will yield $x(k)$ given measurements of the input and output sequencies $\{u(k)\}_{k=0}^{\infty}$, $\{y(k)\}_{k=0}^{\infty}$.

(b) Discuss the solvability of these equations in light of the generalized problem 1 of chapter 5.

18. Consider the nonstationary (time-varying) discrete-time state dynamical equation

$$x(k+1) = A(k)x(k) + B(k)u(k)$$

(a) It is desired to drive $x(0)$ to $x(3)$ for a given $x(0)$ and $x(3)$. Derive a vector-matrix equation whose solution will yield a suitable input sequence $\{u(2), u(1), u(0)\}$.

(b) Suppose

$$A(k) = \begin{bmatrix} k & 0 & 0 \\ 1 & k & 0 \\ 0 & 0 & k-1 \end{bmatrix}, \quad B(k) = \begin{bmatrix} 0 & 0 \\ k & 0 \\ 0 & k-1 \end{bmatrix}$$

and suppose $x(0) = [1 \ 2 \ -1]'$ and $x(3) = [0 \ 0 \ 0]'$. Find the minimum-energy input sequence $\{u(2), u(1), u(0)\}$ which will drive $x(0)$ to $x(3)$.

(c) Characterize the set of all input sequences which will drive $x(0)$ to $x(3)$ for the problem of part (b).

19. Consider the discrete-time nonstationary linear state model

$$x(k+1) = A(k)x(k) + B(k)u(k)$$
$$y(k) = C(k)x(k) + D(k)u(k)$$

(a) Suppose one obtains the measurements $(u(0), y(0))$, $(u(1), y(1))$, and $(u(2), y(2))$. Derive a set of equations whose solution will yield $x(0)$. The equations should depend on the given three measurement points and should be written in matrix form.

(b) Suppose

$$\begin{bmatrix} x_1(k+1) \\ x_2(k+1) \end{bmatrix} = \begin{bmatrix} 0 & 1 \\ \cos(k) & 0.25 \end{bmatrix} \begin{bmatrix} x_1(k) \\ x_2(k) \end{bmatrix} + \begin{bmatrix} 0 \\ 1 \end{bmatrix} \ln(k+1)$$

$$y(k) = [\cos(k) \ \sin(k)] \begin{bmatrix} x_1(k) \\ x_2(k) \end{bmatrix} + [0.9]^k \ln(k+1)$$

Noisy measurements $(u(0), y(0)) = (0, 0.95)$, $(u(1), y(1)) = (0.6931, 0.81)$ are taken. Obtain an estimate of the initial state $x(0)$.

(c) Suppose an additional measurement

$$(u(2), y(2)) = (1.0986, 0.88)$$

is taken. Use the equations of (a) in conjunction with the two measurements of (b), and solve this overspecified set of equations with the Moore Penrose pseudo left inverse to obtain the minimum-error estimate in the least squares sense.

(d) Compare your answers in (b) and (c) with the actual initial state $x(0) = (1, -1)^t$.

20. A discrete-time state model has the form

$$\begin{bmatrix} x_1(k+1) \\ x_2(k+1) \end{bmatrix} = \begin{bmatrix} 0 & 1 \\ 2 & 3 \end{bmatrix} \begin{bmatrix} x_1(k) \\ x_2(k) \end{bmatrix} + \begin{bmatrix} 0 \\ 1 \end{bmatrix} u(k)$$

$$y(k) = \begin{bmatrix} 4 & 5 \end{bmatrix} \begin{bmatrix} x_1(k) \\ x_2(k) \end{bmatrix}$$

Compute the associated difference equation. Can you do so by inspection?

21. Suppose $A = TDT^{-1}$ where $D = \text{diag}[\lambda_1, \lambda_2, ..., \lambda_n]$ is a diagonal matrix. Suppose further that $x(k+1) = Ax(k)$, and $x(0) \neq 0$.

(a) What conditions on the λ_i guarantee that $x(k)$ is asymptotic to the zero vector as k approaches infinity?

(b) What conditions on the λ_i guarantee that $x(k)$ is asymptotic to a finite constant vector as k approaches infinity?

22. A scalar discrete-time system has the response sequence

$$y(k) = a^k y(0) + \sum_{j=0}^{k-1} a^{k-j+1} [u(j)]^p$$

For what values of p is the system linear?

23. A nonlinear discrete-time state model has the form

$$\begin{bmatrix} x_1(k+1) \\ x_2(k+1) \end{bmatrix} = \begin{bmatrix} x_1(k)\, x_2(k) \\ \sin[x_1(k)] + u(k) \end{bmatrix}$$

Construct the linearized state model in the perturbation variables $\Delta x_1(k)$ and $\Delta x_2(k)$ around the nominal (equilibrium) solution $x_1(k) = x_2(k) = 1$ and $u(k) = 0$ for all k.

24. (Moving Average State Model [3]). Many situations require massaging or averaging of raw data prior to display or use in a decision-making process. Averaging borders on low-pass filtering, which suppresses quick deviations in data and helps identify long-term (low-frequency) trends. It is straightforward to develop a simple four-point averager. Let $\{u(k)\}$ be a raw data sequence, and let $\{y(k)\}$ be the sequence in which $y(k)$ is the average of the data points $u(k)$, $u(k-1)$, $u(k-2)$, and $u(k-3)$. Then construct a discrete-time state model representation of the averager of the form

$$x(k+1) = Ax(k) + Bu(k)$$

$$y(k) = Cx(k) + Du(k)$$

where $D = [0.25]$, A is 3×3, B is 3×1, and C is 1×3.

25. (Discrete-Time State Model of a Labor-Management Negotiation [3]). When management and labor representatives negotiate a wage settlement, labor typically submits a wage demand. Management then typically presents a counteroffer less than the wage demand. In a dynamical system perspective, the give-and-take process proceeds by management updating its offer at each step by the addition of some fraction, say, α, of the difference between the wage demand and the offer at the previous step. Also, labor updates its wage demand by subtracting some fraction, say, β, of the difference between

the previous wage demand and offer. Let $x_1(k)$ be the management offer and x_2 the labor demand at each step k. Construct a discrete-time state model representing this process.

REFERENCES

1. Alan Oppenheim and Alan Willsky, *Signals and Systems* (Englewood Cliffs, NJ: Prentice Hall, 1983).

2. Harris McClamroch, *State Models of Dynamic Systems: A Case Study Approach* (New York: Springer-Verlag, 1980).

3. David Luenberger, *Introduction to Dynamic Systems: Theory, Models, and Applications* (New York: Wiley, 1979).

10

EXISTENCE AND
UNIQUENESS OF
STATE TRAJECTORIES

INTRODUCTION

Earlier chapters described several categories of solutions to the state model, zero input state response, zero state system response, etc. Chapters seven and eight communicated the rudiments of numerical simulation. Nowhere was existence and uniqueness of solutions addressed. This chapter illuminates this blind spot by taking up questions of existence and uniqueness of solutions to the nonlinear state dynamical equation

$$\dot{x}(t) = f(x(t),t), \qquad x(t_0) = x_0 \qquad (10.1)$$

where for each t $x(t) \epsilon \mathbf{R}^n$ and $f(\cdot,\cdot):\mathbf{R}^n \times \mathbf{R} \to \mathbf{R}^n$. The usual output equation, $y(t) = g(x(t),t)$, does not affect the solution of equation 10.1 and is therefore dropped from further elaboration. As a notational convenience the second variable, t, in equation 10.1 reflects not only the time dependence of $f(\cdot,\cdot)$ as a time-varying system but also the effect of the unspecified system input $u(\cdot)$ at each t. Note that the linear state dynamics is of course a special case of equation 10.1.

The crux of the existence/uniqueness question rests on equation 10.1. The state trajectory $\phi(\cdot,t_0,x_0)$ is a solution to equation 10.1 over the interval $[a,b]$ if and only if (i) $\phi(t_0,t_0,x_0) = x_0$ (i.e., $\phi(\cdot,t_0,x_0)$ satisfies the given initial condition) and (ii) $\dot{\phi}(t,t_0,x_0) = f(\phi(t,t_0,x_0),t)$ over $[a,b]$ (i.e., $\phi(\cdot,t_0,x_0)$ satisfies the given differential equation). These basic facts allow us to conduct a brief experiment which spotlights the importance of the existence/uniqueness question. The experiment is to numerically solve the differential equation

$$\dot{x}(t) = -\frac{1}{2x(t)}, \quad x(0) = x_0 = 0.1 \qquad (10.2)$$

The integral form of the equation is

$$x(t) = x(0) - \frac{1}{2} \int_0^t x(q)^{-1} dq$$

Applying the Euler formula with a step size of 0.05, yields

$$x(0.05) = 0.1 - \frac{0.05}{2x(0)} = -0.15$$

and

$$x(0.1) = -0.15 - \frac{0.05}{2x(0.05)} = 0.02$$

Unfortunately, these numbers are the proverbial garbage. But did we not apply the formula properly? Yes. Then why garbage? Because a solution simply does not exist for $t > 0.01$. To see this, use separation of variables to solve equation 10.2 analytically as

$$\int_{x_0}^{x(t)} 2x\,dx = -\int_0^t dt$$

to obtain

$$x(t) = \sqrt{x_0^2 - t} \qquad (10.3)$$

Thus, a real solution exists only if $0 \leqslant t \leqslant x_0^2 = 0.01$. Consequently, the preceding number crunching amounts to idle wheel spinning and the question, "When does there exist a solution?" takes on a high level of importance in view of this pathology.

To illustrate the local existence property again, consider the differential equation $\dot{x} = -(2x)^{-1}$ with the new initial condition $x(0) = 1$. The solution as per equation 10.3 is $x(t) = \sqrt{1 - t}$ which has a well-defined real value for $0 \leqslant t \leqslant 1$. For $t > 1$, the solution is complex and hence does not exist. "But what about uniqueness?" one might ask. Shouldn't a solution be unique if it exists? Again, no. Consider the differential equation

$$\dot{x} = 1.5x^{1/3}, \quad x(t_0) = x_0 \qquad (10.4)$$

Again, using separation of variables, one obtains

$$x^2(t) = (t - t_0 + x_0^{2/3})^3$$

Since $x^2(t)$ must be positive, it follows that $t \geqslant t_0 - x_0^{2/3}$ and the solution takes the form

$$x(t) = (t - t_0 + x_0^{2/3})^{1.5} \tag{10.5}$$

Now, clearly, if $x(0) = 0$ is the initial condition, then $x(t) = t^{1.5}$ is a solution. (It satisfies the initial condition and equation 10.4). But let α be an arbitrary constant strictly greater than zero. Then another solution is given by

$$x(t) = \begin{cases} 0, & 0 \leqslant t < \alpha \\ (t - \alpha)^{1.5}, & t \geqslant \alpha \end{cases} \tag{10.6}$$

To see that this is so, merely note that $x(0) = 0$ and that equation 10.4 is satisfied:

$$\dot{x} = 1.5(t - \alpha)^{0.5} = 1.5[(t - \alpha)^{1.5}]^{1/3} = 1.5x(t)^{1/3}$$

Consequently, there are, as a matter of fact, an infinite number of solutions to equation 10.4— i.e., a solution exists, but it is not unique.

A numerical solution to equation 10.4 is also instructive. Again using an Euler scheme with initial condition $x(0) = 0$, one can show that

$$x(0.1) = x(0) + 0.15[x(0)]^{1/3} = 0$$

and

$$x(0.2) = x(0.3) = 0$$

as are all subsequent estimates. This is quite different from the solution given in equation 10.5 and provides another reason for exercising caution in trusting numerical computations. On the other hand, if $x(0) = 0.1$ is the initial condition, we obtain

$$x(0.1) = 0.1 + 0.15(0.1)^{1/3} = 0.1696$$

which is a good approximation to the solution of $x(0.1) = 0.1772$ obtained from equation 10.5.

In conclusion, why embark on a study of the existence and uniqueness of solutions to nonlinear (and hence linear) state dynamical equations? For several reasons: (*i*) simple curiosity, the desire to understand, (*ii*) the need to know that our numerical calculations are not fruitless, and (*iii*) (as we shall see) the uniqueness result provides a tool for showing the equivalence of certain functional representations— e.g., if $z(t)$ is a solution to a differential equation known to have a unique solution and $g(t)h(t)$ is (seemingly) another solution with the same initial condition, then $z(t) = g(t)h(t)$. This fact proves especially useful when we discuss the properties of the state transition matrix in chapter 12.

MATHEMATICAL PRELIMINARIES

Before proceeding with the main thread of the existence and uniqueness discussion, we digress briefly to review some basic mathematical concepts.

Definition 10.1. The *Euclidean norm* of a vector $v = [v_1, \ldots, v_n]^t \in \mathbf{R}^n$ is

$$||v||_2 = \sqrt{v_1^2 + \cdots + v_n^2} \qquad (10.7)$$

Definition 10.2. Let $v_0 \in \mathbf{R}^n$. Then

$$B(v_0;\epsilon) = \{v \in \mathbf{R}^n \mid ||v - v_0||_2 < \epsilon\} \qquad (10.8)$$

is an *open ball* (*open neighborhood*) about v_0 in $\mathbf{R}^n$ of radius $\epsilon > 0$. Of course, the closed ball $\bar{B}(v_0;\epsilon)$ includes all points v in $\mathbf{R}^n$ whose distance from v_0 is less than or equal to ϵ.

As a simple example let $v_0 = (1,1)^t \in \mathbf{R}^2$ and let $\epsilon = 1$. Then $B(v_0;1)$ appears as in Figure 10.1, where $v = (v_1,v_2)^t$ denotes an arbitrary vector in $\mathbf{R}^2$.

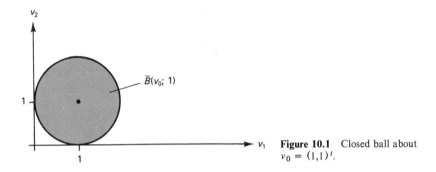

Figure 10.1   Closed ball about $v_0 = (1,1)^t$.

Definition 10.3. Let $f(\cdot,\cdot):\mathbf{R}^n{\times}\mathbf{R} \to \mathbf{R}^n$ satisfy $f(x_0,t_0) = v_0$. Then f, is *continuous at* $(x_0,t_0) \in \mathbf{R}^n{\times}\mathbf{R}$ if for each $\epsilon > 0$, there exists a $\delta > 0$ such that whenever $(x,t) \in B[(x_0,t_0);\delta]$ then $v = f(x,t) \in B(v_0;\epsilon)$.

The idea of continuity is illustrated in Figure 10.2.

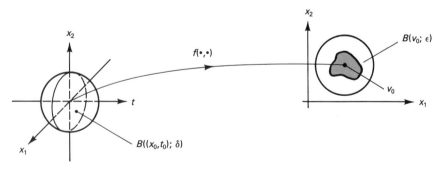

Figure 10.2 Pictorial continuity.

Definition 10.4. A function $f(\cdot,\cdot):\mathbf{R}^n{\times}\mathbf{R} \to \mathbf{R}^n$ is *continuous* on some domain $\mathbf{D} \subset \mathbf{R}^n{\times}\mathbf{R}$ if it is continuous at each point of $\mathbf{D}$.

This definition is of course consistent with our intuitive expectation. Continuity tends to be stronger than necessary for our purposes. Instead, the weaker concept of piecewise continuity proves adequate. To nail down this notion, however, we need the concept of left- and right-hand limits. For fixed x, the function $f(x,t)$ has a well defined *right hand limit* at t' if

$$\lim_{\substack{t \to t' \\ t > t'}} f(x,t)$$

exists. Similarly, $f(x,t)$ has a well-defined *left hand limit* at t' if

$$\lim_{\substack{t \to t' \\ t < t'}} f(x,t)$$

exists.

Definition 10.5. Let S be a subset of **R** which contains at most a finite number of points per unit interval. Then the function $f(x,t)$ is *piecewise continuous in t* if it is continuous at all $t \in S$ *(for an appropriate S)* and has well-defined left- and right-hand limits at each $t' \in S$.

The preceding ideas find application to our general existence and uniqueness results. At the end of the chapter these results are applied to linear time-varying state dynamics in particular. Doing so requires several more definitions.

Definition 10.6. Let $x(\cdot): \mathbf{R} \to \mathbf{R}$ be piecewise continuous. Then the L_∞-norm (also called the *sup-norm*) is defined as

$$\|x(\cdot)\|_\infty = \sup_t |x(t)|$$

If $x(\cdot): \mathbf{D} \subset \mathbf{R} \to \mathbf{R}$, then the restricted sup-norm becomes

$$\|x(\cdot)\|_{\infty,\mathbf{D}} = \sup_{t \in \mathbf{D}} |x(t)|$$

As an example, the sup-norm of $x(t) = (1 - e^{-t})1^t(t)$ is $\|x(\cdot)\|_\infty = 1$.

One can easily extend the notion of a sup-norm to a piecewise continuous vector-valued function $x(\cdot): \mathbf{R} \to \mathbf{R}^n$. Here, the sup-norm is naturally defined as

$$\|x(\cdot)\|_\infty = \max_i \|x_i(\cdot)\|_\infty$$

where $x(t) = [x_1(t),...,x_n(t)]^t$. This vector norm suggests one natural way to norm a matrix. Consider a matrix $A(\cdot): \mathbf{R} \to \mathbf{R}^{n \times n}$, where $A(t) = [a_{ij}(t)]$ and each $a_{ij}(t)$ is a piecewise continuous function. Let $x(\cdot): \mathbf{R} \to \mathbf{R}^n$ also be piecewise continuous. Then $A(t)x(t) = y(t)$ is a vector in $\mathbf{R}^n$. Observe that $y(t) = [y_1(t),...,y_n(t)]^t$ satisfies

$$y_i(t) = \sum_{j=1}^n a_{ij}(t)x_j(t)$$

To obtain an upper bound on $|y_i(t)|$, observe that

$$|y_i(t)| \leqslant \sum_{j=1}^n |a_{ij}(t)| \cdot |x_j(t)| \leqslant \left[\sum_{j=1}^n |a_{ij}(t)|\right] \|x(t)\|_2$$

where $||x(t)||_2 = \sqrt{x_1^2(t) + \ldots + x_n^2(t)}$ is the usual Euclidean vector norm. Such manipulations are standard in system analysis, and the choice of norm must preserve these properties. Hence, it is also natural to define the L_∞ norm of a matrix as follows:

Definition 10.7. Let $A(\cdot):\mathbf{R} \rightarrow \mathbf{R}^{n \times n}$ be an n x n matrix-valued function having piecewise continuous entries, where $A(\cdot) = [a_{ij}(\cdot)]$. Then the L_∞-*norm* or *sup-norm* of $A(\cdot)$ is

$$||A(\cdot)||_\infty = \max_i || \sum_{j=1}^{n} |a_{ij}(\cdot)| ||_\infty$$

The L_∞-norm is said to be a *vector-induced* norm.

As a simple example, consider

$$A(t) = \begin{bmatrix} e^{-t} & 3\sin(t) \\ 2 & 1 \end{bmatrix} 1^+(t)$$

By inspection, $||A(\cdot)||_\infty = 3.216$.

From definition 10.7, it is easily shown that

$$||Ax||_2 \leqslant ||A||_\infty ||x||_2$$

and that if A_1 and A_2 are any two conformable matrices, then

$$||A_1 + A_2||_\infty \leqslant ||A_1||_\infty + ||A_2||_\infty$$

In general, the symbol $||\cdot||$ denotes a norm on a vector space if (*i*) $||A|| = 0$ if and only if $A = [0]$, (*ii*) $||A_1 + A_2|| \leqslant ||A_1|| + ||A_2||$, and (*iii*) $||\alpha A|| = |\alpha| ||A||$ for all scalars α.

The foregoing norms, as well as various others, are pertinent to our investigations throughout the text. The reader may consult some of the references for a fuller discussion [1,2,5–7,11].

LOCAL EXISTENCE AND UNIQUENESS

Is there an obvious property of $f(x,t)$ which guarantees existence of a solution to equation 10.1? Continuity, or at least piecewise continuity, seems reasonable. As it turns out, the continuity of $f(\cdot,\cdot)$ is the key property for guaranteeing the local existence of a solution.

Theorem 10.1 (Peano Existence Theorem). Let $f(\cdot,\cdot):\mathbf{R}^n \times \mathbf{R} \rightarrow \mathbf{R}$ be continuous on a closed, connected bound region $\mathbf{D} \subset \mathbf{R}^n \times \mathbf{R}$. Suppose $(x_0,t_0) \in \mathbf{D}$. Then there exists a solution $\phi(\cdot,t_0,x_0)$ to equation 10.1 defined over some interval $[a,b]$ where $\{(x_0,[a,b])\} \subset \mathbf{D}$ and $t_0 \in [a,b]$.

Several points are important to mention. The solution to equation 10.4 is not unique even though $f(x,t) = 1.5x^{1/3}$ is continuous, something stronger than continuity is needed for uniqueness. Also, the size of $[a,b]$ remains unknown; continuity provides no information; and a and b could represent ϵ-perturbations about t_0— i.e., $a = t_0 - \epsilon$ and $b = t_0 + \epsilon$ for some small positive ϵ. The differential equation 10.2 whose solution is equation 10.3 exhibits this property. Here, $f(x,t) = 0.05x^{-1}$ is continuous for all $x \neq 0$. Equation 10.3 shows that there exists a solution which is very local for small x_0.

Now, what about local uniqueness? Again, is there some property of $f(x,t)$ which guarantees a locally unique solution? Observe from equation 10.4 that $f(x,t) = 1.5\,x^{1/3}$ is not differentiable at $x = 0$. Perhaps if $f(x,t)$ were continuously differentiable, a unique solution would exist locally. In fact, this bit of speculation is true. However, as pointed out in [4], such a strong assumption would exclude far too many functions of engineering interest, such as sawtooth signals which are piecewise continuously differentiable. Fortunately, a weaker condition, called a local Lipschitz condition, implies the local existence of a unique solution.

Definition 10.8. Let $\mathbf{D} \subset \mathbf{R}^n \times \mathbf{R}$ be a connected, closed, bounded set. Then the function $f(x,t)$ satisfies a *local* Lipschitz condition at t_0 on $\mathbf{D}$ with respect to (x,t_0) in $\mathbf{D}$ if there exists a finite constant k such that

$$||f(x^1,t_0) - f(x^2,t_0)||_2 \leqslant k\,||x^1 - x^2||_2 \tag{10.9}$$

for all (x^1,t_0) and (x^2,t_0) in $\mathbf{D}$.

The constant k is called the *Lipschitz constant*. Since $\mathbf{D}$ is a closed, bounded set, and since t_0 is fixed, the Lipschitz condition is local. As an example of a Lipschitz constant and local Lipschitz condition, consider the function $f(x,t) = x^2$ in the region

$$\mathbf{D} = \{(x,t) \epsilon \mathbf{R}^2 \mid -k \leqslant x \leqslant k,\ -k \leqslant t \leqslant k\}$$

for any finite k. Clearly, for each fixed $t \epsilon [-k,k]$, the maximum change Δf in f with respect to changes Δx in x, is given by $\left. \dfrac{\partial f}{\partial x} \right|_{x=k} = 2k$. Heuristically, one reasons that

$$||\Delta f||_2 \leqslant \max_{-k \leqslant x \leqslant k} \left|\left| \frac{\partial f}{\partial x} \right|\right|_2 ||\Delta x||_2$$

and hence one can demonstrate that

$$||f(x^1,t) - f(x^2,t)||_2 \leqslant 2k\,||x^1 - x^2||_2$$

for all (x^1,t) and $(x^2,t) \epsilon \mathbf{D}$. Notice that the function is also continuously differentiable. All continuously differentiable functions satisfy a local Lipschitz condition (see problem 17) but the converse fails.

Theorem 10.2 (Picard's Existence and Uniqueness Theorem). If $f(x,t)$ is continuous on $\mathbf{D} \subset \mathbf{R}^n \times \mathbf{R}$ and satisfies a local Lipschitz condition on $\mathbf{D}$, then for any $(x_0,t_0) \epsilon \mathbf{D}$, there exists a unique solution $\phi(\cdot,t_0,x_0)$ defined over some interval $a < t_0 < b$ with $(x_0,[a,b]) \subset \mathbf{D}$. Moreover, the solution $\phi(\cdot,t_0,x_0)$ depends continuously on t_0 and x_0.

To obtain a graphical interpretation of the Lipschitz condition, consider $f(x,t) = e^{-|t|}\sin(x)$ with $t=0$. Figure 10.3 illustrates how $f(x,0)$ lies within the sector whose region is defined by $|f(x^1,0) - f(x^2,0)| \leqslant |x^1 - x^2|$. Figure 10.4 illustrates a function $f(x,0)$ which has an infinite slope at $x=1$ and hence, the function does not satisfy a Lipschitz condition about this point.

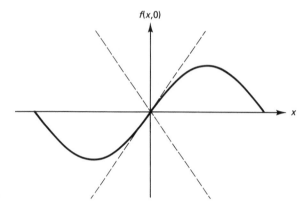

Figure 10.3 A function satisfying a local Lipschitz condition.

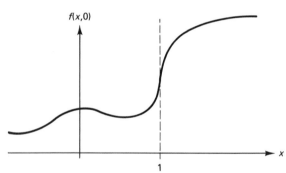

Figure 10.4 A function which fails to satisfy a local Lipschitz condition.

As an example consider

$$\dot{x}(t) = x^2(t)$$

which satisfies a local Lipschitz condition over every closed bounded subset of $\mathbf{R}^2$. Hence, we expect a locally unique solution to exist. Via separation of variables,

$$\int_{x_0}^{x(t)} \frac{dx}{x^2} = \int_{t_0}^{t} dt$$

which leads to

$$x(t) = \frac{x_0}{1 + x_0(t_0 - t)}$$

At $(x_0,t_0) = (1,1)$ we have the unique solution

$$x(t) = \frac{1}{2-t}$$

which unfortunately blows up as t approaches 2. This instant of time is known as the *finite escape time*.

These local existence conditions are interesting but do not provide the kind of information pertinent to either engineering interest or the textual development. Rather, our time functions are defined over semi-infinite intervals $[t_0, \infty)$ and our real concern focuses on global existence and uniqueness— i.e., the existence and uniqueness of solutions to equation 10.1 over these semi-infinite intervals.

GLOBAL EXISTENCE AND UNIQUENESS

In regards to global existence and uniqueness, the objective is to state sufficient conditions for the existence of a unique solution to equation 10.1 rewritten as

$$\dot{x}(t) = f(x(t), t), \quad x(0) = x_0 \tag{10.10a}$$

where

$$x(t) \epsilon \mathbf{R}^n \quad \text{and} \quad t \epsilon \mathbf{R}^+ \triangleq [0, \infty) \tag{10.10b}$$

A simple change of variable makes the statement of the problem and forthcoming existence and uniqueness conditions valid over an arbitrary interval $[t_0, \infty)$. The approach builds on the following assumptions of Desoer [2]:

(*i*) $S \subset \mathbf{R}^+$ contains at most a finite number of points per unit interval,

(*ii*) For each $x \epsilon \mathbf{R}^n$, $f(x, t)$ is continuous at $t \notin S$, (Note, then, that we require continuity only in the variable t.)

(*iii*) For each $t_i \epsilon S$, f(x,t) has finite left- and right-hand limits at $t = t_i$,

(*iv*) $f(\cdot, \cdot): \mathbf{R}^n \times \mathbf{R} \to \mathbf{R}^n$ satisfies a global Lipschitz condition; i.e., there exists a piecewise continuous function $k(\cdot): \mathbf{R}^+ \to \mathbf{R}^+$ such that

$$||f(x^1, t) - f(x^2, t)||_2 \leqslant k(t) ||x^1 - x^2||_2 \tag{10.11}$$

for all $t \epsilon \mathbf{R}^+$ and all points $x^1, x^2 \epsilon \mathbf{R}^n$.

What are the immediate implications of these assumptions? First of all, for each fixed t, the function $f(x, t)$ is continuous at x. This follows from (*iv*). Second, if $x(t)$ is a continuous function, then $f(x(t), t)$ is a piecewise continuous function in t whose points of discontinuity lie in S. This follows from (*i*), (*ii*), (*iii*), and (*iv*). Third, $f(x(t), t)$ is integrable with respect to t and the derivative of its integral equals $f(x(t), t)$ except possibly at $t \epsilon S$.

Theorem 10.3. Suppose the preceding assumptions (*i*) th. ugh (*iv*) hold. Then for each $x_0 \epsilon \mathbf{R}^n$ and $t_0 \epsilon \mathbf{R}^+$, there exists a unique continuous function

$$\phi(\cdot, t_0, x_0): \mathbf{R}^+ \to \mathbf{R}^n$$

such that

$$(a) \quad \dot{\phi}(t, t_0, x_0) = f(\phi(t, t_0, x_0), t)$$

and

(b) $\phi(t_0,t_0,x_0) = x_0$

for all $t \in \mathbf{R}^+$ and $t \notin S$.

This theorem formally states that assumptions (i) through (iv) are sufficient to guarantee the existence of a unique solution to equations 10.10. The beauty of these assumptions stems from their sufficiency in actually constructing a solution [2]. This can be accomplished as follows. Abbreviate $\phi(t,t_0,x_0)$ by $\phi(t)$, and let $\hat{\phi}^k(t)$ be the kth approximation to $\phi(t)$ defined according to the iteration scheme

$$\hat{\phi}^0(t) = x_0$$

$$\hat{\phi}^1(t) = x_0 + \int_0^t f(\hat{\phi}^0(q),q)\,dq$$

$$\vdots$$

$$\hat{\phi}^{k+1}(t) = x_0 + \int_0^t f(\hat{\phi}^k(q),q)\,dq$$

Using the global Lipschitz condition, it can be shown that this iteration formula converges uniformly to a solution of the differential equation $\dot{x} = f(x,t)$, $x(0) = x_0$. One can then show, via the noted Bellman-Gronwall lemma, that the solution is a unique solution, denoted by $\phi(t)$. For more details, consult [2], pp. 34-41.

Note the following information: (1) $\phi(t,t_0,x_0)$ is continuous for all $t \in \mathbf{R}^+$ and differentiable for all $t \notin S$. (2) $\phi(\cdot,t_0,x_0)$ represents the state trajectory. (3) "Unique $\phi(\cdot,t_0,x_0)$" means that if ϕ_1 and ϕ_2 both satisfy

(i) $\dot{\phi}_i = f(\phi_i,t)$ $(i=1,2)$

and

(ii) $\phi_1(t_0,t_0,x_0) = \phi_2(t_0,t_0,x_0)$

then

$$\phi_1(t,t_0,x_0) = \phi_2(t,t_0,x_0)$$

for all $t \in \mathbf{R}^+$. This uniqueness theorem and specifically this meaning of uniqueness run through a number of proofs in succeeding chapters.

EXAMPLE 10.1

In this example we demonstrate the existence of a unique solution for the state equations

$$\begin{bmatrix} \dot{x}_1(t) \\ \dot{x}_2(t) \end{bmatrix} = \begin{bmatrix} x_2(t) \\ q\,\sin[x_1(t)] \end{bmatrix} \tag{10.12}$$

of the inverted pendulum. Specifically, we show that

$$f(x,t) = \begin{bmatrix} x_2(t) \\ q\sin[x_1(t)] \end{bmatrix}$$

satisfies the global Lipschitz condition of equation 10.11.

Let $x^1 = [x_1^1, x_2^1]^t$ and $x^2 = [x_1^2, x_2^2]^t$ be two arbitrary vectors in $\mathbf{R}^2$. Consider the following equality and inequality, where an expression on the left side of equation 10.11 is "massaged" into the right side:

$$||f(x^1,t) - f(x^2,t)||_2 = \left|\left| \begin{bmatrix} x_2^1 - x_2^2 \\ q[\sin(x_1^1) - \sin(x_1^2)] \end{bmatrix} \right|\right|_2$$

$$\leqslant \max[|q|,1] \left|\left| \begin{bmatrix} x_2^1 - x_2^2 \\ 2\cos[0.5(x_1^1 + x_1^2)]\sin[0.5(x_1^2 - x_1^2)] \end{bmatrix} \right|\right|_2$$

Observe that

$$|2\cos[0.5(x_1^1 + x_1^2)]\sin[0.5(x_1^1 - x_1^2)]|$$

$$\leqslant 2|\sin[0.5(x_1^1 - x_1^2)]| \leqslant 2|0.5(x_1^1 - x_1^2)|$$

which results because $[\sin(x)/x] \leqslant 1$. Hence,

$$||f(x^1,t) - f(x^2,t)||_2 \leqslant \max[|q|,1] \left|\left| \begin{bmatrix} x_2^1 - x_2^2 \\ x_1^1 - x_1^2 \end{bmatrix} \right|\right|_2$$

which is the necessary global Lipschitz condition. Note that $\max[|q|,1]$ plays the role of the piecewise continuous function $k(t)$ in equation 10.11. The conclusion of the example is that for each initial condition the nonlinear state equations for the inverted pendulum have a unique solution.

As mentioned earlier, the Lipschitz condition in conjunction with assumptions (i) through (iii) allows one to construct the unique solution to equation 10.10. The proof of Theorem 10.3 proceeds along such paths. This theorem, however, can also be viewed as a corollary to Wintner's global existence theorem [8,4].

As a final point, note that the global Lipschitz condition of equation 10.11 can be restricted. Instead of allowing x^1 and $x^2 \epsilon \mathbf{R}^n$ we can restrict our set to a connected set $\mathbf{D} \subset \mathbf{R}^n$. For example $\mathbf{D}$ could be $\{x \epsilon \mathbf{R}^n | x_1 \geqslant 0, x_2 \geqslant 0,...,x_n \geqslant 0\}$. Theorem 10.3 would then assert the existence of a unique solution in that region, provided the solution remains in that region for all time.

EXAMPLE 10.2

Determine the existence and possible uniqueness of a solution to the normalized state model

$$\begin{bmatrix} \dot{x}_1(t) \\ \dot{x}_2(t) \end{bmatrix} = \begin{bmatrix} x_2(t) \\ -[x_1(t)]^{-2} \end{bmatrix} + \begin{bmatrix} 0 \\ 1 \end{bmatrix} 2 \times 1^+(t) \tag{10.13}$$

for the vertical ascent of a rocket, where $x_1(t)$ is the rocket position above the surface of the earth, $x_2(t)$ is the rocket velocity, and $2 \times 1^+(t)$ is the normalized thrust.

Our concern is with positions above the surface of the earth and with the rocket's vertical ascent. Hence, for proper thrust, our state variables will be in the region

$$\mathbf{D} = \{(x_1, x_2) \mid x_1 \geqslant 1, x_2 \geqslant 0\}$$

The implicit question is, does the function

$$f(x,t) = \begin{bmatrix} x_2(t) \\ -[x_1(t)]^{-2} + 2 \times 1^+(t)] \end{bmatrix} \tag{10.14}$$

satisfy a global Lipschitz condition in the said region for $t \geqslant 0$? To answer the question, it is necessary to show that, for this region,

$$\left\| \begin{matrix} x_2^1 - x_2^2 \\ -\left[\dfrac{1}{x_1^1}\right]^2 + \left[\dfrac{1}{x_1^2}\right]^2 \end{matrix} \right\|_2 \leqslant k(t) \left\| \begin{matrix} x_1^1 - x_1^2 \\ x_2^1 - x_2^2 \end{matrix} \right\|_2 \tag{10.15}$$

for some piecewise continuous $k(t)$. The crux of this task is to show that there exists a piecewise continuous $\hat{k}(t)$ such that

$$\left| -\left[\frac{1}{x_1^1}\right]^2 + \left[\frac{1}{x_1^2}\right]^2 \right| \leqslant \hat{k}(t) \left| x_1^1 - x_1^2 \right|$$

Observe that

$$\left| -\left[\frac{1}{x_1^1}\right]^2 + \left[\frac{1}{x_1^2}\right]^2 \right| = \left| \frac{(x_1^1)^2 - (x_1^2)^2}{(x_1^1)^2 (x_1^2)^2} \right|$$

$$= \left| \frac{(x_1^1 + x_1^2)(x_1^1 - x_1^2)}{(x_1^1)^2(x_1^2)^2} \right| \leqslant 2|x_1^1 - x_1^2|$$

in the region $\mathbf{D}$ ($\hat{k}(t) = 2$), since in $\mathbf{D}$,

$$\frac{x_1^1 + x_1^2}{(x_1^1)^2 (x_1^2)^2} \leqslant 2$$

with maximum value at $x_1^1 = x_1^2 = 2$. Unfortunately, this shows only a region-restricted Lipschitz condition, and the function does not satisfy the precise definition of the global Lipschitz condition (equation 10.11), which requires satisfaction for *all* $x^1, x^2 \in \mathbf{R}^2$. Our experience, however, indicates that the solution is unique; one can show this rigorously by building up locally unique solutions over the entire region and then connecting them.

EXISTENCE AND UNIQUENESS OF THE LINEAR STATE DYNAMICS

This section focuses on the special case of equation 10.1 or equations 10.10, where

$$\dot{x}(t) = f(x(t),t) = A(t)x(t) \tag{10.16}$$

The essential hypothesis is that the $n \times n$ matrix $A(t) = [a_{ij}(t)]$ has entries $a_{ij}(t)$ which are piecewise continuous, where $a_{ij}(\cdot):\mathbf{R}^+ \to \mathbf{R}$. Again, the possible points of discontinuity are the elements of S, and the results are valid for any semi-infinite interval $[t_0, \infty)$. If each $a_{ij}(\cdot)$ is piecewise continuous, then the matrix $A(\cdot):\mathbf{R}^+ \to \mathbf{R}^{n \times n}$ is said to be piecewise continuous. From our early theorems, we suspect that there exists a local solution. In fact, piecewise continuity, coupled with the linearity of equation 10.16, implies the validity of the global Lipschitz condition on $\mathbf{R}^{n \times n} \times \mathbf{R}^+$ and, hence, a unique solution.

Proposition 10.1. Let equation 10.16 hold under the hypothesis that $A(\cdot)$ is piecewise continuous on $\mathbf{R}^+$. Then $f(x,t) = A(t)x$ satisfies the global Lipschitz condition on $\mathbf{R}^{n \times n} \times \mathbf{R}^+$.

Proof. Define the sets $\mathbf{D}_j = [j-1, j)$ for $j=1,2,3,...$ Observe that

$$\bigcup_{j=1}^{\infty} \mathbf{D}_j = \mathbf{R}^+$$

Since each entry of $A(\cdot)$ is piecewise continuous the L_∞ norm of $A(\cdot)$ is well defined over each $\mathbf{D}_j$. In particular, for each $t \in \mathbf{D}_j$ and arbitrary x^1 and x^2 in $\mathbf{R}^n$,

$$||f(x^1,t) - f(x^2,t)||_2 = ||A(t)x^1 - A(t)x^2||_2 \leqslant ||A(\cdot)||_{\infty, \mathbf{D}_j} ||x^1 - x^2||_2$$

which follows by linearity and the Holder inequality. Now, the piecewise continuity of $A(\cdot)$ guarantees that

$$||A(\cdot)||_{\infty, \mathbf{D}_j} < \infty$$

for each $j = 0,1,2, ...$ Hence, equation 10.16 satisfies a Lipschitz condition over each region $\mathbf{D}_j$. Next, demonstrate satisfaction of a global Lipschitz condition over $\mathbf{R}^+$, define the piecewise continuous function

$$k(t) = \begin{cases} 0, & t < 0 \\ ||A(\cdot)||_{\infty, \mathbf{D}_j} & t \in \mathbf{D}_j \end{cases}$$

Then equation 10.16 satisfies the desired global Lipschitz condition.

$$||A(t)x^1 - A(t)x^2||_2 \leqslant k(t)||x^1 - x^2||_2 \qquad \blacksquare$$

Theorem 10.4. If $A(\cdot)$ is piecewise continuous, then for each initial condition $x(0)$, a solution designated $\phi(\cdot,0,x_0)$ to equation 10.16 exists and is unique.

Proof. The theorem follows directly from proposition 10.1. $\qquad \blacksquare$

Recalling that the proof to Proposition 10.1 holds for any semi-infinite interval $[t_0, \infty)$, under a simple change of variable, we obtain the final theorem of this chapter.

Theorem 10.5. If $A(\cdot)$ and $B(\cdot)$ are piecewise continuous over $[t_0,\infty)$, then a solution $\phi(\cdot,t_0,x_0,u)$ to

$$\dot{x}(t) = A(t)x(t) + B(t)u(t), \quad x(t_0) = x_0$$

exists and is unique over $[t_0,\infty)$.

Proof. The proof of the theorem is left as an exercise.   ∎

PROBLEMS

1. Discuss the existence and/or uniqueness of a solution to $\dot{x}(t) = 1.25x^{0.2}$.

2. Plot the solutions of equation 10.6 for $\alpha = 1,2,3,4$.

3. Show that $\dot{x}(t) = \exp[-x(t)]$ satisfies a Lipschitz condition in the region defined by

  (i) $0 < \epsilon \leqslant t_0 \leqslant t$
    and
  (ii) $0 \leqslant x \leqslant K < \infty$.

4. In equation 10.15, find the function $k(t)$.

5. Show that $f(x(t),t) = t\cos(x(t))$ satisfies a global Lipschitz condition.

6. Investigate the existence and/or uniqueness of a solution to the scalar nonlinear differential equation

$$\dot{x} = t\ell n(x) - x, \quad x(0) = x_0 > 0$$

with $t > 0$.

7. A certain "switched" (time-varying) system has the state model

$$\dot{x} = A_1 x + Bu, \quad x(t_0) = x_0$$
$$y = Cx$$

for $0 \leqslant t < 1$ and switches to the state model

$$\dot{x} = A_2 x + Bu$$
$$y = Cx$$

for $t \geqslant 1$. Choose the most appropriate of the following statements:
  (a) There exists a unique solution only for $0 \leqslant t < 1$,
  (b) There exists a unique solution only for $1 \geqslant t$,
  (c) There exists a unique solution for $0 \leqslant t < 1$ or $1 \leqslant t$ but not both,
  (d) There exists a unique solution for all $t \geqslant 0$,
  (e) The uniqueness of the solution cannot be determined.

8. For the differential equation $\dot{x}(t) = t[x(t)]^{0.5}$, $x(t_0) = x_0$, which of the following statements is true?

  (a) No solution exists anywhere.
  (b) There exists a solution everywhere, but it is not necessarily unique.
  (c) There exists a unique solution everywhere.
  Justify your choice.

9. Let the matrices $A(\cdot)$ and $B(\cdot)$ be piecewise continuous over the interval $[t_0,\infty)$, Show that the function

$$f(x(t), u(t), t) = A(t)x(t) + B(t)u(t)$$

satisfies the global Lipschitz condition. First explain what is meant by $A(\cdot)$ being piece-wise continuous on $[t_0, \infty]$. What can be said about the existence and uniqueness of a solution for

$$\dot{x}(t) = A(t)x(t) + B(t)u(t), \quad x(t_0) = x_0$$

10. Determine the existence and uniqueness of the state dynamical equations

$$\begin{bmatrix} \dot{x}_1(t) \\ \dot{x}_2(t) \end{bmatrix} = \begin{bmatrix} x_2(t) \\ \left(\dfrac{g}{\ell}\right) \sin[x_1(t)] + 3u(t) \end{bmatrix}$$

11. Consider the nonlinear differential equation

$$\dot{x}(t) = x^3(t) + 1, \quad x(0) = 2$$

(a) Does this equation have a unique solution? Justify your answer.
(b) Use an Euler predictor and *three* iterations of a trapezoidal corrector to compute an approximate value for $x(0.1)$.

12. Consider the nonlinear differential equation

$$\begin{bmatrix} \dot{x}_1 \\ \dot{x}_2 \end{bmatrix} = \begin{bmatrix} x_2 \\ 1 - \cos(x_1) \end{bmatrix} + \begin{bmatrix} 0 \\ 1 \end{bmatrix} \cos(2t) \, 1^+(t)$$

Does this equation have a unique solution? Justify your answer in an organized, clear manner. (*Hint*: $\cos(a) - \cos(b) = -2\sin[0.5(a+b)]\sin[0.5(a-b)]$.)

13. Show that the initial value problem

$$\dot{x}(t) = tx(t)^{-2}$$

with $x(0) = 0$ has a unique solution.

14. Show that the initial value problem

$$\dot{x}(t) = 2t^{-3}x(t)$$

with $x(0) = 0$ does not have a unique solution. How many solutions does it have? Observe that this is a linear time-varying differential equation.

15. Suppose that there exists a unique solution to $\dot{x} = f(x,t)$, $x(t_0) = x_0$, $t \geq t_0$. Suppose also that due to a modeling error, the differential equation perturbs to $\dot{z} = f(z,t) + p(t)$, $z(t_0) = x_0 + \delta x_0$, $t \geq t_0$. If (i) $t \in [t_0, t_0 + T]$, (ii) the perturbation p(t) is bounded according to $||p(t)||_2 \leq \epsilon_1$, and (iii) the variation in the initial condition is bounded by $||\delta x_0|| \leq \epsilon_0$, construct a bound or $||z(t) - x(t)||_2$ that is valid over $[t_0, t_0 + T]$. (*Hint*: consult [2] pp. 40, 41.)

16. Suppose $\dot{x}(t) = f(x,t)$ does not satisfy a global Lipschitz condition on $\mathbf{R}^n \times \mathbf{R}^+$. Which of the following statements is true?

(a) There does not exist a local solution.
(b) There does not exist a locally unique solution.
(c) There does not exist a global solution.
(d) There does not exist a globally unique solution.
(e) It cannot be determined whether there is a solution.
Find counterexamples to each answer you did not choose.

17. Show that if $f(\cdot,\cdot): \mathbf{R}^n \times \mathbf{R} \to \mathbf{R}^n$ is continuous and differentiable in a bounded, closed, convex domain **D**, then f satisfies a Lipschitz condition on **D**. (See [10], p. 22). Show further that the Lipschitz constant k can be taken to be $\sup_{\mathbf{D}} \|\partial f / \partial x\|_2$. Note that $\mathbf{D}$ is convex if the line segment joining any two points in n-space lies entirely within $\mathbf{D}$.

18. An alternative definition of $\|A(\cdot)\|_\infty$ is

$$\|A(\cdot)\|_\infty = \max_{i,j} \|a_{ij}(\cdot)\|_\infty$$

Show that with this norm (which is not vector-induced) the proof of proposition 10.1 fails. (*Hint*: Consider

$$A(t) = \begin{bmatrix} 1 & 1 \\ 1 & 1 \end{bmatrix}$$

with $x^1 = [1\ 1]^t$ and $x^2 = [0\ 0]^t$.)

19. Use the iteration formula following theorem 10.3 to construct an approximate solution to

$$\dot{x} = 1/x^2, \quad x(0) = 1$$

Stop at ϕ^2 and show that this is a good approximation to the actual solution at $t = 0.2$. Then evaluate at $t = 0.5$ and check your answer using a numerical simulation.

REFERENCES

1. E. A. Coddington and N. Levinson, *Theory of Ordinary Differential Equations* (New York: McGraw-Hill, 1955), chapters 1 and 2.

2. C. A. Desoer, *Notes for a Second Course on Linear Systems* (New York: Van Nostrand Reinhold, 1970), pp. 32–41.

3. R. K. Miller and A. N. Michel, *Ordinary Differential Equations* (New York: Academic Press, 1982).

4. R. A. Struble, *Nonlinear Differential Equations* (New York: McGraw-Hill, 1962).

5. J. K. Hale, *Ordinary Differential Equations* (New York: McGraw-Hill, 1969).

6. H. L. Royden, *Real Analysis* (New York: Macmillan, 1963).

7. L. O. Chua and P. M. Lin, *Computer-Aided Analysis of Electronic Circuits: Algorithms and Computational Techniques* (Englewood Cliffs, NJ: Prentice Hall, 1975).

8. A. Wintner, "The Non-Local Existence Problem of Ordinary Differential Equations," *American Journal of Mathematics* Vol. 67, (1945), pp. 277–284.

9. Garret Birkhoff and Gian-Carlo Roto, *Ordinary Differential Equations* (Waltham, MA: Blaisdell, 1969).

11

THE SOLUTION SPACE OF $\dot{x}(t) = A(t)x(t)$ AND THE STATE TRANSITION MATRIX

INTRODUCTION

Whenever the entries of $A(t)$ are piecewise continuous and bounded over $[t_0, \infty)$, the solution to $\dot{x}(t) = A(t)x(t), x(t_0) = x_0$ exists and is unique. Indeed, an infinite number of initial-condition-dependent unique solutions exist. Accordingly, is there an identifiable structure to the world of all possible solutions, i.e., to the solution space? Is it possible to characterize this solution space, perhaps with "basis solutions" similar to the way basis vectors characterize Euclidean n-space, $\mathbf{R}^n$? To answer these questions, we

    (i) define a set of "vector basis solutions" characterizing the solution space of $\dot{x}(t) = A(t)x(t)$;

    (ii) organize these "basis solutions" into a matrix defined as the state transition matrix;

    (iii) sketch the very basic properties of the state transition matrix; and

    (iv) develop the exponential form of the state transition matrix.

NOTATION AND MATHEMATICAL PRELIMINARIES

In linear algebra, a set of Euclidean vectors $\{v_1, v_2,...,v_n\}$, is *linearly independent* if and only if, whenever

$$\alpha_1 v_1 + \alpha_2 v_2 +...+ \alpha_n v_n = \theta$$

where θ is the zero vector, then the scalars $\alpha_1 = \alpha_2 =...= \alpha_n = 0$. The usual Euclidean basis vectors for $\mathbf{R}^n$ are

$$\eta_i = [0,...,0,1,0,...0]^t$$
$$\underset{\underset{\textit{ith entry}}{\uparrow}}{}$$

We extend these ideas in the following definition.

Definition 11.1. The vector-valued functions $\phi_1(\cdot), \phi_2(\cdot), \ldots, \phi_n(\cdot)$, continuous over $[t_0, t_1]$, are *linearly independent* if and only if, whenever

$$\alpha_1\phi_1(\cdot) + \alpha_2\phi_2(\cdot) +...+ \alpha_n\phi_n(\cdot) \equiv \theta \tag{11.1}$$

then $\alpha_1 = \alpha_2 =...= \alpha_n = 0$.

In this definition, θ is the zero function and the triple bar means that the quantity on the left is identically equal to the zero function. Simply stated, *independent functions can sum to the zero function if and only if the scaling weights α_i are zero*. As an example, consider the set of scalar-valued functions $\{\phi_1(t) = 1, \phi_2(t) = t, \phi_3(t) = t^2\}$. These are clearly continuous and independent over $[0,1]$. Consequently, no linear multiple of ϕ_1 could ever equal ϕ_2 or ϕ_3 over $[0,1]$. Figure 11.1 sketches these three functions, which are linearly independent,

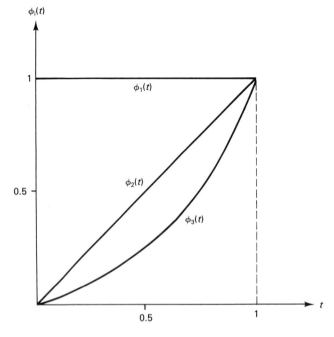

Figure 11.1 Sketch of three independent scalar valued functions.

not just over $[0,1]$, but over any finite subinternal of the real line. Another set of vector-valued functions that are independent over any finite subinterval of the real line is

$$\left\{ \phi_1(t) = \begin{bmatrix} 1 \\ 0 \end{bmatrix}, \phi_2(t) = \begin{bmatrix} 0 \\ 1 \end{bmatrix}, \phi_3(t) = \begin{bmatrix} e^{-t} \\ 0 \end{bmatrix}, \phi_4(t) = \begin{bmatrix} 0 \\ e^{-t} \end{bmatrix} \right\} \qquad (11.2)$$

A set of functions is a *basis* for a given space of functions if (i) the element functions of the set are independent and (ii) every function in the space can be expressed as a linear combination of the elements in the basis set. For example, the functions of equation 11.2 are a basis for the space of functions of the form

$$\phi(t) = \begin{bmatrix} a_0 + a_1 e^{-t} \\ b_0 + b_1 e^{-t} \end{bmatrix}$$

The idea is completely analogous to the notion of a basis in $\mathbf{R}^n$.

As a final notational point, recall that the solution trajectory for $\dot{x}(t) = A(t)x(t)$, $x(t_0) = x_0$ has the designation $\phi(t,t_0,x_0)$, where, by definition, $\phi(t_0,t_0,x_0) = x_0$.

SOLUTION SPACE OF $\dot{x}(t) = A(t)x(t)$

The following theorem completely characterizes the solution space of the homogeneous differential equation $\dot{x}(t) = A(t)x(t)$.

Theorem 11.1. Let $A(t)$ be $n \times n$ with piecewise continuous bounded entries over $[t_0,t]$. The equation $\dot{x}(t) = A(t)x(t)$ has exactly n linearly independent vector solutions $\{\phi_1(t,t_0,\eta_1), \phi_2(t,t_0,\eta_2), \ldots, \phi_n(t,t_0,\eta_n)\}$ satisfying $\phi_i(t_0,t_0,\eta_i) = \eta_i$. Furthermore, every linear combination of these solutions is a solution, and any arbitrary solution is a linear combination of these solutions.

Proof. The uniqueness theorem of the previous chapter guarantees that a unique solution $\phi_i(t,t_0,\eta_i)$ exists for each initial condition $x(t_0) = \eta_i$. The theorem is then proved by verifying the following three points:

(i) The solutions $\phi_i(\cdot,t_0,\eta_i)$ are linearly independent,

(ii) $\sum_{i=1}^{n} \alpha_i \phi_i(\cdot,t_0,\eta_i) \triangleq \psi(\cdot,t_0,x_0)$ is a solution,

(iii) Any arbitrary solution, say, $\psi(\cdot,t_0,x_0)$ is a linear combination of the solutions $\phi_i(\cdot,t_0,\eta_i)$.

Proof of (i). The argument structure here is proof by contradiction; i.e., assume that the solutions $\phi_i(t,t_0,\eta_i)$, $i=1,\ldots,n$ are linearly dependent and reason to the conclusion that the η_i, $i=1,\ldots,n$ are linearly dependent. This patently false conclusion implies that the assumption of linearly dependent solutions is incorrect: the solutions $\phi_i(t,t_0,\eta_i)$ are indeed linearly independent.

Accordingly, suppose that over $[t_0,t_1]$

$$\alpha_1\phi_1(\cdot,t_0,\eta_1) + ... + \alpha_n\phi_n(\cdot,t_0,\eta_n) = \theta \tag{11.3}$$

but $\alpha_i \neq 0$ for at least one i. Then evaluating equation 11.3 at t_0 yields

$$\alpha_1\eta_1 + \alpha_2\eta_2 + ... + \alpha_n\eta_n = \theta$$

But, since at least one $\alpha_i \neq 0$, the set $\{\eta_1,\eta_2, . . . ,\eta_n\}$ must be linearly dependent, an obviously false conclusion.

Proof of (ii). Let an arbitrary linear combination be defined according to

$$\psi(t,t_0,x_0) = \sum_{i=1}^{n} \alpha_i\phi_i(t,t_0,\eta_i)$$

where $x_0 = \alpha_1\eta_1 + ... + \alpha_n\eta_n$ by construction. Clearly, if $\psi(t,t_0,x_0)$ satisfies $\dot{x}(t) = A(t)x(t)$, it is a solution. Differentiating $\psi(t,t_o,x_o)$ and using the linearity of the derivative operator results in the following series of equations:

$$\dot{\psi}(t,t_0,x_0) = \sum_{i=1}^{n} \alpha_i\,\dot{\phi}_i(t,t_0,\eta_i) = \sum_{i=1}^{n} \alpha_i\,A(t)\phi_i(t,t_0,\eta_i)$$

$$= A(t) \sum_{i=1}^{n} \alpha_i\,\phi_i(t,t_0,\eta_i) = A(t)\psi(t,t_0,x_0)$$

Proof of (iii). Let $\psi(t,t_0,x_0)$ designate an arbitrary solution. Certainly, $\psi(t,t_0,x_0)$ satisfies $\dot{x}(t) = A(t)x(t)$ with initial condition $x(t_0) = x_0$. Then let

$$\phi(t,t_0,x_0) = \sum_{i=1}^{n} \alpha_i\,\phi_i(t,t_0,\eta_i)$$

where the α_i are chosen so that

$$x_0 = \alpha_1\eta_1 + ... + \alpha_n\eta_n$$

Now since $\phi(t_0,t_0,x_0) = x_0$, if it also satisfies $\dot{x}(t) = A(t)x(t)$, then $\phi(t,t_0,x_0) = \psi(t,t_0,x_0)$ by Theorem 10.5, the uniqueness theorem—i.e., an arbitrary solution is a linear combination of the ϕ_i's. But then, by the same arguments used to verify (ii), $\phi(t,t_0,x_0)$ satisfies the given differential equation.   ∎

All solutions to $\dot{x}(t) = A(t)x(t)$ now have convenient expressions in terms of the basis solutions, just as all vectors in $\mathbf{R}^n$ have simple representations in terms of the Euclidean basis $\{\eta_1, . . . ,\eta_n\}$. Hence, from the arguments of the preceding proof, one simply sets $(\alpha_1, . . . ,\alpha_n) = x_0^t$. The solution to $\dot{x}(t) = A(t)x(t)$, $x(t_0) = x_0$ takes the form

$$\phi(t,t_0,x_0) = \sum_{i=1}^{n} \alpha_i\,\phi_i(t,t_0,\eta_i)$$

This simple observation underlies the definition and development of the state transition matrix in the next section.

THE STATE TRANSITION MATRIX

Is there a convenient "matrix" expression for the solution of $\dot{x}(t) = A(t)x(t), x(t_0) = x_0$? The observation at the end of the last section suggests one.

Definition 11.2. The *state transition matrix* of the differential equation $\dot{x}(t) = A(t)x(t)$ is

$$\Phi(t,t_0) = [\phi_1(t,t_0,\eta_1) \mid \phi_2(t,t_0,\eta_2) \mid \cdots \mid \phi_n(t,t_0,\eta_n)] \tag{11.4}$$

where the $\phi_i(t,t_0,\eta_i)$ are the previously defined basis solutions.

Note that for pedagogical reasons the state transition matrix has a form which appears to be linked to the usual Euclidean basis $\{\eta_1,...,\eta_n\}$ as initial conditions. In the next chapter, our discussion of the fundamental matrix allows us to explicitly free $\Phi(t,t_0)$ from this apparent dependence.

There are three immediately obvious and very basic properties of the state transition matrix:

(*i*) $\Phi(t_0,t_0) = I$,

(*ii*) $\Phi(t,t_0)$ satisfies the matrix differential equation

$$\dot{M}(t) = A(t)M(t), M(t_0) = I$$

where both $A(t)$ and $M(t)$ are $n \times n$,

(*iii*) $\Phi(t,t_0)$ is uniquely defined. (Why?)

Proposition 11.1. The solution to $\dot{x}(t) = A(t)x(t)$ with $x(t_0) = x_0$ is

$$x(t) = \Phi(t,t_0)x_0 \tag{11.5}$$

for all t.

Proof. The proof relies on the uniqueness theorem; i.e., $x(t)$ uniquely satisfies the given differential equation with initial condition $x(t_0) = x_0$. The goal is to show that $\Phi(t,t_0)x_0$ does also.

Step 1. By construction $\Phi(t_0,t_0) = I$; hence, $x(t_0) = \Phi(t_0,t_0)x_0$—i.e., the initial values coincide.

Step 2. Since $\dot{\Phi}(t,t_0) = A(t)\Phi(t,t_0)$, $[\Phi(t,t_0)x_0] = A(t)[\Phi(t,t_0)x_0]$—i.e., $[\Phi(t,t_0)x_0]$ and $x(t)$ satisfy the same differential equation.

Step 3. By the uniqueness theorem, $x(t) = \Phi(t,t_0)x_0$.   ∎

From Chapter 2, for constant A, the solution to $\dot{x}(t) = Ax(t), x(t_0) = x_0$ is

$$x(t) = e^{A(t-t_0)}x_0 \triangleq \Phi(t,t_0)x_0 \tag{11.6}$$

where the rightmost identification follows from Proposition. 11.1 Unfortunately, solutions to $\dot{x}(t) = A(t)x(t), x(t_0) = x_0$ are not generally given as

$$x(t) = \exp[\int_{t_0}^{t} A(q)dq]x_0 \tag{11.7}$$

The development of sufficient conditions guaranteeing the validity of equation 11.7 follows shortly. But first...

AN EXAMPLE, THE HARD WAY

Recall a trick formula for the solution of the scalar differential equation

$$\dot{x}(t) = -p(t)x(t) + q(t) \tag{11.8}$$

The formula is

$$x(t) = e^{-\rho(t)}[\int e^{\rho(\tau)}q(\tau)d\tau + K] \tag{11.9}$$

where $\rho(t) = \int p(t)dt$ and K is a constant.

With this formula in mind, consider the homogeneous state dynamics

$$\begin{bmatrix} \dot{x}_1 \\ \dot{x}_2 \end{bmatrix} = \begin{bmatrix} 2t & 1 \\ 1 & 2t \end{bmatrix} \begin{bmatrix} x_1 \\ x_2 \end{bmatrix}; \begin{bmatrix} x_1(0) \\ x_2(0) \end{bmatrix} = \begin{bmatrix} \alpha_1 \\ \alpha_2 \end{bmatrix} \tag{11.10}$$

The objectives are: (*i*) find a general solution to these equations in the form $x(t)$, and (*ii*) use the solution to (*i*) to find the state transition matrix. The solution requires solving two simultaneous differential equations in two unknowns— i.e., x_1 cannot be found independently of x_2. Some cleverness becomes necessary. Adding the two differential equations in 11.10 together produces a differential equation in a new variable $z_1 = x_1 + x_2$. Similarly, subtracting the two produces a differential equation in the new variable $z_2 = x_1 - x_2$. Each equation is solvable by the trick formula. The foregoing manipulation is equivalent to executing the time-invariant nonsingular state transformation

$$\begin{bmatrix} z_1 \\ z_2 \end{bmatrix} = \begin{bmatrix} 1 & 1 \\ 1 & -1 \end{bmatrix} \begin{bmatrix} x_1 \\ x_2 \end{bmatrix} \tag{11.11}$$

The resulting new state dynamical equations are

$$\begin{bmatrix} \dot{z}_1 \\ \dot{z}_2 \end{bmatrix} = \begin{bmatrix} 2t+1 & 0 \\ 0 & 2t-1 \end{bmatrix} \begin{bmatrix} z_1 \\ z_2 \end{bmatrix}; \begin{bmatrix} z_1(0) \\ z_2(0) \end{bmatrix} = \begin{bmatrix} \alpha_1 + \alpha_2 \\ \alpha_1 - \alpha_2 \end{bmatrix}$$

As indicated earlier, z_1 and z_2 are decoupled; i.e., each is solvable independently of the other via the trick formula.

Applying equation 11.9 to $\dot{z}_1 = (2t+1)z_1$ produces

$$z_1(t) = (\alpha_1 + \alpha_2)\exp(t^2 + t) \tag{11.12}$$

Similarly, applying equation 11.9 to $\dot{z}_2 = (2t-1)z_2$ produces

$$z_2(t) = (\alpha_1 - \alpha_2)\exp(t^2 - t) \tag{11.13}$$

The inverse of the state transformation of equation 11.11 is

$$\begin{bmatrix} x_1 \\ x_2 \end{bmatrix} = \begin{bmatrix} 0.5 & 0.5 \\ 0.5 & -0.5 \end{bmatrix} \begin{bmatrix} z_1 \\ z_2 \end{bmatrix} \tag{11.14}$$

Substituting equations 11.12 and 11.13 into equation 11.14 yields the general form of the state trajectory of the differential equation 11.10:

$$\phi(t,0,[\alpha_1,\alpha_2]^t) = \begin{bmatrix} x_1(t) \\ x_2(t) \end{bmatrix} = 0.5e^{t^2} \begin{bmatrix} (\alpha_1 + \alpha_2)e^t + (\alpha_1 - \alpha_2)e^{-t} \\ (\alpha_1 + \alpha_2)e^t - (\alpha_1 - \alpha_2)e^{-t} \end{bmatrix} \tag{11.15}$$

This fulfills the earlier objective (i) and is the pedagogical basis for constructing the state transition matrix. Specifically, if $[\alpha_1 \, \alpha_2]^t = [1 \; 0]^t = \eta_1$, then

$$\phi_1(t,0,\eta_1) = 0.5e^{t^2} \begin{bmatrix} e^t + e^{-t} \\ e^t - e^{-t} \end{bmatrix}$$

Similarly, if $[\alpha_1 \, \alpha_2]^t = [0 \; 1]^t = \eta_2$, then

$$\phi_2(t,0,\eta_2) = 0.5e^{t^2} \begin{bmatrix} e^t - e^{-t} \\ e^t + e^{-t} \end{bmatrix}$$

in which case

$$\Phi(t,0) = [\phi_1(t,0,\eta_1) \mid \phi_2(t,0,\eta_2)] = 0.5e^{t^2} \begin{bmatrix} e^t + e^{-t} & e^t - e^{-t} \\ e^t - e^{-t} & e^t + e^{-t} \end{bmatrix}$$

Clearly, $\Phi(0,0) = I$, and the general form of the state trajectory given in equation 11.15 has the specific form

$$x(t) = \Phi(t,0) \begin{bmatrix} \alpha_1 \\ \alpha_2 \end{bmatrix}$$

as expected. Of course it is not necessary to construct the general form 11.15 first; rather it is only necessary to compute the two solutions with initial conditions η_1 and η_2.

EXPONENTIAL FORM OF THE STATE TRANSITION MATRIX

Under certain conditions, the state transition matrix has an exponential form:

$$\Phi(t,t_0) = \exp \begin{bmatrix} \int_{t_0}^{t} A(q)dq \end{bmatrix} \tag{11.16}$$

The goal is to describe sufficient conditions for the validity of equation 11.16. The related proofs are left as problems at the end of the chapter.

Definition 11.3.   Two matrices A and B commute if and only if $AB = BA$.

Theorem 11.2.   If $A(t)$ and

$$\hat{A}(t) = \int_{t_0}^{t} A(q)dq \tag{11.17}$$

commute for all t, then the state transition matrix of $\dot{x}(t) = A(t)x(t)$ has the exponential form of equation 11.16.   ∎

Computing $\hat{A}(t)$ and then checking whether or not it commutes with $A(t)$ is very tedious although straightforward. The following theorem cites some simple sufficient conditions for this property.

Theorem 11.3. $A(t)$ and $\hat{A}(t)$ (in equation 11.17) commute if any of the following holds:

(i) $A(\cdot)$ is constant. (What is the state transition matrix in this case?)

(ii) $A(t) = \alpha(t)M$ where $\alpha(\cdot):\mathbf{R} \to \mathbf{R}$ and M is a constant matrix,

(iii) $A(t) = \displaystyle\sum_{i=1}^{k} \alpha_i(t)M_i$, where $\alpha_i(\cdot):\mathbf{R} \to \mathbf{R}$ and the M_i's are constant matrices which satisfy the condition $M_iM_j = M_jM_i$ for all i and j.   ∎

Corollary 11.3a.   If $A(t)$ has the form of condition (iii) of theorem 11.3 then

$$\Phi(t,t_0) = \prod_{i=1}^{k} \exp\left[M_i \int_{t_0}^{t} \alpha_i(q)\,dq \right] \tag{11.18}$$

AN EXAMPLE, THE EASY WAY

Now let us reconsider $\dot{x}(t) = A(t)x(t)$ when

$$A(t) = \begin{bmatrix} 2t & 1 \\ 1 & 2t \end{bmatrix}$$

It is easy to confirm that $A(t)$ commutes with its integral $\hat{A}(t)$. More simply, however, $A(t)$ has a decomposition $A(t) = \alpha(t)M_1 + M_2$ according to

$$\begin{bmatrix} 2t & 1 \\ 1 & 2t \end{bmatrix} = 2t \begin{bmatrix} 1 & 0 \\ 0 & 1 \end{bmatrix} + \begin{bmatrix} 0 & 1 \\ 1 & 0 \end{bmatrix} = \alpha(t)M_1 + M_2$$

Since M_1 and M_2 commute, this is precisely condition (iii) of theorem 11.3. Now, by equation 11.18

$$\Phi(t,t_0) = \exp[M_1(t^2 - t_0^2)]\exp[M_2(t - t_0)]$$

But $\exp[M_1(t^2 - t_0^2)] = e^{t^2 - t_0^2}I$, where I is the 2×2 identity matrix. And further, expanding $\exp[M_2(t - t_0)]$ in a Taylor series indicates that

$$\exp[M_2(t - t_0)] = \begin{bmatrix} \cosh(t - t_0) & \sinh(t - t_0) \\ \sinh(t - t_0) & \cosh(t - t_0) \end{bmatrix}$$

Hence,

$$\Phi(t,t_0) = e^{t^2 - t_0^2} \begin{bmatrix} \cosh(t - t_0) & \sinh(t - t_0) \\ \sinh(t - t_0) & \cosh(t - t_0) \end{bmatrix} \qquad (11.19)$$

which again is the state transition matrix of equation 11.10.

THE STATE TRANSITION MATRIX FOR PERIODIC A(t)

Historically, periodically time-varying system models have found important application in many areas of engineering and science. For example, Hill [1], in 1886, explained the lunar perigee in terms of the periodically varying gravitational forces of the sun and moon. In more recent times, the study of circuits with time-varying elements [2–4] has found application to filters and parameteric amplifiers [5]. Reference [6] provides a comprehensive look at the various contributions in the field.

The particular focus here rests on the structure of the state transition matrix of a periodically time-varying system. The so called Floquet theory [7] underlies the explanation. The theory says that knowledge of the solution of the zero-input state dynamics for a periodically time varying matrix $A(t)$ completely determines the solution for successive periods. To glimpse the implied meaning of this theory, let $\Phi(t,t_0)$ be the state transition matrix of $\dot{x}(t) = A(t)x(t)$, where $A(t)$ varies periodically with fundamental period $T_0 \neq 0$— i.e., $A(t) = A(t + T_0)$ and for any other T such that $A(t) = A(t + T)$, $kT_0 = T$ for some integer k. Floquet theory implies the existence of an $\hat{A}$ such that $x(t + T_0) = \hat{A}x(t)$. Deriving this relationship requires the following lemma.

Lemma 11.1. If $A(t)$ is periodic with fundamental period T_0, then $\Phi(t + T_0,t_0)$ is a fundamental matrix of $A(t)$, i.e., $\dot{\Phi}(t + T_0,t_0) = A(t)\Phi(t + T_0,t)$ and $\det[\Phi(t + T_0,t)] \neq 0$.

Proof. Since $\Phi(t,t_0)$ is the associated state transition matrix of $A(t)$, $\dot{\Phi}(t,t_0) = A(t)\Phi(t,t_0)$. Substituting $t + T_0$ for t produces

$$\dot{\Phi}(t + T_0,t_0) = A(t + T_0)\Phi(t + T_0,t_0)$$

Since $A(t) = A(t + T_0)$,

$$\dot{\Phi}(t + T_0,t_0) = A(t)\Phi(t + T_0,t_0)$$

This relationship together with the nonsingularity of $\Phi(t + T_0,t_0)$, implies that $\Phi(t + T_0,t_0)$ is a fundamental matrix. ∎

Since $\Phi(t + T_0,0)$ and $\Phi(t,0)$ are both fundamental matrices, there exists a constant nonsingular matrix $\hat{A}$ such that

$$\Phi(t + T_0,0) = \hat{A}\Phi(t,0) \qquad (11.20)$$

By the properties of the state transition matrix (see theorem 12.1, the semigroup property),

$$x(t + T_0) = \Phi(t + T_0,t)x(t) = \Phi(t + T_0,0)\Phi(0,t)x(t)$$

$$= \Phi(t + T_0,0)[\Phi(t,0)]^{-1}x(t) = \hat{A}x(t) \qquad (11.21)$$

so that

$$x(t + T_0) = \hat{A}\,x(t)$$

Now, from equation 11.20, $\hat{A} = \Phi(T_0,0)$, and this implies that $\Phi(kT_0,0) = \hat{A}^k$. This relationship with equation 11.21 yields the discrete-system-like relationship

$$x(t + kT_0) = \hat{A}^k x(t) \qquad (11.22)$$

This verifies the following proposition:

Proposition 11.2. If $\dot{x}(t) = A(t)x(t)$ when $A(t)$ is periodic with fundamental period T_0, then there exists an $\hat{A}$ such that

$$x(t + kT_0) = \hat{A}^k x(t)$$

In other words, knowledge of the state trajectory over one period is sufficient to determine the state trajectory over all other periods. A little thought should augur two underlying properties. First, the structure of the original system $\dot{x}(t) = A(t)x(t)$ is not time-invariant, but the relationship of equation 11.22 is. This hints at a possible exponential structure for $\Phi(t,t_0)$ viz.,

$$\Phi(t + T_0,t) \approx \exp(RT_0) = \hat{A}$$

for an appropriate matrix R. Second, observe that all values of $x(t + kT_0)$ depend only on $x(t)$ over $[t,t + T_0]$. Hence, $\Phi(t,t_0)$ should also possess a periodic structure, i.e.,

$$\Phi(t + t_0,t) \approx P(t)$$

for some periodic matrix $P(t)$. Floquet theory combines these two proportionalities to assert, for the state transition matrix, the structure

$$\Phi(t,t_0) = P(t)\exp[R(t-t_0)]P^{-1}(t_0) \qquad (11.23)$$

where (i) R is a (possibly complex) matrix defined by $\Phi(T_0,0) = \exp(RT_0)$ and

$$P(t) = \Phi(t,0)\exp(-Rt) \qquad (11.24)$$

is a *nonsingular* periodic matrix with period T_0,

Before formally proving these ideas from Floquet theory, let us consider a brief example.

EXAMPLE 11.1.

Consider a diagonal system $\dot{x}(t) = A(t)x(t)$, where

$$A(t) = (\cos(t) + \lambda_1) \begin{bmatrix} 1 & 0 \\ 0 & 0 \end{bmatrix} + (-\sin(t) + \lambda_2) \begin{bmatrix} 0 & 0 \\ 0 & 1 \end{bmatrix}$$

is periodic with period $T_0 = 2\pi$. From theorem 11.3 and equation 11.18, the state transition matrix is

$$\Phi(t,t_0) = \begin{bmatrix} e^{[\sin(t) - \sin(t_0)]} & 0 \\ 0 & e^{[\cos(t) - \cos(t_0)]} \end{bmatrix} \begin{bmatrix} e^{\lambda_1(t - t_0)} & 0 \\ 0 & e^{\lambda_2(t - t_0)} \end{bmatrix}$$

Step 1: Compute R via $\Phi(T_0,0) = \exp(RT_0)$. Since $T_0 = 2\pi$ and

$$\Phi(2\pi,0) = \begin{bmatrix} 1 & 0 \\ 0 & 1 \end{bmatrix} \begin{bmatrix} e^{\lambda_1 2\pi} & 0 \\ 0 & e^{\lambda_2 2\pi} \end{bmatrix}$$

it follows that

$$R = \begin{bmatrix} \lambda_1 & 0 \\ 0 & \lambda_2 \end{bmatrix}$$

Step 2: As per equation 11.24,

$$P(t) = \Phi(t,0) e^{-Rt} = \begin{bmatrix} e^{\sin(t)} & 0 \\ 0 & e^{\cos(t) - 1} \end{bmatrix} \begin{bmatrix} e^{\lambda_1 t} & 0 \\ 0 & e^{\lambda_2 t} \end{bmatrix} \begin{bmatrix} e^{-\lambda_1 t} & 0 \\ 0 & e^{-\lambda_2 t} \end{bmatrix}$$

$$= \begin{bmatrix} e^{\sin(t)} & 0 \\ 0 & e^{\cos(t) - 1} \end{bmatrix}$$

Clearly, $P(t) = P(t + 2\pi)$, so $P(t)$ is periodic with period T_0.

At this point, we can set forth the foundational proofs of Floquet theory. In addition, we show that a linear periodically time-varying system is equivalent to a linear time-invariant system under the coordinate transformation $x(t) = P(t)z(t)$. Hence, analysis of these systems can be carried out (for most practical purposes) via linear time-invariant methods, a powerful set of tools.

First note that R is a well-defined matrix (see [8], p. 126). Thus, $P(t)$ in equation 11.24 is also well defined.

Proposition 11.3. $P(t)$ is periodic with period T_0.

Proof.   By equation 11.20,

$$P(t + T_0) = \Phi(t + T_0,0) \exp[-R(t + T_0)] = \Phi(t,0)\hat{A} \exp[-R(T_0 + t)]$$

But since $\hat{A} = \exp(RT_0)$, it follows that

$$P(t + T_0) = \Phi(t,0)\exp(-Rt) = P(t)$$

as was to be shown.   ∎

Combining the above propositions and discussion yields the main theorem.

Theorem 11.4. The state transition matrix for the periodically time-varying system $\dot{x}(t) = A(t)x(t)$, $A(t + T_0) = A(t)$ for all t, has the structure

$$\Phi(t,t_0) = P(t)\exp[R(t-t_0)]P^{-1}(t_0)$$

where R is as above and $P(t)$ is given in equation 11.24.

The next stage of the development of the Floquet theory proves that $\dot{x}(t) = A(t)x(t)$ is equivalent to the linear time-invariant system $\dot{z}(t) = Rz(t)$ under the time-varying coordinate transformation $x(t) = P(t)z(t)$. From our knowledge of time-varying coordinate transformations, one might attempt to show that

$$\dot{z}(t) = Rz(t) = [P^{-1}(t)A(t)P(t) + P^{-1}(t)\dot{P}(t)]z(t)$$

However, a more straightforward approach is to show that for all $x(0)$, the solution to $\dot{z}(t) = Rz(t)$, $z(0) = x(0)$ implies that $x(t) = P(t)z(t)$ is a solution to $\dot{x}(t) = A(t)x(t)$, $x(0)$ given. To implement this approach, consider that since R is constant, $z(t) = exp(Rt)x(0)$. But then,

$$x(t) = \Phi(t,0)x(0) = P(t)e^{Rt}x(0) = P(t)z(t)$$

implies that $P(t)z(t)$ is a solution, as stated.

Thus, properties of the state trajectory $x(t)$ are related to those properties of $z(t)$ (easily deduced by linear time-invariant analysis) through $P(t)$.

PROBLEMS

1. Find a set of basis vectors for functions of the form

$$\phi(t) = \begin{bmatrix} a_3t^3 + a_2t^2 + a_1t + a_0 \\ b_3t^3 + b_2t^2 + b_1t + b_0 \end{bmatrix}$$

How many such vectors do you need?

2. Construct a counterexample to the claim that if $\phi_1(t)$ and $\phi_2(t)$ are independent over $[t_0,t_1]$, they are independent over $[t_0',t_1']$ for $t_0 < t_0' < t_1' < t_1$. Then prove that if $\phi_1(t)$ and $\phi_2(t)$ are independent over $[t_0',t_1']$, they must be independent over $[t_0,t_1]$.

3. By using the Taylor Series expansion, show that

$$\exp\left\{ \begin{bmatrix} 0 & b \\ b & 0 \end{bmatrix} t \right\} = \begin{bmatrix} \cosh[b(t-t_0)] & \sinh[b(t-t_0)] \\ \sinh[b(t-t_0)] & \cosh[b(t-t_0)] \end{bmatrix}$$

4. Consider the equation $\dot{x}(t) = A(t)x(t)$, where

$$A(t) = \begin{bmatrix} 0 & 0 & 0 \\ 2t & 0 & 0 \\ 1 & 2t & 0 \end{bmatrix}$$

(a) What is $\Phi(t,0)$?

(b) Evaluate the expression found in (a) using the Taylor series definition of the matrix exponential.

5. Consider the differential equation

$$\begin{bmatrix} \dot{x}_1 \\ \dot{x}_2 \end{bmatrix} = \begin{bmatrix} -t & 0 \\ 0 & -1 \end{bmatrix} \begin{bmatrix} x_1(t) \\ x_2(t) \end{bmatrix}$$

(a) How many independent solutions does the differential equation have?
(b) Find the solutions $\phi_1(\cdot,0,\eta_1)$ and $\phi_2(\cdot,0,\eta_2)$. (*Hint*: Use the theory of elementary differential equations to obtain the solutions.)
(c) What is the state transition matrix $\Phi(t,0)$ in terms of your answer to part (b)?
(d) Determine $x(2)$ when $x(0) = [1\ 1]^t$.

6. Let $\dot{x}(t) = A(t)x(t)$, where

$$A(t) = \begin{bmatrix} 0 & -t \\ 0 & -t \end{bmatrix}$$

(a) Find the state transition matrix $\Phi(t,t_0)$.
(b) If $x(1) = [1\ 1]^t$, find $x(2)$.

7. Find the state transition matrix of $\dot{x}(t) = A(t)x(t)$, where

$$A(t) = \begin{bmatrix} 2t & 0 \\ -\cos(t) & 2t \end{bmatrix}$$

8. Two matrices A and B commute if and only if $AB = BA$. Show that if for all t, $A(t)$ and $M(t) = \int_{t_0}^{t} A(q)dq$ commute, then the state transition matrix of $\dot{x}(t) = A(t)x(t)$ is given by

$$\Phi(t,t_0) = \exp \int_{t_0}^{t} A(q)dq$$

9. Show that $A(t)$ and $\int_0^t A(q)dq$ commute if at least one of the following holds:
(a) $A(\cdot)$ is constant. (What is the state transition matrix in this case?)
(b) $A(t) = \alpha(t)M$, where $\alpha(\cdot):\mathbf{R} \to \mathbf{R}$ and M is a constant matrix.
(c) $A(t) = \sum_{i=1}^{k} \alpha_i(t)M_i$, where $\alpha_i:\mathbf{R} \to \mathbf{R}$ and the M_i's are constant matrices which satisfy the conditions $M_iM_j = M_jM_i$ for all i and j.

10. Suppose $A(t)$ is of the form in problem 9, part (c). Show that

$$\Phi(t,t_0) = \prod_{i=1}^{k} \exp[M_i \int_{t_0}^{t} \alpha_i(q)dq]$$

(*Hint*: Be careful of order.)

11. Given $\dot{x}(t) = A(t)x(t)$, where

$$A(t) = \begin{bmatrix} -2t & 0 & 0 \\ 2 & -2t & 0 \\ -1 & 1 & -2t \end{bmatrix}$$

compute the state transition matrix. In solving this problem, state all formulas that you use. Also, verify those conditions which guarantee the validity of the formula(s). Your work should be clear and organized, and should proceed in logical steps.

12. If $\dot{x}(t) = A(t)x(t)$, find the state transition matrix $\Phi(t,t_0)$ when

$$A(t) = \begin{bmatrix} 0 & \omega(t) \\ -\omega(t) & 0 \end{bmatrix}$$

(*Hint*: First compute $\Phi(t,0)$ by defining $\beta(t) = \displaystyle\int_0^t \omega(q)dq$.

13. Find the state transition matrix for the system $\dot{x}(t) = A(t)x(t)$ when

$$A(t) = \begin{bmatrix} \sin(t) & e^{-t} \\ e^{-t} & \sin(t) \end{bmatrix} \begin{bmatrix} \cos(t) & -e^{-t} \\ -e^{-t} & \cos(t) \end{bmatrix}$$

14. Use problem 9, part (c), and the Taylor series definition of the matrix exponential to construct the state transition matrix for

(a)

$$A = \begin{bmatrix} \lambda_1 & 1 \\ 0 & \lambda_1 \end{bmatrix}$$

(b)

$$A = \begin{bmatrix} \lambda_1 & 1 & 0 \\ 0 & \lambda_1 & 1 \\ 0 & 0 & \lambda_1 \end{bmatrix}$$

15. Suppose $\dot{x}(t) = A(t)x(t)$, $x(t_0) = x_0$, and

$$A(t) = \begin{cases} \begin{bmatrix} \lambda_1 & 0 & 0 \\ 0 & \lambda_2 & 1 \\ 0 & 0 & \lambda_2 \end{bmatrix}, & t < 1 \\\\ \begin{bmatrix} \lambda_2 & 0 & 0 \\ 1 & \lambda_2 & 0 \\ 0 & 0 & \lambda_1 \end{bmatrix}, & t \geqslant 1 \end{cases}$$

   (a) Find $\Phi(t,t_0)$ when $t_0, t \geqslant 1$.
   (b) Find $\Phi(t,t_0)$ when $t_0, t < 1$.
   (c) Find $\Phi(t,t_0)$ when $t_0 < 1$.
   (*Hint*: For part (c), remember that the entries of $\Phi(t,t_0)$ must be continuous everywhere and that $x(t) = \Phi(t,t_0)x(t_0)$ for all t.)

16. Consider a periodically time-varying system whose matrix $A(t)$ is

$$A(t) = \begin{bmatrix} -1 & \cos(t) \\ \cos(t) & -1 \end{bmatrix}$$

   (a) Use the results of problems 8, 9, and 11 to compute the state transition matrix $\Phi(t,t_0)$.
   (b) Find $P(t)$ and R, and formulate

$$\Phi(t,t_0) = P(t) e^{R(t-t_0)} P^{-1}(t_0)$$

17. Show that if $A(t_1) A(t_2) = A(t_2) A(t_1)$ for all t_1 and t_2, then the solution to $\dot{x}(t) = A(t)x(t)$ can be expressed as a matrix exponential.

REFERENCES

1. G. W. Hill, "On the Part of the Moon's Motion which is a Function of the Mean Motions of the Sun and Moon," Acta Mathematica, Vol. 8, 1886, pp. 1–36.

2. C. A. Desoer, "Steady-state Transmission through a Network Containing a Single Time-varying Element," IRE Transactions on Circuit Theory, Vol. CT-6, 1959, pp. 244–252.

3. A. Fettweis, "Steady-state Analysis of Circuits Containing a Periodically-operated Switch," IRE Transactions on Circuit Theory, Vol. CT-6, 1959, pp. 252–260.

4. L. E. Franks and I. W. Sandberg, "An Alternative Approach to the Realization of Network Transfer Functions: The N-path Filter," Bell Systems Technical Journal, Vol. 39, 1960, pp. 1321–1350.

5. D. P. Howson and R. B. Smith, *Parametric Amplifier* (Maidenhead, Berkshire, England: McGraw-Hill, 1970).

6. J. A. Richards, *Analysis of Periodically Time-Varying Systems* (New York: Springer-Verlag, 1983).

7. G. Floquet, "Sur les equations defferentiales lineáires," *Annals L'Ecole Normale Super.*, Vol. 12, 1983.

8. C. A. Desoer, *Notes for a Second Course on Linear Systems* (New York: Van Nostrand, 1970).

12

THE FUNDAMENTAL MATRIX
AND PROPERTIES OF
THE STATE TRANSITION MATRIX

INTRODUCTION

The state transition matrix represents a set of basis solutions (also called a *fundamental set of solutions*) for $\dot{x}(t) = A(t)x(t)$. Just as different bases exist for $\mathbf{R}^n$, different sets of basis solutions exist for the solution space of $\dot{x}(t) = A(t)x(t)$. The term *fundamental matrix* refers to that matrix fabricated from any fundamental set of solutions, i.e., from any set of basis solutions. In this chapter, the first objective is to briefly develop the notion of a fundamental matrix and point out its connection with the state transition matrix. Then, using the fundamental matrix concept as background, the second objective is to develop some important properties of the state transition matrix. These properties, their proofs and the associated manipulations illustrate the basic kind of reasoning and necessary perspective found throughout the theoretical investigations of the text.

MATHEMATICAL PRELIMINARIES

There are two new introductory concepts. The first is the notion of the matrix trace.

Definition 12.1. Let $A(\cdot) = [a_{ij}(\cdot)]$ be $n \times n$. Then the (matrix) *trace* of $A(\cdot)$, denoted tr$[A(\cdot)]$, is

$$\text{tr}[A(\cdot)] = \sum_{i=1}^{n} a_{ii}(\cdot) \tag{12.1}$$

The trace is simply the sum of the diagonal entries of the matrix.

The second is a statement of the Leverrier-Souriau-Faddeeva-Frame formula [1,2] stated here without proof.

Definition 12.2. The *Leverrier-Souriau-Faddeeva-Frame formula* for computing the coefficients a_i of the characteristic polynomial of A, $\pi_A(\lambda) = \det[\lambda I - A] = \lambda^n + a_1\lambda^{n-1} + a_2\lambda^{n-2} + \ldots + a_n$, is as follows:

(i) $N_1 = I$ $a_1 = -\text{tr}[A]$

(ii) $N_2 = N_1A + a_1I$ $a_2 = -\dfrac{1}{2}\,\text{tr}[N_2A]$

(iii) $N_3 = N_2A + a_2I$ $a_3 = -\dfrac{1}{3}\,\text{tr}[N_3A]$

$$\vdots$$

and in general,

$$N_n = N_{n-1}A + a_{n-1}I \qquad a_n = -\frac{1}{n}\,\text{tr}[N_nA] \tag{12.2}$$

where

$$[0] = N_nA + a_nI$$

As an example of the use of this algorithm, consider the matrix

$$A = \begin{bmatrix} 1 & 1 \\ 1 & -1 \end{bmatrix}$$

Then from the formula, $\pi_A(\lambda) = \det[\lambda I - A] = \lambda^2 - 2$, $-$ i.e., $a_1 = 0$ and $a_2 = -2$. Using the above algorithm

$$a_1 = -\,\text{tr}[A] = 0$$

and

$$a_2 = -0.5\,\text{tr}[N_2A] = -0.5\,\text{tr}[A^2] = -0.5\,\text{tr}\begin{bmatrix} 2 & 0 \\ 0 & 2 \end{bmatrix} = -2$$

which are the correct coefficients. The important point, for this chapter is that $a_1 = -\text{tr}[A]$. Although of theoretical interest, the algorithm itself is numerically unstable.

THE FUNDAMENTAL MATRIX

The fundamental matrix has columns which represent a set of basis solutions for the solution space of $\dot{x}(t) = A(t)x(t)$.

Definition 12.3. Any $n \times n$ matrix $M(t)$ satisfying the matrix differential equation

$$\dot{M}(t) = A(t)M(t), \ M(t_0) = M_0 \tag{12.3}$$

where $\det[M_0] \neq 0$ is a *fundamental matrix* (of solutions).

Clearly, the state transition matrix satisfies equation 12.3 with $M_0 = I$; therefore, it is a fundamental matrix. Most properties of a fundamental matrix are also properties of the state transition matrix.

Proposition 12.1. If $\det[M_0] \neq 0$, then $\det[M(t)] \neq 0$ for all t.

Proof. The proof is by the method of contradiction. Suppose there exists a t_1, such that $\det[M(t_1)] = 0$.

Step 1. Choose a vector $v = [v_1,...,v_n]^t \neq \theta$ such that $M(t_1)v = \theta$. This is possible since $\det[M(t_1)] = 0$ implies the linear dependence of the columns of $M(t_1)$. That is, a nonzero linear combination of columns of $M(t_1)$ sums to zero, or mathematically, $M(t_1)v = \theta$ with $v \neq \theta$.

Step 2. Define a state trajectory $x(t) = M(t)v$. Observe that

$$\dot{x}(t) = A(t)x(t), \quad x(t_1) = \theta$$

Observe further that the state trajectory $z(\cdot) \equiv \theta$ also satisfies

$$\dot{z}(t) = A(t)z(t), \quad z(t_1) = \theta$$

Then, by the uniqueness theorem, $x(t) = z(t)$ over the range for which $A(t)$ is piecewise continuous.

Step 3. Note that $z(t_0) = x(t_0) = M(t_0)v = \theta$, which implies that $\det[M(t_0)] = 0$, a contradiction. Therefore, $\det[M(t)] \neq 0$ for all t. ∎

Corollary 12.1a. If $\det[M(t_0)] \neq 0$, then $M(t)$ is nonsingular for all t.

Corollary 12.1b. If $M(t)$ is a fundamental matrix of $\dot{M}(t) = A(t)M(t)$, then the state transition matrix of $\dot{x}(t) = A(t)x(t)$ is given by

$$\Phi(t,t_0) = M(t)[M(t_0)]^{-1} \tag{12.4}$$

Equation 12.4 frees the state transition matrix from its dependence on initial conditions as given in definition 11.2.

PROPERTIES OF THE STATE TRANSITION MATRIX

The semigroup property of the state transition matrix describes how segments of state trajectories piece together over disjoint time intervals. For example, if $x(t) = \Phi(t,t_1)x(t_1)$ and $x(t_1) = \Phi(t_1,t_0)x(t_0)$, then does $x(t) = \Phi(t,t_0)x(t_0) = \Phi(t,t_1)\Phi(t_1,t_0)x(t_0)$ for all possible choices of t, t_1 and t_0? Surely, it must, and that is what the semigroup property asserts.

Theorem 12.1 (Semigroup Property).   For all t_1, t_0, and t,

$$\Phi(t,t_0) = \Phi(t,t_1)\Phi(t_1,t_0) \tag{12.5}$$

Proof.   The gist of the proof rests on the left-and right-hand sides of equation 12.5 satisfying the same matrix differential equation and the same initial condition. The uniqueness theorem then guarantees that the equality holds.

Step 1.   Define $M \triangleq \Phi(t_1,t_0)$ as the initial condition of the left-hand side of equation 12.5. Then $\dot{\Phi}(t,t_0) = A(t)\Phi(t,t_0)$ with initial condition $\Phi(t_1,t_0)$.

Step 2.   With respect to the right-hand side of equation 12.5, observe that

(i) $[\Phi(t,t_1)\Phi(t_1,t_0)]_{t=t_1} = \Phi(t_1,t_0) = M$, the required initial condition.
(ii) $d[\Phi(t,t_1)\Phi(t_1,t_0)]/dt = \dot{\Phi}(t,t_1)\Phi(t_1,t_0) = A(t)[\Phi(t,t_1)\Phi(t_1,t_0)]$.

Thus, since the left-and right-hand sides satisfy the same initial condition as well as the same differential equation, they are identical. But then the theorem is proven. ∎

The inverse property states that a simple interchange of arguments in the state transition matrix produces the inverse matrix.

Theorem 12.2 (Inverse Property). $\Phi(t,t_0)$ is nonsingular for all t and t_0, and $[\Phi(t,t_0)]^{-1} = \Phi(t_0,t)$.

Proof.   *Nonsingularity.*   Since $\Phi(t,t_0)$ is a fundamental matrix, it is nonsingular for all t and t_0 by Corollary 12.1a.
Inversion.   By the semigroup property, $\Phi(t,t_0) = \Phi(t,t_1)\Phi(t_1,t_0)$ for arbitrary choices of t_0, t_1, and t. In particular, for $t = t_0$ and $t_1 = t$, we obtain

$$I = \Phi(t_0,t)\Phi(t,t_0)$$

Multiplying on the right by $[\Phi(t,t_0)]^{-1}$ produces the desired result.   ∎

Another property establishes a relationship between the determinant of $\Phi(t,t_0)$ and the trace of $A(t)$.

Theorem 12.3 (Liouville formula [3]).

$$\det[\Phi(t,t_0)] = \exp[\int_{t_0}^{t} \mathrm{tr}[A(q)]dq] \tag{12.6}$$

Proof.   The structure of this proof is rather clever. The objective is to construct a differential equation in $\det[\Phi(t,t_0)]$ whose solution is $\exp\left[\int_{t_0}^{t} \mathrm{tr}\,[A(q)]dq\right]$.
What differential equation has the correct form? Clearly, one such equation is

$$\frac{d}{dt}\det[\Phi(t,t_0)] = \mathrm{tr}[A(t)]\det[\Phi(t,t_0)] \tag{12.7}$$

Fabrication of this differential equation uses the concept of an Euler approximation, the properties of the state transition matrix, and the fact that $a_1 = -\mathrm{tr}[A]$, where

$$\det[\lambda I - A] = \lambda^n + a_1\lambda^{n-1} + \ldots + a_n$$

Step 1. Recall that the state transition matrix satisfies $\dot{\Phi}(t,t_0) = A(t)\Phi(t,t_0)$. For sufficiently small step size h, the forward Euler formula implies that

$$\Phi(t+h,t_0) \approx \Phi(t,t_0) + hA(t)\Phi(t,t_0)$$

$$= h[h^{-1}I + A(t)]\Phi(t,t_0)$$

Taking determinants on both sides produces

$$\det[\Phi(t+h,t_0)] = h^n \det[\frac{1}{h}I + A(t)]\det[\Phi(t,t_0)]$$

Step 2. Applying the algorithm of Leverrier et al. with suitable adjustment for sign results in

$$h^n \det[\frac{1}{h}I + A(t)] = h^n[(\frac{1}{h})^n + a_1(t)(\frac{1}{h})^{n-1} + \ldots + a_n(t)]$$

$$= 1 + a_1(t)h + a_2(t)h^2 + \ldots + a_n(t)h^n$$

$$\approx 1 + h\,\mathrm{tr}\,[A(t)]$$

again for suitably small h. This in turn yields the approximation

$$\det[\Phi(t+h,t_0)] \approx [1 + h\,\mathrm{tr}[A(t)]]\det[\Phi(t,t_0)]$$

Step 3. Subtracting $\det[\Phi(t,t_0)]$ from both sides, defining

$$\det[\Phi(t+h,t_0)] - \det[\Phi(t,t_0)] \triangleq \Delta\det[\Phi(t,t_0)]$$

and dividing by the step size h produces

$$\frac{\Delta\det[\Phi(t,t_0)]}{h} \approx \mathrm{tr}[A(t)]\det[\Phi(t,t_0)]$$

Letting h approach zero, the continuity and piecewise differentiability of $\Phi(t,t_0)$, imply

$$\frac{d}{dt}\det[\Phi(t,t_0)] = \mathrm{tr}[A(t)]\det[\Phi(t,t_0)]$$

which is the differential equation whose solution is $\exp\left[\displaystyle\int_{t_0}^{t} \mathrm{tr}\,[A(q)]\,dq\right]$.   ∎

Corollary 12.3.   For all t, $\det[\Phi(t,t_0)] \neq 0$.

Exercise. If $M(t)$ is any fundamental matrix, find $\det[M(t)]$.

EXAMPLES

An example in the previous chapter showed that if

$$\begin{bmatrix} \dot{x}_1(t) \\ \dot{x}_2(t) \end{bmatrix} = \begin{bmatrix} 2t & 1 \\ 1 & 2t \end{bmatrix} \begin{bmatrix} x_1(t) \\ x_2(t) \end{bmatrix}$$

then the state transition matrix is

$$\Phi(t,t_0) = \exp(t^2 - t_0^2) \begin{bmatrix} \cosh(t-t_0) & \sinh(t-t_0) \\ \sinh(t-t_0) & \cosh(t-t_0) \end{bmatrix}$$

To illustrate the semigroup property, a simple calculation shows that

$$\Phi(1,0) = 2.718 \begin{bmatrix} 1.543 & 1.175 \\ 1.175 & 1.543 \end{bmatrix}$$

$$\Phi(2,1) = 20.086 \begin{bmatrix} 1.543 & 1.175 \\ 1.175 & 1.543 \end{bmatrix}$$

and

$$\Phi(2,0) = 54.598 \begin{bmatrix} 3.762 & 3.627 \\ 3.627 & 3.762 \end{bmatrix}$$

A further calculation verifies that

$$\Phi(2,1)\Phi(1,0) = 54.598 \begin{bmatrix} 1.543 & 1.175 \\ 1.175 & 1.543 \end{bmatrix}^2 = \Phi(2,0)$$

as expected.

To illustrate the inverse property, convert the coshs and sinhs to their exponential form to produce

$$[\Phi(t,t_0)]^{-1} = \Phi(t_0,t) = 0.5e^{t_0^2 - t^2} \begin{bmatrix} e^{(t_0 - t)} + e^{-(t_0 - t)} & e^{(t_0 - t)} - e^{-(t_0 - t)} \\ e^{(t_0 - t)} - e^{-(t_0 - t)} & e^{(t_0 - t)} + e^{-(t_0 - t)} \end{bmatrix}$$

A simple matrix multiplication then verifies the correctness of the inverse matrix.

Finally, to illustrate the Liouville Formula, by direct calculation we obtain

$$\det[\Phi(t,t_0)] = \exp[t^2 - t_0^2] \exp[t^2 - t_0^2] [\cosh^2(t - t_0) - \sinh^2(t - t_0)]$$

$$= \exp[2(t^2 - t_0^2)] = \exp \int_{t_0}^{t} 4q \, dq$$

which is the requisite equality.

PROBLEMS

1. Use the algorithm of Leverrier et al. to verify that the characteristic polynomial of

$$A = \begin{bmatrix} 0 & 1 & 0 \\ 0 & 0 & 1 \\ -a_3 & -a_2 & -a_1 \end{bmatrix}$$

is $\pi_A(\lambda) = \lambda^3 + a_1\lambda^2 + a_2\lambda + a_3$.

2. Let $\dot{x}(t) = A(t)x(t)$, where

$$A(t) = \begin{bmatrix} 0 & \sin(t) \\ 0 & 0 \end{bmatrix}$$

(a) Compute a fundamental matrix $M(t)$ of solutions, when

$$M(0) = \begin{bmatrix} 1 & 1 \\ 1 & -1 \end{bmatrix}$$

(b) Using the result of part (a), compute $\Phi(t,0)$.

(c) Using the result of part (b) together with the semigroup and inverse properties, compute $\Phi(t,t_0)$.

(d) Verify your answer to (c) by computing $\Phi(t,t_0)$ via theorem 11.2 and the Taylor series expansion of the matrix exponential.

3. Let $M(t)$ be a fundamental matrix. Using equation 12.4 and the Liouville formula, construct an expression for $\det[M(t)]$. Now prove that $\det[M(t)] \neq 0$ for all t.

4. Prove the semigroup property of the state transition matrix from memory. That is, show that for any t, t_1, and t_0,

$$\Phi(t,t_0) = \Phi(t,t_1)\Phi(t_1,t_0)$$

5. Prove that $[\Phi(t,t_0)]^{-1} = \Phi(t_0, t)$ from memory.

6. Suppose that the matrix

$$\Phi(t,0) = \begin{bmatrix} \exp(-t)\cos(t) & \exp(-t)\sin(t) \\ -\exp(-t)\sin(t) & \exp(-t)\cos(t) \end{bmatrix}$$

is the state transition matrix of $\dot{x}(t) = A(t)x(t)$.

(a) Find $A(t)$.

(b) What is $\Phi^{-1}(t,0)$? Can you answer this by inspection?

7. Which of the following are not valid state transition matrices?

(a)

$$\begin{bmatrix} \exp(t_0-t) & 0 \\ \sin(t-t_0) & \exp(t_0-t) \end{bmatrix}$$

(c)

$$\begin{bmatrix} \exp[t_0-t] & 0 \\ 0 & \cos[t_0-t] \end{bmatrix}$$

(b)

$$\begin{bmatrix} \exp[-2(t-t_0)] & 1-\exp[-2(t-t_0)] \\ 0 & 1 \end{bmatrix}$$

(d)

$$\begin{bmatrix} \exp(t_0-t) & t_0-t \\ 0 & \exp(t_0-t) \end{bmatrix}$$

8. (a) Let $P(t)$ be an $n \times n$ nonsingular matrix for all t. Show that $[\dot{P}^{-1}(t)] = -P^{-1}(t)\dot{P}(t)P^{-1}(t)$. (*Hint*: Differentiate $I = P(t)\,P^{-1}(t)$.)

(b) Show that the state transition matrix of the so-called *adjoint equation*

$$\dot{\lambda}(t) = -A^t(t)\lambda(t)$$

is given by $\Phi^t(\tau,t)$, where $\Phi(t,\tau)$ is the state transition matrix of $\dot{x}(t) = A(t)x(t)$.

(c) Let $\lambda(t_0) = \lambda_0$ and $x(t_0) = x_0$. Show that

$$\lambda^t(t)x(t) = constant$$

(d) If λ_0 and x_0 are orthogonal, what does (c) say about the corresponding trajectories $\lambda(t)$ and $x(t)$?

(These ideas find use in the theory of optimal control. See, for example, [4], pp. 111–114.)

9. A particular time-dependent process is constrained to the two-dimensional time-dependent surface in $\mathbf{R}^3$ defined by

$$[\lambda_1(t), \lambda_2(t), \lambda_3(t)] \begin{bmatrix} x_1(t) \\ x_2(t) \\ x_3(t) \end{bmatrix} = 0$$

where $[x_1(0), x_2(0), x_3(0)] = [0\ 1\ 1]$. Determine the set of all $\{\lambda_1(t), \lambda_2(t), \lambda_3(t)\}$ which lie on the surface, given that $\dot{x}(t) = A(t)x(t)$, where

$$A(t) = \begin{bmatrix} 0 & -t & 0 \\ 0 & -t & 0 \\ 0 & 0 & -1 \end{bmatrix}$$

10. Consider the differential equation

$$\dot{x}(t) = [A(t) + B(t)]x(t) \qquad\qquad \text{P12.1}$$

where A and B are matrices having continuous entries. Let $\Phi_A(t,t_0)$ be the state transition matrix of $\dot{x}(t) = A(t)x(t)$. Let $M(t) = \Phi_A(t,0)^{-1}B(t)\Phi_A(t,0)$. Show that the state transition matrix of equation P12.1 is of the form $\Phi_A(t,0)\Phi_M(t,t_0)$, where Φ_M denotes the state transition matrix of $\dot{x} = M(t)x$. (*Hint*: Use the substitution $x(t) = \Phi_A(t,0)y(t)$.)

11. Let the state transition matrix of a particular system have the form

$$\Phi(t,0) = \exp[\lambda_1 t] \begin{bmatrix} \cos(t^2) & \sin(t^2) \\ -\sin(t^2) & \cos(t^2) \end{bmatrix}$$

(a) Compute $\Phi(t,t_0)$.

(b) What is the matrix $A(t)$ for the system?

(c) If the system has no input and $x^t(2) = [-0.7814\ -2.3841]$, what does $x(1)$ equal?

12. Suppose $\dot{x}(t) = A(t)x(t)$, where

$$A(t) = \begin{bmatrix} A_{11}(t) & A_{12}(t) \\ 0 & A_{22}(t) \end{bmatrix}$$

Show that the associated matrix has the form

$$\Phi(t,t_0) = \begin{bmatrix} \Phi_{11}(t,t_0) & \Phi_{12}(t,t_0) \\ 0 & \Phi_{22}(t,t_0) \end{bmatrix}$$

where $\dot{\Phi}_{ii}(t,t_0) = A_{ii}(t)\Phi_{ii}(t,t_0)$, for $i = 1,2$. Find an expression for $\Phi_{12}(t,t_0)$.

13. If $\dot{x}(t) = A(t)x(t)$, where

$$A(t) = \begin{bmatrix} 1 & e^{-t} \\ -e^{-t} & 1 \end{bmatrix}$$

then what is $\det[\Phi(t,0)]$?

14. The value at $t = 2$ of the fundamental matrix of solutions to $\dot{x}(t) = Ax(t)$ is

$$M(2) = \begin{bmatrix} 1.5431 & -0.3679 \\ 1.1752 & 0.3679 \end{bmatrix}$$

It is further known that

$$M(0) = \begin{bmatrix} 1 & 1 \\ 0 & 1 \end{bmatrix}$$

What is the value of the state transition matrix $\Phi(4,0)$?

15. (*Review Problem*). Suppose $\dot{x}(t) = A(t)x(t)$, where

$$A(t) = \begin{bmatrix} \lambda_1 & 2t & 0 \\ 0 & \lambda_1 & 2t \\ 0 & 0 & \lambda_1 \end{bmatrix}$$

Compute $\Phi(t,t_0)$.

16. Using the fact that $\Phi(t,t_0) = M(t)M(t_0)^{-1}$, construct proofs of the semigroup and inverse properties.

REFERENCES

1. C. A. Desoer, *Notes for a Second Course on Linear Systems* (New York: Van Nostrand, 1972).

2. V. N. Faddeeva, *Computational Methods of Linear Algebra* (New York: Dover, 1959).

3. L. S. Pontryagin, *Ordinary Differential Equations* (Reading, MA: Addison-Wesley, 1962).

4. George Leitmann, *The Calculus of Variations and Optimal Control: An Introduction* (New York: Plenum Press, 1981).

13

CLOSED-FORM SOLUTION
OF THE TIME-VARYING
STATE EQUATIONS

INTRODUCTION

The complete closed-form solution of the time-varying state dynamical equations

$$\dot{x}(t) = A(t)x(t) + B(t)u(t), \quad x(t_0) = x_o \qquad (13.1a)$$

$$y(t) = C(t)x(t) + D(t)u(t) \qquad (13.1b)$$

equals the sum of the zero-input response and the zero-state response. The associated discussions of this chapter proceed in two phases: (*i*) an heuristic derivation based on [1], of the form of the solution, and (*ii*) a proof that the result of the heuristic derivation is, in fact, the unique solution to the time-varying state dynamical equation, 13.1a. The development culminates one aspect of our investigation into state space theory and sparks inquiries into other numerical, theoretical, and physical aspects.

HEURISTIC DERIVATION OF SOLUTION

The heuristic derivation builds on the input depicted in Figure 13.1. Although pictured as a scalar-valued function, the input should be thought of as a multidimensional vector.

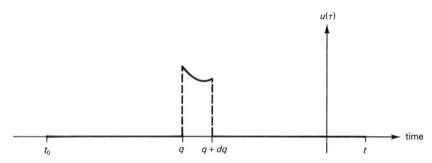

Figure 13.1   Special input for heuristic derivation.

The input, $u(\cdot)$, is defined over the interval $[t_0,t]$ as follows: $u(\tau) = 0$ for all τ in $[t_0,t]$ *except* over the "small" subinterval $[q,q + dq]$ as pictured. The objective of the first four steps of the derivation is to derive an approximate expression for $x(t)$, i.e., $x(\cdot)$ at t given the input. The last step uses linearity (superposition) and continuity to compute $x(t)$ for an arbitrary input defined over $[t_0,t_1]$.

Step 1.   Since $u(\tau) = 0$ over $[t_0,q]$,

$$x(q) = \Phi(q,t_0)x_o \tag{13.2}$$

Step 2.   For sufficiently small dq, the forward Euler formula, implies

$$x(q + dq) \approx x(q) + [A(q)x(q) + B(q)u(q)]dq$$

Substituting equation 13.2 into this equation yields

$$x(q + dq) \approx [\Phi(q,t_0) + A(q)\Phi(q,t_0)dq]x_o + B(q)u(q)dq \tag{13.3}$$

which is an approximate expression for the state vector at time $q + dq$.

Step 3.   Recall that $\Phi(t,t_0)$ satisfies $\dot{\Phi}(t,t_0) = A(t)\Phi(t,t_0)$. Hence, another application of the forward Euler formula yields

$$\Phi(q + dq,t_0) \approx \Phi(q,t_0) + A(q)\Phi(q,t_0)dq$$

Substituting this into equation 13.3 produces

$$x(q + dq) \approx \Phi(q + dq,t_0)x_o + B(q)u(q)dq$$

Step 4.   Since $u(\tau) = 0$ over $[q + dq,t]$.

$$x(t) = \Phi(t,q + dq)x(q + dq)$$

$$\approx \Phi(t,q + dq)\Phi(q + dq,t_0)x_o + \Phi(t,q + dq)B(q)u(q)dq$$

Hence by the semigroup property (theorem 12.1) of the previous chapter,

$$x(t) \approx \Phi(t,t_0)x_o + \Phi(t,q + dq)B(q)u(q)dq \tag{13.4}$$

This expression is an approximation for $x(t)$ given the initial condition x_o and the small input segment sketched in Figure 13.1.

Step 5. Now let $u(\cdot)$ be an arbitrary *admissible input*. View $u(\cdot)$ as a sum of disjoint segments defined over $[q_i, q_i + dq]$. Since the system is linear, each segment of $u(\cdot)$ contributes to the value of $x(t)$ in a linear fashion — i.e., the principle of superposition applies. In particular,

$$x(t) \approx \Phi(t, t_0)x_o + \sum_i \Phi(t, q_i + dq)B(q_i)u(q_i)dq$$

Since the various quantities satisfy the appropriate (piecewise) continuity conditions, one can take the limit as dq collapses to an infinitesimal. The sum then becomes the usual Riemann integral, q_i becomes q and the approximation converges to an equality, i.e.,

$$x(t) = \Phi(t, t_0)x_o + \int_{t_0}^{t} \Phi(t, q)B(q)u(q)dq$$

This quantity then, appears to represent a valid expression for the solution of equation 13.1a. Substituting it into equation 13.1b produces the output response. For a different, more formal derivation, see problem 12 at the end of this chapter.

THEOREMS AND PROOFS

The task now is to verify the result of the heuristic derivation.

Theorem 13.1.   The complete solution to the time-varying state dynamical equation 13.1a is given by

$$x(t) = \Phi(t, t_0)x_o + \int_{t_0}^{t} \Phi(t, q)B(q)u(q)dq \tag{13.5}$$

where

(i) $\Phi(t, t_0)x_o$ is the zero-input state response, and

(ii) $\int_{t_0}^{t} \Phi(t, q)B(q)u(q)dq$ is the zero-state state response.

Proof.   As in many of the previous proofs the uniqueness theorem 10.4 underpins our arguments.

Step 1. At $t = t_0$, the right-hand side of equation 13.5 clearly reduces to x_o.

Step 2. Does the right-hand side of equation 13.5 satisfy the same differential equation as $x(t)$? To answer this question, observe that

$$\dot{\Phi}(t, t_0)x_o + \frac{\partial}{\partial t} \int_{t_0}^{t} \Phi(t, q)B(q)u(q)dq$$

$$= A(t)\Phi(t, t_0)x_o + \Phi(t, t)B(t)u(t) + \int_{t_0}^{t} \frac{\partial}{\partial t} \Phi(t, q)B(q)u(q)dq \tag{13.6}$$

But $\Phi(t,t) = I$ and $\dfrac{\partial}{\partial t}\Phi(t,q) = A(t)\Phi(t,q)$. Hence, equation 13.6 equals

$$A(t)[\Phi(t,t_0)x_o + \int_{t_0}^{t}\Phi(t,q)B(q)u(q)dq] + B(t)u(t)$$

i.e., the right-hand side of equation 13.5 satisfies the given differential equation. But then, by the uniqueness theorem, equation 13.5 is indeed the solution to the time-varying state dynamical equation 13.1a.   ∎

Corollary 13.1.   The response of the time-varying state model given by equations 13.1(a) and (b) is

$$y(t) = C(t)\Phi(t,t_0)x_o + C(t)\int_{t_0}^{t}\Phi(t,q)B(q)u(q)dq + D(t)u(t) \qquad (13.7)$$

This completes the time-varying solution theory. And what about the time-invariant case? Is there any special structure to the response there? The answer, of course, is yes. Recall the time-invariant state model,

$$\dot{x}(t) = A x(t) + Bu(t)$$
$$y(t) = Cx(t) + Du(t) \qquad (13.8)$$

where A, B, C, and D are constant. The special solution structure then depends on the special structure of the state transition matrix. First recall condition (i) of theorem 11.3 stated next as a proposition.

Proposition 13.1.   The constant matrix A commutes with its integral.

Theorem 13.2.   The state transition matrix of the time invariant state model is

$$\Phi(t,t_0) = \Phi(t-t_0,0) = \exp[A(t-t_0)] \qquad (13.9)$$

Proof.   With the aid of proposition 13.1, it is straightforward to show that $\exp[A(t-t_0)]$ satisfies the matrix differential equation $\dot{M}(t) = A(t)M(t)$, $M(t_0) = I$.   ∎

Since the state transition matrix of the time-invariant state dynamical equation depends only on the elapsed time $t-t_0$, it is usually written simply as $\Phi(t)$ or $\Phi(t-t_0)$. The remainder of the text adopts this convention. theorem 13.2 suggests the following modification of theorem 13.1.

Theorem 13.3.   The complete solution to the time-invariant state dynamics has the form

$$x(t) = \exp[A(t-t_0)]x_o + \int_{t_0}^{t}\exp[A(t-q)]Bu(q)dq \qquad (13.10)$$

Corollary 13.3.   The system response of the time-invariant state model is

$$y(t) = Cx(t) + Du(t)$$

where $x(t)$ given by equation 13.10.

This essentially completes the discussion of the solution to the linear-time varying and time-invariant state models. One last proposition, however, proves helpful in the evaluation of equation 13.10.

Proposition 13.2.   For any t and q,

$$\exp[A(t-q)] = \exp[At]\exp[-Aq] \qquad (13.11)$$

Proof.   The proof follows directly from the semigroup property (theorem 12.1) of the state transition matrix.

The import of the proposition is to allow us to use the form

$$x(t) = \exp[A(t-t_0)]x_o + \exp[At]\int_{t_0}^{t}\exp[-Aq]Bu(q)dq \qquad (13.12)$$

which makes the evaluation of the integral more straightforward.

The foregoing theorems thus lead to an explicit analytic expression for computing $x(t)$. Aside from the mathematical elegance involved, some engineering payoff should accrue from this endeavor. In fact, theoretically, the theorems round out our development and will aid both our understanding and our forthcoming investigations of controllability and observability. Numerically speaking, however, the payoff is more difficult to evaluate and to some extent questionable [2]. The next section deals with this sobering issue.

NUMERICAL CONSIDERATIONS

Does the numerical evaluation of equation 13.5 offer a more efficient vehicle for calculating $x(t)$ at any t than does a direct simulation of the underlying differential equation? The debate rages. To gain a perspective on the issue, focus on the solution of equation 13.5. The systems of primary concern in this text have state transition matrices with an exponential form. For simplicity, let us further restrict our attention to the state trajectory of equation 13.12. To efficiently and accurately evaluate the integral in that equation, it is necessary to have fast, accurate, and reliable numerical algorithms to evaluate both $\Phi(t) = \exp(At)$ and $\int_{t_o}^{t}\exp(-Aq)Bdq$. The following example taken from [3] illustrates some of the pitfalls in evaluating $\Phi(t)$.

EXAMPLE 13.1.

Consider $\Phi(t) = \exp(At)$ and find $\Phi(1) = \exp(A)$. From the usual Taylor series,

$$\exp(A) = I + A + \frac{A^2}{2!} + \frac{A^3}{3!} + \cdots + \frac{A^k}{k!} + \cdots \qquad (13.13)$$

For sufficiently large k, the term, $1/k!$, will effectively annihilate A^k. Keeping this in mind, note that a digital computer may compute $\exp(A)$ by successively adding terms until the floating-point sum remains unchanged. To characterize this, define the partial sums

$$S_k(A) = \sum_{j=0}^{k} \frac{A^j}{j!}$$

Since a digital computer has a finite word length, it has finite precision. Define $fl[S_k(A)]$ as the matrix of floating-point numbers obtained by computing $S_k(A)$ using floating-point arithmetic. Suppose, for some integer K,

$$fl[S_K(A)] = fl[S_{K+1}(A)]$$

Then the digital computer approximation to $\exp(A)$ becomes

$$\exp(A) = fl[S_K(A)]$$

To see the unsavoriness of this Taylor series approach, suppose

$$A = \begin{bmatrix} -49 & 24 \\ -64 & 31 \end{bmatrix}$$

An IBM 370 using "short" arithmetic, i.e., a computer with a relative accuracy of $16^{-5} \approx 0.95 \times 10^{-6}$, found K to be 59 using the procedure above [3]. The result was

$$\exp[A] = fl[S_{59}(A)] = \begin{bmatrix} -22.25880 & -1.432766 \\ -61.49931 & -3.474280 \end{bmatrix}$$

The correct answer to six decimal places is

$$e^A = \begin{bmatrix} -0.735759 & 0.551819 \\ -1.471518 & 1.103638 \end{bmatrix}$$

Clearly, then something went amiss in the numerical calculation. In particular, $A^{16}/16!$ and $A^{17}/17!$ have entries with magnitudes between 10^6 and 10^7 but of opposite sign. Since "short" arithmetic calculations have a relative accuracy of only 10^5, the sum

$$\left[\frac{A^{16}}{16!} + \frac{A^{17}}{17!} \right]$$

has a potential absolute error larger than the final result! The conclusion: computing $\exp(At)$ via the Taylor series formula on a digital computer is a numerically unsound approach.

Fortunately, various other methods [3] for computing the matrix exponential exist and have varying levels of speed, accuracy, and reliability. A discussion of several of these, like the Padé approximation and the reduction to real Schur form, picks up in chapter 14.

Once one knows the integrand, $\exp(-Aq)Bdq$, numerically integrating equation 13.12 is more straightforward. Numerical integration of a known integrand, commonly called *numerical quadrature* [4], proceeds via any of the formulas discussed in chapter 7 or with more sophisticated techniques [5]. Since the integrand is known explicitly, implicit formulas like Simpson's rule or any of the Newton-Cotes formulas [4] become simple to implement.

Calculating the matrix exponential and implementing a numerical quadrature procedure is the most transparent means of evaluating equation 13.12. A very clever, different approach appears in [4]. Here, $\exp(At)$, the integral of $\exp(At)B$ and various other integrals of matrix exponentials are formulated as the exponential of a very cleverly defined block matrix. Hence, the computation of the integral reduces again to the problem of finding a fast, accurate, reliable algorithm for calculating the matrix exponential. More on this technique is presented in chapter 14.

Notice that we have deliberately avoided touching on a comparison of evaluating equation 13.12 with performing a direct simulation of the underlying state dynamics. There is as yet no clear-cut answer in the literature. Suffice it to say that depending on the structure of A— i.e., the location of its eigenvalues (see chapter 16), its degree of sparsity, etc.— it may be advantageous to execute a simulation [6].

PROBLEMS

1. Compute the zero-state step response for the system

$$\dot{x} = \begin{bmatrix} 2t & 1 \\ 1 & 2t \end{bmatrix} x + \begin{bmatrix} 0 \\ 1 \end{bmatrix} u(t)$$
$$y = [1 \;\; -1]x$$

2. Consider the system

$$\dot{x}(t) = \begin{bmatrix} 0 & 1 \\ 0 & 0 \end{bmatrix} x(t) + \begin{bmatrix} 1 & 1 \\ 0 & 1 \end{bmatrix} u(t), \quad x(0) = \begin{bmatrix} 3 \\ 0 \end{bmatrix}$$
$$y(t) = [1 \;\; -1] x(t)$$

   (a) Find an expression for the state response when

$$u(t) = \begin{bmatrix} -1 \\ 1 \end{bmatrix} 6t1^+(t)$$

   (b) Find an expression for the complete response $y(t)$

3. Consider the scalar differential equation

$$\ddot{y} = u - \dot{u} + \ddot{u} \qquad y(0) = \dot{y}(0) = 1$$

   (a) Derive the controllable canonical form of this equation.

   (b) Find the initial state vector $x(0)$, given the above initial conditions; assume that $u(t) = 0$ for all t.

   (c) Find the state transition matrix for the state model of part (a).

(d) Given the initial condition of part (b), find the zero-input state response and evaluate at $t = 1$.

(e) Find the zero-state state-response given that $u(t) = t\,1^+(t)$.

4. Given the continuous-time system

$$\dot{x}(t) = \begin{bmatrix} 0 & 0 & 0 \\ 1 & 0 & 0 \\ 1 & 2 & 0 \end{bmatrix} x(t) + \begin{bmatrix} 0 \\ 1 \\ 0 \end{bmatrix} u(t), \quad x(1) = \begin{bmatrix} 1 \\ 0 \\ 0 \end{bmatrix}$$

$$y(t) = x(t)$$

compute $x(2)$ when $u(t) = 1^+(t)$.

5. For the differential equation

$$\begin{bmatrix} \dot{x}_1 \\ \dot{x}_2 \end{bmatrix} = \begin{bmatrix} 0 & 0 \\ -2t & 0 \end{bmatrix} \begin{bmatrix} x_1 \\ x_2 \end{bmatrix} + \begin{bmatrix} 1 \\ 0 \end{bmatrix} u, \quad x(1) = \begin{bmatrix} 1 \\ -1 \end{bmatrix}$$

the state transition matrix is

$$\Phi(t, t_0) = \begin{bmatrix} 1 & 0 \\ t_0^2 - t^2 & 1 \end{bmatrix}$$

Compute $x(2)$, given that $u(t) = 1^+(t)$.

6. Compute the system response at $t = 1$ to the system

$$\dot{x} = \begin{bmatrix} 0 & 2 & 0 \\ 0 & 0 & 3 \\ 0 & 0 & 0 \end{bmatrix} x + \begin{bmatrix} 0 \\ 0 \\ 1 \end{bmatrix} u$$

$$y = [1 \ 1 \ 1]x$$

given that $x(0) = [-1 \ -2 \ 1]'$ and $u = 1^+(t)$.

7. Reproduce the heuristic derivation of the solution to the time-varying state dynamics from memory.

8. Consider the *RC*-circuit in Figure P13.1.

(a) Construct the state model of the circuit, and compute the form of the solution for $v_C(t)$.

(b) Construct a discrete-time system of the form $x(k+1) = ax(k) + bu(k)$ so that $x(k) = v_C(k\Delta t)$ for $u(t) = 1^+(k)$ for all k.

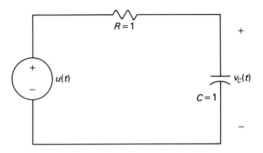

9. Given the state model

$$\dot{x} = \begin{bmatrix} 0 & 1 \\ 0 & 0 \end{bmatrix} x + \begin{bmatrix} 0 \\ 1 \end{bmatrix} u, \quad x(1) = \begin{bmatrix} 3 \\ -2 \end{bmatrix}$$

construct an input of the form $\alpha\, 1^+(t)$ such that $x(2) = \begin{bmatrix} 0 \\ 0 \end{bmatrix}$. (*Remark:* This problem motivates the controllability ideas discussed later in the text.)

10. Consider problem 9 again, with $x(1) = [1\ \ 1]^t$. Does there exist an appropriate input of the form $\alpha\, 1^+(t)$? What about an input of the form $\alpha\,\exp(\beta t)$?

11. Consider the state model

$$\begin{bmatrix} \dot{x}_1(t) \\ \dot{x}_2(t) \end{bmatrix} = \begin{bmatrix} 0 & 1 \\ 1 & 0 \end{bmatrix} \begin{bmatrix} x_1(t) \\ x_2(t) \end{bmatrix} + \begin{bmatrix} 0 \\ 1 \end{bmatrix} u(t); \quad x(0) = \theta$$

$$y(t) = x(t)$$

The input $u(t)$ is a staircase function defined as $u(t) = kT$ for $kT \leqslant t < (k+1)T$, $k = 0,1,2,...$, and T some fixed real number. Also, $u(t) = 0$ for all $t < 0$. Find a discrete-time state model of the form

$$\begin{bmatrix} \hat{x}_1(k+1)T \\ \hat{x}_2(k+1)T \end{bmatrix} = \hat{A}(kT) \begin{bmatrix} \hat{x}_1(kT) \\ \hat{x}_2(kT) \end{bmatrix} + \hat{B}(kT)u(kT)$$

$$\hat{y}(kT) = \hat{C}(kT) \begin{bmatrix} \hat{x}_1(kT) \\ \hat{x}_2(kT) \end{bmatrix} + \hat{D}(kT)u(kT)$$

whose solution at the time instants $k = 0,1,2,...$, exactly equals the response of the continuous-time model for $t \geqslant 0$, i.e., $y(kT) = \hat{y}(kT)$ for $k = 0,1,2,...$.

12. As an alternative to the heuristic derivation given in the text, suppose $\dot{x}(t) = A(t)x(t) + B(t)u(t)$. Postulate a solution of the form $x(t) = \Psi(t)\alpha(t)$, where $\Psi(t)$ is a fundamental matrix and $\alpha(t)$ is to be determined. Substitute $\Psi(t)\alpha(t)$ into the state dynamical equations and determine $\alpha(t)$. Show that $\Psi(t)\alpha(t)$ has the form of equation 13.5.

REFERENCES

1. C. A. Desoer, *Notes for a Second Course on Linear Systems* (New York: Van Nostrand, 1970).

2. Alan J. Laub, "Numerical Linear Algebra Aspects of Control Design Computations," *IEEE Transactions on Automatic Control*, Vol. AC-30, No. 2, February 1985, pp. 97–108.

3. C. B. Moler and C. F. VanLoan, "Nineteen Dubious Ways to Compute the Exponential of a Matrix," *SIAM Review*, Vol. 20, No. 4, October 1978, pp. 801–836.

4. G. E. Forsythe, M. A. Malcolm, and C. B. Moler, *Computer Methods for Mathematical Computations* (Englewood Cliffs, NJ: Prentice Hall, 1977).

5. Charles F. VanLoan, "Computing Integrals Involving the Matrix Exponential," *IEEE Transactions on Automatic Control*, Vol. AC-23, No. 3, June 1978, pp. 395–404.

6. Wahne Enright, "On the Efficient and Reliable Numerical Solution of Large Linear Systems of ODE's," *IEEE Transactions on Automatic Control*, Vol. AC-24, No. 6, December 1979, pp. 905–908.

14

SOME TECHNIQUES FOR COMPUTING THE MATRIX EXPONENTIAL AND ITS INTEGRAL

INTRODUCTION

This chapter begins an intensive focus on the properties and characteristics of the time invariant state model. The chapter looks at several methods for evaluating the exponential $\exp(At)$ of a matrix, the quantity intrinsic to all state trajectory computations. The Taylor series method receives a reprise which consolidates some previous discussions and adds some new dimensions. Next the Padé approximation method, the "scaling and squaring" method, and a coordinated mixture of the two methods are explored. The chapter ends with a brief snapshot of matrix decomposition methods which find widespread numerical and theoretical use and a look at a certain integral whose integrand contains $\exp(At)$.

As pointed out in [1], all methods have a dubious aspect with regard to floating-point calculations. Keep in mind that the aim of each method is to provide a viable means for numerically evaluating the mapping $A \rightarrow \exp(At)$. This mapping has an intrinsic sensitivity to perturbations. Suppose $A + \Delta A$ represents a perturbation on A. We can think of A and $A + \Delta A$ as two matrices separated from each other by some distance d_1. Analogously, $\exp(A)$ and $\exp(A + \Delta A)$ will differ by some other distance, say, d_2. The ratio d_2/d_1, in some sense measures the intrinsic sensitivity of the mapping $A \rightarrow \exp(At)$. This is important because A and

$fl(A)$ are different and, generally speaking, there exists a matrix ΔA such that $fl(A) = A + \Delta A$. This is precisely the round-off phenomenon.

Round-off error is commonly characterized by the so-called machine epsilon. The *machine epsilon* is the smallest positive real number ϵ such that $fl(1 + \epsilon) > fl(1)$. It measures the distance between the floating-point representation of the number 1 and the next largest number represented in the computer. Algorithms which compute $\exp(At)$ should not magnify the round-off already present in A. Algorithms which do not impart further sensitivity to perturbations are said to be *stable*. An algorithm is *reliable* if it tells the user when it is magnifying the round-off error or introducing any other types of excessive error. *Generality,* or the applicability of an algorithm to a broad class of matrices, is a third characteristic, and *efficiency,* a term referring to the needed computer time necessary to solve a problem, is a fourth characteristic of the algorithms referred to in this and various other chapters.

THE TAYLOR SERIES METHOD

Throughout our examination of state variables, the Taylor series has appeared and reappeared many times. Example 13.1 recounted a Taylor series illustration involving so-called catastrophic cancellation in the numerical (floating-point) calculation of $\exp(At)$. As pointed out in [1], the difficulty centers on the "truncated arithmetic," not on the truncation of the infinite series, and a larger word length (say 16 binary digits) would have led to an accurate answer. In addition to catastrophic cancellation, efficiency demands that the terms $A^k/k!$ in the series rapidly approach zero. This is clearly so for $\exp(0.5) = 1 + 0.5 + 0.25/2 + 0.125/6 + \ldots + 0.5^k/k! + \ldots$, and the reason is because $|0.5|$ is close to zero. Early truncation of the matrix Taylor series of $\exp(A)$ requires that A also be close to the zero matrix in some appropriate sense. But what is that sense? Observe that $0.5x$ scales x by half. Now the matrix-vector product $Ax = y$ produces a scaled and rotated vector y from x (see Figure 14.1).

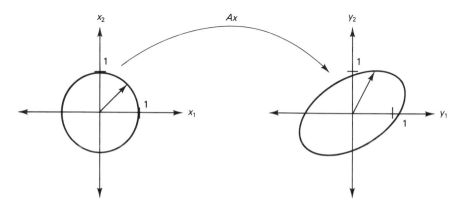

Figure 14.1 Mapping of the region $||x||_2 \leqslant 1$ by the matrix A.

A is close to zero if the product Ax does not "stretch" x much. The *spectral norm* of a matrix measures the maximum stretch x receives whenever multiplied by A.

Definition 14.1.   The *spectral norm* of a matrix A is

$$\| A \| = \max_{\| x \|_2 = 1} \| Ax \|_2$$

where $\| x \|_2$ is the usual Euclidean norm of a vector $x \in R^n$.

EXAMPLE 14.1.

Suppose $x = (x_1 \ x_2)^t$ with $\| x \|_2 = 1$, λ_1 and λ_2 are real scalars such that $|\lambda_1| > |\lambda_2| \geqslant 0$, and

$$A = \begin{bmatrix} \lambda_1 & 0 \\ 0 & \lambda_2 \end{bmatrix}$$

Then

$$Ax = \begin{bmatrix} \lambda_1 x_1 \\ \lambda_2 x_2 \end{bmatrix}$$

and $(\| A \|_2)^2 = \lambda_1^2 x_1^2 + \lambda_2^2 x_2^2 = \lambda_1^2 [x_1^2 + (\lambda_2^2/\lambda_1^2) x_2^2]$. Since $[x_1^2 + (\lambda_2^2/\lambda_1^2) x_2^2] \leqslant x_1^2 + x_2^2 = 1$, with equality whenever $x_2 = 0$, the spectral norm of A is

$$\| A \| = \max_{\| x \|_2 = 1} \| Ax \|_2 = \lambda_1$$

That is, the maximum stretch x can undergo is determined by λ_1.

As a brief aside, and as a precursor of material of subsequent chapters, it is possible to interpret the spectral norm of A in terms of the structure of A. Specifically, if one knows the so-called eigenvalues of A, i.e., the zeros of the characteristic polynomial, then the magnitude of the largest eigenvalue of A equals the spectral norm of A provided that A is symmetric. (See problem 7 for a counterexample when A is not symmetric.) Chapter 15 defines and discusses the so-called eigenstructure of A. More generally, the largest singular value of A (the positive square root of the largest eigenvalue of $A^t A$) also equals the spectral norm for arbitrary A. A discussion of singular values and the very important singular value decomposition of a matrix takes place in chapter 19. The eigenvalues and singular values of a matrix can be accurately and reliably computed using widely available software [2].

If the spectral norm of a matrix A is sufficiently close to zero, then (*i*) the Taylor series form of the exponential has terms which quickly tend to zero, thus improving the efficiency of the numerical computation of $\exp(A)$, and (*ii*) the round-off error does not accumulate from term to term, thus improving the accuracy of the computation. More importantly, however, assuming no round-off error, the spectral norm aids in determining a suitable number, say K, of terms needed in a truncated Taylor series to produce an approximation good to d significant figures

[3]. To determine the proper number of terms, we merely split the Taylor series into two parts: the part approximating $\exp(At)$ for fixed t, say M, and a remainder matrix, say R:

$$\exp(At) = M + R = \sum_{k=0}^{K} \frac{A^k t^k}{k!} + \sum_{k=K+1}^{\infty} \frac{A^k t^k}{k!}$$

An accuracy of d significant figures requires that

$$|r_{ij}| \leqslant 10^{-d} |m_{ij}|$$

for all terms r_{ij} of R and m_{ij} of M. Using the fact that $\|A^k\| \leqslant \|A\|^k$, one concludes that

$$|r_{ij}| \leqslant \sum_{k=K+1}^{\infty} \frac{\|A\|^k t^k}{k!} \qquad (14.1)$$

The goal is to find a simple (conservative) bound for $|r_{ij}|$. To obtain this bound, define a quantity ϵ equal to the ratio of the term $k = K + 2$ and $k = K + 1$ in equation 14.1. Taking this ratio and simplifying produces

$$\epsilon = \frac{\|A\| t}{K + 2}$$

Using ϵ, equation 14.1 implies that

$$|r_{ij}| \leqslant \frac{\|A\|^{K+1} t^{K+1}}{(K+1)!} (1 + \epsilon + \epsilon^2 + \ldots)$$

If K is sufficiently large, so that $\epsilon < 1$, then $1 + \epsilon + \epsilon^2 + \cdots = (1-\epsilon)^{-1}$ and

$$|r_{ij}| \leqslant \frac{\|A\|^{K+1} t^{K+1}}{(K+1)!} (1 - \epsilon)^{-1}$$

Hence, if a relative accuracy of d significant figures is needed for a given t, then it is sufficient to choose K such that

$$\frac{\|A\|^{K+1} t^{K+1}}{(K+1)!} (1 - \epsilon)^{-1} \leqslant 10^{-d} |m_{ij}|$$

for all i and j. The flowchart in Figure 14.2 blocks out an algorithm utilizing this bound. Of course as pointed out in [1], if round-off is present in A, there are cases where this method will fail to give the required accuracy.

Several summary comments end this section. First, the typical A-matrix does not have spectral norm close to zero. However, the associated numerical difficulties in computing $\exp(At)$ in this case are alleviated by the scaling and squaring method to be discussed shortly. This method allows one to satisfactorily use the Taylor series for numerical computation. Second, despite the numerical questions surrounding the Taylor series, it has provided much insight into the theoretical pylons of state variables and will continue to do so. Third, we remind the reader that if one knows $\exp[AT]$ for some T, then $\exp[A(nT)] = [\exp(AT)]^n$. As usual, one must beware of blind applications of this formula: in some cases a small amount of round-off in $\exp[At]$ will lead to erroneous results for $\exp[A(nT)]$. A discussion of this phenomenon takes place shortly.

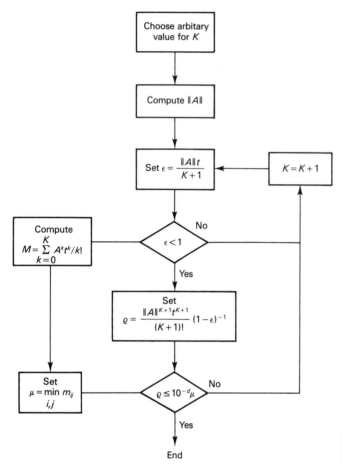

Figure 14.2 Flowchart for suitable number of terms K in truncated Taylor series of $\exp(At)$.

THE PADÉ METHOD

Basic calculus courses typically devote time to the study of infinite series of single variables, say in z. One learns that many infinite series have rational function equivalents. For example, if $|z| < 1$, then

$$\sum_{j=1}^{\infty} z^j = z \sum_{j=0}^{\infty} z^j = z(1-z)^{-1}$$

The Padé approximation to $\exp(At)$ represents a rational function approximation to the infinite Taylor series. It is valid in a neighborhood of the origin—i.e., when the spectral norm of A, $\|A\|$, is small, just as with the Taylor series.

Definition 14.2. The $p-q$ Padé approximation to $\exp(A)$ is given by

$$P_{pq}(A) = [D_{pq}(A)]^{-1} N_{pq}(A) \qquad (14.2a)$$

where

$$N_{pq}(A) = \sum_{j=0}^{p} \frac{(p+q-j)!\,p!}{(p+q)!\,j!\,(p-j)!}\, A^j \qquad (14.2b)$$

and

$$D_{pq}(A) = \sum_{j=0}^{q} \frac{(p+q-j)!\,q!}{(p+q)!\,j!\,(q-j)!}\, (-A)^j \qquad (14.2c)$$

Of course, if one desires to compute $\exp(At)$ for known t, then one simply substitutes At for A in the preceding formulas. The matrix $N_{pq}(A)$ plays the role of a numerator polynomial in the rational function approximation, and $D_{pq}(A)$ plays the role of the denominator.

EXAMPLE 14.2

Let

$$A = \begin{bmatrix} 1 & 0 \\ 1 & 1 \end{bmatrix} \begin{bmatrix} -1 & 0 \\ 0 & -2 \end{bmatrix} \begin{bmatrix} 1 & 0 \\ -1 & 1 \end{bmatrix} = \begin{bmatrix} -1 & 0 \\ 1 & -2 \end{bmatrix}$$

Then

$$\exp(At) = \begin{bmatrix} e^{-t} & 0 \\ e^{-t} - e^{-2t} & e^{-2t} \end{bmatrix} = \begin{bmatrix} 0.3678\ldots & 0 \\ 0.2325\ldots & 0.1353\ldots \end{bmatrix}$$

Our goal is to find the 2-2 Padé approximation. Using equations 14.2, we have

$$N_{22}(A) = I + 0.5\,A + \frac{A^2}{12} = \begin{bmatrix} \frac{7}{12} & 0 \\ \frac{1}{4} & \frac{1}{3} \end{bmatrix}$$

and

$$D_{22}(A) = I - 0.5\,A + \frac{A^2}{12} = \begin{bmatrix} \frac{19}{12} & 0 \\ \frac{-3}{4} & \frac{7}{3} \end{bmatrix}$$

Since $P_{22}(A) = [D_{22}(A)]^{-1} N_{22}(A)$, it is necessary to compute the inverse of $D_{22}(A)$:

$$[D_{22}(A)]^{-1} = \begin{bmatrix} \frac{12}{19} & 0 \\ \frac{27}{133} & \frac{3}{7} \end{bmatrix}$$

The resulting Padé approximation to $\exp(A)$ is

$$P_{22}(A) = \begin{bmatrix} \frac{7}{19} & 0 \\ \frac{30}{133} & \frac{1}{7} \end{bmatrix} \approx \begin{bmatrix} 0.36842 & 0 \\ 0.22556 & 0.14286 \end{bmatrix}$$

The Padé approximation found in the example is reasonably accurate for the low-order approximation used. Again, however, round-off error prevents reliable use of the method for arbitrary A's. As $q = p$ becomes large, the (diagonal) Padé approximation, $N_{qq}(A)$ and $D_{qq}(A)$, tends to the Taylor series for $\exp(0.5A)$ and $\exp(-0.5A)$, respectively [1]. Thus, the method becomes prone to the catastrophic cancellation error illustrated in the previous chapter in example 13.1. In addition, $D_{qq}(A)$ may be poorly conditioned with respect to matrix inversion. Like arguments apply when $p \neq q$.

On the positive side, if the spectral norm of A is not too large, then the Padé approximation produces reasonable results. Further, if $q = p$ (the so-called diagonal Padé approximation) is used, then the entries of $P_{qq}(At)$ remain bounded as $t \to \infty$. This is a compelling reason for choosing $p = q$.

SCALING AND SQUARING IN CONJUNCTION WITH THE DIAGONAL PADÉ APPROXIMATION

The scaling and squaring method utilizes the identity

$$\exp[A] = [\exp(\frac{A}{m})]^m \tag{14.3}$$

The idea is to first choose $m = 2^k$ (a power of 2) so that $\exp(A/m)$ can be reliably computed by known methods and then form $[\exp(A/m)]^m$ by repeated squaring. (Again, if $\exp(At)$ is desired, simply substitute At for A in the discussion.) For example, if $m = 2^3$, then

$$\left[\exp\left[\frac{A}{8} \right] \right]^8 = \left[\left[\left[\exp\left[\frac{A}{8} \right] \right]^2 \right]^2 \right]^2$$

The method depends on a sound means for computing $\exp(A/m)$ and on the presumption that round-off error will not catastrophically accumulate during the squaring process. For sufficiently large m, the spectral norm of A/m becomes sufficiently small for a reliable calculation based on the Taylor series or Padé method. In particular, we seek to choose m such that $\dfrac{\|A\|}{m} \leq 1$.

Under reasonable conditions, repeated squaring produces a satisfactory answer. It is claimed in [1,4] that scaling and squaring with diagonal Padé approximation is one of the most efficient means for evaluating $\exp(At)$ for a given t. Analytically speaking, the quality of the method depends on the accuracy of the approximation matrix

$$M_{pq} = \left[P_{pq}\left[\frac{A}{m} \right] \right]^m$$

where $m = 2^k$ for some k. The question of course is, How closely does M_{pq} approximate $\exp(A)$? Answering this question requires the notion of the sup-norm.

Definition 14.3.   The sup-norm (or L_∞-norm) of an $n \times n$ matrix A is defined as

$$|| A ||_\infty = \max_i \sum_{j=1}^{n} |a_{ij}| \tag{14.4}$$

In other words, we sum the absolute values of the entries in each row of A and choose the largest value as the norm.

According to [1,4], if $m = 2^k$ and

$$\frac{|| A ||_\infty}{m} \leqslant 0.5$$

then there exists an $n \times n$ perturbation matrix E which commutes with A such that

$$M_{pq} = \exp(A + E) \approx \exp(A)$$

provided that $|| A ||_\infty$ is sufficiently small. The relative smallness of E depends on the choice of p and q according to

$$|| E ||_\infty \leqslant \epsilon || A ||_\infty$$

where

$$\epsilon = 2^{3-(p+q)} \frac{p!\, q!}{(p + q)!\, (p + q + 1)!}$$

Again, according to [1,4], $p = q$ (the diagonal Padé approximation) is optimal. The following algorithm often proves effective:

Step 1: Preliminaries. Compute $|| A ||_\infty$ and choose an error tolerance $\delta > 0$.

> *Comment*: The goal of the algorithm is to compute $M_{qq} = \exp(A + E)$, where $|| E ||_\infty < \delta || A ||_\infty$.

Step 2: Initializations.

(i) $j = \max\{0, 1 + \lfloor \log_2(|| A ||_\infty) \rfloor\}$, where $\lfloor \cdot \rfloor$ denotes the integer part of.

(ii) $A \leftarrow \dfrac{A}{2^j}$,

(iii) Find the smallest nonnegative integer q such that

$$2^{3-(2q)} = \frac{(q!)^2}{(2q)!\, (2q + 1)!} \leqslant \delta$$

(iv) $D \leftarrow I, N \leftarrow I, Q \leftarrow I$

Step 3: Construction of the Padé Approximate M_{qq}, denoted by M. For $k = 1, \ldots, q$,

(i) $\beta \leftarrow \dfrac{(2q-k)!\, q!}{(2q)!\, k!\, (q-k)!}$

Comment: β takes on values equal to the various coefficients in $N_{qq}(A)$ and $D_{qq}(A)$ of the Padé approximation.

(ii) $Q \leftarrow AQ, \ N \leftarrow N + \beta Q, \ D \leftarrow D + (-1)^k \beta Q$

(iii) Apply Gaussian elimination (chapter 6) to solve $DM = N$ for M.

Step 4: Squaring Phase. Square M j times; i.e., for $k = 1,...,j$, $M \leftarrow M^2$.

Comment: M is the desired approximation.

To conclude this section, we shave away some of the glamor of this very effective method. Figure 14.3 displays a curve resembling a camel hump. For some stable matrices, A, $\| \exp(At) \|$ grows before it decays. Such matrices exhibit the so-called *hump phenomenon*. It is possible, although not necessarily the case, for the relative error between M and $P_{qq}(A/m)$ to be very large when one tries to pass through the hump. For a more thorough discussion of this situation, see [1,3].

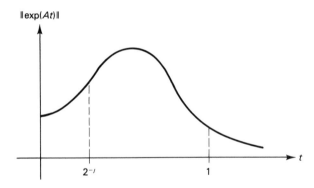

$\| \exp(At) \|$

Figure 14.3   Hypothetical spectral norm curve of $\exp(At)$.

MATRIX DECOMPOSITION METHODS

Matrix decomposition methods are methods which factor the A-matrix into a product of matrices which permits a straightforward computation of $\exp(At)$. For example, if A is diagonal, say, $A = diag(\lambda_1, \ldots, \lambda_n)$, then $\exp(At) = diag(\lambda_1 t,..,\lambda_n t)$, a straightforward calculation. The idea is to factor A via a similarity transformation, i.e.,

$$A = PMP^{-1} \tag{14.5}$$

observing all the while that

$$\exp(At) = P \exp(Mt) P^{-1} \tag{14.6}$$

Provided that M leads to an easily computed exponential (loosely speaking, M should be as close to diagonal as possible), and provided that P is well conditioned with respect to inversion, such methods prove most effective, especially for large matrices and repeated calculation of $\exp(At)$ for various t's. As might be expected, these two objectives are often in conflict [1]. The eigenvalue-eigenvector method discussed in chapter 15 constructs a diagonal M; but the associated P can often be ill conditioned with respect to inversion. The so-called Schur decomposition method, discussed shortly, emphasizes a well-conditioned P-matrix.

Definition 14.4.   An $n \times n$ matrix A is *orthogonal* if $A^t A = I$, i.e., $A^t = A^{-1}$.

Definition 14.5.   The *Schur decomposition* of an $n \times n$ matrix A is

$$A = QMQ^t$$

where Q is orthogonal and M is triangular.

As per equation 14.6, $\exp(At) = Q \exp(Mt) Q^t$, and the difficulty resides with the computation of the exponential of the triangular matrix M. In reference [5], Parlett discusses the exponential of triangular matrices. If M is upper triangular, then $\exp(Mt)$ must also be upper triangular. The diagonal entries of $\exp(Mt)$ are simply the exponential of the diagonal entries of M. Difficulty arises in computing the off-diagonal terms. Since a discussion of this method is beyond the intended purpose of this text, the reader is directed to [5] for further details. However, the Schur decomposition is numerically satisfactory, and the reader should be aware of its existence and the structure of its factorization. Moreover, there are very good "canned" programs available for the computation of the factorization [2,6].

INTEGRALS INVOLVING THE MATRIX EXPONENTIAL

In this section, we show that a certain integral involving the exponential of a matrix can be cleverly evaluated using matrix exponential techniques. This integral often arises in the solution of state equations. Specifically, suppose the input to the state equation $\dot{x} = Ax + Bu$ is constant. Then

$$\int_0^t \exp(Aq) \, B \, dq \tag{14.7}$$

is germane to the solution of the equation. We seek to show that this integral can be found by taking the exponential of a certain matrix.

Accordingly, consider the matrix exponential

$$\exp\left\{ t \begin{bmatrix} A & B \\ 0 & 0 \end{bmatrix} \right\} = \begin{bmatrix} M_{11}(t) & M_{12}(t) \\ 0 & I \end{bmatrix} \tag{14.8}$$

for appropriate matrices $M_{11}(t)$ and $M_{12}(t)$ to be defined. (Also see problem 12 of chapter 12.) Equation 14.8 is the state transition matrix for the state model

$$\dot{x} = \begin{bmatrix} A & B \\ 0 & 0 \end{bmatrix} x$$

Hence, the right hand side of equation 14.8 must satisfy the matrix differential equation

$$\begin{bmatrix} \dot{M}_{11}(t) & \dot{M}_{12}(t) \\ 0 & 0 \end{bmatrix} = \begin{bmatrix} A & B \\ 0 & 0 \end{bmatrix} \begin{bmatrix} M_{11}(t) & M_{12}(t) \\ 0 & I \end{bmatrix}$$

Of particular interest is the $M_{12}(t)$-term, which must satisfy

$$\dot{M}_{12}(t) = AM_{12}(t) + B$$

This equation has a solution with the same form as the solution of the time-invariant state dynamical equation 13.10; thus,

$$M_{12}(t) = \exp(At)\, M_{12}(0) + \int_0^t \exp[A(t-q)]\, B\, dq$$

From equation 14.8 and the properties of the matrix exponential applied thereto, $M_{12}(0) = [0]$. Therefore,

$$M_{12}(t) = \int_0^t \exp[A(t-q)]B\, dq$$

If we can effectively compute the matrix exponential of equation 14.8, then we can effectively compute $M_{12}(t)$ which is the desired integral. Any of the methods so far elaborated upon, as well as others to come are effective in evaluating equation 14.8.

PROBLEMS

1. Computation of the Taylor series and Padé approximations require evaluations of polynomials in A. Let

$$p(A) = \beta_q A^q + \beta_{q-1} A^{q-1} + \cdots + \beta_1 A + \beta_0 I$$

The *Horner method* suggests evaluating $p(A) \triangleq P$ as follows:

(i) $P \leftarrow \beta_q A + \beta_{q-1} I$
(ii) For $k = q-2, q-3, \ldots, 0$, $P \leftarrow AP + \beta_k I$

    (a) Verify that this algorithm in fact yields $p(A)$.

    (b) For $q = p = 3$, compute the coefficients of the Padé matrix polynomial $N_{qq}(A)$ and $D_{qq}(A)$. Evaluate the polynomials using the Horner method when

$$A = \begin{bmatrix} -1 & 0 \\ 1 & -2 \end{bmatrix}$$

    (c) Consult reference [3], p. 393, for a more efficient means of evaluating polynomials in A.

2. Many system-related problems require evaluating A to a power, say $A^p = Q$. The question of efficient computation arises naturally. An approach to this problem is the following:

(i) Decompose p in a binary expansion

$$p = \sum_{k=0}^{K} \beta_k 2^k$$

such that $\beta_K \neq 0$; e.g., if $p = 7$, then $p = 2^2 + 2^1 + 2^0$.

(ii) Let $P \leftarrow A$ and $Q \leftarrow A^{\beta_0}$. For $k = 1,2,...,K$, $P \leftarrow P^2$ and if $\beta_q = 1$, then $Q \leftarrow Q * P$

   (a) For $p = 7$, evaluate A^7 using this algorithm when

$$A = \begin{bmatrix} -1 & 0 \\ 1 & -2 \end{bmatrix}$$

   (b) Write a Fortran program which executes the algorithm. Evaluate A^{57} using your program when

$$A = \begin{bmatrix} -3 & 2 \\ -4 & 3 \end{bmatrix}$$

   (*Hint*: $A^{57} = A$.)

   (c) Repeat part (b) for

$$A = \begin{bmatrix} 1 & 1 \\ 2 & 1 \end{bmatrix} \begin{bmatrix} 0.99 & 0 \\ 0 & 0.1 \end{bmatrix} \begin{bmatrix} -1 & 1 \\ 2 & -1 \end{bmatrix} = \begin{bmatrix} -0.79 & 0.89 \\ -1.78 & 1.88 \end{bmatrix}$$

3. Compute the 2-2 and 3-3 Padé approximations for $\exp(A)$ when A is

   (a)

$$\begin{bmatrix} -3 & 2 \\ -4 & 3 \end{bmatrix}$$

   (b)

$$\begin{bmatrix} -1 & 1 \\ 1 & -2 \end{bmatrix}$$

   (c)

$$\begin{bmatrix} -1 & 1 & 0 \\ 0 & -1 & 1 \\ 0 & 0 & -1 \end{bmatrix}$$

4. Use the scaling-and-squaring method with the Padé method to evaluate $\exp(At)$ when $t = 1$ and A is

   (a)

$$\begin{bmatrix} 1 & 1 \\ -1 & 1 \end{bmatrix} \begin{bmatrix} 63 & 0 \\ 0 & -57 \end{bmatrix} \begin{bmatrix} 0.5 & -0.5 \\ 0.5 & 0.5 \end{bmatrix} = \begin{bmatrix} 3 & -60 \\ -60 & 3 \end{bmatrix}$$

   (b)

$$\begin{bmatrix} -64 & 64 & 0 \\ 0 & -64 & 64 \\ 0 & 0 & -64 \end{bmatrix}$$

   (*Hint:* Look at your answer to problem 3, part (c).)

   (c)

$$\begin{bmatrix} -49 & 24 \\ -64 & 31 \end{bmatrix}$$

5. Let M be a nonsingular matrix. Show that there exists a matrix A such that $M = \exp(A)$. Is A unique?

6. The so called *Caley-Hamilton* method for computing the matrix exponential is as follows:

(i) Let the characteristic polynomial of A be given by

$$\pi_A(\lambda) = \lambda^n + \alpha_{n-1}\lambda^{n-1} + \cdots + \alpha_1\lambda + \alpha_0$$

(ii) The Caley-Hamilton theorem implies that

$$\exp(At) = \sum_{j=0}^{n-1} \xi_j(t)\, A^j$$

for appropriate analytic functions $\xi_j(t)$.

(iii) The analytic functions $\xi_j(t)$ are given by

$$\xi_j(t) = \sum_k \frac{\beta_{kj}\, t^k}{k!}$$

where

$$\beta_{kj} = \begin{cases} \delta_{kj} & \text{if } k < n \\ \alpha_j & \text{if } k = n \\ \alpha_0\beta_{k-1,n-1} & \text{if } k > n,\, j = 0 \\ \alpha_j\beta_{k-1,n-1} + \beta_{k-1,j-1} & \text{if } k > n,\, j > 0 \end{cases}$$

Use this method to compute the matrix exponential for those matrices given in problem 4.

7. Suppose

$$A = \begin{bmatrix} 0 & 1 \\ 0 & 0 \end{bmatrix}$$

(a) Compute the eigenvalues of A.

(b) Compute $A^t A$ and then the singular values of A.

(c) What is the spectral norm of A? Is it equal to the largest eigenvalue of A?

8. Show that the spectral norm of

$$A = \begin{bmatrix} 2 & 1 \\ 1 & 2 \end{bmatrix}$$

is 3.

REFERENCES

1. Cleve Moler and Charles Van Loan, "Nineteen Dubious Ways to Compute the Exponential of a Matrix," *SIAM Review*, Vol. 20, No. 4, October 1978, pp. 80–836.

2. B. T. Smith, J. M. Boyle, Y. Ikebe, V. C. Klema, and C. B. Moler, *Matrix Eigensystem Routines: EISPACK Guide*, 2d ed., (New York: Springer-Verlag, 1970).

3. M. L. Liou, "A Novel Method of Evaluating Transient Response," *Proceedings of the IEEE*, Vol. 54, 1966, pp. 20–30.

4. Gene H. Golub and Charles Van Loan, *Matrix Computations* (Baltimore: The John Hopkins University Press, 1983).

5. B. N. Parlett, "A Recurrence among the Elements of Functions of Triangular Matrices," *Linear Algebra Applications*, Vol. 14, 1976, pp. 117–121.

6. The IMSL Library, IMSL Inc., Houston, Texas, 1980.

15

AN EIGENVALUE-EIGENVECTOR
METHOD FOR COMPUTING
$\Phi(t) = \exp(At)$

INTRODUCTION

This chapter sketches the well-known eigenvalue-eigenvector method for computing the matrix exponential assuming distinct eigenvalues or at the very least a nondefective A-matrix [1,2]. Earlier, we briefly examined the effects of finite-word-length arithmetic on such a calculation. The application of the QR-algorithm and other orthogonal transformation-like algorithms to eigenvalue-eigenvector problems has resulted in numerically acceptable and accurate computations for a broad variety of large matrices. Thus, an understanding of eigenvalue-eigenvector methods for constructing $\exp[At]$, the time-invariant state transition matrix, becomes quite meaningful in an applications context.

The essential goal of the chapter is to develop a state transformation T such that a given A-matrix has the decomposition $A = T D T^{-1}$, where D is a diagonal matrix. This decomposition, then leads to the structure, $\exp[At] = T \exp[Dt] T^{-1}$ where if D has the diagonal form, $D = \text{diag}[\lambda_1, \lambda_2, \ldots, \lambda_n]$, then $\exp[Dt] = \text{diag}[e^{\lambda_1 t}, e^{\lambda_2 t}, \ldots, e^{\lambda_n t}]$. It turns out that the λ_i are eigenvalues of A and the columns of T are associated eigenvectors. This decomposition is possible whenever A has a *full set of eigenvectors* in which case A is said to be *nondefective*. In other words,

words, if A is $n \times n$ and *nondefective*, it must have n linearly independent eigenvectors over the field of complex scalars.

The case of computing the state transition matrix when A is defective (loosely speaking A has repeated eigenvalues) requires a delicate conceptual and numerical development. The burden rests on the numerical computation of the *Jordan form* $A = TJT^{-1}$, where J is a block diagonal matrix and T has some columns which are generalized eigenvectors of A. Our avenue for developing this concept runs out of a Laplace transform frequency domain characterization of the state transition matrix presented in a later chapter [1,3]. For the numerical aspects of the Jordan form, see [2,4].

MATHEMATICAL PRELIMINARIES

Recall that the *characteristic polynomial* of the matrix A has the form

$$\pi_A(\lambda) = \det(\lambda I - A) = \lambda^n + a_1 \lambda^{n-1} + \cdots + a_{n-1}\lambda + a_n$$

$$= (\lambda - \lambda_1)(\lambda - \lambda_2) \cdots (\lambda - \lambda_n) \qquad (15.1)$$

The zeros of this polynomial are the (possibly complex) numbers λ_i, called the *eigenvalues* of A. The set $\sigma(A) = \{\lambda_1, \ldots, \lambda_n\}$ has the tag, *spectrum of A* and refers to the set of eigenvalues of A. Associated with each λ_i is a *right eigenvector* in $\mathbf{C}^n$, given by $e_i \neq \theta$, satisfying

$$Ae_i = \lambda_i e_i \qquad (15.2)$$

for each i. Also associated with each λ_i of A is a *left eigenvector* in $\mathbf{C}^n$, given by $w_i \neq \theta$, which for each i satisfies

$$w_i^* A = \lambda_i w_i^* \qquad (15.3)$$

where w_i^* designates the complex conjugate transpose of the (possibly complex) n-vector w_i.

Since the zeros of a polynomial are unique, the eigenvalues of A are unique. On the other hand the set of right or left eigenvectors of A is not unique, as verified by the following proposition.

Proposition 15.1.   Let $\xi \neq 0$ be any complex number. If e_i is any right eigenvector of A associated with λ_i, then $\tilde{e}_i = \xi e_i$ is also a right eigenvector of A associated with λ_i. A similar statement holds for left eigenvectors w_i.

Proof.   Since $\xi \neq 0$, $\xi e_i \neq \theta$. Hence, $\xi[Ae_i] = \xi[\lambda_i e_i]$, from which it follows that $A[\xi e_i] = \lambda_i[\xi e_i]$, showing that $\tilde{e}_i = \xi e_i$ is the stated eigenvector.   ∎

It is often convenient to require eigenvectors to have unit length. Proposition 15.1 legitimizes this choice. Also, assuming the matrix A is real, $\pi_A(\lambda)$ must have real coefficients. If a specific zero, say λ_i, is not real, then there exists a λ_j such that $\lambda_j = \bar{\lambda}_i$. Furthermore, the definition of an eigenvector, together with Proposition 15.1 implies that it is always possible to choose $e_j = \bar{e}_i$ and $w_j = \bar{w}_i$. In sum, we have the following proposition.

Proposition 15.2. For a real $n \times n$ matrix A, nonreal eigenvalues of A occur in complex conjugate pairs and it is always possible to choose the corresponding eigenvectors as complex conjugates of each other.

The convention is to assume that eigenvectors corresponding to complex conjugate pairs of eigenvalues are complex conjugates of each other.

EXAMPLE 15.1

Find the right eigenvectors of the matrix

$$A = \begin{bmatrix} a & 1 \\ 0 & b \end{bmatrix} \tag{15.4}$$

Computing the characteristic polynomial, $\pi_A(\lambda) = (\lambda - a)(\lambda - b)$, leads to the assignment $\lambda_1 = a$ and $\lambda_2 = b$. To compute e_1, observe that since

$$\begin{bmatrix} a & 1 \\ 0 & b \end{bmatrix} \begin{bmatrix} x \\ y \end{bmatrix} = \lambda_1 \begin{bmatrix} x \\ y \end{bmatrix} = \begin{bmatrix} ax \\ ay \end{bmatrix}$$

it follows that $ax + y = ax$ or $y = 0$. Choosing $x = 1$ produces $e_1 = [1 \ 0]^t$.

To compute e_2, observe that

$$\begin{bmatrix} a & 1 \\ 0 & b \end{bmatrix} \begin{bmatrix} x \\ y \end{bmatrix} = \begin{bmatrix} bx \\ by \end{bmatrix}$$

in which case $(b - a)x = y$. Again, choosing $x = 1$, gives $y = b - a$. Thus, $e_2 = [1 \ b - a]^t$. It turns out that whenever $a = b$ for this particular A-matrix, $e_1 = e_2$ and A does not have a full set of eigenvectors. Any matrix which does not have a full set of eigenvectors is said to be *defective*. If $a \neq b$, then the 2×2 matrix A has two linearly independent eigenvectors and is nondefective.

Exercise. Compute the left eigenvectors of the preceding matrix A.

It is easy to conjecture that a matrix with nondistinct eigenvalues is defective. However, such a conjecture is untrue. Simply note that the $n \times n$ identity matrix has all n eigenvalues equal to 1 yet has the full set $\{\eta_1, \eta_2, \ldots, \eta_n\}$ of eigenvectors.

For the remainder of this chapter, we will assume that A is nondefective.

AN EIGENVALUE-EIGENVECTOR FACTORIZATION OF EXP(At)

Theorem 15.1. If the $n \times n$ matrix A is nondefective, then the state transformation $T = [e_1 e_2 \cdots e_n]$ effects the decomposition $A = TDT^{-1}$, where $D = \text{diag}[\lambda_1, \lambda_2, \ldots, \lambda_n]$ is a diagonal matrix whose diagonal entries are the eigenvalues of A.

Proof.   Since A is nondefective, it has a full set $\{e_1, e_2, ..., e_n\}$, of eigenvectors which are linearly independent over the field of complex scalars. Thus, T^{-1} exists. By the definition of a right eigenvalue-eigenvector pair, $AT = TD$, from which it follows that $A = TDT^{-1}$. Finally, observe that T in fact is a state transformation associated with the equivalent state variables $z = Tx$.   ∎

In view of theorem 15.1, $e^{At} = T e^{Dt} T^{-1}$, which can be demonstrated by considering each term in a Taylor series expansion of $\exp(At) = \exp(TDT^{-1}t)$. Since D is diagonal,

$$
e^{Dt} = \begin{bmatrix} e^{\lambda_1 t} & & & \\ & e^{\lambda_2 t} & & \\ & & \ddots & \\ & & & e^{\lambda_n t} \end{bmatrix} \tag{15.5}
$$

EXAMPLE 15.2

Let a and b be real scalars, and compute $\exp(At)$, where

$$
A = \begin{bmatrix} a & b \\ -b & a \end{bmatrix} \tag{15.6}
$$

Step 1.   Compute the eigenvalues of A. Note that

$$
\pi_A(\lambda) = \det\left[\lambda I - A\right] = \det\begin{bmatrix} \lambda - a & -b \\ b & \lambda - a \end{bmatrix} = 0
$$

from which it follows that the eigenvalues of A are the zeros of $(\lambda - a)^2 + b^2 = 0$, i.e., $\lambda_1 = a + jb$ and $\lambda_2 = a - jb$.

Step 2.   Compute the associated right eigenvectors e_1 and e_2. The eigenvector e_1 satisfies $Ae_1 = \lambda_1 e_1$, hence,

$$
\begin{bmatrix} a & b \\ -b & a \end{bmatrix} \begin{bmatrix} \alpha_1 \\ \alpha_2 \end{bmatrix} = (a + jb) \begin{bmatrix} \alpha_1 \\ \alpha_2 \end{bmatrix}
$$

so that $\alpha_2 = j\alpha_1$. Let us choose $\alpha_1 = 1$ forcing $\alpha_2 = j$. In consequence,

$$
e_1 = \begin{bmatrix} \alpha_1 \\ \alpha_2 \end{bmatrix} = \begin{bmatrix} 1 \\ j \end{bmatrix}
$$

Since e_2 can always be chosen as $\bar{e}_1$, the complex conjugate of e_1,

$$
e_2 = \begin{bmatrix} 1 \\ -j \end{bmatrix}
$$

Step 3.   Form the state transformation T, and compute T^{-1}. By definition,

$$T = \begin{bmatrix} e_1 & | & e_2 \end{bmatrix} = \begin{bmatrix} 1 & 1 \\ j & -j \end{bmatrix}$$

so that

$$T^{-1} = \frac{1}{2} \begin{bmatrix} 1 & -j \\ 1 & j \end{bmatrix}$$

After suitable arithmetic manipulation,

$$e^{At} = T e^{Dt} T^{-1} = e^{at} \begin{bmatrix} \cos(bt) & \sin(bt) \\ -\sin(bt) & \cos(bt) \end{bmatrix} \tag{15.7}$$

If a matrix A has complex conjugate eigenvalues and is not of the form of equation 15.6, the associated matrix exponential is generically much uglier. With a little cleverness, however, it is still possible to utilize the result of the preceding example to obtain meaningful forms for the state transition matrix of A. To develop this approach, suppose A is 2×2, has complex conjugate pairs of eigenvalues, and is nondefective with eigenvalues $\lambda_1 = a + jb$ and $\lambda_2 = a - jb$. Then, from the foregoing development,

$$A = \begin{bmatrix} e_1 & | & e_2 \end{bmatrix} \begin{bmatrix} a+jb & 0 \\ 0 & a-jb \end{bmatrix} \begin{bmatrix} e_1 & | & e_2 \end{bmatrix}^{-1}$$

$$= \begin{bmatrix} e_1 & | & \bar{e}_1 \end{bmatrix} \begin{bmatrix} a+jb & 0 \\ 0 & a-jb \end{bmatrix} \begin{bmatrix} e_1 & | & \bar{e}_1 \end{bmatrix}^{-1} \tag{15.8}$$

Now consider the transformation matrices

$$M = 0.5 \begin{bmatrix} 1 & -j \\ 1 & j \end{bmatrix}$$

and

$$M^{-1} = \begin{bmatrix} 1 & 1 \\ j & -j \end{bmatrix}$$

It is a simple matter to show that $A = TDT^{-1} = (TM)(M^{-1}DM)(M^{-1}T^{-1})$ $= (TM)(M^{-1}DM)(TM)^{-1}$ has the elegant structure

$$A = \begin{bmatrix} \text{Re}(e_1) & | & \text{Im}(e_1) \end{bmatrix} \begin{bmatrix} a & b \\ -b & a \end{bmatrix} \begin{bmatrix} \text{Re}(e_1) & | & \text{Im}(e_1) \end{bmatrix}^{-1} \tag{15.9}$$

where $\text{Re}(\cdot)$ and $\text{Im}(\cdot)$ denote the real and imaginary parts, respectively, and where all matrices in the factorization of equation 15.9 are now *real*. Using the results of example 15.2 and the result of problem 1 at the end of the chapter,

$$e^{At} = \begin{bmatrix} \text{Re}(e_1) & | & \text{Im}(e_1) \end{bmatrix} e^{at} \begin{bmatrix} \cos(bt) & \sin(bt) \\ -\sin(bt) & \cos(bt) \end{bmatrix} \begin{bmatrix} \text{Re}(e_1) & | & \text{Im}(e_1) \end{bmatrix}^{-1} \tag{15.10}$$

This indicates that e^{At} can be computed without resort to complex arithmetic, a welcome feature. Of course, this development generalizes in the obvious way to A-matrices having several pairs of complex conjugate eigenvalues. In such cases, the transformation matrix M becomes block diagonal with appropriate 2×2 blocks of the form

$$0.5 \begin{bmatrix} 1 & -j \\ 1 & j \end{bmatrix}$$

for adjacent pairs of complex conjugate eigenvectors, and 1's in all remaining diagonal entries. Thus, $M^{-1}DM$ is a block diagonal matrix, termed $\tilde{D}$, having the property set out in the following proposition.

Proposition 15.3. Suppose $\tilde{D} = \text{diag} \begin{bmatrix} D_1, \ldots, D_n \end{bmatrix}$ is a block diagonal matrix with square blocks, D_i. Then

$$e^{\tilde{D}t} = \text{block} - \text{diag} \begin{bmatrix} e^{D_1 t}, \ldots, e^{D_n t} \end{bmatrix} \tag{15.11}$$

The proof of proposition 15.3 is left as an exercise for the reader. To illustrate the ideas it embraces consider the following example.

EXAMPLE 15.3

Evaluate e^{At}, where

$$A = \begin{bmatrix} -1 & -4 & 0 \\ 1 & -1 & 0 \\ 4 & 2 & -3 \end{bmatrix}$$

Step 1. The eigenvalues of A are the roots of the equation

$$\det(\lambda I - A) = \det \begin{bmatrix} \lambda + 1 & 4 & 0 \\ -1 & \lambda + 1 & 0 \\ -4 & -2 & \lambda + 3 \end{bmatrix} = (\lambda + 3) \begin{bmatrix} (\lambda + 1)^2 + 4 \end{bmatrix} = 0$$

Thus, the spectrum of A is

$$\sigma(A) = \{\lambda_1 = -3, \lambda_2 = -1 - 2j, \lambda_3 = -1 + 2j\}$$

Since there are three distinct eigenvalues, A is nondefective and has a full set of eigenvectors as follows.

Step 2. To calculate the right eigenvectors of A, consider that with regard to e_1,

$$\begin{bmatrix} -1 & -4 & 0 \\ 1 & -1 & 0 \\ 4 & 2 & -3 \end{bmatrix} \begin{bmatrix} \alpha_1 \\ \alpha_2 \\ \alpha_3 \end{bmatrix} = -3 \begin{bmatrix} \alpha_1 \\ \alpha_2 \\ \alpha_3 \end{bmatrix}$$

with solution $e_1 = [0 \ 0 \ 1]^t$. With regard to e_2, observe that $Ae_2 = \lambda_2 e_2$ requires that

$$\begin{bmatrix} -1 & -4 & 0 \\ 1 & -1 & 0 \\ 4 & 2 & -3 \end{bmatrix} \begin{bmatrix} \alpha_1 \\ \alpha_2 \\ \alpha_3 \end{bmatrix} = [-1 - 2j] \begin{bmatrix} \alpha_1 \\ \alpha_2 \\ \alpha_3 \end{bmatrix}$$

Solving for α_1, α_2, and α_3 leads to one possible solution, viz.,

$$e_2 = \begin{bmatrix} -2j \\ 1 \\ 2.5 - j1.5 \end{bmatrix}$$

Since $\lambda_2 = \bar{\lambda}_3$ — i.e., λ_2 and λ_3 are complex conjugates of each other— we choose $e_3 = \bar{e}_2$. Then, utilizing the techniques which led to equations 15.9 and 15.10, we define the transformation

$$\tilde{T} = \left[e_1 \mid \text{Re}(e_2) \mid \text{Im}(e_2) \right]$$

In this case

$$\tilde{T} = \begin{bmatrix} 0 & 0 & -2 \\ 0 & 1 & 0 \\ 1 & 2.5 & -1.5 \end{bmatrix}$$

and

$$\tilde{T}^{-1} = \begin{bmatrix} -0.75 & -2.5 & 1 \\ 0 & 1 & 0 \\ -0.5 & 0 & 0 \end{bmatrix}$$

with

$$\tilde{D} = \tilde{T}^{-1}A\tilde{T} = \begin{bmatrix} -3 & 0 & 0 \\ 0 & -1 & -2 \\ 0 & 2 & -1 \end{bmatrix}$$

a *block diagonal* matrix. Using equation 15.10, we have

$$e^{At} = \tilde{T} \begin{bmatrix} e^{-3t} & 0 & 0 \\ 0 & e^{-t}\cos(2t) & -e^{-t}\sin(2t) \\ 0 & e^{-t}\sin(2t) & e^{-t}\cos(2t) \end{bmatrix} \tilde{T}^{-1}$$

or, in the *unnecessarily* reduced form,

$$e^{At} = \begin{bmatrix} e^{-t}\cos(2t) & -2e^{-t}\sin(2t) & 0 \\ 0.5e^{-t}\sin(2t) & e^{-t}\cos(2t) & 0 \\ q(t) & p(t) & e^{-3t} \end{bmatrix}$$

where $q(t) = -0.75e^{-3t} + 1.25 \, e^{-t} \sin(2t) + 0.75 \, e^{-t}\cos(2t)$ and $p(t) = -2.5e^{-3t} + 2.5 \, e^{-t} \cos(2t) - 1.5 \, e^{-t}\sin(2t)$.

The culmination of the development, then, is the form of the state trajectory, which becomes

$$x(t) = Te^{Dt} T^{-1} x(t_0) + T \int_{t_0}^{t} e^{D(t-\tau)} [T^{-1}B] u(\tau) d\tau \qquad (15.12)$$

with appropriate modification for complex conjugate eigenvalues. Of course, a similar expression results for the system response $y(t)$.

The following proposition formalizes an obvious property which proves very useful.

Proposition 15.4. Let $T = [e_1,...,e_n]$ be a matrix whose columns are the right eigenvectors of a nondefective matrix A. Then $[T^{-1}]*$ is a matrix whose columns are associated left eigenvectors.

Proof. Since $AT = TD$ where $D = \text{diag}[\lambda_1,...,\lambda_n]$, it follows that $T^{-1}A = DT^{-1}$. But each row of T^{-1} satisfies equation 15.3, the defining equation for a left eigenvector. Hence, the complex conjugate transpose of each row of T^{-1} is a left eigenvector of A.   ∎

NUMERICAL CONSIDERATIONS

Computationally, the preceding techniques find implementation in a numerically stable way via "canned" algorithms available, for example in the EISPACK [5,6] and IMSL [7] libraries. The particular algorithms are generally based on a Hessenberg reduction of A and then a QR transformation which decomposes the Hessenberg-reduced A-matrix as QR where Q is a unitary matrix and R an upper triangular matrix [8,9]. Q has orthonormal columns, implying that $Q'Q = I$. These types of calculations are carried out using orthogonal transformations and thus do not increase the inherent sensitivity of the eigenvalues of A to perturbations ΔA.

Sensitivity of the eigenvalues to perturbations has the potential for being serious in any number of common circumstances. The potential difficulty rests with the fact that the A-matrix of a typical process such as a manufacturing process or a power system is often an approximate linearization of some underlying nonlinear state dynamics and has inaccurate entries as well. This means that A only approximates the real plant A-matrix. Coupled with the fact that the previously mentioned eigenvalue routines compute the eigenvalues not of A but of a "neighboring matrix," $A + \Delta A$, one may very well determine eigenvalues far different from those of the plant. To get a handle on how sensitive some matrix eigenvalues are to perturbations in the entries consider an example found in Wilkinson [10, p. 90] of the following 20×20 matrix:

$$A = \begin{bmatrix} 20 & 20 & 0 & 0 & & & & \\ 0 & 19 & 20 & 0 & & & & \\ 0 & 0 & 18 & 20 & & & & \\ & & & & \cdot & & & \\ & & & & & \cdot & & \\ & & & & & & \cdot & \\ & & & & & & 2 & 20 \\ & & & & & & 0 & 1 \end{bmatrix}$$

Clearly, the eigenvalues are 1,2,...,20, which are well separated, not clustered. If this matrix is perturbed by some δ in row 20, column 1, then the new eigenvalues $\lambda_i(\delta)$ of the matrix are proportional to the old eigenvalues λ_{iold}, according to

$$\lambda_i(\delta) = \lambda_{iold} + \xi_i\delta$$

where, unfortunately,

$$\xi_{20} = -\xi_1 \approx 4\cdot10^7$$

and

$$\xi_{10} = -\xi_{11} \approx 4\cdot10^{12}$$

One concludes that the eigenvalues of A can be highly sensitive to perturbations in the data. This potential difficulty should be borne in mind when analyzing or designing control systems. (See [10,11] for further details.)

PROBLEMS

1. Show that for any nonsingular transformation T, $\exp[TMT^{-1}] = T\exp[M]T^{-1}$.

2. Consider the parallel RLC circuit shown in Figure P15.2.

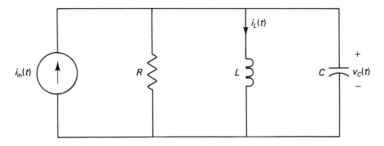

Figure P15.2.   Figure for Problem 2.

Suppose both $v_C(t)$ and $i_L(t)$ are desired outputs.

(a) Write a state model for this circuit.

(b) Suppose $R = 1\Omega$, $L = 1H$, and $C = 0.5F$. Find an expression for the state transition matrix. Your result should contain only real matrices.

(c) Find the resulting state trajectory when the input is a step function and the initial conditions are $i_L(0) = 1A$ and $v_C(0) = 1V$.

3. Consider the state dynamical equation $\dot{x}(t) = Ax(t)$.

(a) Find an expression for the state transition matrix.

(b) Find $x(0)$ given that $x(1) = [1\ 1\ -1]^t$ when

(i)

$$A = \begin{bmatrix} -1 & 0 & 0 \\ 0 & -2 & 5 \\ 0 & -2 & 4 \end{bmatrix}$$

(ii)

$$A = \begin{bmatrix} -3 & 2 & 2 \\ 0 & -1 & 0 \\ -6 & 6 & 4 \end{bmatrix}$$

4. Given the time-varying system

$$\dot{x}(t) = A_1(t)x(t), \quad x(t_0) = x_0 \text{ for } 0 \leqslant t_0, t \leqslant 1$$

and

$$\dot{x}(t) = A_2(t)x(t) \quad \text{for } 1 < t$$

where

$$A_1(t) = \begin{bmatrix} -6 & 6 \\ -4 & 4 \end{bmatrix}$$

and

$$A_2(t) = \begin{bmatrix} 1/t & 1 \\ 1 & 1/t \end{bmatrix}$$

construct an expression for the state transition matrix of the time-varying state equations.

5. Given $\dot{x}(t) = A(t)x(t)$, compute the associated state transition matrix when

$$A(t) = \begin{bmatrix} 2t-2 & 2 \\ -1 & 2t \end{bmatrix}$$

and find $x(-1)$ when the input is identically zero and $x(0) = [1\ 1]^t$. In solving this problem, state all formulas that you use. Also, verify those conditions which guarantee the validity of the formulas. Your work should be clear and organized and proceed in logical steps.

6. Given the homogeneous state model $\dot{x} = Ax$, find a closed-form expression for the state transition matrix $\Phi(t - t_o)$ when

(i)

$$A = \begin{bmatrix} -6 & -12 & 0 & 0 \\ 2 & 4 & 0 & 0 \\ 0 & 0 & -1 & 1 \\ 0 & 0 & 0 & -1 \end{bmatrix}$$

(ii)

$$A = \begin{bmatrix} 0 & 0 & 0 & 0 & 0 & 0 & 0 \\ 1 & 0 & 0 & 0 & 0 & 0 & 0 \\ -1 & 2 & 0 & 0 & 0 & 0 & 0 \\ 0 & 0 & 0 & 1 & -1 & 0 & 0 \\ 0 & 0 & 0 & 1 & 1 & 0 & 0 \\ 0 & 0 & 0 & 0 & 0 & 0 & 1 \\ 0 & 0 & 0 & 0 & 0 & -2 & -2 \end{bmatrix}$$

7. Let the A-matrix of a state model be

$$A = \begin{bmatrix} a & b \\ b & a \end{bmatrix}$$

for real scalars a and b.

(a) What are the eigenvalues of A?
(b) What is a set of right eigenvectors of A?
(c) From your results in (b), form the matrix $T = [e_1 \; e_2]$.
 Compute T^{-1}. What are the left eigenvectors of A?
(d) Compute an expression for the state transition matrix. (*Hint*: Are there cosh's and sinh's in your answer?)

8. Compute the state transition matrix associated with the time-invariant state model $\dot{x}(t) = Ax(t)$, where

$$A = \begin{bmatrix} a-b & 2b & 0 \\ -b & a+b & 0 \\ 0 & 0 & -1 \end{bmatrix}$$

for real scalars a and b.

9. Consider the time-varying switched state model description

$$\dot{x}(t) = \begin{bmatrix} a & b \\ -b & a \end{bmatrix} x(t), \qquad t_0 \leq t < 1$$

and

$$\dot{x}(t) = \begin{bmatrix} a & b \\ b & a \end{bmatrix} x(t), \qquad 1 \leq t$$

What is the state transition matrix?

10. A certain switched time-varying system $\dot{x}(t) = A(t)x(t)$ has a matrix $A(t)$ defined as

$$A(t) = \begin{bmatrix} a-b & 2b \\ -b & a+b \end{bmatrix}$$

for $0 \leq t_0, t \leq \pi$, where $a = 0$ and $b = 1$. For $\pi < t$,

$$A(t) = \begin{bmatrix} -1 & 0 \\ (t+1)^{-1} & -1 \end{bmatrix}$$

What is the associated state transition matrix?

11. Derive equation 2.33 in Chapter 2 using the eigenvalue-eigenvector method of computing $\exp[At]$.

12. Let $\dot{x} = Ax$, and suppose A has real eigenvalues λ_1, λ_2, and λ_3, where

$$A = \begin{bmatrix} 0 & 1 & 0 \\ 0 & 0 & 1 \\ -a_3 & -a_2 & -a_1 \end{bmatrix}$$

Find eigenvectors for A in terms of λ_i.

13. Consider the state model

$$\dot{x} = \begin{bmatrix} -2 & 0 & 2 \\ 2 & 0 & -2 \\ -1 & 1 & 2 \end{bmatrix} x + \begin{bmatrix} 1 \\ -1 \\ 1 \end{bmatrix} u$$

(a) Find the state transition matrix $\Phi(t)$. *Hint*: Consider the state transformation $x = Tz$, where

$$T = \begin{bmatrix} 1 & -1 & 1 \\ 0 & 1 & -1 \\ 0 & 0 & 1 \end{bmatrix}$$

(b) If $u(t) = 0$ for all t, compute $x(2)$ given that $x(1) = [1\ 0\ 0]^t$.

14. Consider the state model

$$\begin{bmatrix} \dot{x}_1(t) \\ \dot{x}_2(t) \end{bmatrix} = \begin{bmatrix} 0 & 1 \\ -1 & 0 \end{bmatrix} \begin{bmatrix} x_1(t) \\ x_2(t) \end{bmatrix} + \begin{bmatrix} 0 \\ 1 \end{bmatrix} u(t), \quad x(0) = \begin{bmatrix} 1 \\ 1 \end{bmatrix}$$

$$y(t) = x(t)$$

The input $u(t)$ is a staircase function defined as $u(t) = kT$ for $kT \leqslant t < (k+1)T$, $k = 0,1,2,...$, and T some fixed real number; also, $u(t) = 0$ for all $t < 0$. Find a discrete-time state model of the form

$$\begin{bmatrix} \hat{x}_1((k+1)T) \\ \hat{x}_2((k+1)T) \end{bmatrix} = \hat{A}(kT) \begin{bmatrix} \hat{x}_1(kT) \\ \hat{x}_2(kT) \end{bmatrix} + \hat{B}(kT)u(kT)$$

$$\hat{y}(kT) = \hat{C}(kT) \begin{bmatrix} \hat{x}_1(kT) \\ \hat{x}_2(kT) \end{bmatrix} + \hat{D}(kT)u(kT)$$

whose solution for $k = 0,1,2...$ exactly equals the response of the preceding continuous-time system for $t \geqslant 0$—i.e., $y(kT) = \hat{y}(kT)$ for $k = 0,1,2,...$.

15. (Two useful properties). Show that

(a)   (b)

$$\det[A] = \prod_{i=1}^{n} \lambda_i.$$ $$\text{trace }[A] \triangleq \sum_{i=1}^{n} a_{ii} = \sum_{i=1}^{n} \lambda_i.$$

(*Hint*: For (a), consider the definition of the characteristic polynomial. For (b), consider one of the coefficients of the characteristic polynomial computed via Leverrier's algorithm (definition 12.2).)

16. Consider again problem 5 of chapter 11, letting

$$A(t) = t \begin{bmatrix} 0 & -1 \\ 0 & -1 \end{bmatrix}$$

What can you deduce about the more general form

$$A(t) = \alpha(t) [e_1 \ e_2] \begin{bmatrix} \lambda_1 & 0 \\ 0 & \lambda_2 \end{bmatrix} [e_1 \ e_2]^{-1}?$$

REFERENCES

1. C. A. Desoer, *Notes for a Second Course on Linear Systems* (New York: Van Nostrand, 1970).

2. Cleve Moler and Charles Van Loan, "Nineteen Dubious Ways to Compute the Exponential of a Matrix," *SIAM Review*, Vol. 20, No. 4, October 1978, pp. 801–836.

3. W. M. Wonham, *Linear Multivariable Control: A Geometric Approach,* 2d ed. (New York: Springer-Verlag, 1979).

4. G. H. Golub and J. H. Wilkinson, "Ill-conditioned Eigensystems and the Computation of the Jordan Canonical Form," *SIAM Review,* Vol. 18, 1976, pp. 578–619.

5. EISPACK, Eigensystem Package, Argonne National Laboratory.

6. B. S. Garbow, et al., *Matrix Eigensystem Routines—EISPACK Guide Extension,* Lecture Notes in Computer Science, Vol. 51 (New York: Springer-Verlag, 1977).

7. IMSL, International Mathematical and Statistical Libraries, IMSL Inc., Houston.

8. J. Stoer and R. Bulirsch, *Introduction to Numerical Analysis* (New York: Springer-Verlag, 1980), pp. 370–371.

9. J. G. F. Francis, "The QR Transformation I, II," *Computer Journal,* Vol. 4, 1961, pp. 265–271; Vol. 4, 1962, pp. 332–345.

10. J. H. Wilkinson, *The Algebraic Eigenvalue Problem* (London: Oxford University Press, 1965).

11. Chris C. Paige, "Properties of Numerical Algorithms Related to Computing Controllability," *IEEE Transactions Automatic Control,* Vol. AC-26, No. 1, February 1981, pp. 130–138.

16

THE STATE TRANSITION MATRIX
AND
THE SOLUTION OF
THE DISCRETE-TIME STATE MODEL

INTRODUCTION

The concepts of a state transition matrix, a fundamental matrix, the semigroup property, etc., all have analogs for the discrete-time state model (DTSM)

$$x(k+1) = A(k) x(k) + B(k) u(k)$$
$$y(k) = C(k) x(k) + D(k) u(k)$$

This chapter addresses the development of these important counterparts. We begin with a direct derivation of the state transition matrix and its properties.

THE STATE TRANSITION MATRIX AND ITS PROPERTIES

The state transition matrix, $\Phi(t,t_0)$ for a continuous-time state model defines transitions from a state at time t_0 to a state at time t for arbitrary t_0 and t according to the formula $x(t) = \Phi(t,t_0)x(t_0)$. Because the A-matrix of a discrete-time state model is sometimes singular, the arbitrariness of t_0 and t will not carry over unless $A(k)$ is nonsingular for all k.

Definition 16.1. The *state transition matrix* for the discrete-time state model $x(k+1) = A(k)x(k)$ is the matrix $\Phi(n,k)$ such that for any $x(k)$, $x(n) = \Phi(n,k)x(k)$.

To derive the structure of $\Phi(n,k)$, let $x(k+1) = A(k)x(k)$ with $x(k)$ given. Then a simple straightforward "plug and chug" produces

$$x(k+1) = A(k)\,x(k)$$
$$x(k+2) = A(k+1)x(k+1) = A(k+1)A(k)x(k)$$
$$\vdots$$

$$x(n) = A(n-1)\,A(n-2)...A(k+1)\,A(k)\,x(k) = \prod_{j=k}^{n-1} A(j)\,x(k)$$

This derivation prompts the following proposition.

Proposition 16.1. The discrete-time state transition matrix has the structure

$$\Phi(n,k) = \prod_{j=k}^{n-1} A(j) = A(n-1)\,A(n-2)...A(k+1)\,A(k) \qquad (16.1)$$

Proposition 16.2. The state transition matrix satisfies the difference equation

$$\Phi(k+1,j) = A(k)\Phi(k,j), \quad \Phi(j,j) = I, \ k \geq j$$

A straightforward application of equation 16.1 proves this proposition and is left as a problem at the end of the chapter.

Proposition 16.3 (Semigroup property). For $n \geq k \geq j$,

$$\Phi(n,k)\Phi(k,j) = \Phi(n,j)$$

A straightforward application of equation 16.1 also proves this proposition and is left as a problem at the end of the chapter.

Proposition 16.4 (Inverse property). If $A(k)$ is nonsingular for all k,

$$[\Phi(n,k)]^{-1} = \prod_{j=n-1}^{k} [A(j)]^{-1} \qquad (16.2)$$

Using the well-known property of matrix inverses that if A and B are invertible then $(AB)^{-1} = B^{-1}A^{-1}$, it is unchallenging to demonstrate that equation 16.2 is the inverse of equation 16.1 and conversely.

This development of the discrete-time state transition matrix falls directly out of the state model dynamics. By contrast, in the continuous-time case the development proceeded from a discussion of a fundamental basis of solutions generated from initial conditions $x(t_0) = \eta_i$ for $i = 1, \ldots, n$. The next section examines the notion of a fundamental set of solutions and the associated fundamental matrix in the discrete-time context.

FUNDAMENTAL MATRIX

A fundamental matrix depends on the notion of a linear independent set of sequences.

Definition 16.2. A set of q vector sequences $\{x^1(k)\}, \{x^2(k)\}, \ldots, \{x^q(k)\}$ for k $= 0, 1, 2, \ldots$ is *linearly independent* if and only if, whenever

$$\alpha_1 x^1(k) + \alpha_2 x^2(k) + \ldots + \alpha_q x^q(k) = \theta \qquad (16.3)$$

for all k, then $\alpha_1 = \alpha_2 = \ldots = \alpha_q = \theta$.

Of course, it is possible that at some particular k's or even many k's the vectors are linearly dependent; however, they cannot be linearly dependent for all k. The idea corresponds exactly to the continuous-time definition 11.1.

The notion of a fundamental set of solutions follows immediately.

Definition 16.3. A *fundamental set of solutions* for the state dynamical equation $x(k+1) = A(k)x(k)$ (where $A(k)$ is, as usual, $n \times n$) is a set of n linearly independent vector sequences $\{x^1(k)\}, \{x^2(k)\}, \ldots, \{x^n(k)\}$ whose terms satisfy the equation $x^i(k+1) = A(k)x^i(k)$ for all $k \geqslant 0$.

The importance of a fundamental set of solutions derives from the property that an arbitrary solution can be expressed as a linear combination of the fundamental set. Conversely, every linear combination of a fundamental set of solutions is a solution. Proofs of these properties are left as exercises at the end of the chapter.

Organizing a fundamental set of solutions side by side as the columns of a matrix results in a fundamental matrix.

Definition 16.4. A *fundamental matrix* is any matrix whose columns form a fundamental set of solutions. It is denoted by

$$\Psi(k) = [x^1(k) \; x^2(k) \ldots x^n(k)]$$

where the $x^i(k)$ satisfy definition 16.3.

Since each column of a fundamental matrix satisfies $x^i(k+1) = A(k)x^i(k)$ for all $k \geqslant 0$, the following proposition results.

Proposition 16.5. A fundamental matrix satisfies $\Psi(k+1) = A(k)\Psi(k)$.

Exercise. Show that the state transition matrix is a fundamental matrix. (*Hint*: Consider proposition 16.2.)

Exercise. Construct an example with the property that $\det[\Psi(j)] \neq 0$ but $\det[\Psi(j+1)] = 0$. (*Hint:* What must be the structure of the $A(k)$ matrix?)

The following proposition describes an important property of the discrete-time fundamental matrix, and its proof essentially duplicates its continuous-time counterpart, proposition 12.1, and is left as an exercise.

Proposition 16.6. If $A(k)$ is nonsingular for all k, then any associated fundamental matrix $\Psi(k)$ is nonsingular for all k.

The following proposition allows us to link the fundamental matrix and the state transition matrix.

Proposition 16.7. If $A(k)$ is nonsingular for all k, then the state transition matrix and any fundamental matrix of solutions are related by $\Phi(n,k) = \Psi(n)\Psi^{-1}(k)$ for $n \geqslant k$.

Proof. Since $A(k)$ is nonsingular for all k, by proposition 16.6, $\Psi(k)$ is nonsingular for all k and any solution to $x(j+1) = A(j)x(j)$ with $x(k)$ given has the form $x(n) = \Psi(n)\Psi^{-1}(k)x(k)$. This is clear since $\Psi(n)\Psi^{-1}(k)x(k)$ satisfies the proper initial condition and difference equation. But since this is true for all possible initial conditions $x(k)$, it follows that the state transition matrix $\Phi(n,k) = \Psi(n)\Psi^{-1}(k)$. ∎

EXAMPLE 16.1

It is known that a second-order DTSM of the form $x(k+1) = A(k)x(k)$ produces two fundamental sequences $\{x^1(k)\} = \{[1\ 0]'\}$ and $\{x^2(k)\} = \{[1\ 1]', [2\ -1]', [1\ 1]', [4\ -1]',...\}$. Find (a) the associated fundamental matrix $\Psi(k)$, and (b) $A(k)$, and (c) the state transition matrix $\Phi(k,0)$.

Solution to part (a): By definition 16.4, $\Psi(k) = [x^1(k)\ x^2(k)]$. But $x^2(k) = [1 + 0.5k[1 + (-1)^{k+1}]\quad(-1)^k]'$. Therefore,

$$\Psi(k) = \begin{bmatrix} 1 & 1 + 0.5k[1 + (-1)^{k+1}] \\ 0 & (-1)^k \end{bmatrix}$$

Solution to part (b): Here the objective is simply to use the information about the sequence to generate a set of four simultaneous equations in the entries of $A(k)$. Accordingly, let

$$A(k) = \begin{bmatrix} a_{11}(k) & a_{12}(k) \\ a_{21}(k) & a_{22}(k) \end{bmatrix}$$

Substituting the values of $\{x^1(k)\}$ into the state model produces

$$\begin{bmatrix} 1 \\ 0 \end{bmatrix} = \begin{bmatrix} a_{11}(k) & a_{12}(k) \\ a_{21}(k) & a_{22}(k) \end{bmatrix} \begin{bmatrix} 1 \\ 0 \end{bmatrix}$$

for all k, from which it follows that that $a_{11}(k) = 1$ and $a_{21}(k) = 0$. Using the general expression in part (a) for $\{x^2(k)\}$ leads to

$$\begin{bmatrix} 1 + 0.5(k+1)[1 + (-1)^{k+2}] \\ (-1)^{k+1} \end{bmatrix} = \begin{bmatrix} 1 & a_{12}(k) \\ 0 & a_{22}(k) \end{bmatrix} \begin{bmatrix} 1 + 0.5k[1 + (-1)^{k+1}] \\ (-1)^k \end{bmatrix}$$

from which it follows that $a_{22}(k) = -1$ and that

$$0.5k[1 + (-1)^{k+2}] + 0.5[1 + (-1)^{k+2}]$$
$$= 0.5k[1 + (-1)^{k+1}] + a_{12}(k)(-1)^k$$

This reduces to

$$k(-1)^k + 0.5 + 0.5(-1)^k = a_{12}(k)(-1)^k$$

or, equivalently, to

$$a_{12}(k) = k + 0.5[1 + (-1)^k]$$

in which case

$$A(k) = \begin{bmatrix} 1 & k + 0.5[1 + (-1)^k] \\ 0 & -1 \end{bmatrix}$$

Solution to part (c): Since $A(k)$ is nonsingular for all k, by proposition 16.7,

$$\Phi(k,0) = \begin{bmatrix} 1 & 1 + 0.5k[1 + (-1)^{k+1}] \\ 0 & (-1)^k \end{bmatrix} \begin{bmatrix} 1 & -1 \\ 0 & 1 \end{bmatrix}$$

COMPLETE SOLUTION OF THE DISCRETE-TIME STATE MODEL

In chapter 9, the formula of equation 9.29 was asserted to solve the state dynamical equation $x(k+1) = Ax(k) + Bu(k)$ with $x(0)$ given. The material here supports that assertion by heuristically deriving the solution to the time-varying DTSM. Then we verify that the result in fact solves the time-varying dynamics.

Straightforward evaluation of the state dynamics produces

$x(1) = A(0)x(0) + B(0)u(0)$

$x(2) = A(1)x(1) + B(1)u(1)$

$\qquad = A(1)A(0)x(0) + A(1)B(0)u(0) + B(1)u(1)$

$x(3) = A(2)x(2) + B(2)u(2)$

$\qquad = A(2)A(1)A(0)x(0) + A(2)A(1)B(0)u(0) + A(2)B(1)u(1) + B(2)u(2)$

$\qquad = \Phi(3,0)x(0) + \Phi(3,1)B(0)u(0) + \Phi(3,2)B(1)u(1) + \Phi(3,3)B(2)u(2)$

Hence, for $k \geqslant 1$,

$$x(k) = \Phi(k,0)x(0) + \sum_{j=0}^{k-1} \Phi(k,j+1)B(j)u(j) \tag{16.4}$$

At $k = 0$ the sum is undefined (extraneous), and hence the formula reduces to $x(0)$, the given initial condition.

The following proves that equation 16.4 is indeed the solution to the time-varying state dynamics.

Theorem 16.1. The solution to the time-varying discrete-time state dynamical equation $x(k + 1) = A(k)x(k) + B(k)u(k)$ for a given initial condition $x(0)$ and given input sequence $\{u(k), k \geqslant 0\}$ is given by equation 16.4.

Proof. The proof is by induction and entails showing that the formula satisfies (i) the proper initial condition at $k = 0$ and (ii) the state dynamics.

(i) By definition, at $k = 0$ the formula reduces to $x(0)$. Further, at $k = 1$ the right side of equation 16.4 reduces to $x(1)$.

(ii) Suppose the formula is valid up to k. We seek to prove that it is valid at $k + 1$. At $k + 1$ the formula becomes

$$\Phi(k + 1,0)x(0) + \sum_{j=0}^{k} \Phi(k + 1, j + 1)B(j)u(j) \tag{16.5}$$

which can be written as

$$\Phi(k + 1,0)x(0) + \sum_{j=0}^{k-1} \Phi(k + 1, j + 1)B(j)u(j) + B(k)u(k) \tag{16.6}$$

From proposition 16.2, the state transition matrix satisfies the relation

$$\Phi(k + 1, j + 1) = A(k)\Phi(k, j + 1)$$

Thus, equation 16.6 reduces to the form

$$A(k)\Phi(k,0)x(0) + \sum_{j=0}^{k-1} A(k)\Phi(k, j + 1)B(j)u(j) + B(k)u(k)$$

$$= A(k)\left[\Phi(k,0)x(0) + \sum_{j=0}^{k-1} \Phi(k, j + 1)B(j)u(j)\right] + B(k)u(k)$$

$$= A(k)x(k) + B(k)u(k)$$

where the last equality follows because equation 16.4 is assumed valid up through k by the induction hypothesis. But this last equality is by definition equal to $x(k + 1)$, and thus equation 16.4 is valid at $k + 1$.   ∎

Since $y(k) = C(k)x(k) + D(k)u(k)$, it follows from equation 16.4 that

$$y(k) = C(k)\Phi(k,0)x(0) + \sum_{j=0}^{k-1} C(k)\Phi(k, j + 1) B(j)u(j) + D(k)u(k) \tag{16.7}$$

Of course, numerical evaluation of this formula, as well as that of equation 16.4, is simply a programming exercise unless one has a special form for $\Phi(k,0)$. Such a form does exist in the time-invariant case, as discussed next.

THE TIME-INVARIANT DISCRETE-TIME STATE MODEL

Applying the foregoing results to the linear time-invariant DTSM

$$x(k+1) = Ax(k) + Bu(k)$$
$$y(k) = Cx(k) + Du(k)$$

produces equation 9.29, repeated here as

$$x(k) = A^k x(0) + \sum_{j=0}^{k-1} A^{k-1-j} Bu(j) = A^k x(0) + \sum_{j=0}^{k-1} A^j Bu(k-1-j) \quad (16.8)$$

where the state transition matrix now has the form $\Phi(k,j) = \Phi(k-j,0) \triangleq \Phi(k-j) = A^{k-j}$.

As with equation 9.30, if A has the factorization $A = TDT^{-1}$, then equation 16.8 reduces to

$$x(k) = TD^k T^{-1} x(0) + T \sum_{j=0}^{k-1} D^{k-1-j} T^{-1} Bu(j) \quad (16.9a)$$

or

$$x(k) = TD^k T^{-1} x(0) + T \sum_{j=0}^{k-1} D^j T^{-1} Bu(k-1-j) \quad (16.9b)$$

As per chapter 15, whenever A is nondefective, such a factorization exists and equations 16.9 apply.

EXAMPLE 16.2

To illustrate the evaluation of equations 16.9, consider the DTSM

$$x(k+1) = \begin{bmatrix} 0.5 & 1 \\ 0 & 1 \end{bmatrix} x(k) + \begin{bmatrix} 1 \\ 1 \end{bmatrix} 1^+(k), \quad x(0) = \begin{bmatrix} 1 \\ -1 \end{bmatrix} \quad (16.10)$$

where $1^+(k)$ is the step sequence. The objective is to find a closed-form expression for $x(k)$.

Step 1. Factor the A-matrix into TDT^{-1}. By inspection, the eigenvalues of the A-matrix are 0.5 and 1. A little arithmetic in computing the matrix of right eigenvectors and its inverse then produces the factorization

$$\begin{bmatrix} 0.5 & 1 \\ 0 & 1 \end{bmatrix} = \begin{bmatrix} 1 & 2 \\ 0 & 1 \end{bmatrix} \begin{bmatrix} 0.5 & 0 \\ 0 & 1 \end{bmatrix} \begin{bmatrix} 1 & -2 \\ 0 & 1 \end{bmatrix} \quad (16.11)$$

Step 2. Compute the zero-input state response. Using the factorization of equation 16.11, the zero-input response is given by $x(k) = TD^k T^{-1}$ according to

$$x(k) = \begin{bmatrix} 1 & 2 \\ 0 & 1 \end{bmatrix} \begin{bmatrix} 0.5^k & 0 \\ 0 & 1 \end{bmatrix} \begin{bmatrix} 1 & -2 \\ 0 & 1 \end{bmatrix} \begin{bmatrix} 1 \\ -1 \end{bmatrix} = \begin{bmatrix} 3(0.5)^k - 2 \\ -1 \end{bmatrix} \quad (16.12)$$

Step 3. Compute the zero-state state response. From equation 16.22b, the zero-state state response is

$$x(k) = \begin{bmatrix} 1 & 2 \\ 0 & 1 \end{bmatrix} \sum_{j=0}^{k-1} \begin{bmatrix} 0.5^j & 0 \\ 0 & 1 \end{bmatrix} \begin{bmatrix} 1 & -2 \\ 0 & 1 \end{bmatrix} \begin{bmatrix} 1 \\ 1 \end{bmatrix}$$

$$= \begin{bmatrix} 1 & 2 \\ 0 & 1 \end{bmatrix} \begin{bmatrix} \sum_{j=0}^{k-1}(0.5)^j & 0 \\ 0 & k \end{bmatrix} \begin{bmatrix} -1 \\ 1 \end{bmatrix} = \begin{bmatrix} 2(0.5)^k - 2 + 2k \\ k \end{bmatrix}$$

where we have used the formula

$$\sum_{j=0}^{k-1} \lambda^j = \frac{1-\lambda^k}{1-\lambda}, \quad \lambda \neq 1$$

to evaluate the sum.

Step 4. Combine the zero-input and zero-state responses to obtain the complete response

$$x(k) = \begin{bmatrix} 5(0.5)^k - 4 + 2k \\ k - 1 \end{bmatrix}$$

Clearly, the evaluation of the sum in step 3 of the preceding example depends intimately on a suitable expression for A^k. In the case of complex eigenvalues, equations 16.9 would require complex arithmetic. To avoid complex arithmetic, we follow a path similar to that followed in chapter 15. Suppose A is 2×2 and of the form

$$A = \begin{bmatrix} a & b \\ -b & a \end{bmatrix}$$

Then it is possible to show (see problem 11) that

$$A^k = \rho^k \begin{bmatrix} \cos(k\theta) & \sin(k\theta) \\ -\sin(k\theta) & \cos(k\theta) \end{bmatrix}$$

where $\rho = \sqrt{a^2 + b^2}$ and $\theta = \tan^{-1}[b/a]$. Furthermore, as per equation 15.9, if A is 2×2 with complex eigenvalues $a + jb$ and $a - jb$ and right eigenvectors e_1 and e_2, then

$$A^k = [\text{Re}(e_1) | \text{Im}(e_1)] \rho^k \begin{bmatrix} \cos(k\theta) & \sin(k\theta) \\ -\sin(k\theta) & \cos(k\theta) \end{bmatrix} [\text{Re}(e_1) | \text{Im}(e_1)]^{-1} \quad (16.13)$$

These relationships extend naturally to larger dimensional matrices by factoring A into a block diagonal D-matrix having appropriate 2×2 blocks representing complex eigenvalues and 1×1 blocks for real eigenvalues.

In the case of repeated eigenvalues, the binomial expansion proves useful in constructing A^k for certain triangular matrices. Consider, for example, computing A^k where A^k has the form $[\lambda I + Y]^k$. Specifically, suppose

$$A^k = \begin{bmatrix} \lambda_1 & 1 \\ 0 & \lambda_1 \end{bmatrix}^k = \left[\lambda_1 \begin{bmatrix} 1 & 0 \\ 0 & 1 \end{bmatrix} + \begin{bmatrix} 0 & 1 \\ 0 & 0 \end{bmatrix} \right]^k = [\lambda_1 I + Y]^k \qquad (16.14)$$

where, clearly, $\lambda_1 I$ and Y commute, i.e., $(\lambda_1 I)Y = Y(\lambda_1 I)$. The *binomial expansion* for matrices says that if two properly dimensioned square matrices X and Y commute $(XY = YX)$, then

$$(X + Y)^k = X^k + \binom{k}{1} X^{k-1} Y + \binom{k}{2} X^{k-2} Y^2 + \ldots + Y^k \qquad (16.15)$$

where the *binomial coefficient* is given by

$$\binom{k}{j} = \frac{k!}{(k-j)!\, j!} \qquad (16.16)$$

Applying the binomial expansion to equation 16.14 yields

$$A^k = \lambda_1^k I^k + \binom{k}{1} \lambda_1^{k-1} I^{k-1} Y + \binom{k}{2} \lambda_1^{k-2} I^{k-2} Y^2 + \ldots + Y^k$$

But since $Y^k = [0]$ for all $k \geqslant 2$, this expression reduces to

$$A^k = \lambda_1^k I + k\lambda_1^{k-1} Y = \begin{bmatrix} \lambda_1^k & k\lambda_1^{k-1} \\ 0 & \lambda_1^k \end{bmatrix} \qquad (16.17)$$

THE IMPULSE RESPONSE OF THE TIME-INVARIANT DTSM

The impulse response of a linear system defines a very common and very useful input-output representation of the system. In elementary courses, the discrete-time impulse response often refers to the response of a single-input, linear, time-invariant system to an impulse $\delta(k)$ applied to the system when it is in a relaxed state at time zero. Actually, the term "impulse response" in some sense misrepresents the situation for discrete-time systems, since $\delta(k) = 1$ if $k = 0$ and is zero otherwise. Thus, $\delta(k)$ is not an impulse in the traditional sense. Nevertheless, it plays the same role in the discrete-time world as $\delta(t)$ does in continuous time.

The utility of the impulse response representation rests with the property that the response to an arbitrary input is the discrete convolution of the input sequence and the impulse response sequence. As described in a number of elementary texts [3–5], one arrives at this property by showing, first, that input sequences have a representation as the convolutional sum of the impulse with the given input sequence. Then, using linearity and shift invariance, the effect of each term of the input sequence on the response is seen as the impulse response properly weighted and shifted in time. Finally, summing all these contributions leads to the desired convolutional representation.

For MIMO systems, one looks at the impulse response of each port, holding the remaining ports at ground. One then organizes these responses into a matrix. The resulting matrix designates the impulse response matrix.

Definition 16.5. The *impulse response matrix* for the time-invariant DTSM is given by

$$H(k) = [h_1(k) \mid h_2(k) \mid ... \mid h_m(k)] \qquad (16.18)$$

where $h_i(k)$ is the response of the relaxed system (all initial conditions zero) to the impulsive input $\delta(k)\eta_i$ where $\eta_i \in \mathbf{R}^m$ has a 1 in the ith position and zeros elsewhere.

The applicability of this definition comes from the representation of $H(k)$ in terms of the system matrices A, B, C, and D. This permits a straightforward input-output analysis of discrete-time systems.

To express $H(k)$ in terms of the system matrices, consider the response of the DTSM to the input $u(k) = \delta(k)\eta_i$ assuming that $x(0) = \theta$. Since the system structure is causal, it follows immediately that the response for $k < 0$ is zero. At $k = 0$, the response is $h_i(0) = D\,\delta(0)\eta_i = D\eta_i$. For $k = 1$, consider $x(1) = Ax(0) + B\delta(0)\eta_i = B\delta(0)\eta_i$, which implies that the response is $h_i(1) = CB\eta_i$. At $k = 2$, $x(2) = Ax(1) + B\delta(1)\eta_i = AB\eta_i$, and thus $h_i(2) = CAB\eta_i$. In general, $h_i(k) = CA^{k-1}B\eta_i$ for $k \geqslant 1$.

From equation 16.18 and the foregoing discussion, the resulting impulse response matrix has the structure

$$H(k) = \begin{cases} [0] & k < 0 \\ D & k = 0 \\ CA^{k-1}B & k \geqslant 1 \end{cases} \qquad (16.19)$$

The efficient evaluation and the utility of this expression depend on calculating a closed-form expression for A^{k-1} (as discussed in the previous section) and on the property that the response to an arbitrary one-sided sequence $\{u(k)\}$ can be expressed as the convolution sum [3–5]

$$y(k) = \sum_{j=-\infty}^{\infty} H(k-j)u(j) = \sum_{j=0}^{k-1} CA^{k-1-j}Bu(j), \quad k \geqslant 1 \qquad (16.20)$$

and $y(k) = Du(0)$ for $k = 0$. This formula is identical to equation 16.8 when the initial state $x(0) = \theta$.

It is important to note that one can develop this equivalence from a different, but very instructive, perspective. The outline of such a development is briefly sketched as follows and can be found as well in [3–5] in greater detail. Recall that

$$u(k) = \sum_{j=0}^{\infty} \delta(k-j)u(j)$$

By linearity and shift invariance of the system, the response of this sum is simply the sum of the weighted shifted impulse responses (those responses associated with each term $\delta(k - j)u(j)$) and takes the form of equation 16.20.

In addition to supplying a convenient input-output description of the system, the impulse response proves critical in developing conditions for system stability and in identifying external models for unknown systems from input-output measurements.

THE DTSM IN THE Z-DOMAIN

Recall the one-sided or unilateral Z-transform of a vector sequence $\{u(k)\}$ given by

$$Z\{u(k)\} = \sum_{k=0}^{\infty} u(k)z^{-k} \qquad (16.21)$$

where z is both a placemarker for terms in the sequence $\{u(k)\}$ and a complex variable. Since equation 16.21 is a series in the complex variable z, there is an associated region of convergence in the z-plane where the series is well defined. For example, if $u(k) = a^k 1^+(k)$, then $Z\{u(k)\} = z/(z-a)$ with the region of convergence $|z| > |a|$. Many texts [3–5] have extensive expositions of Z-transforms covering the particulars. These discussions also include a listing and examples of the various properties of the Z-transform. Of particular importance here are the properties of linearity, shifting, and convolution. Linearity is that property in virtue of which the scaled sum of sequences has a Z-transform equal to the scaled sum of the individual Z-transforms. The convolution property is such that the Z-transform of the convolution of two sequences equals the product of the two associated Z-transforms. And the *forward shift property* posits that

$$Z\{u(k-n)\} =$$
$$z^{-n}u(z) + u(-1)z^{-n+1} + \dots + u(-n+1)z^{-1} + u(-n) \qquad (16.22)$$

while according to the *backward shift property*,

$$Z\{u(k+n)\} = z^n u(z) - u(0)z^n - u(1)z^{n-1} - \dots - u(n-1)z \quad (16.23)$$

These well-known properties permit us to analyze the Z-transform of the state model and arrive at the ubiquitous transfer function matrix of the DTSM. Taking the Z-transform of the time-invariant DTSM and applying the properties of linearity and shifting produces

$$z Z\{x(k)\} - zx(0) = A Z\{x(k)\} + B Z\{u(k)\}$$
$$Z\{y(k)\} = C Z\{x(k)\} + D Z\{u(k)\}$$

By denoting $Z\{x(k)\}$ as $x(z)$ and similarly $Z\{u(k)\}$ as $u(z)$, etc., this equation simplifies to

$$x(z) = (zI-A)^{-1}Bu(z) + z(zI-A)^{-1}x(0) \qquad (16.24)$$
$$y(z) = Cx(z) + Du(z)$$

Eliminating $x(z)$ produces

$$y(z) = [C(zI-A)^{-1}B + D]u(z) + Cz(zI-A)^{-1}x(0) \qquad (16.25)$$

The *transfer function matrix* thus takes the expected form

$$H(z) = C(zI-A)^{-1}B + D \qquad (16.26)$$

Intuitively, the transfer function matrix should be the Z-transform of the impulse response matrix. To see whether this intuition indeed holds, consider $Z\{H(k)\}$:

$$Z\{H(k)\} = \sum_{k=0}^{\infty} H(k)z^{-k} = D + \sum_{k=1}^{\infty} CA^{k-1}Bz^{-k} = D + z^{-1} \sum_{k=0}^{\infty} CA^k Bz^{-k}$$

For convenience, suppose A is nondefective and $A = T\Lambda T^{-1}$, where $\Lambda = \text{diag}[\lambda_1, \ldots, \lambda_n]$. Then

$$Z\{H(k)\} = D + z^{-1} CT \sum_{k=0}^{\infty} \Lambda^k z^{-k} T^{-1} B$$

$$= D + z^{-1} CT [I - z^{-1} \Lambda]^{-1} T^{-1} B$$

with region of convergence $|z| > \max_i |\lambda_i|$, $i = 1, \ldots, n$. Simplifying then produces

$$Z\{H(k)\} = C[zI - A]^{-1} B + D$$

This coincides with $H(z)$, as intuitively expected. A similar derivation applies to defective A-matrices where one factors the A-matrix into the Jordan form [1,6]. Accordingly, one concludes that *the Z-transform of the impulse response matrix coincides with the transfer function matrix.*

For one-sided sequences, such as the impulse response matrix sequence, the unilateral Z-transform generates the unique representation $Z^{-1}\{H(z)\} = \{H(k)\}$, for $k \geq 0$, for the inverse Z-transform of $H(z)$. This leads to the following theorem.

Theorem 16.2. Let $H(k)$ be the impulse response matrix (sequence) of equation 16.19 and $H(z)$ the transfer function matrix of equation 16.26. Then $Z\{H(k)\} = H(z)$, and conversely, $\{H(k)\} = Z^{-1}\{H(z)\}$.

EXAMPLE 16.3

Compute the transfer function matrix, the impulse response matrix from the transfer function matrix, and the response of the following system when $u(k) = 1^+(k)$ and $x(0) = \theta$:

$$x(k+1) = \begin{bmatrix} \lambda & 0 & 0 \\ 1 & \lambda & 0 \\ 0 & 1 & \lambda \end{bmatrix} x(k) + \begin{bmatrix} 1 \\ 0 \\ 0 \end{bmatrix} u(k)$$

$$y(k) = \begin{bmatrix} 1 & 0 & 0 \\ 0 & 1 & 0 \end{bmatrix} x(k)$$

Step 1. Compute $(zI - A)^{-1}$:

$$(zI - A)^{-1} = \begin{bmatrix} z-\lambda & 0 & 0 \\ -1 & z-\lambda & 0 \\ 0 & -1 & z-\lambda \end{bmatrix}^{-1} = (z-\lambda)^{-3} \begin{bmatrix} (z-\lambda)^2 & 0 & 0 \\ (z-\lambda) & (z-\lambda)^2 & 0 \\ 1 & (z-\lambda) & (z-\lambda)^2 \end{bmatrix}$$

Step 2. Evaluate $C(zI - A)^{-1}B$. Since $D = [0]$,

$$H(z) = C(zI-A)^{-1}B = \begin{bmatrix} (z-\lambda)^{-1} \\ (z-\lambda)^{-2} \end{bmatrix} = (z-\lambda)^{-2} \begin{bmatrix} z-\lambda \\ 1 \end{bmatrix}$$

Step 3. A partial fraction expansion of $H(z)/z$ produces

$$\frac{H(z)}{z} = \frac{\begin{bmatrix} -\lambda^{-1} \\ \lambda^{-2} \end{bmatrix}}{z} + \frac{\begin{bmatrix} \lambda^{-1} \\ -\lambda^{-2} \end{bmatrix}}{z-\lambda} + \frac{\begin{bmatrix} 0 \\ \lambda^{-1} \end{bmatrix}}{(z-\lambda)^2}$$

whereupon

$$H(z) = \begin{bmatrix} -\lambda^{-1} \\ \lambda^{-2} \end{bmatrix} + \frac{z}{z-\lambda}\begin{bmatrix} \lambda^{-1} \\ -\lambda^{-2} \end{bmatrix} + \frac{z}{(z-\lambda)^2}\begin{bmatrix} 0 \\ \lambda^{-1} \end{bmatrix}$$

Step 4. Taking inverse z-transforms yields the impulse response matrix

$$H(k) = \begin{bmatrix} -\lambda^{-1} \\ \lambda^{-2} \end{bmatrix}\delta(k) + \begin{bmatrix} \lambda^{k-1} \\ -\lambda^{k-2} \end{bmatrix}1^{+}(k) + \begin{bmatrix} 0 \\ k\,\lambda^{k-2} \end{bmatrix}1^{+}(k)$$

Step 5. Since $z\{1^{+}(k)\} = z/(z-1) = u(z)$, from step 4 it follows that

$$y(z) = H(z)u(z) = \frac{z}{(z-\lambda)^2(z-1)}\begin{bmatrix} z-\lambda \\ 1 \end{bmatrix}$$

Expanding $Y(z)/z$ by partial fractions produces

$$\frac{y(z)}{z} = \frac{\begin{bmatrix} (1-\lambda)^{-1} \\ (1-\lambda)^2 \end{bmatrix}}{z-1} + \frac{\begin{bmatrix} -(1-\lambda)^{-1} \\ -(1-\lambda)^{-2} \end{bmatrix}}{z-\lambda} + \frac{\begin{bmatrix} 0 \\ -(1-\lambda)^{-1} \end{bmatrix}}{(z-\lambda)^2}$$

Step 6. Taking the inverse Z-transform of $y(z)$ produces the response sequence

$$y(k) = \begin{bmatrix} (1-\lambda)^{-1} \\ (1-\lambda)^{-2} \end{bmatrix}1^{+}(k) - \begin{bmatrix} (1-\lambda)^{-1} \\ (1-\lambda)^{-2} \end{bmatrix}\lambda^{k}1^{+}(k) - \begin{bmatrix} 0 \\ (1-\lambda)^{-1} \end{bmatrix}k\,\lambda^{k-1}1^{+}(k)$$

For discussions of the related topics of discrete-time (digital) control theory and design, the reader is directed to [6,7].

PROBLEMS

1. Prove proposition 16.2 by applying equation 16.1 to $A(k)\Phi(k,j)$.

2. Prove proposition 16.3.

3. Prove proposition 16.4.

4. Find $x(10)$ for the system $x(k+1) = A(k)x(k)$, where $x(0) = [1\ \ 1]^{t}$ and the 2×2 matrix $A(k)$ is

(a)
$$\begin{bmatrix} 0 & \cos(k) \\ 0 & 0 \end{bmatrix}$$

(c)
$$\begin{bmatrix} \cos(2\pi k) & 1 \\ 0 & 1 \end{bmatrix}$$

(b)
$$\begin{bmatrix} a & 0 \\ 0 & b \end{bmatrix}$$

5. Which of the systems defined in problem 4 have invertible state transition matrices? Assume a and b in part (b) are nonzero.

6. Show that if $A(k)$ is nonsingular for all k and $x(k+1) = A(k)x(k)$ with $x(0) \neq \theta$, then for every k, no term in the sequence $\{x(k)\}$ is zero.

7. Prove that an arbitrary linear combination of fundamental solutions of $x(k+1) = A(k)x(k)$ is a solution as well.

8. Prove that if $\{x(k) \mid k = 0,1,2,...\}$ is a solution of $x(k+1) = A(k)x(k)$, then $\{x(k)\}$ is a linear combination of the set of fundamental solutions denoted by $\{x^i(k) = \eta_i \mid k \geqslant 0\}$ for $i = 1, ... , n$.

9. Two fundamental sequences of the second-order state dynamical equation $x(k+1) = A(k)x(k)$ are given by $\{x^1(k)\} = \{[1 \ 0]^t\}$ and

$$\{x^2(k)\} = \left\{ \begin{bmatrix} 0 \\ 1 \end{bmatrix}, \begin{bmatrix} 1 \\ 1 \end{bmatrix}, \begin{bmatrix} 3 \\ 1 \end{bmatrix}, \begin{bmatrix} 6 \\ 1 \end{bmatrix}, \begin{bmatrix} 10 \\ 1 \end{bmatrix}, ... \right\}$$

(a) Find $A(k)$.

(b) Find the associated $\Psi(k)$.

(c) Find the state transition matrix $\Phi(k,0)$.

10. Consider the discrete-time state model

$$x(k+1) = \begin{bmatrix} -0.1 & 0.6 \\ 0.6 & -0.1 \end{bmatrix} x(k) + \begin{bmatrix} 1 \\ -1 \end{bmatrix} u(k)$$

$$y(k) = [1 \ -1] x(k)$$

(a) Find $x(6)$ when $x(0) = [8 \ 8]^t$ and $u(k) = 0$ for all k.

(b) Find $x(6)$ when $x(0) = [0 \ 0]^t$ and $u(k) = (1.4)^k$ for $k \geqslant 0$ and 0 otherwise.

(c) Compute $y(6)$.

11. Recall that the zero-input state response of the time-invariant discrete-time state model

$$x(k+1) = Ax(k) + Bu(k), \quad x(0) = x_0$$

is given by $x(k) = \Phi(k)x_0$, where $\Phi(k) = A^k$ is the discrete-time state transition matrix. Derive a meaningful expression for A^k when, assuming a nonzero b,

$$A = \begin{bmatrix} a & b \\ -b & a \end{bmatrix}$$

is 2×2 and has complex eigenvalues.

(a) What are the eigenvalues λ_1 and λ_2 of A?

(b) What are the right eigenvectors of A?

(c) Derive the form of a transformation T with the property that $A = T\Lambda T^{-1}$, where $\Lambda = \text{diag}[\lambda_1,\lambda_2]$.

(d) Using the result of part (c), derive an expression for A^k which is real. (*Hint*: Express λ_i^k in polar form and use the fact that $\exp(j\theta) = \cos(\theta) + j\sin(\theta)$.)

(e) If $\hat{A}$ is an arbitrary 2×2 matrix with complex eigenvalues $\hat{\lambda}_1 = a + jb$ and $\hat{\lambda}_2 = a - jb$ and with complex eigenvectors $\hat{e}_1$ and $\hat{e}_2$, derive an expression for A^k based on your answer to part (d).

12. Assuming $\lambda \neq 0$, use the binomial expansion to compute an expression for the following matrices:

(a)
$$\begin{bmatrix} \lambda & 0 & 0 \\ 1 & \lambda & 0 \\ 0 & 1 & \lambda \end{bmatrix}$$

(b)
$$\begin{bmatrix} \lambda & 0 & 0 \\ 1 & \lambda & 0 \\ 1 & 1 & \lambda \end{bmatrix}$$

(c)
$$\begin{bmatrix} \lambda & 0 & 0 & 0 \\ 1 & \lambda & 0 & 0 \\ 0 & 0 & \lambda & -1 \\ 0 & 0 & 0 & \lambda \end{bmatrix}$$

13. Consider the discrete-time system model

$$x(k+1) = \begin{bmatrix} 0.5 & 0 \\ 1 & 0.5 \end{bmatrix} x(k) + \begin{bmatrix} 1 \\ 0 \end{bmatrix} u(k)$$

$$y(k) = [0 \quad 1] x(k)$$

(a) From memory, what is the formula for the impulse response of a discrete-time state model?
(b) Compute a closed-form expression for the impulse response of the given model.
(c) Compute $H(4)$, i.e., the impulse response evaluated at $k = 4$.
(d) Compute the frequency response function $H(e^{j\theta})$ of the model.
(e) Find the magnitude of $H(e^{j\theta})$ when $\theta = \pi/2$ and $\theta = 5\pi/2$.

14. Suppose the impulse response of a certain single-input single-output third-order DTSM is $\{0,0,0,1,0,1,0,...\}$. Find a realization $\{A,B,C\}$ of the associated DTSM such that A, B, and C are in the controllable canonical form. Then compute the transfer function matrix $H(z)$.

15. Recall the continuous-time state dynamical equation $\dot{x}(t) = Ax(t) + Bu(t)$, which has the analytic solution given by equation 13.10.
(a) Use equation 13.12 to find matrices $\hat{A}$ and $\hat{B}$ such that

$$x((k+1)T) = \hat{A} x(kT) + \hat{B}u(kT)$$

assuming that $u(t) = u(kT)$ for $kT \leqslant t < (k+1)T$.
(b) Apply the formula of part (a) to the continuous-time state model

$$\begin{bmatrix} \dot{x}_1 \\ \dot{x}_2 \end{bmatrix} = \begin{bmatrix} -1 & 0 \\ -1 & -2 \end{bmatrix} \begin{bmatrix} x_1 \\ x_2 \end{bmatrix} + \begin{bmatrix} 1 \\ 1 \end{bmatrix} u$$

Solve the resulting DTSM for $x(kT)$ (i.e., find a closed-form expression for $x(kT)$), assuming that $x(0) = [1 \quad 1]^t$ and $u(kT) = \exp(-kT)$.

16. For a discrete-time state model in which

$$A = \begin{bmatrix} \lambda & 0 & 0 & 0 \\ 1 & \lambda & 0 & 0 \\ 0 & 0 & \lambda & -1 \\ 0 & 0 & 0 & \lambda \end{bmatrix}, \quad B = \begin{bmatrix} 1 & 0 \\ 0 & 0 \\ 0 & 1 \\ 0 & 0 \end{bmatrix}, \quad C = \begin{bmatrix} 0 & 1 & 0 & 0 \\ 0 & 0 & 0 & 1 \end{bmatrix}$$

(a) Find $(zI - A)^{-1}$.

(b) Find the transfer function matrix $H(z)$.

(c) Execute an appropriate partial fraction expansion and take the inverse Z-transform to obtain the impulse response matrix $H(k)$.

(d) If the input sequence is $u(k) = [1 \quad 1]^t 1^+(k)$, find the response sequence $\{y(k)\}$ using Z-transform methods. Assume $x(0) = \theta$.

17. Repeat problem 16 for the system

$$A = \begin{bmatrix} 1 & 1 \\ -1 & 1 \end{bmatrix}, \quad B = \begin{bmatrix} 1 & 0 \\ 0 & 1 \end{bmatrix}, \quad C = [1 \quad 0]$$

18. Repeat problem 10 using Z-transform methods.

19. Consider the discrete-time state model

$$x(k+1) = \begin{bmatrix} 0 & 1 \\ 0 & 0 \end{bmatrix} x(k) + \begin{bmatrix} 1 & -1 \\ -1 & -1 \end{bmatrix} u(k), x(0) = \begin{bmatrix} -1 \\ -1 \end{bmatrix}$$

Find a general expression for $x(k)$ given that $u(k) = [(-1)^k \quad -(-1)^k]^t$.

20. Given the discrete-time matrix equation

$$x(k+2) = A[x(k+1) + x(k)] + Bu(k)$$

$$y(k) = Cx(k)$$

Derive the z-domain transfer function matrix and make provision for initial conditions $x(0)$ and $x(1)$.

21. A discrete-time system having a state model is completely controllable if and only if, for any final state x_1 and any initial state x_0, there exists an input sequence which will drive x_0 to x_1 in a finite number of steps. Prove or give a counterexample to the statement, "A discrete-time system is completely controllable if and only if every state x_0 can be driven to the zero state in a finite number of steps."

22. The probability that a hypothetical computer will learn a certain hypothetical five-move chess play after seeing it k times is given by the difference equation

$$p(k+2) + 0.3p(k+1) + 0.02p(k) = 1^+(k)$$

for $k \geqslant 0$, where $p(0) = 0.858$ and $p(1) = 0.508$. Recall that $1^+(k) = 1$ for all $k \geqslant 0$ and zero otherwise.

(a) Using Z-transform techniques, solve for $p(k)$.

(b) Write the given difference equation in controllable canonical form.

(c) Investigate controllability, observability, and stability.

23. Show that a sequence $\{u(k)\}$ has a reduced rational Z-transform of degree n of the form

$$u(z) = \frac{b_0 z^n + b_1 z^{n-1} + b_2 z^{n-2} + \dots + b_n}{z^n + a_1 z^{n-1} + a_2 z^{n-2} + \dots + a_n}$$

if and only if $u(k+n) + a_1 u(k+n-1) + \dots + a_n u(k) = 0$ for all $k > 0$. (See [1].)

REFERENCES

1. Alan V. Oppenheim and Alan S. Willsky, *Signals and Systems* (Englewood Cliffs, NJ: Prentice Hall, 1983).

2. Robert A. Gabel and Richard A. Roberts, *Signals and Linear Systems* (New York: Wiley, 1980).

3. Clare D. McGillem and George R. Cooper, *Continuous and Discrete Signal and System Analysis* (New York: Holt, Rinehart and Winston, 1974).

4. David Luenberger, *Introduction to Dynamic Systems: Theory, Models, and Applications* (New York: Wiley, 1979).

5. L. A. Zadeh and C. A. Desoer, *Linear System Theory: The State Space Approach* (New York: McGraw-Hill, 1963).

6. Karl J. Astrom and Bjorn Wittenmark, *Computer Controlled Systems: Theory and Design* (Englewood Cliffs, NJ: Prentice Hall, 1984).

7. Gene F. Franklin and J. D. Powell, *Digital Control of Dynamic Systems* (Reading, MA: Addison-Wesley, 1980).

17

DECOMPOSITION
OF THE TIME-INVARIANT
STATE TRANSITION MATRIX

INTRODUCTION

This chapter applies and expands the eigenvalue-eigenvector development begun in the previous chapter to construct a decomposition of the state transition matrix as a sum of matrix directed modes. These matrix or subspace directions are invariant under multiplication by A. In many ways they reflect the essential nature of the state model dynamics and hence of the actual physical system. In particular, *if the $n \times n$ matrix A has distinct eigenvalues $\lambda_1, ..., \lambda_n$*, the desired decomposition is

$$\Phi(t) = e^{At} = e^{\lambda_1 t} R_1 + e^{\lambda_2 t} R_2 + \cdots + e^{\lambda_n t} R_n \qquad (17.1)$$

where each $n \times n$ matrix R_i takes the form

$$R_i = \left[\xi_{i1} e_i \mid \xi_{i2} e_i \mid \cdots \mid \xi_{in} e_i \right] = e_i \left[\xi_{i1} \mid \xi_{i2} \mid \cdots \mid \xi_{in} \right] \qquad (17.2)$$

for appropriate (possibly complex) scalars $\xi_{ij}, i, j = 1, ..., n$. Each column then is simply a (possibly zero) scalar ξ_{ij} times the ith eigenvector e_i. Since $A e_i = \lambda_i e_i$, it follows that $A R_i = \lambda_i R_i$, meaning that the matrix R_i is invariant under multiplication by A except for a possible scaling or rotation induced by the possibly complex number λ_i.

A harvest of spinoffs grows out of this decomposition. For example, the state trajectory also has an elegant decomposition into directed modes, i.e., eigenvalue oscillations in the direction of associated eigenvectors (see equation 17.5). The decomposition also provides a clear path for developing the notions of controllability, observability, and stability, as well as defining the avenues of interaction among the concepts.

To sketch these avenues of interaction, consider the structure of the system and state responses in the context of the decomposition 17.1. The state response takes the form

$$x(t) = \sum_{i=1}^{n} e^{\lambda_i t} R_i x_0 + \sum_{i=1}^{n} \int_{0}^{t} e^{\lambda_i (t-\tau)} R_i B u(\tau) d\tau \qquad (17.3)$$

As will be seen in chapter 20, the controllability of a certain oscillation or mode, say λ_i, depends on whether or not the product $R_i B$ is zero. If the product is not zero, the mode can be excited by the input. This information affects the *bounded-ness* and hence the *stability* of the state trajectory (see chapters 22 and 24). Also, if $R_i B$ is not zero, it is always possible to construct a state feedback to move this eigenvalue to some other location. Chapter 20 considers these matters.

With regard to the complete system response, the decomposition of equation 17.1 implies

$$y(t) = \sum_{i=1}^{n} e^{\lambda_i t} CR_i x_0 + \sum_{i=1}^{n} \int_{0}^{t} e^{\lambda_i (t-\tau)} CR_i B u(\tau) d\tau \qquad (17.4)$$

This form provides the framework for investigations of system observability through the matrix products CR_i, $i = 1, \ldots, n$. Also, the nature of input-output stability becomes clearer in the context of the matrix products, CR_i and $CR_i B$, $i = 1,\ldots,n$.

DECOMPOSITION OF THE ZERO-INPUT STATE RESPONSE

To better understand the preceding decomposition, we first decompose the zero-input state response into a sum of directed modes. The decomposition takes the form

$$x(t) = e^{A(t-t_0)} x_0 = \xi_1(t_0) e^{\lambda_1 (t-t_0)} e_1 + \cdots + \xi_n(t_0) e^{\lambda_n (t-t_0)} e_n \qquad (17.5)$$

whenever A has distinct eigenvalues. Constructing this decomposition entails two essential steps. First, since $x_0 \in \mathbf{R}^n$ is arbitrary, at $t = t_0$, x_0 must be a (possibly complex) linear combination of (possibly complex) eigenvectors. Hence, the set of eigenvectors must be a basis for $\mathbf{R}^n$ over the field of complex scalars $\mathbf{C}$. The second step involves demonstrating that $e^{At} e_i = e^{\lambda_i t} e_i$, in which case equation 17.5 follows.

The following lemma aids the development.

Lemma 17.1.   Let A have distinct eigenvalues $\{\lambda_1,\ldots,\lambda_n\}$ and corresponding right eigenvectors $\{e_1,\ldots,e_n\}$. Then

$$(A - \lambda_2 I)(A - \lambda_3 I) \cdots (A - \lambda_n I)\alpha_i e_i = \theta \qquad (17.6)$$

for all $i \neq 1$ and for all (possibly complex) scalars α_i.

Proof.   Without loss of generality, assume that $\alpha_i \neq 0$.

Case 1. For pedagogical reasons, let $i = n$. Equation 17.6 then becomes

$$(A - \lambda_2 I) \; \cdots \; (A - \lambda_n I) \alpha_n e_n = \alpha_n (A - \lambda_2 I) \; \cdots \; (A e_n - \lambda_n e_n)$$

By definition, $A e_n - \lambda_n e_n = \theta$; hence, if $i = n$, equation 17.6 holds.

Case 2. Suppose $i \neq n$. Then, for any $j \neq i$, the usual eigenvalue-eigenvector relationship implies that $(A - \lambda_j I) \alpha_i e_i = \alpha_i (A e_i - \lambda_j e_i) = \alpha_i (\lambda_i - \lambda_j) e_i$. Hence,

$$(A - \lambda_2 I) \; \cdots \; (A - \lambda_n I) \alpha_i e_i =$$

$$\alpha_i (\lambda_i - \lambda_n) \; \cdots \; (\lambda_i - \lambda_{i+1})(A - \lambda_2 I) \; \cdots \; (A - \lambda_i I) e_i$$

But this expression equals θ, since $(A - \lambda_i I) e_i = \theta$. Consequently, equation 17.6 results. ∎

Theorem 17.1.   Let A have distinct eigenvalues $\{\lambda_1,\ldots,\lambda_n\}$. Then any collection of associated right (or left) eigenvectors is a basis for $\mathbf{C}^n$ over the field $\mathbf{C}$ of complex scalars. Also, since $\mathbf{R}^n \subset \mathbf{C}^n$, such a collection is a basis for $\mathbf{R}^n$.

Proof.   Let $\{e_1, e_2, \ldots, e_n\}$ designate an arbitrary set of right eigenvectors of A corresponding to the eigenvalues $\{\lambda_1, \lambda_2, \ldots, \lambda_n\}$. Since there are n eigenvectors, and since $\mathbf{C}^n$ is n-dimensional, it suffices to demonstrate the linear independence of the set $\{e_1, \ldots, e_n\}$. The proof proceeds by contradiction: we assume that the set is linearly dependent and deduce that at least one eigenvector is zero. But a zero eigenvector contradicts the definition of an eigenvector and hence the assumption of linear dependence of the set must be false. More formally, if the set $\{e_1, \ldots, e_n\}$ is linearly dependent, then there exist complex scalars $\alpha_1, \ldots, \alpha_n$ not all zero such that

$$\alpha_1 e_1 + \; \cdots \; + \alpha_n e_n = \theta \tag{17.7}$$

Without loss of generality, assume that $\alpha_1 \neq 0$. Then, multiplying both sides of equation 17.7 by $[(A - \lambda_2 I)(A - \lambda_3 I) \cdots (A - \lambda_n I)]$ results in

$$\alpha_1 (A - \lambda_2 I)(A - \lambda_3 I) \; \cdots \; (A - \lambda_n I) e_1 = \theta$$

by Lemma 17.1. Using methods analogous to the proof of that lemma, this further reduces to

$$\alpha_1 (\lambda_1 - \lambda_n)(\lambda_1 - \lambda_{n-1}) \; \cdots \; (\lambda_1 - \lambda_2) e_1 = \theta \tag{17.8}$$

Since $\alpha_1 \neq 0$, and since the eigenvalues are distinct, the eigenvector e_1 must equal zero, a contradiction. ∎

An immediate corollary of Theorem 17.1, justifying the use of the state transformations of the previous chapter, is the following.

Corollary 17.1.   The transformation $T = [e_1, \ldots, e_n]$ is nonsingular whenever $\{\lambda_1, \ldots, \lambda_n\}$ are distinct.

Theorem 17.2. Let A have distinct eigenvalues $\{\lambda_1,...,\lambda_n\}$ and right eigenvectors $\{e_1,...,e_n\}$. Then

$$e^{At}e_i = e^{\lambda_i t}e_i \tag{17.9}$$

Proof. From the Taylor series definition of the matrix exponential,

$$e^{At}e_i = [I + At + \frac{A^2 t^2}{2!} + \cdots + \frac{A^k t^k}{k!} + \cdots]e_i \tag{17.10}$$

Observe that $A^k e_i = A^{k-1}(Ae_i) = \lambda_i A^{k-1} e_i$. Continuing this process leads to $A^k e_i = \lambda_i^k e_i$. Hence, from equation 17.10, we obtain $e^{At}e_i = e^{\lambda_i t}e_i$. ∎

Theorem 17.2 says that if the eigenvalues of A are $\lambda_1,...,\lambda_n$, then the eigenvalues of $\exp(At)$ are $\exp(\lambda_i t), i = 1,...,n$, for all t. Also, the associated eigenvectors coincide—i.e., they are the same for both A and $\exp(At)$.

In view of theorem 17.1, for any initial condition $x(t_0) = x_0$, there exist (possibly complex) scalars $\xi_1(t_0), \xi_2(t_0)$, dots , $\xi_n(t_0)$ such that

$$x_0 = \xi_1(t_0)e_1 + \xi_2(t_0)e_2 + \cdots + \xi_n(t_0)e_n$$

Using theorem 17.2, the resulting zero-input state response becomes

$$x(t) = e^{A(t-t_0)}x_0 = \xi_1(t_0)e^{\lambda_1(t-t_0)}e_1 + \cdots + \xi_n(t_0)e^{\lambda_n(t-t_0)}e_n \tag{17.11}$$

The zero-input state response has thus been decomposed into a sum of *directed modes* $e^{\lambda_i(t-t_0)}e_i$. Figure 17.1 provides a two-dimensional illustration of the decomposition of

$$\begin{bmatrix} x_1(t) \\ x_2(t) \end{bmatrix} = e^{-t}\begin{bmatrix} 1 \\ 0 \end{bmatrix} + 2e^{-2t}\begin{bmatrix} 0 \\ 1 \end{bmatrix} \tag{17.12}$$

for $t \geqslant 0$.

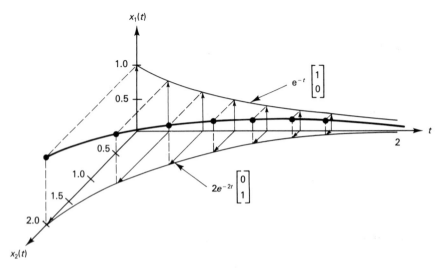

Figure 17.1 Zero-input state response as per equation 17.12.

EXAMPLE 17.1.

Suppose

$$A = \begin{bmatrix} a & b \\ -b & a \end{bmatrix}$$

which has eigenvalues $\lambda_1 = a + jb$ and $\lambda_2 = a - jb$. The corresponding eigenvectors are $e_1 = [1 \ j]^t$ and $e_2 = [1 \ -j]^t$. According to the preceding development,

$$x(t) = e^{At}x_0 = \xi_1 e^{\lambda_1 t}e_1 + \xi_2 e^{\lambda_2 t}e_2 \tag{17.13}$$

where t_0 has been set to zero and explicit t_0-dependencies dropped. Note that ξ_1 and ξ_2 are chosen such that $x_0 = \xi_1 e_1 + \xi_2 e_2$, or more specifically,

$$\begin{bmatrix} \xi_1 \\ \xi_2 \end{bmatrix} = \begin{bmatrix} e_1 & | & e_2 \end{bmatrix}^{-1} x_0$$

Also, since x_0 is a real vector and since $e_2 = \bar{e}_1$, $\xi_1 = \bar{\xi}_2$.

Since ξ_i, e_i, and λ_i are generally not real, the decomposition of equation 17.13 lacks obvious physical meaning. To obtain a more physically meaningful expression, let $\phi_1 = \text{Re}[e_1]$, $\phi_2 = \text{Im}[e_1]$, $\rho_1 = \text{Re}[\xi_1]$, and $\rho_2 = \text{Im}[\xi_1]$. Then $e_1 = \phi_1 + j\phi_2$, $e_2 = \phi_1 - j\phi_2$, $\xi_1 = \rho_1 + j\rho_2$, and $\xi_2 = \rho_1 - j\rho_2$. Substituting these expressions into equation 17.13 and simplifying then produces the more physically meaningful expression

$$x(t) = 2e^{at}[\rho_1\cos(bt) - \rho_2\sin(bt)]\phi_1 \tag{17.14}$$
$$- 2e^{at}[\rho_1\sin(bt) + \rho_2\cos(bt)]\phi_2$$

This form of the state trajectory mirrors the sinusoidal nature intuitively expected.

EXAMPLE 17.2

Decompose $e^{At}x_0$ into directed modes when

$$A = \begin{bmatrix} a & 1 \\ 0 & b \end{bmatrix}, \quad x(0) = \begin{bmatrix} 2 \\ 3 \end{bmatrix}$$

assuming that $a \neq b$.

Step 1. The eigenvalues of A are $\lambda_1 = a$ and $\lambda_2 = b$.

Step 2. The associated eigenvectors are

$$e_1 = \begin{bmatrix} 1 \\ 0 \end{bmatrix}$$

and

$$e_2 = \begin{bmatrix} 1 \\ b-a \end{bmatrix}$$

Step 3. Express $x(0)$ as a linear combination of the eigenvectors, i.e., find $\xi_1(0)$ and $\xi_2(0)$. Using the above expressions

$$\begin{bmatrix} 2 \\ 3 \end{bmatrix} = \begin{bmatrix} e_1 & | & e_2 \end{bmatrix} \begin{bmatrix} \xi_1(0) \\ \xi_2(0) \end{bmatrix} = \begin{bmatrix} 1 & 1 \\ 0 & b-a \end{bmatrix} \begin{bmatrix} \xi_1(0) \\ \xi_2(0) \end{bmatrix}$$

Solving then produces

$$\begin{bmatrix} \xi_1(0) \\ \xi_2(0) \end{bmatrix} = \begin{bmatrix} 1 & \dfrac{-1}{b-a} \\ 0 & \dfrac{1}{b-a} \end{bmatrix} \begin{bmatrix} 2 \\ 3 \end{bmatrix} = \begin{bmatrix} 2 - \dfrac{3}{b-a} \\ \dfrac{3}{b-a} \end{bmatrix}$$

Step 4. The resulting decomposition of the state trajectory is

$$x(t) = e^{At} \begin{bmatrix} 2 \\ 3 \end{bmatrix} = \left(2 - \frac{3}{b-a}\right) e^{at} \begin{bmatrix} 1 \\ 0 \end{bmatrix} + \left(\frac{3}{b-a}\right) e^{bt} \begin{bmatrix} 1 \\ b-a \end{bmatrix}$$

DECOMPOSITION OF THE STATE TRANSITION MATRIX

The decomposition 17.1 of the state transition matrix builds along lines similar to those set out in the previous section. At the initial condition $t = 0$, equation 17.1 must reduce to

$$I = R_1 + R_2 + \cdots + R_n$$

Given that each R_i has the form of equation 17.2, is it possible to find scalars ξ_{ij} to guarantee equality? Certainly, since the eigenvectors are a basis for $\mathbf{R}^n$, it is always possible to find ξ_{ij} $i = 1,\dots,n$, such that the jth column, η_j, of I, has the form

$$\eta_j = \xi_{1j}e_1 + \xi_{2j}e_2 + \cdots + \xi_{nj}e_n = \begin{bmatrix} e_1 & | & e_2 & | & \cdots & | & e_n \end{bmatrix} \begin{bmatrix} \xi_{1j} \\ \xi_{2j} \\ \vdots \\ \xi_{nj} \end{bmatrix}$$

for $j = 1,\dots,n$. Thus, the identity matrix has the form

$$I = \begin{bmatrix} e_1 & | & e_2 & | & \cdots & | & e_n \end{bmatrix} \begin{bmatrix} \xi_{11} & \xi_{12} & \cdots & \xi_{1n} \\ \xi_{21} & \xi_{22} & \cdots & \xi_{2n} \\ \vdots & \vdots & & \vdots \\ \xi_{n1} & \xi_{n2} & \cdots & \xi_{nn} \end{bmatrix}$$

in which case if $T = \begin{bmatrix} e_1 & | & e_2 & | & \cdots & | & e_n \end{bmatrix}$, then the scalars ξ_{ij} are given by

$$T^{-1} = \begin{bmatrix} \xi_{11} & \xi_{12} & \cdots & \xi_{1n} \\ \xi_{21} & \xi_{22} & \cdots & \xi_{2n} \\ \vdots & \vdots & & \vdots \\ \xi_{n1} & \xi_{n2} & \cdots & \xi_{nn} \end{bmatrix} = \begin{bmatrix} w_1^* \\ w_2^* \\ \vdots \\ w_n^* \end{bmatrix} \qquad (17.15)$$

Again, T^{-1} exists, since the linear independence of the eigenvectors guarantees the nonsingularity of T. Recall that the rows of T^{-1} are the complex conjugate transpose of the *left* eigenvectors w_i of A. This observation proves very useful in the forthcoming eigenvalue-eigenvector stability characterizations. At any rate,

$$\Phi(t) = e^{At} = e^{At}I = e^{At}R_1 + e^{At}R_2 + \cdots + e^{At}R_n \qquad (17.16)$$

Using the result of theorem 17.2 and the definition of R_i yields

$$\Phi(t) = e^{At} = e^{\lambda_1 t}R_1 + e^{\lambda_2 t}R_2 + \cdots + e^{\lambda_n t}R_n$$

whenever A has distinct eigenvalues.

EXAMPLE 17.3

Let us consider a decomposition of the state transition matrix for the matrix

$$A = \begin{bmatrix} a & b \\ -b & a \end{bmatrix}$$

From the foregoing discussion,

$$\Phi(t) = e^{At} = e^{\lambda_1 t}R_1 + e^{\lambda_2 t}R_2$$

$$= e^{at}\left[e^{jbt}R_1 + e^{-jbt}R_2\right] = e^{at}\left[e^{jtb}R_1 + e^{-jbt}\overline{R}_1\right]$$

After some manipulation, this becomes

$$\Phi(t) = 2e^{at}\cos(bt)\,\mathrm{Re}[R_1] - 2e^{at}\sin(bt)\,\mathrm{Im}[R_1]$$

In order to tie the decomposition of the state transition matrix together with the solution of the time-invariant state model and to glimpse the notions of unobservability and uncontrollability to come, consider the following example.

EXAMPLE 17.4

Suppose the matrices A, B, C, and D of the usual state model have structures

$$A = \begin{bmatrix} 1 & 0 & 1 \\ 0 & 1 & -1 \\ 1 & 1 & 1 \end{bmatrix} \begin{bmatrix} -1 & 0 & 0 \\ 0 & 0 & 0 \\ 0 & 0 & 1 \end{bmatrix} \begin{bmatrix} 2 & 1 & -1 \\ -1 & 0 & 1 \\ -1 & -1 & 1 \end{bmatrix}, \qquad (17.17)$$

$$B = \begin{bmatrix} 1 & 1 \\ 0 & -1 \\ 1 & 1 \end{bmatrix}, \quad C = [1 \ \ 0 \ \ -1], \quad D = [0]$$

Part 1. Compute the state transition matrix. By inspection,

$$\Phi(t) = \begin{bmatrix} 1 & 0 & 1 \\ 0 & 1 & -1 \\ 1 & 1 & 1 \end{bmatrix} \begin{bmatrix} e^{-t} & 0 & 0 \\ 0 & 1 & 0 \\ 0 & 0 & e^{t} \end{bmatrix} \begin{bmatrix} 2 & 1 & -1 \\ -1 & 0 & 1 \\ -1 & -1 & 1 \end{bmatrix}$$

Part 2. Decompose $\Phi(t)$ into matrix directed modes. From equation 17.17 and the fact that $R_i = e_i w_i^*$, we obtain

$$R_1 = \begin{bmatrix} 1 \\ 0 \\ 1 \end{bmatrix} [2 \ \ 1 \ \ -1] = \begin{bmatrix} 2 & 1 & -1 \\ 0 & 0 & 0 \\ 2 & 1 & -1 \end{bmatrix}$$

$$R_2 = \begin{bmatrix} 0 \\ 1 \\ 1 \end{bmatrix} [-1 \ \ 0 \ \ 1] = \begin{bmatrix} 0 & 0 & 0 \\ -1 & 0 & 1 \\ -1 & 0 & 1 \end{bmatrix}$$

$$R_3 = \begin{bmatrix} 1 \\ -1 \\ 1 \end{bmatrix} [-1 \ \ -1 \ \ 1] = \begin{bmatrix} -1 & -1 & 1 \\ 1 & 1 & -1 \\ -1 & -1 & 1 \end{bmatrix}$$

As a check, note that $R_1 + R_2 + R_3 = I$.

Part 3. Compute the state response when

$$x(0) = [0 \ \ 1 \ \ 1]' \text{ and } u(t) = [1 \ \ 1]' 1^+(t)$$

Recall equation 17.3:

$$x(t) = \sum_{i=1}^{3} e^{\lambda_i t} R_i x(0) + \sum_{i=1}^{3} \int_0^t e^{\lambda_i(t-\tau)} R_i B u(\tau) d\tau$$

To evaluate this formula, first consider $R_i x_0$:

$$R_1 x_0 = \begin{bmatrix} 0 \\ 0 \\ 0 \end{bmatrix}, \quad R_2 x_0 = \begin{bmatrix} 0 \\ 1 \\ 1 \end{bmatrix}, \quad R_3 x_0 = \begin{bmatrix} 0 \\ 0 \\ 0 \end{bmatrix} \tag{17.18}$$

Next, consider $R_i B$:

$$R_1 B = e_1 \left[w_1^* B \right] = \begin{bmatrix} 1 \\ 0 \\ 1 \end{bmatrix} [1 \ \ 0] = \begin{bmatrix} 1 & 0 \\ 0 & 0 \\ 1 & 0 \end{bmatrix} \tag{17.19a}$$

$$R_2 B = e_2 \left[w_2^* B \right] = \begin{bmatrix} 0 \\ 1 \\ 1 \end{bmatrix} [0 \ \ 0] = \begin{bmatrix} 0 & 0 \\ 0 & 0 \\ 0 & 0 \end{bmatrix} \tag{17.19b}$$

$$R_3B = e_3 \left[w_3^* B \right] = \begin{bmatrix} 1 \\ -1 \\ 1 \end{bmatrix} \begin{bmatrix} 0 & 1 \end{bmatrix} = \begin{bmatrix} 0 & 1 \\ 0 & -1 \\ 0 & 1 \end{bmatrix} \qquad (17.19c)$$

In view of equations 17.18 and 17.19, the state response for $t \geqslant 0$ takes the form

$$x(t) = \begin{bmatrix} 0 \\ 1 \\ 1 \end{bmatrix} + \begin{bmatrix} 1 \\ 0 \\ 1 \end{bmatrix} \left[\int_0^t e^{-(t-\tau)} d\tau \right] + \begin{bmatrix} 1 \\ -1 \\ 1 \end{bmatrix} \left[\int_0^t e^{(t-\tau)} d\tau \right]$$

$$= \begin{bmatrix} 0 \\ 1 \\ 1 \end{bmatrix} + (1 - e^{-t}) \begin{bmatrix} 1 \\ 0 \\ 1 \end{bmatrix} + (e^t - 1) \begin{bmatrix} 1 \\ -1 \\ 1 \end{bmatrix}$$

Part 4. Compute the complete system response when $x(0) = [1 \;\; -7 \;\; 3]^t$ and $u(t) = e^{-7t}\cos(\pi t) 1^+(t) [1 \;\; 1]^t$. First, recall equation 17.4:

$$y(t) = \sum_{i=1}^{3} e^{\lambda_i t} CR_i x(0) + \sum_{i=1}^{3} \int_0^t e^{\lambda_i(t-\tau)} CR_i Bu(\tau) d\tau$$

Then, knowing from equations 17.19 that

$$R_1B = \begin{bmatrix} 1 & 0 \\ 0 & 0 \\ 1 & 0 \end{bmatrix}, \quad R_2B = \begin{bmatrix} 0 & 0 \\ 0 & 0 \\ 0 & 0 \end{bmatrix}, \quad R_3B = \begin{bmatrix} 0 & 1 \\ 0 & -1 \\ 0 & 1 \end{bmatrix}$$

and that $C = [1 \;\; 0 \;\; -1]$, a straightforward calculation verifies that

$$CR_1B = [0 \;\; 0], \quad CR_2B = [0 \;\; 0], \quad CR_3B = [0 \;\; 0]$$

The complete response thus reduces to

$$y(t) = \sum_{i=1}^{3} e^{\lambda_i t} CR_i x(0)$$

which is independent of the input $u(t)$. Further, $CR_1 = CR_3 = [0 \;\; 0 \;\; 0]$ and $CR_2 = [1 \;\; 0 \;\; -1]$ implies that

$$y(t) = -2 \, 1^+(t)$$

where $1^+(t)$ indicates that our concern is for $t \geqslant 0$.

As a forethought of the controllability and observability discussions to come, observe that

(1) $CR_1 = CR_3 = [0 \;\; 0 \;\; 0]$ means that the modes containing the frequencies $\lambda_1 = -1$ and $\lambda_3 = 1$ are not observable at the output port.
(2) $R_2B = [0]$ (the zero matrix) means that the mode containing $\lambda_2 = 0, e_2$ is *not* excitable (i.e., is not controllable) from the input port.
(3) The fact that the input is decoupled from the output means that both the impulse response matrix and the transfer function matrix of the system are zero.

ANALOGIES WITH DISCRETE-TIME SYSTEMS

The preceding notions all carry over isomorphically (one to one) to the discrete-time state representation. For example, the discrete-time (time-invariant) state transition matrix has the decomposition

$$\Phi(k) = A^k = \sum_{i=1}^{n} \lambda_i^k R_i \qquad (17.20)$$

provided that A has distinct eigenvalues. Given this formula, the zero-input state response sequence is defined via

$$x(k) = A^k x(0) = \sum_{i=1}^{n} \lambda_i^k R_i x(0) \qquad (17.21)$$

Continuing the obvious analogy, the complete state response becomes

$$x(k) = \sum_{i=1}^{n} \lambda_i^k R_i x(0) + \sum_{i=1}^{n} \left[\sum_{j=0}^{k-1} \lambda_i^{k-1-j} R_i Bu(j) \right] \qquad (17.22)$$

Finally, the complete system response reduces to

$$y(k) = \sum_{i=1}^{n} \lambda_i^k CR_i x(0) + \sum_{i=1}^{n} \left[\sum_{j=0}^{k-1} \lambda_i^{k-1-j} CR_i Bu(j) \right] \qquad (17.23)$$

Theoretically (not numerically) speaking, it is interesting to note a somewhat different means for computing the matrices R_i in the decomposition of equation 17.20 than the eigenvector-based approach described earlier. Consider that

$$e^{At} = e^{\lambda_1 t} R_1 + \cdots + e^{\lambda_n t} R_n \qquad (17.24)$$

Taking the zeroth through $(n-1)$st derivative of equation 17.24 and evaluating at $t = 0$ yields

$$A^k = \lambda_1^k R_1 + \cdots + \lambda_n^k R_n$$

for $k = 0,\ldots,n-1$. Thus, it follows that

$$\left[R_1 \mid R_2 \mid \cdots \mid R_n \right] V = \left[I \mid A \mid \cdots \mid A^{n-1} \right] \qquad (17.25)$$

where

$$V = \begin{bmatrix} I & \lambda_1 I & \lambda_1^2 I & \cdots & \lambda_1^{n-1} I \\ I & \lambda_2 I & \lambda_2^2 I & \cdots & \lambda_2^{n-1} I \\ & & & & \\ & & & & \\ I & \lambda_n I & \lambda_n^2 I & \cdots & \lambda_n^{n-1} I \end{bmatrix} \qquad (17.26)$$

is a so-called block Vandermonde [1] matrix. Consequently,

$$\left[R_1 \mid R_2 \mid \cdots \mid R_n \right] = \left[I \mid A \mid \cdots \mid A^{n-1} \right] V^{-1} \qquad (17.27)$$

The entries of V^{-1} can be identified with the coefficients of certain Lagrange polynomials. To develop this equivalence, let $V = [v_{ij}]$, where $v_{ij} = \lambda_i^{j-1}$ and all the λ_i's are distinct. Then define the jth Lagrange polynomial as

$$p_j^n(\lambda) = \frac{\pi_A(\lambda)}{(\lambda - \lambda_j)\left[\dfrac{d}{d\lambda}\pi_A(\lambda)\right]_{\lambda = \lambda_j}} \tag{17.28}$$

where the characteristic polynomial $\pi_A(\lambda) = (\lambda - \lambda_1)...(\lambda - \lambda_n)$. By this means, it is possible to show [1] that the ijth entry of V^{-1}, say, v_{ij}^{-1}, is given by

$$v_{ij}^{-1} = \xi_{j,i-1}$$

where $\xi_{j,i-1}$ is the coefficient of λ_j^{i-1} in equation 17.28.

This development follows [2], in which the motivation is somewhat different from here. The interested reader is encouraged to read this article and discover a use of feedback control and canonical forms.

CONCLUDING REMARKS

The decomposition developed in this chapter can be generalized to the case of non-distinct eigenvalues. In that case, one must introduce the notion of multiplicity of eigenvalues in the characteristic polynomial and appropriately multiply terms in the decomposition of equation 17.16 by specific powers of t and in that of equation 17.20 by specific powers of k. In fact, this decomposition is constructed quite naturally in the next chapter using frequency domain techniques. The reader can consult [3] and [4] for similar treatments.

PROBLEMS

1. Consider the matrix A of example 15.3. Construct the decomposition of the state transition matrix. Simplify your expression so that all quantities are real.

2. Let $\dot{x} = Ax$, and suppose

$$A = \begin{bmatrix} 0 & 1 & 0 \\ 0 & 0 & 1 \\ -a_3 & -a_2 & -a_1 \end{bmatrix}$$

has eigenvalues $\lambda_1, \lambda_2,$ and λ_3. Find eigenvectors for A in terms of λ_i, and decompose the state trajectory into directed modes for $x(0) = [1\ 0\ 1]^t$.

3. Decompose the state transition matrix into a sum of matrix directed modes when

$$A = \begin{bmatrix} a & b \\ b & a \end{bmatrix}$$

for real scalars a and b.

4. Decompose $\exp[At]$ into a sum of matrix directed modes when

$$A = \begin{bmatrix} 5 & -2 & 0 & 0 \\ 12 & -5 & 0 & 0 \\ 0 & 0 & 1 & 0 \\ 0 & 0 & 0 & 1 \end{bmatrix}$$

5. Let $\dot{x} = Ax + Bu$, where A is as in problem 4 and

$$B = \begin{bmatrix} 1 \\ 3 \\ 0 \\ 0 \end{bmatrix}$$

(a) Using the decomposition of problem 4, evaluate the integral

$$x(t) = \int_0^t \exp[A(t-\tau)]B\sin(2\pi\tau)d\tau$$

(b) If $y = Cx$ and

$$C = [3\ 1\ 0\ 0]$$

evaluate the integral

$$y(t) = C\int_0^t \exp[A(t-\tau)]B\sin(2\pi\tau)d\tau$$

6. Prove that if A has distinct eigenvalues $\{\lambda_1,...,\lambda_n\}$ with right eigenvectors $\{e_1,...,e_n\}$ and left eigenvectors $\{w_1,...,w_n\}$, then

$$R_i = e_i w_i^*$$

7. Prove equation 17.20, utilizing the theorems of the chapter wherever possible.

8. Consider the discrete-time system

$$x(k+1) = Ax(k) + Bu(k)$$
$$y(k) = Cx(k)$$

where

$$A = \begin{bmatrix} 1 & 1 \\ 1 & 2 \end{bmatrix} \begin{bmatrix} 0 & 0 \\ 0 & -3 \end{bmatrix} \begin{bmatrix} 2 & -1 \\ -1 & 1 \end{bmatrix}$$

$$B = \begin{bmatrix} 2 \\ 2 \end{bmatrix}$$

and

$$C = [-2\ \ 2]$$

(a) Compute the state transition matrix by inspection.
(b) Decompose the state transition matrix into a sum of matrix directed modes.
(c) Evaluate the zero-state state response given that $u(k) = \sin(2k)\,1^+(k)$.
(d) Evaluate the zero-input system response given that $x(1) = [1\ \ 7]^t$.

9. Spectral Mapping and Functions of a Matrix.

(a) Suppose $g(\lambda)$ is a polynomial in λ. Show that if λ_i is an eigenvalue of A, then $g(\lambda_i)$ is an eigenvalue of $g(A)$.

(b) Show that if e_i is a right eigenvector of A that is associated with λ_i, then e_i is a right eigenvector of $\tilde{A} = g(A)$.

(c) Show that if $\tilde{A} = g(A)$, then

$$\tilde{A} = TDT^{-1}$$

where $T = [e_1,...,e_n]$ and $D = \text{diag}[g(\lambda_1),...,g(\lambda_n)]$, assuming that A is nondefective.

(d) Compute the eigenvalues of $g(A) = [A^3 + A^2 - A + I]\exp[A]$ when

$$A = \begin{bmatrix} 1 & 0 & -2 \\ 0 & 1 & 0 \\ 1 & 0 & -2 \end{bmatrix}$$

REFERENCES

1. E. W. Cheney, *Introduction to Approximation Theory* (New York: McGraw-Hill, 1966).

2. B. R. Barmish, J. A. Fleming, J. S. Thorp, and J. C. Dunn, "The Transferability of Bounded Initial Regions by Feedback Compensation," *International Journal of Control*, Vol. 20, No. 5, 1974, pp. 801–810.

3. Charles A. Desoer, *Notes for a Second Course on Linear Systems* (New York: Van-Nostrand Reinhold, 1970).

4. C. T. Chen, *Linear System Theory and Design* (New York: Holt, Rinehart, and Winston, 1984).

18

IMPULSE RESPONSE MATRICES, TRANSFER FUNCTION MATRICES, AND THE JORDAN FORM

INTRODUCTION

As mentioned earlier, the state model formulation is an internal system description. Chapter 4 developed "artificial" canonical state model representations of differential equation models. There, derivatives of the output variable $y(t)$ were related to derivatives of the input variable $u(t)$. Such differential equation models depict *external system models*. This chapter furnishes the basic definitions and properties of two other classical and yet very modern external system descriptions: the *impulse response matrix model* of a linear time-varying system, and the *transfer function matrix model* of a linear time-invariant system having a known state model [1]. There is no straightforward generalization of these ideas to nonlinear systems, although the interested reader might consult [2] and [3].

Both the impulse response matrix and the transfer function matrix find wide application in control systems. One important area of application is realization theory [4,5]. Realization theory centers on methods for determining different characterizations of models of a system from input-output measurements.

IMPULSE RESPONSE MATRICES

Our purpose here is to introduce and interpret the definition of the impulse response matrix for a linear, lumped, time-varying system. The development then continues by linking the impulse response matrix with the analytic solution of the time-varying state model.

Definition 18.1. The *impulse response* matrix of a linear, lumped, time-varying system is an $r \times m$ matrix map $H(\cdot,\cdot) : \mathbf{R} \times \mathbf{R} \to \mathbf{R}^{r \times m}$, given by

$$H(t,\tau) = [h_1(t,\tau), \dots, h_m(t,\tau)] \tag{18.1}$$

where each column $h_i(t,\tau)$ represents the response of the system to the impulsive input $u(t) = \delta(t - \tau)\eta_i$, i.e., an impulse applied to the ith input port at time τ. Note that, as before, $\eta_i \in \mathbf{R}^m$ and has zeros everywhere except in the ith row.

Specification of the impulse response depends on two parameters: τ, the time of application of the impulse, and t, the time of observation or the time at which the system response is measured. In an engineering sense, each column of $H(t,\tau)$ is computed by applying an impulse to the ith system input port, grounding the remaining input ports, and measuring the output vector $h_i(t,\tau)$ at each $t \in (-\infty, \infty)$. One then repeats the process for each $\tau \in (-\infty, \infty)$, a tedious process at best.

Observe that if, for a given t and τ, $t < \tau$ and $H(t,\tau) \neq [0]$ (the zero matrix), then the system is noncausal, since a response is observed prior to the application of the input. For causal systems, $H(t,\tau) = [0]$ whenever $t < \tau$.

The impulse response matrix draws its utility from the convolution-like relationship linking a general system input $u(\cdot)$ with a response $y(\cdot)$ via

$$y(t) = \int_{-\infty}^{\infty} H(t,\tau) u(\tau) \, d\tau \tag{18.2}$$

For the linear, lumped, causal, time-varying state model, $H(t,\tau)$ has a familiar structure set forth in the following theorem.

Theorem 18.1. The impulse response matrix for a linear, time-varying, lumped state model $\{A(\cdot), B(\cdot), C(\cdot), D(\cdot)\}$ is

$$H(t,\tau) = \begin{cases} C(t)\Phi(t,\tau)B(\tau) + D(t)\delta(t-\tau) & t \geqslant \tau \\ [0] & t < \tau \end{cases} \tag{18.3}$$

Proof. *Step 1.* Recall that $H(t,\tau)$ is a matrix designated as

$$H(t,\tau) = [h_1(t,\tau), \dots, h_m(t,\tau)]$$

where $h_i(t,\tau)$ is the system response to the impulsive input $u(t) = \eta_i \delta(t-\tau)$. The form of this input causes an excitation at the ith input port while holding the remaining ports at ground.

Step 2. From the general analytic solution of the time-varying state model equation 13.7, the response $h_i(t,\tau)$ for $t \geqslant \tau$ is

$$h_i(t,\tau) = C(t) \int_{-\infty}^{t} \Phi(t,q)B(q)\delta(q-\tau)dq\,\eta_i + D(t)\delta(t-\tau)\,\eta_i \quad (18.4a)$$

or

$$h_i(t,\tau) = [C(t)\Phi(t,\tau)B(\tau) + D(t)\delta(t-\tau)]\eta_i \quad (18.4b)$$

In equation 18.4a, q represents the variable of integration and τ the time of application of the impulse. Also, an implicit requirement is that the state $x(-\infty) = \theta$. The causality requirement of $H(t,\tau) = [0]$ for $\tau > t$ allows us to take the upper limit of integration in equation 18.2 to be t, and that in turn renders t the upper limit of integration in equation 18.4a.

Step 3. Recalling that $I = [\eta_1,...,\eta_m]$, it follows that

$$H(t,\tau) = [h_1(t,\tau),...,h_m(t,\tau)] = [C(t)\Phi(t,\tau)B(\tau) + D(t)\delta(t-\tau)][\eta_1,...,\eta_m]$$

$$= C(t)\Phi(t,\tau)B(\tau) + D(t)\delta(t-\tau)$$

for $t \geqslant \tau$. ∎

Corollary 18.1. If A, B, C, and D are constant matrices, the impulse response matrix for the time-invariant state model is

$$H(t,\tau) = H(t-\tau,0) \triangleq H(t-\tau)$$

$$= \begin{cases} C\exp[A(t-\tau)]B + D\delta(t-\tau), & t \geqslant \tau \\ 0 & t < \tau \end{cases} \quad (18.5)$$

Observe that in the time-invariant case, the impulse response matrix depends only on the difference $t-\tau$, i.e., the elapsed time between the observation of the response and the application of the impulsive input. Notationally, $H(t)$ designates the time-invariant impulse response matrix.

True-False Exercises

1. If the impulse response matrix

$$H(t,q) = \begin{bmatrix} -1 & 2 \\ 2 & -1 \end{bmatrix}$$

for $t = 1$ and $q = 2$, then the system is causal.

2. The impulse response matrix for a linear time-invariant state model is

$$H(t) = \begin{cases} C\exp(At)B + D & t \geqslant 0 \\ [0] & t < 0 \end{cases}$$

THE TRANSFER FUNCTION MATRIX

The derivation of the transfer function matrix $H(s)$ of the linear time-invariant lumped state representation utilizes the one-sided Laplace transform. The reader is cautioned to consult the standard references for the usual subtleties. Once again, recall the time-invariant state model

$$\dot{x}(t) = Ax(t) + Bu(t)$$
$$y(t) = Cx(t) + Du(t) \tag{18.6}$$

With the slight notational misuse of substituting s for t to denote the Laplace transform, the s-plane equivalent of equation 18.6 becomes

$$sx(s) - x(0) = Ax(s) + Bu(s)$$

$$y(s) = Cx(s) + Du(s)$$

Solving for $x(s)$ and then $y(s)$ produces

$$x(s) = (sI - A)^{-1}x(0) + (sI - A)^{-1}Bu(s) \tag{18.7}$$

and

$$y(s) = C(sI - A)^{-1}x(0) + [C(sI - A)^{-1}B + D]u(s) \tag{18.8}$$

Clearly, the term $(sI - A)^{-1}x(0)$ represents the zero-input state response, the term $(sI - A)^{-1}Bu(s)$ the zero-state state response, $C(sI - A)^{-1}x(0)$ the zero-input response, and $[C(sI - A)^{-1}B + D]u(s)$ the zero-state response. Despite all this "consistent" terminology, however, one can think of $u(s)$ as the "two-sided" Laplace transform of an admissible signal $u(t)$, where $u(t) = 0$ for all $t \leqslant T$ for some finite real number T. In such perspective, $[C(sI - A)^{-1}B + D]u(s)$ represents the sum total of the system behavior. Hence, it is natural to single out the quantity $[C(sI - A)^{-1}B + D]$. For the most part, however, we will take the input $u(t) = 0$ for $t < 0$ without loss of generality. This permits one to use the more commonplace one-sided Laplace transformation with explicit initial conditions.

Definition 18.2. The transfer function matrix of the time-invariant state model given by equation 18.6 is

$$H(s) = C(sI - A)^{-1}B + D \tag{18.9}$$

Definition 18.2 is an input-output description, commonly called an *external* description. The rational matrix $(sI - A)^{-1}$ characterizes the internal system structure. In the time domain, this corresponds to the dynamical equation $\dot{x}(t) = Ax(t)$. The distinction between internal and external descriptions arises because modes present in the internal description sometimes disappear from the external description. For example, if $C = [0]$, then all modes are dropped from the system output. If every mode of the internal description is present in the external description, the system is said to be *completely observable*. If every internal system mode can be excited by the input ports, either directly or indirectly, the system is *completely controllable*. Modes not excitable are said to be *uncontrollable*—i.e., they are decoupled from the system input. These skeletal ideas are fleshed out in chapters 20 and 21.

As a simple example, for $\lambda_1 \neq \lambda_2$ consider

$$\begin{bmatrix} \dot{x}_1 \\ \dot{x}_2 \end{bmatrix} = \begin{bmatrix} \lambda_1 & 0 \\ 0 & \lambda_2 \end{bmatrix} \begin{bmatrix} x_1 \\ x_2 \end{bmatrix} + \begin{bmatrix} 0 \\ 1 \end{bmatrix} u$$

$$y = \begin{bmatrix} 1 & 1 \end{bmatrix} \begin{bmatrix} x_1 \\ x_2 \end{bmatrix}$$

Some straightforward arithmetic produces

$$H(s) = C(sI - A)^{-1}B = \frac{s - \lambda_1}{(s - \lambda_1)(s - \lambda_2)} = \frac{1}{s - \lambda_2}$$

Observe that the mode associated with λ_1 disappears. As will be made clear in chapter 20, this mode is uncontrollable. Thus, it will not appear in the reduced transfer function.

Let us return for the present to the transfer function matrix $H(s)$ of equation 18.9. The difficulty in evaluating this quantity lies in the computation of the inverse matrix $(sI - A)^{-1}$.

FADDEEV-LEVERRIER ALGORITHM FOR COMPUTING $(sI - A)^{-1}$

The Faddeev-Leverrier algorithm provides a straightforward recursive (although, numerically unstable) approach to computing $(sI - A)^{-1}$. The algorithm, originally given in definition 12.2, states that

$$(sI - A)^{-1} = \frac{R(s)}{\pi_A(s)} = \frac{N_1 s^{n-1} + N_2 s^{n-2} + \cdots + N_{n-1}s + N_n}{s^n + a_1 s^{n-1} + \cdots + a_{n-1}s + a_n} \qquad (18.10)$$

where $\pi_A(s)$ is the characteristic polynomial of A, $R(s)$ is the adjugate matrix of $sI - A$, and

$$N_1 = I \qquad\qquad a_1 = -\operatorname{tr}[N_1 A]$$

$$N_2 = N_1 A + a_1 I \qquad a_2 = -\frac{1}{2}\operatorname{tr}[N_2 A]$$

$$N_3 = N_2 A + a_2 I \qquad a_3 = -\frac{1}{3}\operatorname{tr}[N_3 A]$$

$$\vdots \qquad\qquad\qquad \vdots \qquad\qquad (18.11)$$

$$N_n = N_{n-1} A + a_{n-1} I \qquad a_n = \frac{-1}{n}\operatorname{tr}[N_n A]$$

$$[0] = N_n A + a_n I$$

Note that the adjugate matrix of $sI - A$ is the transpose of the cofactor matrix of $sI - A$.

The proof of the Faddeev-Leverrier algorithm breaks down into three parts: (*i*) show that $(sI - A)^{-1}$ has the form of equation 18.10, (*ii*) show that the N_i

matrices satisfy equations 18.11, and finally, (iii) show that the coefficients a_i of the characteristic polynomial satisfy equations 18.11.

Part (i). From Cramer's rule (equation 6.5), it is possible to solve the equation

$$(sI - A) v_i(s) = \eta_i$$

for the column vector $v_i(s)$ which must have the form $r_i(s)/\pi_A(s)$, where $r_i(s)$ is a column vector of polynomials in s. Specifically, let $M_i^j(s)$ equal the matrix $(sI - A)$ except in the ith column, which is replaced by η_i. By equation 6.5,

$$r_i^j(s) = \det[M_i^j(s)]$$

where $r_i^j(s)$ is the jth entry of $r_i(s)$ and where $r_i^j(s)$ has degree at most $n-1$. Hence,

$$(sI - A)^{-1} \triangleq [v_i(s),...,v_n(s)] = \frac{1}{\pi_A(s)} [r_1(s),...,r_n(s)] \triangleq \frac{R(s)}{\pi_A(s)}$$

requires $R(s)$ to be an $n \times n$ matrix whose entries are polynomials of degree at most $n-1$, and $\pi_A(s) = \det(sI - A)$ is the characteristic polynomial. Clearly, it is possible to write $R(s)$ as

$$R(s) = N_1 s^{n-1} + N_2 s^{n-2} + \cdots + N_n \tag{18.12}$$

for some constant $n \times n$ matrices N_i.

Part (ii). To show that the N_i matrices satisfy the recursive relations on the left-hand side of equations 18.11, observe that

$$\pi_A(s)I = R(s) (sI - A) \tag{18.13}$$

Since $\pi_A(s) = s^n + a_1 s^{n-1} + \cdots + a_n$, and with $R(s)$ given by equation 18.12, equation 18.13 may be fleshed out as

$$Is^n + a_1 Is^{n-1} + a_2 Is^{n-2} + \cdots + a_n I$$
$$= [N_1 s^{n-1} + N_2 s^{n-2} + N_3 s^{n-3} + \cdots + N_n] (sI - A)$$
$$= N_1 s^n + (N_2 - N_1 A)s^{n-1} + (N_3 - N_2 A)s^{n-2} + \cdots - N_n A$$

By equating coefficients of like powers of s, it follows that the N_i matrices satisfy the left-hand side of equation 18.11.

Part (iii). It remains to show that the formulas on the right-hand side of equation 18.11 indeed produce the coefficients of the characteristic polynomial of A. First, note that $N_1 A = A$, $N_2 A = N_1 A^2 + a_1 A = A^2 + a_1 A$, $N_3 A = N_2 A^2 + a_2 A = A^3 + a_1 A^2 + a_2 A$, and so on. Then, in general,

$$N_k A = A^k + a_1 A^{k-1} + \cdots + a_{k-1} A$$

Taking the trace of both sides yields

$$\text{tr}[N_k A] = \text{tr}[A^k] + a_1 \text{tr}[A^{k-1}] + a_2 \text{tr}[A^{k-2}] + \cdots + a_k \text{tr}[A] \tag{18.14}$$

The idea now is to simplify this equality of traces. Since the trace of a matrix sums the diagonal entries, $\text{tr}[AT] = \text{tr}[TA]$ for compatible square matrices T and A. But then, $\text{tr}[T_1AT_2] = \text{tr}[AT_2T_1]$. Also, from linear algebra [6], every matrix A is similar to an upper triangular matrix, say U. In other words, there exists a nonsingular T such that $U = T^{-1}AT$. For example, if A has distinct eigenvalues, then A can be diagonalized, and thus the transformed A is upper triangular. Consequently, for an appropriate upper triangular matrix U, $\text{tr}[A] = \text{tr}[T^{-1}UT] = \text{tr}[UT^{-1}T] = \sum_{i=1}^{n} u_{ii} = \sum_{i=1}^{n} \lambda_i$, where the λ_i are the eigenvalues of A which coincide with the diagonal entries of U. Furthermore, $\text{tr}[A^k] = \sum_{i=1}^{n} \lambda_i^k$. Therefore, equation 18.14 becomes

$$\text{tr}[N_kA] = d_k + a_1d_{k-1} + \cdots + a_{k-1}d_1 \tag{18.15}$$

where

$$d_k = \text{tr}[A^k] = \sum_{i=1}^{n} \lambda_i^k$$

The right-hand side of equation 18.15 is part of the Newton formulas [6] defined by

$$-ka_k = d_k + a_1d_{k-1} + \cdots + a_{k-1}d_1 \tag{18.16}$$

It then follows that

$$a_k = -\frac{1}{k}\,\text{tr}[N_kA]$$

as was to be shown. ∎

The preceding proof relies heavily on the validity of the Newton formulas 18.16. A formal proof is somewhat detailed, but a third-order example provides a good illustration. Let A be 3×3 with eigenvalues λ_1, λ_2, and λ_3, not necessarily distinct. Then

$$\pi_A(\lambda) = s^3 + a_1s^2 + a_2s + a_3 = (s - \lambda_1)(s - \lambda_2)(s - \lambda_3)$$

$$= s^3 - (\lambda_1 + \lambda_2 + \lambda_3)s^2 + (\lambda_1\lambda_2 + \lambda_1\lambda_3 + \lambda_2\lambda_3)s - \lambda_1\lambda_2\lambda_3$$

Equating coefficients in s^2 leads to

$$-a_1 = (\lambda_1 + \lambda_2 + \lambda_3) = \text{tr}[A]$$

and similarly, from the equality of the coefficients of s, we obtain

$$-2a_2 = -2(\lambda_1\lambda_2 + \lambda_1\lambda_3 + \lambda_2\lambda_3) = (\lambda_1^2 + \lambda_2^2 + \lambda_3^2) - (\lambda_1 + \lambda_2 + \lambda_3)^2$$

$$= \text{tr}[A^2] + a_1\text{tr}[A]$$

Finally, our illustration of the formulas becomes complete after observing that

$$-3a_3 = (\lambda_1^3 + \lambda_2^3 + \lambda_3^3) + a_1(\lambda_1^2 + \lambda_2^2 + \lambda_3^2) + a_2(\lambda_1 + \lambda_2 + \lambda_3)$$

$$= \text{tr}[A^3] + a_1\text{tr}[A^2] + a_2\text{tr}[A]$$

The following example ends our discussion of the Faddeev-Leverrier formulas.

EXAMPLE 18.1

Using Leverrier's algorithm, compute $(sI - A)^{-1}$ when

$$A = \begin{bmatrix} -1 & -4 & 0 \\ 1 & -1 & 0 \\ 4 & 2 & -3 \end{bmatrix}$$

Step 1. Clearly, $N_1 = I$ and $a_1 = -\text{tr}[A] = 5$.

Step 2. $N_2 = A + a_1 I$. Thus,

$$N_2 = \begin{bmatrix} 4 & -4 & 0 \\ 1 & 4 & 0 \\ 4 & 2 & 2 \end{bmatrix}$$

Since $a_2 = -0.5\text{tr}[N_2 A]$, with

$$N_2 A = \begin{bmatrix} -8 & -12 & 0 \\ 3 & -8 & 0 \\ 6 & -14 & -6 \end{bmatrix}$$

$a_2 = -0.5[-22] = 11$.

Step 3. Since $N_3 = N_2 A + a_2 I$, we have

$$N_3 = \begin{bmatrix} -8 & -12 & 0 \\ 3 & -8 & 0 \\ 6 & -14 & -6 \end{bmatrix} + \begin{bmatrix} 11 & 0 & 0 \\ 0 & 11 & 0 \\ 0 & 0 & 11 \end{bmatrix} = \begin{bmatrix} 3 & -12 & 0 \\ 3 & 3 & 0 \\ 6 & -14 & 5 \end{bmatrix}$$

Again, $a_3 = -\frac{1}{3}\text{tr}[N_3 A]$, so that

$$N_3 A = \begin{bmatrix} -15 & 0 & 0 \\ 0 & -15 & 0 \\ 0 & 0 & -15 \end{bmatrix}$$

in which case $a_3 = 15$.

Step 4. Concluding, we have

$$(sI - A)^{-1} = \frac{Is^2 + \begin{bmatrix} 4 & -4 & 0 \\ 1 & 4 & 0 \\ 4 & 2 & 2 \end{bmatrix} s + \begin{bmatrix} 3 & -12 & 0 \\ 3 & 3 & 0 \\ 6 & -14 & 5 \end{bmatrix}}{s^3 + 5s^2 + 11s + 15}$$

Use of the Faddeev-Leverrier algorithm for singular systems is discussed in [7]. Again, however, it is well known that the algorithm is particularly sensitive to round-off error, making it numerically unstable as a general algorithm. Thus, caution with regard to its use is advisable.

Exercise. The Faddeev-Leverrier algorithm applied to a square matrix A provides a method for doing which of the following?

(*a*) Computing the coefficients of the characteristic polynomial of A.

(*b*) Computing A^{-1}.

(*c*) Computing $(sI - A)^{-1}$.

(*d*) Computing at least one of the above in a numerically unstable way.

(*e*) (a), (c), and (d).

(*f*) All of the above.

(*g*) None of above.

EXPANSION BY PARTIAL FRACTIONS: DISTINCT EIGENVALUES

By means of the Faddeev-Leverrier algorithm, it is possible to compute $[R(s)/\pi_A(s)] = (sI - A)^{-1}$. Traditionally, the advantage of the Laplace transform stems from its utility in computing time responses. For elementary signals and systems, these time responses flow out of an expansion of the transfer function by partial fractions and the application of a dictionary of precomputed inverse transforms. The presentation here traces a similar path. This leads to another method for computing the state transition matrix $\Phi(t)$.

The matrix $R(s)$ in equation 18.10 is the adjugate matrix or, equivalently, the transpose of the cofactor matrix of $sI - A$. Each entry of $R(s)$ is therefore a polynomial in s of degree at most $n-1$. (Recall that A is $n \times n$.) *Assuming that the eigenvalues* $\{\lambda_1,...,\lambda_n\}$ *of* A *are distinct*, equation 18.10 admits the partial fraction expansion

$$(sI - A)^{-1} = \frac{R(s)}{\pi_A(s)} = \frac{R_1}{s - \lambda_1} + \frac{R_2}{s - \lambda_2} + \cdots + \frac{R_n}{s - \lambda_n} \qquad (18.17)$$

where $\pi_A(s) = (s - \lambda_1)(s - \lambda_2)...(s - \lambda_n)$ and where

$$R_i = \lim_{s \to \lambda_i} \left[\frac{(s - \lambda_i) R(s)}{\pi_A(s)} \right] \qquad (18.18)$$

is called a *residue matrix*.

The partial fraction expansion given in equation 18.17, together with the R_i of equation 18.18, represents the frequency domain counterpart of the decomposition of $\exp[At]$ into matrix directed modes. To see this, take the inverse Laplace transform of equation 18.17:

$$\mathcal{L}^{-1}[(sI - A)^{-1}] = \sum_{i=1}^{n} \mathcal{L}^{-1} \left[\frac{R_i}{s - \lambda_i} \right] = \sum_{i=1}^{n} R_i e^{\lambda_i t} 1^+(t) = \exp[At] 1^+(t)$$

Because of the unique decomposition of the state transition matrix into matrix directed modes, the residue matrices R_i must precisely equal the matrix directed modes of equation 17.1. In this regard, we have the following propositions.

Proposition 18.1. $\mathcal{L}[\Phi(t)] = (sI - A)^{-1}$.

Residue matrices or matrix directions satisfy some interesting relations.

Theorem 18.2. The residue matrices $R_i, i = 1,...,n$, satisfy the relations

(i)

$$\sum_{i=1}^{n} R_i = I \tag{18.19a}$$

(ii)

$$AR_i = \lambda_i R_i \tag{18.19b}$$

(iii)

$$R_i A = \lambda_i R_i \tag{18.19c}$$

(iv)

$$R_j R_i = \delta_{ji} R_j \tag{18.19d}$$

where $\delta_{ji} = 1$ if $i = j$ and 0 if $i \neq j$.

Proof. The development in chapter 17 makes the proof clear. For example,

$$\Phi(0) = I = \sum_{i=1}^{n} R_i$$

and since each column of R_i is a scalar multiple of e_i, *(ii)* follows immediately. For *(iii)* and *(iv)*, one writes R_i as

$$R_i = [\xi_{i1}e_i | \xi_{i2}e_i | \cdots | \xi_{in}e_i] = \begin{bmatrix} \zeta_{1i} \ w_i^* \\ \zeta_{2i} \ w_i^* \\ \vdots \ \vdots \\ \zeta_{ni} \ w_i^* \end{bmatrix}$$

where

$$[e_1 | e_2 | \cdots | e_n]^{-1} = \begin{bmatrix} \xi_{11} & \xi_{12} & \cdots & \xi_{1n} \\ \xi_{21} & \xi_{22} & \cdots & \xi_{2n} \\ \vdots & & & \vdots \\ \xi_{n1} & \xi_{n2} & \cdots & \xi_{nn} \end{bmatrix} = \begin{bmatrix} w_1^* \\ w_2^* \\ \vdots \\ w_n^* \end{bmatrix}$$

The interested reader should fill in the details.   ∎

The preceding properties are generalized to the case of nondistinct eigenvalues in the next section. But first, an example is illustrative.

EXAMPLE 18.2

Let

$$A = \begin{bmatrix} -1 & -4 & 0 \\ 1 & -1 & 0 \\ 4 & 2 & -3 \end{bmatrix}$$

The characteristic polynomial is

$$\pi_A(s) = \det(sI - A) = (s+3)\left[(s+1)^2 + 4\right]$$

producing eigenvalues $\lambda_1 = -3, \lambda_2 = -1 - 2j$, and $\lambda_3 = -1 + 2j$. From example 18.1,

$$\Phi(s) = \frac{R(s)}{\pi_A(s)} = \frac{\begin{bmatrix} s^2 + 4s + 3 & -4s - 12 & 0 \\ s + 3 & s^2 + 4s + 3 & 0 \\ 4s + 6 & 2s - 14 & s^2 + 2s + 5 \end{bmatrix}}{(s+3)(s+1+2j)(s+1-2j)} \qquad (18.20)$$

The partial fraction expansion takes the form

$$\Phi(s) = \frac{R_1}{s+3} + \frac{R_2}{s+1+2j} + \frac{R_3}{s+1-2j}$$

Using the residue formula 18.18, we obtain

$$R_1 = \begin{bmatrix} 0.0 & 0.0 & 0.0 \\ 0.0 & 0.0 & 0.0 \\ -0.75 & -2.5 & 1.0 \end{bmatrix}, R_2 = \bar{R}_3 = \begin{bmatrix} 0.5 & -j & 0 \\ 0.25j & 0.5 & 0 \\ 0.375 + j0.625 & 1.25 - j0.75 & 0 \end{bmatrix}$$

Next, the relationship

$$\Phi(t) = R_1 e^{-3t} + R_2 e^{-(1+2j)t} + R_3 e^{-(1-2j)t} = \mathcal{L}^{-1}[\Phi(s)]$$

produces

$$\Phi(t) = \begin{bmatrix} e^{-t}\cos(2t) & -2e^{-t}\sin(2t) & 0.0 \\ 0.5e^{-t}\sin(2t) & e^{-t}\cos(2t) & 0.0 \\ q(t) & p(t) & e^{-3t} \end{bmatrix}$$

where

$$q(t) = -0.75e^{-3t} + 0.75e^{-t}\cos(2t) + 1.25e^{-t}\sin(2t)$$
$$p(t) = -2.5e^{-3t} + 2.5e^{-t}\cos(2t) - 1.5e^{-t}\sin(2t)$$

The foregoing example illustrates not only the partial fraction expansion of $\Phi(s) = (sI - A)^{-1}$, but also the computation of $\Phi(t)$.

EXPANSION BY PARTIAL FRACTIONS: MULTIPLE EIGENVALUES

As a foretaste of the structure of the partial fraction expansion of $(sI - A)^{-1}$ when A has multiple eigenvalues, consider the matrix

$$A = \begin{bmatrix} \lambda_1 & 1 & 0 \\ 0 & \lambda_1 & 0 \\ 0 & 0 & \lambda_1 \end{bmatrix}$$

By direct computation,

$$(sI-A)^{-1} = \frac{R(s)}{\pi_A(s)} = \frac{\begin{bmatrix} (s-\lambda_1)^2 & s-\lambda_1 & 0 \\ 0 & (s-\lambda_1)^2 & 0 \\ 0 & 0 & (s-\lambda_1)^2 \end{bmatrix}}{(s-\lambda_1)^3} = \frac{\begin{bmatrix} s-\lambda_1 & 1 & 0 \\ 0 & s-\lambda_1 & 0 \\ 0 & 0 & s-\lambda_1 \end{bmatrix}}{(s-\lambda_1)^2}$$

$R(s)$ and $\pi_A(s)$ have common factors, so a partial fraction expansion of $(sI - A)^{-1}$ should build on a reduced form of $R(s)/\pi_A(s)$. The construction, however, requires some additional mathematics. First, we formally state the Caley-Hamilton theorem, mentioned several times throughout the text.

Theorem 18.3 (Caley-Hamilton Theorem) [6,1]. Every square matrix A satisfies its characteristic polynomial; i.e., $\pi_A(A) = [0]$.

The proof of the Caley-Hamilton theorem can be found in the aforementioned references or can be obtained as a result of lemmas 20.1 and 20.2 (See chapter 20.) The interesting aspect of the characteristic polynomial $\pi_A(\lambda)$ is that $\pi_A(A) = [0]$. In general, a polynomial $p(\lambda)$, such that $p(A) = [0]$ is called an *annihilating polynomial* of the matrix A. Thus, the characteristic polynomial $\pi_A(\lambda)$ annihilates A. Are there any other polynomials which annihilate A? Indeed, there are. Let $p(\lambda) = g(\lambda)\pi_A(\lambda)$. Then, clearly, $p(\lambda)$ is an annihilating polynomial of A. Are there polynomials of degree smaller than $\pi_A(\lambda)$ which annihilate A? To answer this, consider the $n \times n$ identity matrix I. $\pi_I(\lambda) = (\lambda-1)^n$, so that $\psi(\lambda) = \lambda-1$ annihilates I—i.e., $\psi(I) = I - I = [0]$. On the other hand, if A has distinct eigenvalues $\lambda_1,...,\lambda_n$, then $\pi_A(\lambda)$ is the polynomial of least degree which annihilates A. These results suggest the following definition.

Definition 18.3. The polynomial of least degree with leading coefficient 1 which annihilates the matrix A is the *minimal polynomial* of A, denoted by $\psi_A(\lambda)$.

From the preceding discussion, the minimal polynomial of the identity I is $\psi_I(\lambda) = \lambda - 1$, whereas the characteristic polynomial is $\pi_A(\lambda) = (\lambda - 1)^n$. Also, by direct computation, the minimal polynomial of

$$A = \begin{bmatrix} \lambda_1 & 1 & 0 \\ 0 & \lambda_1 & 0 \\ 0 & 0 & \lambda_1 \end{bmatrix}$$

is $\psi_A(\lambda) = (\lambda - \lambda_1)(\lambda - \lambda_1)$, whereas the characteristic polynomial is $\pi_A(\lambda) = (\lambda - \lambda_1)^3$.

These examples suggest that the minimal and characteristic polynomial are related. Clearly, both are annihilating polynomials of A. Also, the minimal polynomial appears to divide the characteristic polynomial. In fact, it is possible to show that the minimal polynomial divides *every* annihilating polynomial of A. To verify this, let $p_A(\lambda)$ be any annihilating polynomial of A. Then, by the usual polynomial division algorithm, there exist polynomials $f(\lambda)$ and $r(\lambda)$ such that

$$p_A(\lambda) = \psi_A(\lambda)f(\lambda) + r(\lambda)$$

where $r(\lambda)$ has degree strictly less than $\psi_A(\lambda)$. Substituting A for λ yields

$$p_A(A) = \psi_A(A)f(A) + r(A)$$

Since $p_A(A) = \psi_A(A) = [0]$, it follows that $r(A) = [0]$. However, $r(\lambda)$ has degree strictly less than $\psi_A(\lambda)$, and by definition, $\psi_A(\lambda)$ is the polynomial of least degree for which $\psi_A(A) = [0]$. Hence, $r(\lambda) \equiv 0$. This means that every annihilating polynomial of A is divisible by the minimal polynomial without remainder; in particular, the minimal polynomial divides the charcteristic polynomial.

It is also possible to show that the minimal polynomial is unique. To do so, let $\rho_A(\lambda)$ be any other minimal polynomial of A. Then $\psi_A(\lambda)$ and $\rho_A(\lambda)$ must differ by a constant (multiplicative) factor. But by definition, this factor must be 1. Hence, $\rho_A(\lambda) = \psi_A(\lambda)$, establishing the uniqueness of the minimal polynomial.

From the foregoing discussion, the minimal polynomial must have the structure

$$\psi_A(s) = (s-\lambda_1)^{m_1}(s-\lambda_2)^{m_2} \cdots (s-\lambda_\sigma)^{m_\sigma} \tag{18.21}$$

where $\lambda_1, \lambda_2, ..., \lambda_\sigma$ denote the distinct eigenvalues of A. But then, clearly, $m_1 + m_2 + ... + m_\sigma \leqslant n$. Thus, the minimal polynomial of A determines the structure of the partial fraction expansion of $(sI - A)^{-1}$. Specifically, for the preceding A-matrix

$$(sI - A)^{-1} = \frac{\begin{bmatrix} s-\lambda_1 & 1 & 0 \\ 0 & s-\lambda_1 & 0 \\ 0 & 0 & s-\lambda_1 \end{bmatrix}}{\psi_A(s)} = \frac{\begin{bmatrix} 0 & 1 & 0 \\ 0 & 0 & 0 \\ 0 & 0 & 0 \end{bmatrix}}{(s-\lambda_1)^2} + \frac{\begin{bmatrix} 1 & 0 & 0 \\ 0 & 1 & 0 \\ 0 & 0 & 1 \end{bmatrix}}{s-\lambda_1}$$

The structure of this partial fraction expansion is formalized in the following theorem.

Theorem 18.4.   Let A be $n \times n$ and $\psi_A(s)$ be as in equation 18.21. Then

$$(sI - A)^{-1} = \frac{R(s)}{\pi_A(s)} = \frac{\hat{R}(s)}{\psi_A(s)} = \sum_{i=1}^{\sigma} \sum_{j=1}^{m_i} \frac{R_i^j}{(s-\lambda_i)^j} \tag{18.22}$$

where, for $j = 1,...,m_i$,

$$R_i^j = \frac{1}{(m_i - j)!} \lim_{s \to \lambda_i} \left[\frac{d^{m_i-j}}{ds^{m_i-j}} \left[(s-\lambda_i)^{m_i} (sI - A)^{-1} \right] \right] \tag{18.23}$$

are the residue matrices analogous to those of equation 18.18.

Proof.   The proof follows Zadeh and Desoer [1]. Since $(sI-A)^{-1} = \hat{R}(s)/\psi_A(s)$ as per equation 18.22 where $\hat{R}(s)$ is a polynomial matrix and $\psi_A(s)$

has the form of equation 18.21, it follows that $(sI - A)^{-1}$ has a pole of order no greater than m_i at $s = \lambda_i$. It is possible then to define the constant matrices

$$R_i^{m_i} \triangleq \lim_{s \to \lambda_i} \left[(s - \lambda_i)^{m_i} \frac{\hat{R}(s)}{\psi_A(s)} \right]$$

for all $i = 1, 2, \ldots, \sigma$. If one then defines polynomials $\xi_i(s) = \psi_A(s)/(s - \lambda_i)^{m_i}$, then, since $\xi_i(\lambda_i) \neq 0$ (by construction),

$$R_i^{m_i} = \frac{\hat{R}(\lambda_i)}{\xi_i(\lambda_i)}$$

and is consistent with equation 18.23.

The idea now is to remove the m_ith-order pole from $\hat{R}(s)/\psi_A(s)$ and repeat the procedure. In particular, $R_i^{m_i} = \hat{R}(\lambda_i)/\xi_i(\lambda_i)$ implies that $\hat{R}(s) - \xi_i(s)R_i^{m_i} = [0]$ at $s = \lambda_i$—i.e., each entry of this matrix has a zero at $s = \lambda_i$. But then, there exists a polynomial matrix $P_1(s)$ such that

$$\hat{R}(s) - \xi_i(s)R_i^{m_i} = (s - \lambda_i)P_1(s)$$

Dividing this expression by the minimal polynomial and rearranging yields

$$\frac{\hat{R}(s)}{\psi_A(s)} = \frac{R_i^{m_i}}{(s - \lambda_i)^{m_i}} + \frac{P_1(s)}{(s - \lambda_i)^{m_i - 1}\xi_i(s)}$$

The second term on the right represents the remainder after removing the m_ith-order pole at λ_i from $\hat{R}(s)/\psi_A(s)$.

At this point, we repeat the procedure on

$$\frac{P_1(s)}{(s - \lambda_i)^{m_i - 1}\xi_i(s)}$$

which has an $(m_i - 1)$th-order pole at λ_i. As such since $\xi_i(\lambda_i) \neq 0$, it is possible to define the constant matrices

$$R_i^{m_i - 1} = \lim_{s \to \lambda_i} \left[\frac{(s - \lambda_i)^{m_i - 1} P_1(s)}{(s - \lambda_i)^{m_i - 1}\xi_i(s)} \right] = \frac{P_1(\lambda_i)}{\xi_i(\lambda_i)}$$

$$= \lim_{s \to \lambda_i} \frac{d}{ds} \left[\frac{(s - \lambda_i)^{m_i} R_i^{m_i}}{(s - \lambda_i)^{m_i}} + \frac{(s - \lambda_i)^{m_i} P_1(s)}{(s - \lambda_i)^{m_i - 1}\xi_i(s)} \right]$$

consistent with equation 18.23. Therefore, there exists a polynomial matrix $P_2(s)$ such that $P_1(s) - \xi_i(s)R_i^{m_i - 1} = (s - \lambda_i)P_2(s)$, and

$$\frac{P_1(s)}{(s - \lambda_i)^{m_i - 1}\xi_i(s)} = \frac{R_i^{m_i - 1}}{(s - \lambda_i)^{m_i - 1}} + \frac{P_2(s)}{(s - \lambda_i)^{m_i - 2}\xi_i(s)}$$

Repeating the procedure until the pole at $s = \lambda_i$ is exhausted results in

$$\frac{\hat{R}(s)}{\psi_A(s)} = \sum_{j=1}^{m_i} \frac{R_i^j}{(s - \lambda_i)^j} + \frac{\hat{P}(s)}{\xi_i(s)}$$

for an appropriate matrix $\hat{P}(s)$.

Finally, iterating this procedure through each of the remaining eigenvalues of A yields equations 18.22 and 18.23. ∎

There are two immediate corollaries to this partial fraction expansion.

Corollary 18.4a.   The state transition matrix is given by

$$\Phi(t) = \sum_{i=1}^{\sigma} \sum_{j=1}^{m_i} R_i^j \frac{t^{j-1}}{(j-1)!} e^{\lambda_i t} \tag{18.24}$$

for $t \geqslant 0$.

Proof.

$$\Phi(t) = \exp[At]1^+(t) = \sum_{i=1}^{\sigma} \sum_{j=1}^{m_i} R_i^j \mathcal{L}^{-1}\left[\frac{1}{(s-\lambda_i)^j}\right]$$

$$= \sum_{i=1}^{\sigma} \sum_{j=1}^{m_i} R_i^j \frac{t^{j-1}}{(j-1)!} e^{\lambda_i t} 1^+(t) \qquad ∎$$

For the next corollary, recall the time-invariant system impulse response matrix $H(t)$ of equation 18.5.

Corollary 18.4b. $H(s) = \mathcal{L}[H(t)] = C(sI-A)^{-1}B + D$.

For an appropriately defined (restricted) function class, say, $\{H(t)\}$, there is a one-to-one correspondence between impulse response matrices and transfer function matrices. For example, if $\mathcal{L}(\cdot)$ represents the one-sided Laplace transform, $\{H(t)\}$ contains functions which are zero for $t < 0$, and the usual other assumptions hold, then the one-sided Laplace transform is one to one.

The residue matrices for the partial fraction expansion for the case of nondistinct eigenvalues also satisfy some interesting properties. For a nondistinct eigenvalue λ_i, a cyclic structure among the R_i^j is noted and theorem 18.2 has the following analog.

Theorem 18.5.   For $i = 1,2,...,\sigma$, the residue matrices satisfy the following relations:

(i)

$$\sum_{i=1}^{\sigma} R_i^1 = I \tag{18.25a}$$

(ii)

$$R_i^{j+1} = R_i^j(A - \lambda_i I), \quad j=1,2,...,m_i - 1 \tag{18.25b}$$

(iii)

$$R_i^{m_i}(A - \lambda_i I) = [0] \tag{18.25c}$$

(*iv*)

$$R_i^1 R_j^1 = R_i \delta_{ij} \tag{18.25d}$$

Proof. Establishing equations 18.25a–c requires transforming equation 18.22 into a form from which those equations follow directly. First observe that, from equation 18.22, it follows that for all $s \neq \lambda_i, i = 1,\ldots,\sigma$,

$$I = \sum_{i=1}^{\sigma} \left[\sum_{j=1}^{m_i} \frac{R_i^j}{(s - \lambda_i)^j} (sI - A) \right] \tag{18.26}$$

The bracketed term on the right of equation 18.26 can be rewritten as

$$\left[\frac{R_i^1}{s - \lambda_i} + \frac{R_i^2}{(s - \lambda_i)^2} + \cdots + \frac{R_i^{m_i}}{(s - \lambda_i)^{m_i}} \right] \left[(s - \lambda_i)I + (\lambda_i I - A) \right]$$

$$= R_i^1 + \frac{\left[R_i^1(\lambda_i I - A) + R_i^2 \right]}{s - \lambda_i} + \cdots$$

$$+ \frac{\left[R_i^j(\lambda_i I - A) + R_i^{j+1} \right]}{(s - \lambda_i)^j} + \cdots + \frac{R_i^{m_i}(\lambda_i I - A)}{(s - \lambda_i)^{m_i}} \tag{18.27}$$

Notice that each of the $(m_i + 1)$ matrices in the numerators on the right-hand side of equation 18.27 is constant. Suppose, for notational convenience, we define each of these matrices as P_i^j, $j = 1,\ldots,m_i + 1$, where $P_i^1 = R_i^1, P_i^{j+1} = R_i^j(\lambda_i I - A) + R_i^{j+1}$ for $j = 1,\ldots,m_i$ and $P_i^{m_i+1} = R_i^{m_i}(\lambda_i I - A)$. In this notation, equation 18.26 has the form

$$I = \sum_{i=1}^{\sigma} \sum_{j=0}^{m_i} \frac{P_i^{j+1}}{(s - \lambda_i)^j} \tag{18.28}$$

which must hold for all $s \neq \lambda_i, i = 1,\ldots,\sigma$. Since both sides represent rational functions in s, equating coefficients implies

$$\sum_{i=1}^{\sigma} P_i^1 = \sum_{i=1}^{\sigma} R_i^1 = I$$

(establishing equation 18.25a) and that $P_i^{j+1} = [0]$ for all $j \geqslant 1$ and all i. In particular,

$$P_i^{j+1} = R_i^j(\lambda_i I - A) + R_i^{j+1} = [0]$$

(establishing equation 18.25b) and

$$P_i^{m_i+1} = R_i^{m_i}(\lambda_i I - A) = [0]$$

establishing equation 18.25c.

To establish equation 18.25d, consider equation 18.27 again in light of the fact that $P_i^{j+1} = [0]$ for $j \geqslant 1$ and for all i. Simply put,

$$\left[\sum_{j=1}^{m_i} \frac{R_i^j}{(s - \lambda_i)^j} \right] (sI - A) = R_i^1 \tag{18.29}$$

or equivalently,

$$\left[\sum_{j=1}^{m_i} \frac{R_i^j}{(s - \lambda_i)^j} \right] = R_i^1 (sI - A)^{-1} = R_i^1 \sum_{k=1}^{\sigma} \left[\sum_{j=1}^{m_k} \frac{R_k^j}{(s - \lambda_k)^j} \right]$$

Rearranging yields that, for all $s \neq \lambda_i$, $i = 1, \dots, \sigma$,

$$(R_i^1 - I) \sum_{j=1}^{m_i} \frac{R_i^j}{(s - \lambda_i)^j} + R_i^1 \sum_{k=1}^{\sigma} \left[\sum_{j=1}^{m_k} \frac{R_k^j}{(s - \lambda_k)^j} \right] = 0$$

from which it follows that

$$\left[R_i^1 \right]^2 = R_i^1$$

and

$$R_i^1 R_k^1 = [0] = R_i \delta_{ik} \qquad \blacksquare$$

EXAMPLE 18.3

Suppose

$$A = \begin{bmatrix} -2 & 1 & 1 & -1 & 0 \\ 2 & -2 & -2 & 2 & 1 \\ -2 & 1 & 1 & -1 & 0 \\ 1 & -1 & -1 & 1 & 1 \\ 1 & -1 & -1 & 1 & 0 \end{bmatrix}$$

Calculate

(i) $(sI - A)^{-1}$.

(ii) The minimal polynomial of A.

(iii) The partial fraction expansion of $(sI - A)^{-1}$.

(iv) The state transition matrix $\Phi(t)$.

In addition to these computations, demonstrate that parts (i)–(iv) of the theorem 18.5 hold.

Part (i). From Leverrier's algorithm,

$$\pi_A(s) = s^5 + 2s^4 + s^3 = s^3(s + 1)^2 \tag{18.30}$$

and

$$R(s) = Is^4 + N_2 s^3 + N_3 s^2 + N_4 s \tag{18.31}$$

where

$$N_2 = \begin{bmatrix} 0 & 1 & 1 & -1 & 0 \\ 2 & 0 & -2 & 2 & 1 \\ -2 & 1 & 3 & -1 & 0 \\ 1 & -1 & -1 & 3 & 1 \\ 1 & -1 & -1 & 1 & 2 \end{bmatrix}$$ (18.32a)

$$N_3 = \begin{bmatrix} 0 & 0 & 0 & 0 & 0 \\ 3 & -2 & -3 & 3 & 2 \\ -1 & 0 & 1 & 0 & 0 \\ 2 & -2 & -2 & 3 & 2 \\ 1 & -1 & -1 & 1 & 1 \end{bmatrix}$$ (18.32b)

and

$$N_4 = \begin{bmatrix} 0 & 0 & 0 & 0 & 0 \\ 1 & -1 & -1 & 1 & 1 \\ 0 & 0 & 0 & 0 & 0 \\ 1 & -1 & -1 & 1 & 1 \\ 0 & 0 & 0 & 0 & 0 \end{bmatrix}$$ (18.32c)

From equation 18.22,

$$(sI - A)^{-1} = \frac{R(s)}{s^3(s+1)^2}$$

with $R(s)$ defined by equations 18.31 and 18.32.

Part (ii). "Plugging and chugging" away at the computation of the minimal polynomial of A might proceed as follows:

(a) Try $A(A + I)$:

$$A(A + I) = \begin{bmatrix} 1 & -1 & -1 & 1 & 0 \\ 1 & -1 & -1 & 1 & 1 \\ 1 & -1 & -1 & 1 & 0 \\ 1 & -1 & -1 & 1 & 1 \\ 0 & 0 & 0 & 0 & 0 \end{bmatrix}$$

This fails the test.

(b) Try $A^2(A + I)$:

$$A^2(A + I) = \begin{bmatrix} -1 & 1 & 1 & -1 & 0 \\ 0 & 0 & 0 & 0 & 0 \\ -1 & 1 & 1 & -1 & 0 \\ 0 & 0 & 0 & 0 & 0 \\ 0 & 0 & 0 & 0 & 0 \end{bmatrix}$$

(c) Checking $A^2(A + I)^2$, yields

$$\psi_A(A) = A^2(A + I)^2 = [0]$$

Hence, $\psi_A(s) = s^2(s + 1)^2$ is the minimal polynomial. With this information, one can proceed to the partial fraction expansion of $(sI - A)^{-1}$.

Note that a more direct way to compute $\hat{R}(s)/\psi_A(s)$ would be to combine $Is^4 + N_2s^3 + N_3s^2 + N_4s$ into the single polynomial matrix $R(s)$. Cancelling all factors common to $R(s)$ and $\pi_A(s)$ then yields $\hat{R}(s)/\psi_A(s)$.

Part (iii). The partial fraction expansion of $(sI - A)^{-1}$ is

$$(sI - A)^{-1} = \frac{Is^3 + N_2s^2 + N_3s + N_4}{s^2(s + 1)^2} \tag{18.33}$$

$$= \frac{R_1^1}{s} + \frac{R_1^2}{s^2} + \frac{R_2^1}{s + 1} + \frac{R_2^2}{(s + 1)^2}$$

The objective is to compute the R_i^j in terms of the N_i. To accomplish this, observe that for $s = 0$,

$$R_1^2 = \left[\frac{Is^3 + N_2s^2 + N_3s + N_4}{(s + 1)^2}\right]_{s=0} = N_4 \tag{18.34a}$$

where N_4 is given by equation 18.32c and

$$R_1^1 = \frac{d}{ds}\left[\frac{Is^3 + N_2s^2 + N_3s + N_4}{(s + 1)^2}\right]_{s=0}$$

$$= \frac{3Is^2 + 2N_2s + N_3}{(s + 1)^2}\bigg|_{s=0} - 2\frac{Is^3 + N_2s^2 + N_3s + N_4}{(s + 1)^3}\bigg|_{s=0}$$

$$= N_3 - 2N_4 = \begin{bmatrix} 0 & 0 & 0 & 0 & 0 \\ 1 & 0 & -1 & 1 & 0 \\ -1 & 0 & 1 & 0 & 0 \\ 0 & 0 & 0 & 1 & 0 \\ 1 & -1 & -1 & 1 & 1 \end{bmatrix} \tag{18.34b}$$

Also, for the pole $s = -1$,

$$R_2^2 = \frac{Is^3 + N_2s^2 + N_3s + N_4}{s^2}\bigg|_{s=-1} = -I + N_2 - N_3 + N_4$$

$$= \begin{bmatrix} -1 & 1 & 1 & -1 & 0 \\ 0 & 0 & 0 & 0 & 0 \\ -1 & 1 & 3 & -1 & 0 \\ 0 & 0 & 0 & 2 & 0 \\ 0 & 0 & 0 & 0 & 2 \end{bmatrix} \tag{18.34c}$$

and finally

$$R_2^1 = \frac{d}{ds}\left[\frac{Is^3 + N_2s^2 + N_3s + N_4}{s^2}\right]_{s=-1}$$

$$= \left[\frac{3Is^2 + 2N_2s + N_3}{s^2}\right]_{s=-1} - 2\left[\frac{Is^3 + N_2s^2 + N_3s + N_4}{s^3}\right]_{s=-1}$$

$$= I - N_3 + 2N_4 = \begin{bmatrix} 1 & 0 & 0 & 0 & 0 \\ -1 & 1 & 1 & -1 & 0 \\ 1 & 0 & 0 & 0 & 0 \\ 0 & 0 & 0 & 0 & 0 \\ -1 & 1 & 1 & -1 & 0 \end{bmatrix} \qquad (18.34d)$$

Part (iv). The partial fraction decomposition of equation 18.33 with the R_i^j matrices given in equations 18.34 leads to the state transition matrix

$$\Phi(t) = \mathcal{L}^{-1}\{(sI-A)^{-1}\} = \qquad (18.35)$$

$$[R_1^1 + R_1^2 t + R_2^1 e^{-t} + R_2^2 te^{-t}]1^+(t)$$

The forms $R_1^2 t$ and $R_2^2 t\, e^{-t}$ appear because of the presence of multiple eigenvalues, whose multiplicity in the minimal polynomial $\psi_A(s) = s^2(s+1)^2$ is greater than unity.

Part (v). The last task is to show that the properties of the matrices R_i^j hold. Two facts are utilized in the demonstration:

(i)

$$\pi_A(s) = s^5 + a_1 s^4 + a_2 s^3 + a_3 s^2 + a_4 s + a_5$$

$$= s^5 + 2s^4 + s^3$$

i.e., $a_1 = 2, a_2 = 1$, and $a_3 = a_4 = a_5 = 0$, and from Leverrier's algorithm

(ii)

$$N_4 A + a_4 I = N_4 A = [0].$$

With these points in mind, note that

$$R_1^1(A - (0)I) = R_1^1 A = (N_3 - 2N_4)A$$

$$= N_3 A = N_4 - a_4 I = N_4 = R_1^2$$

as required by theorem 18.5. Also,

$$R_2^2[A - (-1)I] = R_2^1(A + I)$$

$$= (I - N_3 + 2N_4)A + (I - N_3 + 2N_4)$$

$$= N_2 - 2I - (N_4 - a_4 I) + I - N_3 + 2N_4$$

$$= N_2 - I - N_3 + N_4 = R_2^2$$

Hence, $R_2^1(A + I) = R_2^2$, as required. Finally, equation 18.25a, requires

$$I = R_1^1 + R_2^1 = N_3 - 2N_4 + I - N_3 + 2N_4 = I$$

Observe in the example that computation of the residue matrices R_i^j proceeds more smoothly via the matrices N_i rather than by direct numerical calculations. References [8] and [9] provide discussions of various methods for computing the residues. In [9] the focus lies on a programmable method.

Exercise. A particular eigenvalue λ_1 of a time-invariant state model has an associated residue matrix $R_1^1 = I$. Also,

$$[A - \lambda_1 I] = \begin{bmatrix} 0 & 0 & 0 & 0 \\ 1 & 0 & 0 & 0 \\ 0 & 1 & 0 & 0 \\ 0 & 0 & 1 & 0 \end{bmatrix}$$

Which of the following is the contribution of the residue matrix R_1^4 to the state transition matrix?

(a) $\begin{bmatrix} 0 & 0 & 0 & 0 \\ 0 & 0 & 0 & 0 \\ 0 & 0 & 0 & 0 \\ 1 & 0 & 0 & 0 \end{bmatrix} t^3 exp(\lambda_1 t)$ (b) $\begin{bmatrix} 0 & 0 & 0 & 0 \\ 0 & 0 & 0 & 0 \\ 1 & 0 & 0 & 0 \\ 0 & 1 & 0 & 0 \end{bmatrix} t^3 exp(\lambda_1 t)$

(c) $\begin{bmatrix} 0 & 0 & 0 & 0 \\ 0 & 0 & 0 & 0 \\ 0 & 0 & 0 & 0 \\ 1 & 0 & 0 & 0 \end{bmatrix} t^4 exp(\lambda_1 t)$ (d) $\begin{bmatrix} 0 & 0 & 0 & 0 \\ 0 & 0 & 0 & 0 \\ 1 & 0 & 0 & 0 \\ 0 & 1 & 0 & 0 \end{bmatrix} t^4 exp(\lambda_1 t)$

(e) Cannot be determined   (f) None of the above

THE JORDAN FORM

Using Laplace transform methods, the development thus far has led to the general form of the time-invariant state transition matrix

$$\Phi(t) = \exp[At] = \sum_{i=1}^{\sigma} \sum_{j=1}^{m_i} R_i^j \frac{t^{j-1}}{(j-1)!} e^{\lambda_i t} 1^+(t) \tag{18.35}$$

where $\psi_A(\lambda) = (\lambda - \lambda_1)^{m_1}(\lambda - \lambda_2)^{m_2} \cdots (\lambda - \lambda_\sigma)^{m_\sigma}$ is the minimal polynomial of A having distinct eigenvalues $\lambda_1, \lambda_2, ..., \lambda_\sigma$. Now, is there a parallel time-domain derivation of this formula? Indeed, there is. To obtain it, recall that the decomposition here smacks of the decomposition constructed in chapter 17 for the case where A had distinct eigenvalues and a full set of eigenvectors. Here, the number of independent eigenvectors of A is $n - \sum_{i=1}^{\sigma} (m_i - 1)$; thus, A cannot be diagonalized. But isn't there some *almost* diagonal equivalent form of A whose matrix exponential would yield equation 18.35? Again, the answer is that there is. The special form is called the *Jordan form of A*. An example of a matrix in Jordan form is

$$
A = \begin{bmatrix} -1 & 1 & | & 0 \\ 0 & -1 & | & 0 \\ -- & -- & + & -- \\ 0 & 0 & | & -2 \end{bmatrix}
$$

This matrix is not diagonal and cannot be diagonalized. However, it is block diagonal and can be viewed as almost diagonal. Using the Laplace transform method or the Taylor series approach, we obtain

$$
\Phi(t) = \exp[At] = \begin{bmatrix} e^{-t} & te^{-t} & | & 0 \\ 0 & e^{-t} & | & 0 \\ -- & -- & + & -- \\ 0 & 0 & | & e^{-2t} \end{bmatrix} = \begin{bmatrix} \exp\begin{bmatrix} -1 & 1 \\ 0 & -1 \end{bmatrix} t & | & 0 \\ -- & + & -- \\ 0 & | & \exp[-2t] \end{bmatrix}
$$

Since the matrix exponential of a block-diagonal matrix is a block-diagonal matrix of matrix exponentials, is it possible to construct a transformation T such that $A = TJT^{-1}$? J, of course, must be block diagonal, say, of the form $J = $ block-diag$[J_1,...,J_p]$, where p is equal to the number of independent eigenvectors. Each block J_i will have the form

$$
J_i = \begin{bmatrix} \lambda_i & 1 & 0 & & & \\ 0 & \lambda_i & 1 & \cdots & & \\ & & & \cdots & & \\ & & \cdot & & & \\ 0 & & \cdot & & \lambda_i & 1 \\ 0 & & \cdots & & 0 & \lambda_i \end{bmatrix} \qquad (18.36)
$$

It then follows that

$$
\Phi(t) = \exp[At] = T\exp[Jt]T^{-1} \qquad (18.37)
$$

where

$$
\exp[Jt] = \text{block-diag}[\exp(J_1t),...,\exp(J_pt)] \qquad (18.38)
$$

The key to evaluating equation 18.38 lies in knowing a simple or standardized formula for $\exp[J_it]$. Assuming J_i is $n_i \times n_i$, and using either the Laplace transform method or the Taylor series approach,

$$
\exp[J_i t] =
\begin{bmatrix}
e^{\lambda_i t} & te^{\lambda_i t} & \cdots & \dfrac{t^{n_i-1}}{(n_i-1)!}\,e^{\lambda_i t} \\[2mm]
0 & e^{\lambda_i t} & \cdots & \dfrac{t^{n_i-2}}{(n_i-2)!}\,e^{\lambda_i t} \\[2mm]
\vdots & \vdots & \ddots & \vdots \\[2mm]
0 & 0 & \cdots & e^{\lambda_i t}
\end{bmatrix}
\tag{18.39}
$$

Specifying the structure of T completes the decomposition, but not the justification. An inductive justification can be found in [10] and other developments in [6]. From past experience, it is a sure bet that the decomposition depends on the eigenvectors of A. Suppose A has p independent eigenvectors $\{e_1^1, e_2^1,\ldots,e_p^1\}$ associated with the eigenvalues $\lambda_1,\ldots,\lambda_p$. Then it is possible to show that there exists a basis

$$
\left\{ e_1^1,\ldots,e_1^{n_1},e_2^1,\ldots,e_2^{n_2},\ldots,e_p^1,\ldots,e_p^{n_p} \right\}
\tag{18.40}
$$

for $\mathbf{C}^n(\mathbf{R}^n)$ over the field of complex scalars. The vectors $e_i^j, j \geq 2$, are termed *generalized eigenvectors*. They depend on e_i^1, or, equivalently, e_i^1 generates the $e_i^j, j \geq 2$, according to the formulas

$$
Ae_i^2 = e_i^1 + \lambda_i e_i^2,\ Ae_i^3 = e_i^2 + \lambda_i e_i^3,\ldots,\ Ae_i^{n_i} = e_i^{n_i-1} + \lambda_i e_i^{n_i}
\tag{18.41}
$$

Thus, if one defines the matrix T according to

$$
T = \left[e_1^1,\ldots,e_1^{n_1},e_2^1,\ldots,e_2^{n_2},\ldots,e_p^1,\ldots,e_p^{n_p} \right]
\tag{18.42}
$$

then $AT = TJ$, and J must be block diagonal with blocks of the form given by equation 18.36. By assumption, the set of 18.40 is a basis for $\mathbf{R}^n$, which means that T is nonsingular. It then follows that $A = TJT^{-1}$. Before proving the existence of such a basis, a couple of examples illustrate the decomposition into Jordan form and generate some experience in regard to the computation involved.

EXAMPLE 18.4

This example illustrates the computation of the Jordan decomposition of the matrix

$$
A =
\begin{bmatrix}
-1 & 0 & 0 & 1 & 0 \\
1 & -1 & 0 & -1 & 0 \\
-1 & 0 & 0 & 2 & 0 \\
0 & 1 & 0 & 0 & -1 \\
1 & 0 & 0 & -1 & -1
\end{bmatrix}
$$

Executing some straightforward arithmetic leads to a characteristic polynomial $\pi_A(\lambda) = \det[\lambda I - A] = (\lambda + 1)^3 \lambda^2$. Thus, A has only two distinct eigenvalues, $\lambda_1 = -1$ and $\lambda_2 = 0$, and the vital question becomes how many independent eigenvectors there are. Solving $Ae_i = \lambda_i e_i$ for $\lambda_i = -1$ and $\lambda = 0$, respectively, produces only two independent eigenvectors

$$e_1^1 = \begin{bmatrix} 0 \\ 1 \\ 0 \\ 0 \\ 1 \end{bmatrix}$$

and   (18.43)

$$e_2^1 = \begin{bmatrix} 0 \\ 0 \\ 1 \\ 0 \\ 0 \end{bmatrix}$$

Thus, there are two generalized eigenvectors, say, e_1^2 and e_1^3, attached to $\lambda_1 = -1$ and one generalized eigenvector, e_2^2, attached to $\lambda_2 = 0$. These eigenvectors must all satisfy

$$\begin{bmatrix} -1 & 0 & 0 & 1 & 0 \\ 1 & -1 & 0 & -1 & 0 \\ -1 & 0 & 0 & 2 & 0 \\ 0 & 1 & 0 & 0 & -1 \\ 1 & 0 & 0 & -1 & -1 \end{bmatrix} \begin{bmatrix} a \\ b \\ c \\ d \\ e \end{bmatrix} = \lambda_i \begin{bmatrix} a \\ b \\ c \\ d \\ e \end{bmatrix} + e_i^{j-1} \qquad (18.44)$$

where e_i^j is represented by $[a \; b \; c \; d \; e]^t$. For example, to compute e_1^2, we use equation 18.44 to obtain the following equations:

(i) $-a + d = -a$, implying that $d = 0$.
(ii) $a - b = -b + 1$, implying that $a = 1$.
(iii) $-a + 2d = -c$, implying that $c = 1$.
(iv) $b - e = 0$, implying the simple choice $b = e = 0$.

The resulting vector is $e_1^2 = [1 \; 0 \; 1 \; 0 \; 0]^t$. Similarly, one can use equation 18.44 to construct e_1^3 as follows:

(i) $-a + d = -a + 1$, implying that $d = 1$.
(ii) $a - b - d = -b$, implying that $a = 1$.
(iii) $-a + 2d = -c + 1$, implying that $c = 0$.
(iv) $b - e = -d$, implying a simple choice of $b = 0$ and $e = 1$.

Thus, $e_1^3 = [1 \; 0 \; 0 \; 1 \; 1]^t$. Of course, other choices are possible, but all lead to the same result. Then, using equation 18.44 again results in

$e_2^2 = [1\ 0\ 0\ 1\ 0]^t$. With $T = [e_1^1, e_1^2, e_1^3, e_2^1, e_2^2]$, the Jordan decomposition of A takes the form

$$
A = \begin{bmatrix} 0 & 1 & 1 & 0 & 1 \\ 1 & 0 & 0 & 0 & 0 \\ 0 & 1 & 0 & 1 & 0 \\ 0 & 0 & 1 & 0 & 1 \\ 1 & 0 & 1 & 0 & 0 \end{bmatrix} \begin{bmatrix} -1 & 1 & 0 & 0 & 0 \\ 0 & -1 & 1 & 0 & 0 \\ 0 & 0 & -1 & 0 & 0 \\ 0 & 0 & 0 & 0 & 1 \\ 0 & 0 & 0 & 0 & 0 \end{bmatrix} \begin{bmatrix} 0 & 1 & 0 & 0 & 0 \\ 1 & 0 & 0 & -1 & 0 \\ 0 & -1 & 0 & 0 & 1 \\ -1 & 0 & 1 & 1 & 0 \\ 0 & 1 & 0 & 1 & -1 \end{bmatrix}
$$

It then follows that the associated state transition matrix is

$$\Phi(t) = \exp[At]$$

$$
= \begin{bmatrix} 0 & 1 & 1 & 0 & 1 \\ 1 & 0 & 0 & 0 & 0 \\ 0 & 1 & 0 & 1 & 0 \\ 0 & 0 & 1 & 0 & 1 \\ 1 & 0 & 1 & 0 & 0 \end{bmatrix} \begin{bmatrix} e^{-t} & te^{-t} & 0.5t^2e^{-t} & 0 & 0 \\ 0 & e^{-t} & te^{-t} & 0 & 0 \\ 0 & 0 & e^{-t} & 0 & 0 \\ 0 & 0 & 0 & 1 & t \\ 0 & 0 & 0 & 0 & 1 \end{bmatrix} \begin{bmatrix} 0 & 1 & 0 & 0 & 0 \\ 1 & 0 & 0 & -1 & 0 \\ 0 & -1 & 0 & 0 & 1 \\ -1 & 0 & 1 & 1 & 0 \\ 0 & 1 & 0 & 1 & -1 \end{bmatrix}
$$

EXAMPLE 18.5

This example sketches one of the vagaries in constructing the Jordan decomposition. Let

$$
A = \begin{bmatrix} 1 & 2 & 0 & 1 \\ 0 & 1 & 0 & 0 \\ 0 & -1 & 1 & 0 \\ 0 & 0 & 0 & 1 \end{bmatrix}
$$

Some straightforward number grinding reveals a characteristic polynomial of $\pi_A(\lambda) = (\lambda - 1)^4$ and a minimal polynomial $\psi_A(\lambda) = (\lambda - 1)^2$. It follows that there must be at least one Jordan 2×2 block in the Jordan decomposition of A. The remaining blocks are either a second 2×2 Jordan block or two 1×1 blocks containing the value 1. The choice depends on how many independent eigenvectors A has: if A has two independent eigenvectors, there are two 2×2 Jordan blocks; if A has three independent eigenvectors, there are two 1×1 blocks. To find out which alternatives are obtained, note that the number of independent eigenvectors associated with $\lambda = 1$ must equal the dimension of the null space of $(\lambda I - A) = (I - A)$. Specifically,

$$
(I - A) = \begin{bmatrix} 0 & -2 & 0 & -1 \\ 0 & 0 & 0 & 0 \\ 0 & 1 & 0 & 0 \\ 0 & 0 & 0 & 0 \end{bmatrix}
$$

has a two-dimensional null space spanned by the vectors $v_1 = [1\ 0\ 0\ 0]^t$ and $v_2 = [0\ 0\ 1\ 0]^t$. Thus, there are two 2×2 Jordan blocks associated with A.

Now, which linear combination of v_1 and v_2 yields e_1^1, and which linear combination yields e_1^2? The answer is obtained by a trial-and-error procedure. First, let e_1^1 have the general form $e_1^1 = [\alpha \ 0 \ \beta \ 0]^t$. The recursive relationship among the eigenvectors of a Jordan block requires that

$$
\begin{bmatrix} 1 & 2 & 0 & 1 \\ 0 & 1 & 0 & 0 \\ 0 & -1 & 1 & 0 \\ 0 & 0 & 1 & 1 \end{bmatrix} \begin{bmatrix} a \\ b \\ c \\ d \end{bmatrix} = \begin{bmatrix} a \\ b \\ c \\ d \end{bmatrix} + \begin{bmatrix} \alpha \\ 0 \\ \beta \\ 0 \end{bmatrix}
\tag{18.45}
$$

where $e_1^2 = [a \ b \ c \ d]^t$. Then, equation 18.45 implies

$$
a + 2b + d = a + \alpha
\tag{18.46a}
$$

and

$$
-b + c = c + \beta
\tag{18.46b}
$$

The choices $e_1^1 = [1 \ 0 \ 0 \ 0]^t$ and $e_1^2 = [0 \ 0 \ 0 \ 1]$ satisfy the constraints of equations 18.46. A similar procedure leads to the choices $e_2^1 = [0 \ 0 \ 1 \ 0]^t$ and $e_2^2 = [0 \ -1 \ 0 \ 2]^t$. The resulting Jordan decomposition of A is

$$
A = \begin{bmatrix} 1 & 0 & 0 & 0 \\ 0 & 0 & 0 & -1 \\ 0 & 0 & 1 & 0 \\ 0 & 1 & 0 & 2 \end{bmatrix} \begin{bmatrix} 1 & 1 & 0 & 0 \\ 0 & 1 & 0 & 0 \\ 0 & 0 & 1 & 1 \\ 0 & 0 & 0 & 1 \end{bmatrix} \begin{bmatrix} 1 & 0 & 0 & 0 \\ 0 & 2 & 0 & 1 \\ 0 & 0 & 1 & 0 \\ 0 & -1 & 0 & 0 \end{bmatrix}
$$

The associate state transition matrix is

$$
\Phi(t) = \begin{bmatrix} 1 & 0 & 0 & 0 \\ 0 & 0 & 0 & -1 \\ 0 & 0 & 1 & 0 \\ 0 & 1 & 0 & 2 \end{bmatrix} \begin{bmatrix} e^t & te^t & 0 & 0 \\ 0 & e^t & 0 & 0 \\ 0 & 0 & e^t & te^t \\ 0 & 0 & 0 & e^t \end{bmatrix} \begin{bmatrix} 1 & 0 & 0 & 0 \\ 0 & 2 & 0 & 1 \\ 0 & 0 & 1 & 0 \\ 0 & -1 & 0 & 0 \end{bmatrix}
$$

There are, of course, other choices for the transformation matrix. For example, A also has the decomposition

$$
A = \begin{bmatrix} 1 & 0 & 0 & -1 \\ 0 & 1 & 0 & -1 \\ -1 & 0 & 1 & 1 \\ 0 & -1 & 0 & 2 \end{bmatrix} \begin{bmatrix} 1 & 1 & 0 & 0 \\ 0 & 1 & 0 & 0 \\ 0 & 0 & 1 & 1 \\ 0 & 0 & 0 & 1 \end{bmatrix} \begin{bmatrix} 1 & 1 & 0 & 1 \\ 0 & 2 & 0 & 1 \\ 1 & 0 & 1 & 0 \\ 0 & 1 & 0 & 1 \end{bmatrix}
$$

As mentioned earlier, a rigorous proof of the existence of the Jordan form fills numerous pages of [6] and [10]. Only two further points need to be made. First, the (generalized) eigenvectors $\{e_i^1, e_i^2, \ldots, e_i^{n_i}\}$ associated with the ith Jordan block span an *A-invariant space*—i.e., for any vector $v \in \text{span}\{e_i^1, \ldots, e_i^{n_i}\}$, $Av \in \text{span}\{e_i^1, \ldots, e_i^{n_i}\}$. And second, each span$\{e_i^1, \ldots, e_i^{n_i}\}$ contains only one right eigenvector, namely, e_i^1, to within a multiplicative constant—i.e., $e_j^1 \notin \text{span } \{e_i^1, \ldots, e_i^{n_i}\}$ for $j \neq i$. To prove the latter assertion, suppose the contrary, i.e., suppose the linear combination

$$\xi_1 e_i^1 + \xi_2 e_i^2 + \cdots + \xi_{n_i} e_i^{n_i}$$

is an eigenvector with ξ_j not all zero. Then it must satisfy the usual eigenvector-eigenvalue equation,

$$A \sum_{j=1}^{n_i} \xi_j e_i^j = \lambda \sum_{j=1}^{n_i} \xi_j e_i^j$$

Using the relationship among the eigenvectors in the set given in equation 18.41, it follows that

$$\xi_1 \lambda_i e_i^1 + \xi_2 (e_i^1 + \lambda_i e_i^2) + \cdots + \xi_{n_i} (e^{n_i - 1} + \lambda_i e_i^{n_i})$$

$$= \lambda \xi_1 e_i^1 + \lambda \xi_2 e_i^2 + \cdots + \lambda \xi_{n_i} e_i^{n_i}$$

Equating the coefficients of each e_i^j requires that

$$\xi_1 \lambda_i + \xi_2 = \lambda \xi_1$$
$$\xi_2 \lambda_i + \xi_3 = \lambda \xi_2$$
$$\vdots \qquad\qquad (18.47)$$
$$\xi_{n_i - 1} \lambda_i + \xi_{n_i} = \lambda \xi_{n_i - 1}$$
$$\xi_{n_i} \lambda_i = \lambda \xi_{n_i}$$

It immediately follows that $\lambda = \lambda_i$. For if this were not so, then $\xi_{n_i} \lambda_i = \lambda \xi_{n_i}$ would require $\xi_{n_i} = 0$. Further, each ascending equation would then require $\xi_j = 0$ for $j = n_i - 1, \ldots, 1$, contradicting the hypothesis that not all the ξ_j were zero.

Substituting λ_i for λ, and starting at the top of equation 18.47 and descending to the bottom implies that $\xi_2 = 0$, $\xi_3 = 0, \ldots, \xi_{n_i} = 0$. It follows that $\xi_1 e_i^1$ is the eigenvector (to within a multiplicative constant) contained in the span $\{e_i^1, e_i^2, \ldots, e_i^{n_i}\}$.

DISCRETE-TIME SYSTEM COUNTERPARTS

Theorems akin to those developed in preceding sections run through the theory of discrete-time systems. Our purpose here is to sketch their counterparts. Accordingly, recall the discrete-time stationary state dynamical equations

$$x(k + 1) = Ax(k) + Bu(k)$$

$$y(k) = Cx(k) + Du(k)$$

The Z-transforms of $\{x(k)\}_{k=0}^{\infty}$ and $\{x(k + 1)\}_{k=0}^{\infty}$ are

$$x(z) = \sum_{k=0}^{\infty} x(k) z^{-k} \triangleq Z[\{x(k)\}]$$

and

$$Z[\{x(k+1)\}] = \sum_{k=0}^{\infty} x(k+1)z^{-k} = x(1) + x(2)z^{-1} + \cdots$$

$$= zx(z) - zx(0)$$

From this, it is straightforward to arrive at

$$x(z) = z(zI - A)^{-1}x(0) + (zI - A)^{-1}Bu(z)$$

Observe that if the input sequence is identically zero, then

$$x(z) = z(zI - A)^{-1}x(0)$$

so that the state transition matrix is

$$\Phi(k) = A^k = Z^{-1}\{z(zI - A)^{-1}\}$$

Finally, since $y(k) = Cx(k) + Du(k)$,

$$y(z) = Cz(zI - A)^{-1}x(0) + [C(zI - A)^{-1}B + D]u(z)$$

Letting $x(0) = \theta$ allows us to define the obvious.

Definition 18.4.   The *transfer function matrix* of the discrete-time stationary linear state model is

$$H(z) = C(zI - A)^{-1}B + D$$

The crux of constructing this transfer function matrix is again the computation of $(zI - A)^{-1}$. As with the continuous-time case,

$$(zI - A)^{-1} = \frac{R(z)}{\pi_A(z)} = \frac{N_1 z^{n-1} + N_2 z^{n-2} + \cdots + N_{n-1}z + N_n}{z^n + a_1 z^{n-1} + \cdots + a_{n-1}z + a_n} \tag{18.48}$$

where the N_i and the a_i can be computed using Leverrier's algorithm. In constructing the partial fraction expansion, one first expands $R(z)/z\pi_A(z)$ and then computes the appropriate form of $R(z)/\pi_A(z)$ by multiplying through by z.

As a final point, discrete-time system A-matrices obviously have a Jordan form that is computed as described earlier. The distinction between the continuous- and discrete-time cases lies in the computation of the state transition matrix. In the latter, $\Phi(k) = TJ^k T^{-1}$, where

$$J^k = \text{block-diag}[J_1^k, \dots, J_p^k]$$

and for $k \geqslant 0$,

$$J_i^k = \begin{bmatrix} \lambda_i^k & k\lambda_i^{k-1} & \dfrac{k(k-1)}{2!}\lambda_i^{k-2} & \dfrac{k(k-1)(k-2)}{3!}\lambda_i^{k-3} & \cdots \\[2ex] 0 & \lambda_i^k & k\lambda_i^{k-1} & \dfrac{k(k-1)}{2!}\lambda_i^{k-2} & \cdots \\[2ex] & & & \vdots & \\[1ex] 0 & 0 & 0 & 0 & \cdots\,\lambda_i^k \end{bmatrix}$$

For example,

$$
\begin{bmatrix} \lambda & 1 & 0 & 0 \\ 0 & \lambda & 1 & 0 \\ 0 & 0 & \lambda & 1 \\ 0 & 0 & 0 & \lambda \end{bmatrix}^4 = \begin{bmatrix} \lambda^4 & 4\lambda^3 & 6\lambda^2 & 4\lambda \\ 0 & \lambda^4 & 4\lambda^3 & 6\lambda^2 \\ 0 & 0 & \lambda^4 & 4\lambda^3 \\ 0 & 0 & 0 & \lambda^4 \end{bmatrix}
$$

The remaining features of A-matrices, of course, again duplicate their continuous-time counterparts.

CONCLUDING REMARKS

This chapter has introduced a great many ideas—ideas that have borne further fruit through the efforts of many researchers. For example, numerical aspects of computing $H(s)$ find expression in [11,12,13], while advanced theories of control based on special factorizations of the transfer function matrix $H(s)$ have been developed in other works. A good summary of this development can be found in [14], where numerous references to the literature also occur.

We close with a couple of comments on the development in [15]. First, computation of $\exp[At]$ by the inverse Laplace transform of $(sI - A)^{-1}$ has several numerical difficulties. For example, the number of flops needed is proportional to n^4 (A is $n \times n$), which seriously increases the effects of round-off. Second, the storage requirements are on the order of n^3, which is significantly more than those of other methods, such as the Padé technique. Multiple eigenvalues compound the storage problem even further, since one must numerically distinguish between distinct and nondistinct eigenvalues.

In the case of the Jordan form, the problem of confluent eigenvalues, the determination of the size of the Jordan blocks, and computation of the generalized eigenvectors combine to make construction of the Jordan form numerically unstable. In [15], Parlett discusses an algorithm which clusters neighboring eigenvalues into upper triangular blocks. This technique limits round-off magnification to pertinent upper triangular blocks.

PROBLEMS

1. Suppose A is $n \times n$ and nonsingular. Show that $A^{-1} = -(a_n)^{-1}N_n$, where N_n is as defined in Leverrier's algorithm.

2. Use the result of problem 1 to compute A^{-1} when

$$
A = \begin{bmatrix} -1 & -4 & 0 \\ 1 & -1 & 0 \\ 4 & 2 & -3 \end{bmatrix}
$$

3. Compute the nine cofactors of $(sI - A)$ for the matrix A of problem 2. What is the cofactor matrix? What is its transpose? Compare your result with equation 18.10.

4. Compute (*i*) $(sI - A)^{-1}$, (*ii*) the associated partial fraction expansion, and (*iii*) the state transition matrices for each of the following A-matrices:

(**a**) $\begin{bmatrix} -1 & 1 & 0 \\ 0 & -1 & 1 \\ 0 & 0 & -1 \end{bmatrix}$ (**b**) $\begin{bmatrix} 0 & -1 & 3 \\ 2 & -3 & 6 \\ 1 & -1 & 2 \end{bmatrix}$ (**c**) $\begin{bmatrix} 0 & 0 & 0 & 0 & 0 \\ 0 & -2 & -1 & 1 & 0 \\ 0 & 2 & 1 & -1 & 0 \\ 0 & 0 & 0 & -1 & 1 \\ 0 & 0 & 0 & 0 & -1 \end{bmatrix}$

(*Hint*: For part (b), consider extending the ideas of a partial fraction expansion of a scalar transfer function having complex eigenvalues.)

5. A particular state model has a 1×2 matrix C, a 2×2 matrix B, and a 2×2 matrix A. Experimental measurements produce zero-state system responses as follows:

(*i*)

$$\text{If } u(t) = \begin{bmatrix} 1 \\ 0 \end{bmatrix} 1^+(t), \text{ then } y(t) = 3 - 3e^{-t} - 2te^{-t}$$

(*ii*)

$$\text{If } u(t) = \begin{bmatrix} 0 \\ t \end{bmatrix} 1^+(t), \text{ then } y(t) = 3 - t - 3e^{-t} - 2te^{-t}$$

(**a**) Working strictly in the time domain, determine the impulse response matrix.
(**b**) Determine the transfer function matrix. (*Note*: This can be done independently of part (a).)
(**c**) Determine A when $C = [2 \ -1]$ and $B = \begin{bmatrix} 1 & 0 \\ 1 & -1 \end{bmatrix}$. (*Hint*: Consider derivatives of $H(t) = C \exp(At)B$.)

6. Consider the two-port shown in Figure P18.6. The z-parameter transfer function matrix of this two-port has the form

$$\begin{bmatrix} V_1(s) \\ V_2(s) \end{bmatrix} = \begin{bmatrix} z_{11}(s) & z_{12}(s) \\ z_{21}(s) & z_{22}(s) \end{bmatrix} \begin{bmatrix} I_1(s) \\ I_2(s) \end{bmatrix}$$

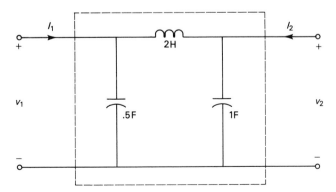

Figure P18.6 Figure for Problem 6.

(a) Compute each of the z_{ij}'s and, thereby, the transfer function matrix of the circuit.

(b) Another often more useful transfer function takes the form

$$\begin{bmatrix} V_1(s) \\ I_1(s) \end{bmatrix} = \begin{bmatrix} a(s) & b(s) \\ c(s) & d(s) \end{bmatrix} \begin{bmatrix} V_2(s) \\ -I_2(s) \end{bmatrix}$$

Compute $a(s)$, $b(s)$, $c(s)$, and $d(s)$ and, thereby, this transfer function matrix (see [16], pp. 133–134).

(c) Compute the impulse response matrices for parts (a) and (b).

7. Suppose that in the frequency domain the state response of a particular time-invariant state model has the form

$$x(s) = \frac{1}{s(s+2)} \begin{bmatrix} s+1 & 1 \\ 1 & s+1 \end{bmatrix} \begin{bmatrix} x_1(0) \\ x_2(0) \end{bmatrix} + \frac{1}{s+2} \begin{bmatrix} 1 \\ -1 \end{bmatrix} u(s)$$

(a) If the system response satisfies

$$y(s) = \begin{bmatrix} x_1(s) - x_2(s) + u(s) \\ x_1(s) + x_2(s) + u(s) \end{bmatrix}$$

compute the transfer function matrix of the system.

(b) Compute the impulse response matrix for the system.

(c) What are the C- and D-matrices for the system.

(d) Construct the A-matrix for the linear time-invariant state model which gives rise to the given frequency domain representation.

(e) Compute the B-matrix for the same state model.

8. Suppose $\dot{x} = Ax$, where

$$A = \begin{bmatrix} -1 & 0 & 0 & 0 \\ 1 & -1 & 0 & 0 \\ 0 & 0 & -1 & 1 \\ 0 & 0 & 0 & -2 \end{bmatrix}$$

(a) Find the characteristic polynomial $\pi_A(s)$ of A.

(b) Compute $(sI - A)^{-1} = R(s)/\pi_A(s)$.

(c) Find the minimal polynomial $\psi_A(s)$ of A. (Justify your answer.)

(d) Find $\hat{R}(s)$, where $(sI - A)^{-1} = \hat{R}(s)/\psi_A(s)$.

(e) What is the form (structure) of the partial fraction expansion of the matrix in part (d)? Give the definition of the residue matrices.

(f) Compute the required residue matrices R_i^j.

(g) Compute the state transition matrix.

9. Consider the state model

$$\dot{x} = Ax + Bu$$
$$y = Cx$$

where

$$A = \begin{bmatrix} -1 & 1 & 1 \\ 0 & -1 & 0 \\ 0 & 0 & 0 \end{bmatrix}, \quad B = \begin{bmatrix} 0 \\ 1 \\ 0 \end{bmatrix}, \quad C = \begin{bmatrix} 1 & 0 & -1 \\ 0 & 0 & 1 \end{bmatrix}$$

(a) Compute the characteristic polynomial of A. What are the system eigenvalues?

(b) Compute $(sI - A)^{-1}$ by Leverrier's algorithm.

(c) Compute the transfer function matrix.

(d) Compute a matrix partial fraction expansion of $(sI - A)^{-1}$.

(e) Construct an expression for the state transition matrix in terms of matrix directed modes.

(f) If $x(0) = [1\ 0\ 1]'$, find the zero-input state response. Evaluate at $t = 1$.

(g) Find the system response if $x(0) = [1\ 0\ 1]'$ and $u(t) = 2\delta(t-1)$.

10. Consider the discrete-time system

$$x(k+1) = Ax(k) + Bu(k), \qquad x(0) = x_0$$
$$y(k) = Cx(k) + Du(k)$$

Let $x(z)$, $u(z)$, and $y(z)$ respectively denote the Z-transform of the sequences $\{x(k)\}$, $\{u(k)\}$, and $\{y(k)\}$.

(a) Derive an expression for the zero-input state response, and determine the state transition matrix $\Phi(k) = A^k, k \geq 0$.

(b) Derive an expression for the zero-state state response.

(c) Given your answer to part (b), determine the form of the transfer function matrix.

(d) Suppose the matrices A, B, C, and D are given by

$$A = \begin{bmatrix} 1 & 0 & -2 \\ 0 & 1 & 0 \\ 1 & 0 & -2 \end{bmatrix}, \quad B = \begin{bmatrix} 1 \\ 1 \\ 1 \end{bmatrix}, \quad C = [0\ 0\ 1],\ D = [1]$$

where A is known to have eigenvalues $\lambda_1 = 1, \lambda_2 = -1$, and $\lambda_3 = 0$.

(i) Find the right eigenvectors of A.

(ii) Find the left eigenvectors of A.

(iii) Decompose the discrete-time state transition matrix into matrix directed modes.

(e) Construct a partial fraction expansion of $(zI - A)^{-1}$ for the given matrices A, B, C, and D. (*Hint*: Use the result of part (d).)

(f) Compute the transfer function matrix of the discrete-time system.

11. Suppose the state trajectory of $\dot{x} = Ax + Bu, y = [1\ 2]x$ is given by

$$x(t) = \begin{bmatrix} e^{-t} & e^{-t} - e^t \\ 0 & e^t \end{bmatrix} x(0) + [e^t - 1] \begin{bmatrix} 1 \\ -1 \end{bmatrix}$$

when $u(t) = 1^+(t)$.

(a) Find the impulse response of the system.

(b) Find the A-matrix of the system.

(c) Find the B-matrix of the system.

12. Let $\psi(t)$ be a fundamental matrix for the usual linear time-varying state model. Determine an expression for the impulse response matrix.

13. Suppose

$$\dot{x} = Ax + Bu$$
$$y = Cx$$

where

$$A = \begin{bmatrix} -2 & 0 & 0 & 1 \\ 0 & -1 & 0 & 0 \\ -1 & -1 & -1 & 2 \\ 0 & 0 & 0 & -1 \end{bmatrix}, \quad B = \begin{bmatrix} 1 \\ 1 \\ 1 \\ 0 \end{bmatrix}$$

It is known that the eigenvalues of A are $\lambda_1 = -2, \lambda_2 = \lambda_3 = \lambda_4 = -1$ and that $\psi_A(s) = (s + 2)(s + 1)^2$. It is also known that in a partial fraction expansion of $(sI - A)^{-1}$

$$
R_2^1 = \begin{bmatrix} 0 & 0 & 0 & 1 \\ 0 & 1 & 0 & 0 \\ -1 & 0 & 1 & 1 \\ 0 & 0 & 0 & 1 \end{bmatrix}
$$

(a) Without directly computing $(sI - A)^{-1}$, compute the remaining matrices R_i^j.
(b) Compute the state transition matrix.
(c) If the transfer function of the system is

$$
H(s) = \frac{2}{s + 2} - \frac{1}{(s + 1)^2}
$$

determine the impulse response matrix.
(d) If $C = [a \; b \; c \; d]$ with $d = 0$, determine the C-matrix of the system.

14. Use the Laplace transform method to find the state transition matrices for the following A-matrices which are in Jordan form:

$$
\textbf{(a)} \quad \begin{bmatrix} \lambda & 1 \\ 0 & \lambda \end{bmatrix} \qquad \textbf{(b)} \quad \begin{bmatrix} \lambda & 1 & 0 \\ 0 & \lambda & 1 \\ 0 & 0 & \lambda \end{bmatrix} \qquad \textbf{(c)} \quad \begin{bmatrix} \lambda & 1 & 0 & 0 \\ 0 & \lambda & 1 & 0 \\ 0 & 0 & \lambda & 1 \\ 0 & 0 & 0 & \lambda \end{bmatrix}
$$

15. Repeat problem 14 assuming the A-matrices are for a discreter-time system.

16. For each of the following A-matrices, find the eigenvalues, the right eigenvectors, the generalized eigenvectors, and the factorization $A = TJT^{-1}$, where J is in Jordan form. Determine the corresponding state transition matrix assuming a continuous-time system

$$
\textbf{(a)} \quad \begin{bmatrix} -3 & 4 \\ -1 & 1 \end{bmatrix} \qquad \textbf{(b)} \quad \begin{bmatrix} -2 & 1 & 0 \\ 2 & -2 & 1 \\ 0 & -2 & -2 \end{bmatrix} \qquad \textbf{(c)} \quad \begin{bmatrix} -2 & 1 & 1 & 1 \\ -1 & -1 & 1 & 1 \\ 0 & 1 & -1 & 0 \\ -1 & 0 & 1 & 0 \end{bmatrix}
$$

17. Repeat problem 16 assuming a discrete-time system.

18. Consider problem 15 again.
(a) Determine conditions on λ for which $\Phi(k) \to 0$ as $k \to \infty$.
(b) Determine conditions on λ for which the sequence $\{\Phi(k)x(0)\}$ satisfies $\|\Phi(k)x(0)\|_\infty < \infty$ for all possible initial conditions $x(0)$.
(*Note*: The L_∞-norm or sup-norm of a vector sequence $\{z(k)\}$ where $z(k) \in \mathbf{R}^n$ is $\|z(k)\|_\infty = \max_k \|z_i(k)\|_\infty$, where $z_i(k)$ is the ith entry of $z(k)$. Further, $\|z_i(k)\|_\infty = \sup_k |z_i(k)|$. This problem foreshadows the notion of stability in which trajectories like $\Phi(k)x(0)$ must remain finite for all $k \geqslant 0$.)

19. Let $u(k) = 1^+(k)$, $B = [0 \; 0 \; 0 \; 1]'$, and

$$
J = \begin{bmatrix} \lambda & 1 & 0 & 0 \\ 0 & \lambda & 1 & 0 \\ 0 & 0 & \lambda & 1 \\ 0 & 0 & 0 & \lambda \end{bmatrix}
$$

(a) Work strictly in the time domain to develop a closed-form expression for the zero-state state response

$$x(k) = \sum_{j=1}^{k} J^j Bu(k-j)$$

when $\lambda = 1$.

(b) Repeat the calculation for $\lambda \neq 1$.

(*Hint*: Consult chapter 16 and consider each term in $x(k)$ separately.)

20. Repeat problem 19 when $u(k) = \lambda^k 1^+(k)$.

21. Show that the transfer function matrix of the system

$$\begin{bmatrix} \dot{x}_1 \\ \dot{x}_2 \end{bmatrix} = \begin{bmatrix} A_{11} & A_{12} \\ 0 & A_{22} \end{bmatrix} \begin{bmatrix} x_1 \\ x_2 \end{bmatrix} + \begin{bmatrix} B_1 \\ 0 \end{bmatrix}$$

$$y = [C_1 \ C_2] \begin{bmatrix} x_1 \\ x_2 \end{bmatrix}$$

has the form

$$H(s) = C_1(sI - A_{11})^{-1}B_1$$

(*Note*: This result is the basis for showing that uncontrollable modes do not appear in the transfer function matrix.)

22. Suppose two *n*th-order state models (A, B, C) and $(\hat{A}, \hat{B}, \hat{C})$ have the same transfer function $H(s)$. Show that they are zero-state equivalent, i.e., that there exists a non-singular state transformation matrix T such that $A = T^{-1}\hat{A}T$, $B = T^{-1}\hat{B}$, etc.

23. Suppose two systems (A, B, C) and $(\hat{A}, \hat{B}, \hat{C})$ have the property that

$$CA^iB = \hat{C}\hat{A}^i\hat{B}$$

for all $i \geqslant 0$. Show that the two state models are zero-state equivalent.

REFERENCES

1. L. A. Zadeh and C. A. Desoer, *Linear System Theory: The State Space Approach* (New York: McGraw-Hill, 1963).

2. Wilson J. Pugh, *Nonlinear System Theory: The Volterra/Wiener Approach* (Baltimore: The John Hopkins University Press, 1981).

3. M. Schetzen, *The Volterra and Wiener Theories of Nonlinear Systems* (New York: Wiley, 1980).

4. C. Bruni, A. Isidori, and A. Ruberti, "A Method of Realization Based on the Moments of the Impulse Response Matrix," *IEEE Transactions on Automatic Control*, Vol. AC-14, April 1969, pp. 203–204.

5. L. M. Silverman, "Realization of Linear Dynamical Systems," *IEEE Transactions on Automatic Control*, Vol. AC-16, December 1971, pp. 554–567.

6. F. R. Gantmacher, *The Theory of Matrices*, Vol. 1 (New York: Chelsea Publishing Co., 1960).

7. B. G. Mertzios, "Leverrier's Algorithm for Singular Systems," *IEEE Transactions on Automatic Control*, Vol. AC-29, No. 7, July 1984, pp. 652–653.

8. J. F. Mahoney and B. D. Sivazlian, "Partial Fractions Expansion: A Review of Computational Methodology and Efficiency," *Journal of Computational and Applied Mathematics*, Vol. 9, 1983, pp. 247–269.

9. J. J. Bongiorno, Jr., "A Recursive Algorithm for Computing the Partial Fraction Expansion of Rational Functions Having Multiple Poles," *IEEE Transactions on Automatic Control*, Vol. AC-29, No. 7, July 1984, pp. 650–651.

10. I. M. Gel'fand, *Lectures on Linear Algebra* (New York: Interscience, 1961).

11. Cleve Moler and Charles Van Loan, "Nineteen Dubious Ways to Compute the Exponential of a Matrix," *SIAM Review*, Vol. 20, No. 4, October 1978, pp. 801–836.

12. P. Misra and R. V. Patel, "Computation of Transfer Function Matrices of Linear Multivariable Systems," *Automatica*, Vol. 23, No. 5, 1987, pp. 635–640.

13. P. R. Cappello and A. J. Laub, "Systolic Computation of Multivariable Frequency Response," *IEEE Transactions on Automatic Control*, Vol 33, No. 6, June 1988, pp. 550–558.

14. F. Callier and C. Desoer, *Multivariable Feedback Systems* (New York: Springer-Verlag, 1982).

15. B. N. Parlett, "A Recurrence among the Elements of Functions of Triangular Matrices," *Linear Algebra Applications*, 14 (1976), pp. 117–121.

16. Gabor C. Temes and Jack W. LaPatra, *Introduction to Circuit Synthesis and Design* (New York: McGraw-Hill, 1977).

19

THE SINGULAR VALUE DECOMPOSITION AND STATE SPACE APPLICATIONS

INTRODUCTION

Terms such as matrix rank, left and right inverse, the condition of a matrix, the range space of a matrix, and the null space of a matrix, define structures intrinsic to system theory. Their numerical computation therefore becomes critical to a viable theory. The singular value decomposition provides a numerically reliable, stable, accurate factorization of a (rectangular) matrix that permits straightforward computation of these and various other entities pertinent to state variable analysis.

THE SINGULAR VALUE DECOMPOSITION OF A MATRIX

The *singular value decomposition* (SVD) of a matrix, say, $Q:\mathbf{R}^q \rightarrow \mathbf{R}^n$, is a factorization of Q into a special structure. This special structure clearly displays the structure of the domain space as seen from the map Q, the structure of the range space as seen by Q, and how the matrix map stretches and rotates vectors from its domain space to vectors in its range space.

Recall from chapter 14 that an orthogonal $n \times n$ matrix, say, U, is a matrix that satisfies $U^t U = I$. In other words, the columns of U form an orthonormal set of basis vectors for $\mathbf{R}^n$. For a given $n \times q$ $(n \leqslant q)$ matrix Q, let $\{\sigma_1,...,\sigma_n\}$ denote the *singular values* of Q, i.e., *the positive square roots of the eigenvalues of QQ^t.* Note that the assumption that $n \leqslant q$ is only for convenience.

Theorem 19.1. Let Q be an $n \times q$ $(n \leqslant q)$ matrix with rank$[Q] = p \leqslant n$ (i.e., $Q \in \mathbf{R}_p^{n \times q}$, where the subscript p denotes those $n \times q$ matrices of rank p). Then there exist orthogonal matrices $U \in \mathbf{R}^{n \times n}$ and $V \in \mathbf{R}^{q \times q}$ such that

$$Q = U \Sigma V^t = [U_1 \ U_2] \begin{bmatrix} S & 0 \\ 0 & 0 \end{bmatrix} \begin{bmatrix} V_1^t \\ V_2^t \end{bmatrix} \tag{19.1}$$

where $S = \text{diag} \ (\sigma_1,..., \sigma_p)$ is $p \times p$ with $\sigma_1 \geqslant ... \geqslant \sigma_p > \sigma_{p+1} = ... = \sigma_n = 0$, U_1 is $n \times p$, U_2 is $n \times (n-p)$, V_1^t is $p \times q$, etc.

The proof of theorem 19.1 can be found in [1–3]. Here, we explain some interesting features of the decomposition of Q. First, the columns of U are known as the *left singular vectors* of Q. These form an orthonormal set of eigenvectors for QQ^t that correspond to the squared singular values $(\sigma_1)^2,...,(\sigma_n)^2$ in that order. The columns of V make up an orthonormal set of eigenvectors of Q^tQ that correspond to the squared singular values $(\sigma_1)^2,...,(\sigma_n)^2$, $(\sigma_{n+1})^2 = 0,...,(\sigma_q)^2 = 0$, in that order. These vectors are called the *right singular vectors* of Q. This terminology essentially defines the structure of the matrices in the decomposition of Q. Before elaborating, a simple example will help interpret the mathematics.

EXAMPLE 19.1

Suppose

$$Q = \begin{bmatrix} 0.72 & 1.04 \\ 1.46 & 0.72 \end{bmatrix}$$

Then the SVD of Q is given by

$$Q = U_1 S V_1^t = [u_1 \ u_2] \begin{bmatrix} \sigma_1 & 0 \\ 0 & \sigma_2 \end{bmatrix} \begin{bmatrix} v_1^t \\ v_2^t \end{bmatrix} = \begin{bmatrix} 0.6 & -0.8 \\ 0.8 & 0.6 \end{bmatrix} \begin{bmatrix} 2 & 0 \\ 0 & 0.5 \end{bmatrix} \begin{bmatrix} 0.8 & 0.6 \\ 0.6 & -0.8 \end{bmatrix}$$

Because rank$[Q] = 2$ and Q is 2×2, the matrices U_2 and V_2 mentioned in theorem 19.1 are absent.

To tie Q to the definitions presented after the theorem, observe that

$$QQ^t = \begin{bmatrix} 1.6 & 1.8 \\ 1.8 & 2.65 \end{bmatrix}$$

The associated characteristic polynomial is

$$\pi_{QQ^t}(\lambda) = \lambda^2 - 4.25\lambda + 1 = (\lambda - 4)(\lambda - 0.25)$$

The positive square roots of $\lambda_1 = 4$ and $\lambda_2 = 0.25$ are, of course, the singular values $\sigma_1 = 2$ and $\sigma_2 = 0.5$.

Observe further that

$$[QQ^t][u_1 \ u_2] = \begin{bmatrix} 1.6 & 1.8 \\ 1.8 & 2.65 \end{bmatrix} \begin{bmatrix} 0.6 & -0.8 \\ 0.8 & 0.6 \end{bmatrix} = \begin{bmatrix} 2.4 & -0.2 \\ 3.2 & 0.15 \end{bmatrix}$$

$$= \begin{bmatrix} 0.6 & -0.8 \\ 0.8 & 0.6 \end{bmatrix} \begin{bmatrix} 4 & 0 \\ 0 & 0.25 \end{bmatrix} = [\lambda_1 u_1 \ \lambda_2 u_2]$$

That is, the columns of U (the left singular vectors of Q) are the right eigenvectors of QQ^t. A similar calculation shows that the columns of V (the right singular vectors of Q) are the right eigenvectors of Q^tQ.

Geometrically, the SVD of Q makes transparent the structure of the mapping by Q of the set of points $\|x\|_2 = 1$ to the hyperellipsoid defined by $\{y \mid y = Qx\}$. Figure 19.1 illustrates this structure. Notice that the singular values $\sigma_1 = 2$ and $\sigma_2 = 0.5$ specify the lengths of the semiaxes of the ellipsoid, while $u_1 = (0.6 \ 0.8)^t$ and $u_2 = (-0.8 \ 0.6)^t$ specify the direction of these axes. The latter vectors, u_1 and u_2, determine an orthonormal basis for the image space or range space of the mapping Q. Notice further that the spectral norm (see chapter 14) of Q, $\|Q\|$, is clearly given by $\sigma_1 = 2$ and that the rank of Q equals the number of nonzero singular values, which is 2.

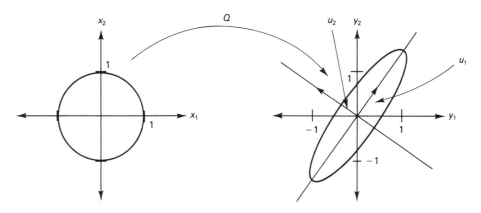

Figure 19.1 Illustration of how the SVD specifies the structure of the linear mapping Q.

Example 19.1 alludes to several aspects of a linear map Q that are pertinent to state space analysis. Recall the discussions in chapters 5 and 9 which dealt with the range space, null space, rank, etc., of a matrix. The *range space* or *image space*, Im[Q] of a mapping Q is given by the span of the columns of U_1. This means that the SVD provides the user with an orthonormal basis for the image space Im[Q] = Im[U_1].

Recall also that the *null space* of Q, denoted $N[Q]$, equals the set of all z satisfying $Qz = 0$. An orthonormal basis for the null space of a matrix also drops out of the SVD: $N[Q] = \text{Im}[V_2]$. Two other spaces of importance are the space orthogonal to the image of Q, given by $\text{Im}[Q]^{\perp} = \text{Im}[U_2]$, and the space orthogonal to $N[Q]$, i.e., $N[Q]^{\perp} = \text{Im}[V_1]$. Figure 19.2 presents a picture of these spaces in $\mathbf{R}^3$.

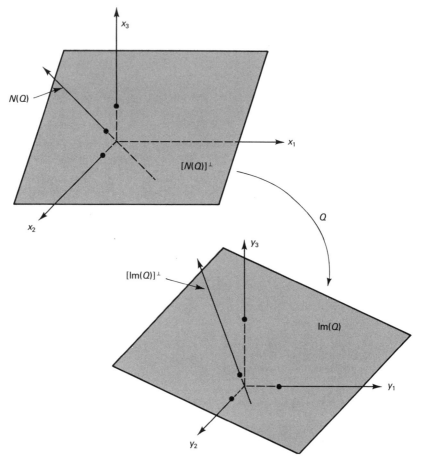

Figure 19.2   Mapping and SVD structure of a hypothetical 3×3 matrix Q.

The figure is based on a hypothetical 3×3 matrix map whose null space has dimension 1. The upper picture of $\mathbf{R}^3$ represents the domain space of the map, and the line perpendicular to the shaded plane represents $N[Q]$. The shaded plane defined by $N[Q]^{\perp}$ represents $N[Q]^{\perp}$. All vectors in $N[Q]$ map to zero. The shaded plane defined by $N[Q]^{\perp}$ maps one to one onto $\text{Im}[Q]$, represented by the shaded plane in the lower picture. The latter is spanned by the vectors $\{u_1, u_2\}$ where $U_1 = [u_1 \ u_2]$. Further, the vector perpendicular to this plane in the range space is given by the span of u_3 where $U_2 = [u_3]$. The following corollary to theorem 19.1 formalizes these ideas.

Corollary 19.1.   Suppose the SVD of Q is given by equation 19.1 and, as per theorem 19.1, $\sigma_1 \geqslant \ldots \geqslant \sigma_p > \sigma_{p+1} = \ldots = \sigma_n = 0$, then rank $[Q] = p$, the columns of $V_2 = [v_{p+1}, ..., v_n]$ provide an orthonormal basis for $N[Q]$, the columns of $U_1 = [u_1, ..., u_p]$ provide an orthonormal basis for $Im[Q]$, and $||Q||$ (the spectral norm of Q) equals σ_1.

The proof of these asssertions are left as exercises at the end of the chapter.

Before closing the section, some numerical considerations are in order. First, computing the singular values of Q by first computing $Q^t Q$ or QQ^t, and then finding the associated eigenvalues, etc., can often lead to erroneous results in the presence of finite-word-length arithmetic. To see this, let $|\mu| < \sqrt{\epsilon}$ where ϵ is the machine epsilon, which is the smallest number such that $fl(1 + \epsilon) = fl(1)$. By construction, $fl(1 + \mu^2) = 1$. Define

$$Q^t = \begin{bmatrix} 1 & \mu & 0 \\ 1 & 0 & \mu \end{bmatrix}$$

which obviously has rank 2. However, in the world of finite word lengths, one must observe that

$$fl(Q^t Q) = fl \begin{bmatrix} 1 + \mu^2 & 1 \\ 1 & 1 + \mu^2 \end{bmatrix} = \begin{bmatrix} 1 & 1 \\ 1 & 1 \end{bmatrix}$$

which means that $fl(Q^t Q)$ has rank 1.

This brief exercise suggests some examination of the SVD itself. Does the SVD truly give the SVD of Q? Diplomatically speaking, the answer is almost. For a given Q, the SVD routines available in say, [4], compute the exact SVD for a matrix $Q + \Delta Q$, where ΔQ is a perturbation matrix, and where $||\Delta Q|| / ||Q||$ is not an unreasonable multiple of the machine epsilon. Thus, a very reasonable approximation is obtained to the SVD of Q. One can be even more precise and say that the number of nonzero singular values of $Q + \Delta Q$ is at least as many as that of Q up to the machine epsilon [5]. Analytically, an upper bound can be placed on the singular-value deviations: if $\tilde{\sigma}_1 \geqslant ... \geqslant \tilde{\sigma}_n$ are the singular values of $Q + \Delta Q$ and $\sigma_1 \geqslant ... \geqslant \sigma_n$ those of Q, then

$$|\sigma_i - \tilde{\sigma}_i| \leqslant ||\Delta Q||$$

for $i = 1, ..., n$, where $||Q|| \triangleq \sigma_{max}(Q) = \sigma_1$. In a very practical sense, the singular values of a matrix are relatively insensitive to matrix perturbations and/or noisy data in the entries. The technique used to compute the SVD via a program in LINPACK [4] is numerically stable and reliable.

DETERMINATION OF RANK

Determination of the rank of a matrix is critical to various theorems of state space analysis. For example, the system $\dot{x} = Ax + Bu$ is said to be controllable (i.e., there exists an input which will drive any given initial state to a desired final state in a finite time) if and only if $rank[Q] = n$, where A is $n \times n$, B is $n \times m$, and $Q = [B \ AB \ ... \ A^{n-1}B]$. Such properties spotlight *rank* making it worthy of serious numerical consideration. Accordingly, suppose that, de facto, the rank of a certain matrix Q is 2. The *effective rank* of Q may still be 1 because the columns of Q are almost dependent on each other. For example, let

$$\dot{x} = \begin{bmatrix} 1 & \epsilon \\ 0 & 1 \end{bmatrix} x + \begin{bmatrix} 1 \\ 1 \end{bmatrix} u$$

where ϵ is a small number. Then $Q = [B \ AB]$ satisfies

$$Q = \begin{bmatrix} 1 & 1+\epsilon \\ 1 & 1 \end{bmatrix}$$

Thus, de facto, the vectors $(1 \ 1)^t$ and $(1+\epsilon \ 1)^t$ are independent, although they are almost dependent vectors. One can then reasonably conclude that the effective rank is 1. Practically speaking, the system would not be controllable.

Because the singular values are "well conditioned" with respect to matrix perturbations, the SVD of Q provides a useful means of determining the numerical and effective rank of a matrix. These in turn lead naturally to the existence and determination of the left and right inverses of a matrix.

Theorem 19.2 specifies the minimum distance between a matrix Q of rank p and the set of lower rank matrices $\{M \,|\, \text{rank}[M] = k < p\}$. The theorem says that this distance equals the singular value σ_{k+1} of Q.

Theorem 19.2.   Let the SVD of $Q \in \mathbf{R}_p^{n \times q}$ be given by theorem 19.1. Define the rank-k matrix Q_k as

$$Q_k = \sum_{j=1}^{k} \sigma_j u_j v_j^t$$

where $k < p = \text{rank}[Q]$. Then for all $n \times q$ matrices M such that $\text{rank}[M] = k$,

$$\min_M \|Q - M\| = \|Q - Q_k\| = \sigma_{k+1}$$

For a proof of this result, see [1].

If $k = p$, then $\|Q - Q_p\| = \sigma_{p+1} = 0$, i.e., the distance of Q from the set of rank p matrices (of which Q is a member) is zero. The distance of Q from the set of rank $p-1$ matrices is σ_p, the smallest nonzero singular value of Q, and so on for rank $p-j$ matrices. This result specifies a means for computing the numerical and effective ranks of a matrix Q: given the machine ϵ, pick a $\delta > \epsilon$ which is a modest multiple of ϵ; then call the effective rank of Q the number of nonzero singular values greater than δ. The formula, however, belies the complexity of the engineering situation: because of noise in the entries of the plant matrices A and B, a matrix, say Q', must double for the actual system controllability matrix Q. It is possible to show that

$$\|Q - Q'\| \leq \alpha_1 \|Q'\|$$

for the accuracy measure α_1. An SVD of Q' would compute the singular values of another matrix $\tilde{Q}$, close to Q', that satisfies

$$\|Q' - \tilde{Q}\| \leq \alpha_2 \|Q'\|$$

for another accuracy measure α_2. Finally, as per [6],

$$\|Q - \tilde{Q}\| \leqslant (\alpha_1 + \alpha_2) \|Q'\| \leqslant \left(\frac{\alpha_1 + \alpha_2}{1 - \alpha_2}\right) \tilde{\sigma}_1$$

Note that whenever $\tilde{\sigma}_p > [(\alpha_1 + \alpha_2)/(1 - \alpha_2)]\tilde{\sigma}_1$ and $\tilde{\sigma}_{p+1} \leqslant [(\alpha_1 + \alpha_2)/(1 - \alpha_2)]\tilde{\sigma}_1$, the effective rank of Q is p, since σ_p could not be zero. However, $\sigma_j, j \geqslant p + 1$, could be zero, thus numerically determining the mathematical rank of the matrix.

   The effective rank of a matrix does not always adequately measure the physical phenomena of importance or interest. Toward that end we define another term called the *useful rank* of the matrix, which is tied to the particular application in which the matrix participates. The effective rank is always greater than or equal to the useful rank. For example, in controllable and observability studies (see chapters 20 and 21) the rank of a particular matrix determines the dimensions of certain important subspaces, specifically the controllability subspace and the unobservable subspace. The effective rank corresponds to the effective (numerical) dimension of these spaces, while the useful rank determines the useful dimension of the spaces with respect to a particular application. For instance, suppose rank[Q] determines the dimension of the controllable subspace. And suppose further that the singular values of Q are $\sigma_1 = 100$, $\sigma_2 = 10$, $\sigma_3 = 1$, $\sigma_4 = 10^{-6}$, and $\sigma_5 = 10^{-7}$, with a machine ϵ of 10^{-10}. Then, from theorem 19.2, the distance from Q to the set of rank 3 matrices is 10^{-6}. So, assuming noisy entries in Q on the order of 10^{-5} or less, the useful rank of Q would be 3 whereas the effective rank would be 5. Even without noise, the useful rank would be 3.

   According to [2,5], σ_p is sensitive to scale. A normalized measure of this nearness property is $\sigma_p/\|Q\| = \sigma_p/\sigma_1$. This number is called the *condition number* $\kappa(Q)$ with respect to pseudo inversion, a notion shortly to be defined. $\kappa(Q)$ measures the distance from a normalized Q to a normalized set of rank-deficient matrices.

THE MOORE-PENROSE RIGHT, LEFT, AND PSEUDO INVERSES

   The equation $Qx = b$ underlies many engineering problems, as seen in chapters 5–8. Oftentimes the equation is overdetermined or underdetermined and corrupted by noisy data, making it an inconsistent set of equations. Solving $Qx = b$ directly then becomes impossible, leading to the equivalent problem

$$\min_x \|Qx - b\|_2 \tag{19.2}$$

where one minimizes the Euclidean norm of $Qx - b$ over all possible x. If one then defines a pseudo inverse

$$Q^+ = V \begin{bmatrix} S^{-1} & 0 \\ 0 & 0 \end{bmatrix} U^t \tag{19.3}$$

where the SVD of Q is

$$U \begin{bmatrix} S & 0 \\ 0 & 0 \end{bmatrix} V^t$$

one obtains

$$x^* = Q^+ b$$

as that x which minimizes equation 19.2 [1]. Further, Q^+ can be shown to be that matrix which satisfies

$$\min_{M \in \mathbf{R}^{n \times q}} \| QM - I_p \|_F$$

where $p = \min(n,q)$ and $\| M \|_F$ denotes the so-called *Frobenius norm* of M. The Frobenius norm of M is defined as the square root of the sum of the entries squared, i.e., the square root of $\text{tr}[MM^t]$. In other words, Q^+ is that unique matrix such that QQ^+ is closest to I_p if $q \geqslant n$ in the Euclidean sense. If $n \geqslant q$, then $Qx = b$ is overdetermined and $Q^+ Q$ is closest to I_q in the least squares sense [1].

Definition 19.1. The *Moore-Penrose pseudo inverse* of Q, denoted Q^+, is that unique matrix satisfying the following conditions:

(i) $QQ^+ Q = Q$
(ii) $Q^+ QQ^+ = Q^+$
(iii) $(QQ^+)^t = QQ^+$ (symmetry condition)
(iv) $(Q^+ Q)^t = Q^+ Q$ (symmetry condition)

It is a simple matter (see exercise 12) to show that Q^+ as defined in equation 19.3 satisfies these conditions. This brings us to a discussion of the right and left inverses of a full-rank rectangular matrix. Recall that a *right inverse* of an $n \times q$ matrix Q, $\text{rank}[Q] = n \leqslant q$, is any matrix Q^{-R} satisfying $QQ^{-R} = I$.

Proposition 19.1. If Q has a right inverse Q^{-R}, then the Moore-Penrose pseudo right inverse is given by

$$Q^{-R} = Q^t [QQ^t]^{-1} = Q^+$$

Proof. Since $\text{rank}[Q] = n$, $[QQ^t]^{-1}$ exists and $QQ^{-R} = QQ^t [QQ^t]^{-1} = I$. It remains to show only that $Q^+ = Q^t [QQ^t]^{-1}$. In fact, it is sufficient to show that $Q^t [QQ^t]^{-1}$ satisfies definition 19.1. Exercise 13 does just this.   ∎

Now recall that a $q \times n$ matrix R, where $q \geqslant n$ and $\text{rank}[R] = n$, has a *left inverse* denoted R^{-L} which satisfies $R^{-L} R = I$. Accordingly, we have the following proposition.

Proposition 19.2. If R has a left inverse R^{-L}, the Moore-Penrose pseudo right inverse is given by

$$R^+ = (R^t R)^{-1} R^t = R^{-L}$$

The proof of proposition 19.2 is left as an exercise.

AN APPLICATION YET TO COME

In the next chapter our focus turns to a subspace of the state space termed the *controllable subspace* of the state dynamical equation $\dot{x} = Ax + Bu$. Briefly, the controllable subspace is the image of the $n \times nm$ controllable matrix

$$Q = [B \; AB \ldots A^{n-1}B] = [U_1 \; U_2] \begin{bmatrix} S & 0 \\ 0 & 0 \end{bmatrix} \begin{bmatrix} V_1^t \\ V_2^t \end{bmatrix}$$

(if computed in a numerically stable way) where the right-hand side is the SVD of Q. As discussed earlier, the columns of U_1 are an orthonormal basis for the image of Q and, hence, for the controllable subspace associated with the pair (A,B) defining the state dynamics.

Im[Q] has the property that any state $x(0) \, \epsilon \, \text{Im}[Q]$ can be driven to any other vector in Im[Q] by some input in finite time (see chapter 20). It is often convenient to use a state transformation $Tz = x$ to extract the controllable subspace from the x-coordinate frame and display it more clearly in a new z-coordinate frame. This is accomplished with the SVD under the specific state transformation $Uz = [U_1 \; U_2]z = x$. Using this numerically stable state transformation, it is possible to show that in the z-coordinates

$$\dot{z} = [U^t A U]z + U^t Bu$$

has the very elegant structure

$$\begin{bmatrix} \dot{z}_1 \\ \dot{z}_2 \end{bmatrix} = \begin{bmatrix} U_1^t A U_1 & U_1^t A U_2 \\ 0 & U_2^t A U_2 \end{bmatrix} \begin{bmatrix} z_1 \\ z_2 \end{bmatrix} + \begin{bmatrix} B_1 \\ 0 \end{bmatrix} u \qquad (19.4a)$$

where

$$\begin{bmatrix} B_1 \\ 0 \end{bmatrix} = \begin{bmatrix} U_1^t \\ U_2^t \end{bmatrix} B \qquad (19.4b)$$

The z_1 part of the new coordinates now embodies the controllable part of the state space. In general, the structure serves a useful design purpose. Since the controllable part of the state space resides with z_1, a designer can construct control inputs to realize design objectives in a more straightforward manner than in the original x-coordinate frame. The designer can then reinterpret these control inputs in the x-coordinates by using the inverse state transformation, $U^t x = z$. This form is called the *Kalman controllable form* and is developed in detail in chapter 20.

EXAMPLE 19.2

Suppose a system has state dynamics

$$\dot{x} = \begin{bmatrix} -3 & 2 & 2 \\ 0 & -1 & 0 \\ -6 & 6 & 4 \end{bmatrix} x + \begin{bmatrix} 1 & 1 & 2 \\ 0 & 1 & 1 \\ 2 & 0 & 2 \end{bmatrix} u \qquad (19.5)$$

Then the controllability matrix $Q = [B \; AB \; A^2 B]$ takes the form

$$Q = \begin{bmatrix} 1 & 1 & 2 & | & 1 & -1 & 0 & | & 1 & 1 & 2 \\ 0 & 1 & 1 & | & 0 & -1 & -1 & | & 0 & 1 & 1 \\ 2 & 0 & 2 & | & 2 & 0 & 2 & | & 2 & 0 & 2 \end{bmatrix}$$

The SVD of Q is

$$Q = U \Sigma V^t = [U_1 \; U_2] \begin{bmatrix} S & 0 \\ 0 & 0 \end{bmatrix} \begin{bmatrix} V_1^t \\ V_2^t \end{bmatrix} \tag{19.6}$$

where

$$[U_1 \mid U_2] = \begin{bmatrix} -0.588 & -0.458 & | & -0.667 \\ -0.196 & -0.719 & | & 0.667 \\ -0.784 & 0.523 & | & 0.333 \end{bmatrix} \tag{19.7a}$$

$$\begin{bmatrix} S & | & 0 \\ - & - & - \\ 0 & | & 0 \end{bmatrix} = \begin{bmatrix} 5.92 & 0 & | & 0 \\ 0 & 3 & | & 0 \\ - & - & - & | & - \\ 0 & 0 & | & 0 \end{bmatrix} \tag{19.7b}$$

(the zero block in the lower right hand corner represents a 7×7 zero matrix, and the other zero blocks are dimensioned accordingly), and

$$\begin{bmatrix} V_1^t \\ V_2^t \end{bmatrix} = \begin{bmatrix} -0.365 & -0.133 & -0.497 & -0.365 & 0.133 & -0.232 & -0.365 & -0.133 & -0.497 \\ 0.196 & -0.392 & -0.196 & 0.196 & 0.392 & 0.588 & 0.196 & -0.392 & -0.196 \\ \hline -0.372 & -0.307 & 0.188 & 0.284 & 0.321 & -0.573 & 0.284 & -0.321 & 0.188 \\ 0.831 & -0.103 & -0.877 & -0.791 & 0.109 & -0.497 & -0.791 & -0.109 & -0.877 \\ 0 & 0.272 & 0.324 & -0.372 & 0.788 & 0.760 & -0.843 & 0.212 & 0.564 \\ 0 & -0.636 & 0.594 & -0.948 & -0.145 & 0.721 & -0.395 & 0.145 & -0.176 \\ 0 & -0.215 & 0.409 & -0.571 & -0.161 & -0.288 & 0.736 & 0.161 & -0.178 \\ 0 & -0.272 & -0.324 & 0.372 & 0.212 & -0.760 & 0.842 & 0.788 & -0.564 \\ 0 & -0.354 & -0.327 & -0.360 & 0.677 & 0.815 & -0.170 & -0.677 & 0.776 \end{bmatrix} \tag{19.7c}$$

In the z-coordinate frame, under the state transformation $Uz = x$,

$$
\begin{bmatrix} \dot{z}_1 \\ \dot{z}_2 \end{bmatrix} = \begin{bmatrix} 0.692 & -1.13 & -9.54 \\ -0.462 & -0.692 & 3.53 \\ 1.57 \cdot 10^{-7} & -1.05 \cdot 10^{-7} & -2 \cdot 10^{-6} \end{bmatrix} \begin{bmatrix} z_1 \\ z_2 \end{bmatrix} \tag{19.8}
$$

$$
+ \begin{bmatrix} -2.156 & -0.784 & -2.94 \\ 0.588 & -1.177 & -0.589 \\ -0.001 & 0 & -0.001 \end{bmatrix} u
$$

where for all practical purposes the bottom rows of the transformed matrices A and B are zero.

CONCLUDING REMARKS

This chapter has merely touched on the many facets of the SVD. Other notions and uses for it include the construction of projections of maps onto invariant subspaces, the computation of bases for the intersection of two or more subspaces, the calculation of "angles" between subspaces, and the calculation of robustness measures for feedback controllers. The interested reader is directed to the references.

PROBLEMS

1. Let σ_i, u_i, and v_i, respectively, be the ith singular value, the ith left singular vector, and the ith right singular vector of A. Using the orthogonality of the matrices U and V in the SVD of A, show that

  (*i*) $Av_i = \sigma_i u_i$
  (*ii*) $A^t u_i = \sigma_i v_i$
  (*Note*: In general, $A[V_1 \ V_2] = [U_1 \ U_2]\Sigma$.)

2. Show that $\|Av_i\|_2 = \sigma_i$.

3. Given theorem 19.1, prove corollary 19.1; i.e., prove that

  (**a**) rank$[Q] = p$
  (**b**) span$[u_1,...,u_p] = \text{Im}[Q]$
  (**c**) span$[v_{p+1},...,v_n] = N[Q]$
  (**d**) $\|Q\| = \sigma_1$

4. Let

$$
Q = \begin{bmatrix} 1 & 0 & 0 & 0 \\ 0 & 1 & 0 & 0 \end{bmatrix}
$$

  (**a**) By inspection, find an orthonormal basis for $N[Q]$.
  (**b**) Show by direct computation that any vector in $\mathbf{R}^4$ can be expressed as a linear combination of vectors in $\text{Im}[Q^t]$ and $N[Q]$.

(c) Show by direct computation that any vector in $\mathbf{R}^3$ can be expressed as a linear combination of vectors in $\text{Im}[Q]$ and $N[Q']$.

(d) Show that $\text{rank}[Q] = \text{rank}[Q']$.

5. Generalize the result of problem 1 to show that if Q is $n \times q$, then

(a) For any $x \in \mathbf{R}^n$, there exists an $x^1 \in \text{Im}[Q]$ and an $x^2 \in N[Q']$ such that $x = x^1 + x^2$.

(b) For any $y \in \mathbf{R}^q$, there exists a $y^1 \in \text{Im}[Q']$ and a $y^2 \in N[Q]$ such that $y = y^1 + y^2$.
(*Note*: Results (a) and (b) are often written as $\mathbf{R}^n = \text{Im}[Q] \oplus N[Q']$ and $\mathbf{R}^q = \text{Im}[Q'] \oplus N[Q]$, which are *direct sum decompositions* of $\mathbf{R}^n$ and $\mathbf{R}^q$, respectively.)

(c) How would the SVD of Q be useful in constructing a direct sum decomposition?

(d) Show that $\text{rank}[Q] = \text{rank}[Q']$.

6. The Moore-Penrose pseudo inverse becomes less obscure if one realizes that

(a) $Q|_{\text{Im}[Q']}$ (i.e, Q restricted to $\text{Im}[Q']$) is 1:1 and onto (bijective) from $\text{Im}[Q'] \subset \mathbf{R}^q$ to $\text{Im}[Q] \subset \mathbf{R}^n$.

(b) $Q'|_{\text{Im}[Q]}$ is a bijection from $\text{Im}[Q]$ to $\text{Im}[Q']$.

Prove these results.

7. Using the results of problem 6, show that

(a) $\text{Im}[Q'] = \text{Im}[Q'Q]$

(b) $N[Q] = N[Q'Q]$

(c) $\text{Im}[Q] = \text{Im}[QQ']$

(d) $N[Q'] = N[QQ']$

(e) $\text{rank}[Q] = \text{rank}[Q'] = \text{rank}[QQ'] = \text{rank}[Q'Q]$

8. Show that for all $\sigma \in \mathbf{R}$ with $\sigma \neq 0$,

$$\det \begin{bmatrix} \sigma I^n & Q \\ Q' & \sigma I_q \end{bmatrix} = \sigma^{n-q} \det[\sigma^2 I_q - Q'Q]$$

$$= \sigma^{q-n} \det[\sigma^2 I_n - QQ']$$

9. Oftentimes it is useful to know the square root of the sum of the squares of the entries of a matrix. This is called the *Frobenious norm* of the matrix and is denoted by $\|Q\|_F$, i.e.,

$$\|Q\|_F^2 = \sum_{i=1}^{n} \sum_{j=1}^{q} (q_{ij})^2$$

(a) Show that $\|Q\|_F^2 = \text{trace}[QQ']$.

(b) Use part (a) to prove that $\|Q\|_F^2 = (\sigma_1)^2 + \ldots + (\sigma_p)^2$ when Q has the SVD given by theorem 19.1.

10. Prove theorem 19.2 as follows:

(a) Show that $U^t Q_k V = \text{diag}(\sigma_1, \ldots, \sigma_k, 0, \ldots, 0)$.

(b) Using the result of part (a), show that $\text{rank}[Q_k] = k$ and $\|Q - Q_k\| = \sigma_{k+1}$.

(c) Let $N[M]$ equal $\text{span}\{m_1, \ldots, m_{n-k}\}$, and show that $\text{span}\{m_1, \ldots, m_{n-k}\} \cap \text{span}\{v_1, \ldots, v_{k+1}\} \neq 0$.

(d) With z in the intersection of the two spans, show that

$$Qz = \sum_{i=1}^{k+1} \sigma_i (v_i^t z) u_i$$

and that

$$\|Q - M\|^2 \geq (\sigma_{k+1})^2$$

11. Suppose the singular values of a 4×5 matrix Q are 1.5, 1.4, 1.0, and 0.01. What can you deduce about the effective rank of Q and the relative orthogonality of its rows and columns.

12. Show that Q^+ as defined in equation 19.3 satisfies the four Moore-Penrose conditions of definition 19.1.

13. If $Q^{-R} = Q'[QQ']^{-1}$ is a right inverse of Q, show that Q^{-R} satisfies the four conditions of definition 19.1 and hence that proposition 19.1 is true.

14. Prove proposition 19.2.

15. Consider the matrix equation $Qx = b$ given by

$$\begin{bmatrix} -1 & 1 & 1 & \cdots & 1 \\ 0 & -1 & 1 & \cdots & 1 \\ & & \vdots & & \\ & & & & 1 \\ 0 & 0 & 0 & & -1 \end{bmatrix} \begin{bmatrix} 1 \\ (0.5) \\ \vdots \\ (0.5)^{n-1} \end{bmatrix} = \begin{bmatrix} -(0.5)^{n-1} \\ -(0.5)^{n-1} \\ \vdots \\ -(0.5)^{n-1} \end{bmatrix}$$

    (a) Prove that adding $(0.5)^{n-1}$ to every element in the first row of Q produces a singular matrix. (*Hint*: First let $n = 2$ and then let $n = 3$.)

    (b) Show that the nth singular value σ_n of Q behaves as $(0.5)^n$.

    (c) Comment on the solution of this equation. (Consult [2,5].)

16. Consider the state dynamical equation

$$\dot{x} = \begin{bmatrix} 0 & 0 & 0 \\ 0 & 0 & 2 \\ 0 & 2 & 0 \end{bmatrix} x + \begin{bmatrix} 0 \\ 1 \\ 1 \end{bmatrix} u$$

Compute the controllability matrix $Q = [B \; AB \; A^2B]$, find the associated SVD, and put the system into the form of equations 19.4.

17. Consider the matrix A of problem 11, and let $y = [0 \; -1 \; 1]x$. Construct the observability matrix

$$R = \begin{bmatrix} C \\ CA \\ CA^2 \end{bmatrix}$$

and compute the associated SVD. The unobservable subspace is the null space of R. What is an orthonormal basis for the unobservable subspace? Pick a vector in this subspace. Show by direct computation that the associated zero-input response is identically zero.

18. Show that a square singular $n \times n$ matrix A is arbitrarily close to a full-rank matrix. (*Hint*: Consider the SVD of A, and construct an arbitrarily small perturbation matrix which would result in a full-rank matrix.)

REFERENCES

1. Gene H. Golub and Charles F. Van Loan, *Matrix Computations* (Baltimore, MD: The John Hopkins University Press, 1983).

2. Alan J. Laub, "Linear Multivariable Control: Numerical Considerations," *American Mathematical Society Short Course on Control Theory* (Providence, RI: August 1978).

3. Frank M. Callier and Charles A. Desoer, *Multivariable Feedback Systems* (New York: Springer-Verlag, 1982).

4. J. Dongarra, J. R. Bunch, C. B. Moler, and G. W. Stewart, *LINPACK Users Guide* (Philadelphia: SIAM Publications, 1979).

5. Virginia C. Klema and Alan J. Laub, "The Singular Value Decomposition: Its Computation and Some Applications," *IEEE Transactions on Automatic Control,* Vol. AC-25, No. 2, April 1980, pp. 164–176.

6. Chris C. Paige, "Properties of Numerical Algorithms Related to Computing Controllability," *IEEE Transactions on Automatic Control,* Vol. AC-26, No. 1, February 1981, pp. 130–138.

20

THE CONTROLLABILITY
OF THE TIME-INVARIANT
STATE MODEL

INTRODUCTION

The controllability ideas spawned in chapters 5 and 9 mature in this chapter. In chapter 5, the controllability problem centered on finding impulsive inputs to drive an arbitrarily specified $x_0 = x(0^-)$ to an arbitrarily specified $x_1 = x(0^+)$. This is possible whenever $Q = [B\ AB \ldots A^{n-1}B]$ has rank n, where B is $n \times m$ and A is $n \times n$. Recall, the rank of a constant matrix equals the number of independent columns, which in turn equals the number of independent rows. For pedagogical reasons, the perspective of chapter 5 was deliberately narrow. As a general rule, impulsive inputs are inappropriate for control purposes: instantaneous changes of states are often undesirable. Realistically speaking, it is impossible and, in any event, often unnecessary to drive an arbitrary x_0 to an arbitrary x_1 for many models representing physical systems.

Accordingly, the middle part of this chapter centers on the more realistic problem of characterizing those states which are mutually transferable to one another by a continuous input defined over some finite interval $[t_0, t_1]$. The set of states which are mutually transferable to one another defines the *controllable subspace* of the state space. The first part of the chapter describes the basic definitions and equivalences, while the third part sketches the relationship of controllability to directed modes. Various equivalences are then stated, proven, and illustrated. Analogies with discrete-time systems follow this, and the chapter climaxes with a discussion of the important problem of pole placement.

A number of texts have similar developments, and the interested reader should browse through [1–8] to obtain a feel for the other perspectives.

BASIC DEFINITIONS AND EQUIVALENCES

Given the continuous-time state dynamical equation, $\dot{x} = Ax + Bu$, what constitutes a controllable state? The following definition answers this question.

Definition 20.1.   A state $x_0 = x(t_0) \epsilon \mathbf{R}^n$ is *controllable* over $[t_0, t_1]$ if there exists an input $u(\cdot)$ defined over $[t_0, t_1]$ such that

$$\theta = \Phi(t_1 - t_0)x_0 + \int_{t_0}^{t_1} \Phi(t_1 - q)Bu(q)\,dq \qquad (20.1)$$

Such an input $u(\cdot)$ is said to drive or transfer x_0 to θ. Because of the assumed time-invariant structure, a simple change in the variable of integration and/or a translation of the interval $[t_0, t_1]$ demonstrates that if x_0 is controllable over $[t_0, t_1]$, it is controllable over every finite interval. For time-varying systems, controllability is dependent on the interval, and definition 20.1 holds with equation 20.1 changed to its time-varying counterpart. Finally, the definition of a controllable state leads to the notion of system controllability.

Definition 20.2.   The state model $\dot{x} = Ax + Bu$ is said to be *controllable* (often called *completely controllable*) if and only if every state $x_0 \epsilon \mathbf{R}^n$ is controllable.

In the vernacular, the locution, *the pair (A,B) is controllable,* refers to *system controllability*. The notion of system controllability has several equivalent characterizations.

Theorem 20.1.   The following are equivalent:

(*i*) (A,B) is controllable over $[t_0, t_1]$.
(*ii*) For each $x_0 = x(t_0)$, there exists a $u(t)$, $t_0 \leqslant t \leqslant t_1$, which will drive x_0 to θ.
(*iii*) For each x_0 and x_1 in $\mathbf{R}^n$, there exists a $u(t)$, $t_0 \leqslant t \leqslant t_1$, which will drive x_0 to x_1.
(*iv*) For each x_1 in $\mathbf{R}^n$, there exists a $u(\cdot)$ defined over $[t_0, t_1]$ which will drive $\theta = x(t_0)$ to $x_1 = x(t_1)$.

Rather than prove the theorem, the following discussion argues the theorem's validity in a colloquial fashion. Clearly, the first two equivalences follow by definition. To see the equivalence between (*ii*) and (*iii*), observe that from (*iii*), for each pair (x_0, x_1), there exists a $u(\cdot)$ such that

$$x_1 = \Phi(t_1 - t_0) x_0 + \int_{t_0}^{t_1} \Phi(t_1 - q) Bu(q) dq$$

Rewriting produces

$$\theta = \Phi(t_1 - t_0) [x_0 - \Phi(t_0 - t_1) x_1] + \int_{t_0}^{t_1} \Phi(t_1 - q) Bu(q) dq \qquad (20.2)$$

Now, define $\hat{x}_0 = [x_0 - \Phi(t_0 - t_1) x_1]$. System controllability implies that x_0 and x_1 can take on any value in $\mathbf{R}^n$. Thus, as x_0 and x_1 range over all possible values in $\mathbf{R}^n$, so does $\hat{x}_0$. In other words, the existence of an input for each pair (x_0, x_1) in $\mathbf{R}^n$, where $u(\cdot)$ transfers x_0 to x_1, implies the existence of an input to transfer each $\hat{x}_0$ in $\mathbf{R}^n$ to θ. Proof of the converse proceeds in a similar manner.

Finally, the equivalence between (ii) and (iv) obtains because equation 20.2 implies that θ is driven to $-\tilde{x}_0$ by some $u(\cdot)$, where $\tilde{x}_0 = \Phi(t_1 - t_0)\hat{x}_0$. Since this is true for each $-\tilde{x}_0$ in $\mathbf{R}^n$, the equivalence follows.

As indicated earlier, system controllability is not always possible. Hence, it is more practical to characterize those states which are controllable.

Definition 20.3. The set of all controllable states is the *controllable subspace*.

As the development will verify, states in the controllable subspace are mutually transferable to one another. First, however, it is important to point out that if two states are controllable, so is their linear combination.

Proposition 20.1. The controllable subspace is a linear subspace—i.e., if x_0 and x_1 lie in the controllable subspace, then for real scalars α_0 and α_1, $\hat{x} = \alpha_0 x_0 + \alpha_1 x_1$ also does.

Proof. The details of the proof are straightforward and left as an exercise. The general method is to use the solution formula 13.10 for the time-invariant state dynamics to show that if u_1 drives θ to x_1 and u_2 drives θ to x_2, then $u_1 + u_2$ drives θ to $x_1 + x_2$. ∎

Exercise. Let the state space of a particular system be $\mathbf{R}^3$. Suppose $[1 \ 0 \ 1]^t$ and $[0 \ 2 \ 0]^t$ are known to be controllable states. Then which of the following states is (are) definitely also controllable:

      (a) $[1 \ 0 \ -1]^t$ (b) $[1 \ 1 \ 1]^t$ (c) $[0 \ 0 \ 1]^t$

      (d) $[1 \ 0 \ 0]^t$ (e) two of the above (f) none of the above

CHARACTERIZATION OF THE CONTROLLABLE SUBSPACE

Our immediate task centers on characterizing the controllable subspace of the pair (A, B) in terms of the column space of the matrix $Q = [B \ \ AB \ \ A^2B \ \ ... \ \ A^{n-1}B]$. Recall that the column space of Q is the span of the columns of Q, i.e., col-sp$[Q] = \{v \in \mathbf{R}^n | v = \alpha_1 q_1 + ... + \alpha_p q_p\}$, where q_i denotes the ith column of Q, α_i

are scalars, and Q is $n \times p$ for $p = mn$. The development of chapter 5, suggests that the pair (A,B) is completely controllable if and only if rank$[Q] = n$. This follows as a corollary to the main theorem of this section. The main theorem will show that for the state dynamics, $\dot{x} = Ax + Bu$, the state $\hat{x}$ is controllable if and only if $\hat{x} \epsilon$ col-sp$[Q]$. The proofs of this theorem and its corollary require a series of propositions and lemmas together with the development of the so-called Kalman controllable canonical form. A technique for constructing an input which will drive x_0 to x_1 arises out of this development. In particular, showing that $\hat{x} \epsilon$ col-sp$[Q]$ is controllable requires the construction of an input which will drive $\hat{x}$ and θ and, in effect, $\hat{x}$ to any arbitrary $\bar{x}$. We begin with a lemma describing an invariant-rank condition on a nested sequence of matrices.

Lemma 20.1. Define the matrix $Q_k = [B \quad AB \quad ... \quad A^{k-1}B]$. If for some k, rank$[Q_{k+1}]$ equals rank$[Q_k]$, then rank$[Q_j] = $ rank$[Q_k]$ for all $j \geq k$.

Proof. The lemma is trivially true if $j = k$ or $j = k + 1$. Hence, suppose $j > k + 1$. The condition that rank$[Q_k] = $ rank$[Q_{k+1}]$ means that

$$\text{rank}[B \quad AB \quad ... \quad A^{k-1}B] = \text{rank}[Q_k \quad A^kB]$$

where $Q_{k+1} = [Q_k \quad A^kB]$. Thus, the columns of Q_{k+1} are linearly dependent on the columns of Q_k. Mathematically, then, there exist matrices $M_0, M_1, ..., M_{k-1}$ of appropriate dimension such that

$$A^kB = [B]M_0 + [AB]M_1 + ... + [A^{k-1}B]M_{k-1} \tag{20.3}$$

Multiplying both sides of 20.3 by A gives

$$A^{k+1}B = [AB]M_0 + [A^2B]M_1 + ... + [A^kB]M_{k-1} \tag{20.4}$$

This equation means that the columns of $A^{k+1}B$ are linearly dependent on the columns of $Q_{k+1} = [B \quad AB \quad ... \quad A^kB]$. Now, observe that $Q_{k+2} = [Q_{k+1} \quad A^{k+1}B]$. Thus, from equation 20.4, the columns of Q_{k+2} are linearly dependent on the columns of Q_{k+1}, which are in turn linearly dependent on the columns of Q_k. Therefore,

$$\text{rank}[Q_k] = \text{rank}[Q_{k+1}] = \text{rank}[Q_{k+2}]$$

Inductively continuing the argument gives the result that rank$[Q_j] = $ rank$[Q_k]$ for all $j \geq k$ whenever rank$[Q_k] = $ rank$[Q_{k+1}]$, which is what was to be proven. ∎

Lemma 20.2. For any integer $p > n$, rank$[B \ AB ... A^{p-1}B] = $ rank$[B \ AB ... A^{n-1}B]$.

Proof. Consider again the matrix Q_k. In regard to rank, only two possibilities occur:

(i) rank$[Q_{k+1}] \geq $ rank$[Q_k] + 1$, or
(ii) rank$[Q_{k+1}] = $ rank$[Q_k]$.

Since each Q_k has n rows, its maximum rank is n; i.e., rank$[Q_k] \leq n$ for all k. In view of the preceding two possibilities, the rank of each matrix in the sequence $\{Q_1, Q_2, ..., Q_n\}$ either exceeds the rank of the previous matrix by 1 or equals it. By lemma 20.1, if the ranks of two successive matrices ever coincide, the rank remains fixed for all succeeding terms. Hence, $\text{rank}[Q_n]$ is either n or $\text{rank}[Q_{n-1}]$ $= \text{rank}[Q_n] < n$. This implies the result of the lemma.   ∎

The following corollary states a fundamental consequence and property of col-sp$[Q]$.

Corollary 20.2.   Let $Q = [B \quad AB \quad ... \quad A^{n-1}B]$. Then for any $x_0 \epsilon \text{col-sp}[Q]$, $Ax_0 \epsilon \text{col-sp}[Q]$.

Proof.   Since $Q = Q_n$, as defined earlier, the proof of lemma 20.1 implies the desired result.   ∎

The corollary to lemma 20.2 says that the column space of Q is *A-invariant;* i.e., multiplication of any vector in col-sp$[Q]$ by A results in another vector in the column space. Since A relates the state vector to its derivative, this corollary implies that all changes in state which lie in col-sp$[Q]$ must take place within the confines of col-sp$[Q]$. Intuitively speaking, motion begun inside the col-sp$[Q]$ remains within col-sp$[Q]$. More formally, if $x_0 \epsilon \text{col-sp}[Q]$, then $x(t) = e^{At}x_0$ is a member of col-sp$[Q]$.

The foregoing lemmas and the A-invariance of col-sp$[Q]$ allow the development of a state transformation which extracts and makes explicit the controllable part of the pair (A,B). Specifically, there exists a state transformation $x = [U_1 \ U_2]z$ such that

$$\begin{bmatrix} \dot{z}_1 \\ \dot{z}_2 \end{bmatrix} = \begin{bmatrix} \tilde{A}_{11} & \tilde{A}_{12} \\ 0 & \tilde{A}_{22} \end{bmatrix} \begin{bmatrix} z_1 \\ z_2 \end{bmatrix} + \begin{bmatrix} B_1 \\ 0 \end{bmatrix} u$$

in which all the controllable aspects of x project onto the z_1-part of z. This equivalent system plays an integral part in the characterization of the controllable subspace. We begin by defining two matrices U_1 and U_2 as follows. Let $\text{rank}[Q] = p$, let U_1 be an $n \times p$ matrix whose columns form a basis for the column space of Q, and let U_2 be an $n \times (n-p)$ matrix whose columns in conjunction with those of U_1 form a basis for $\mathbf{R}^n$, i.e., col-sp$[U_1 \ U_2] = \mathbf{R}^n$. Several important propositions regarding $[U_1 \ U_2]$ emerge from this definition.

Proposition 20.2. $[U_1 \ U_2]^{-1}$ exists.

Proof.   The result is obvious, since the $n \times n$ matrix $[U_1 \ U_2]$ has n linearly independent columns which are a basis for $\mathbf{R}^n$.   ∎

Proposition 20.3. $AU_1 = U_1 \tilde{A}_{11}$ for an appropriate $p \times p$ matrix $\tilde{A}_{11}$.

Proof.   From corollary 20.2 of lemma 20.2, $AU_1 \epsilon \text{col-sp}[Q]$. Each column of AU_1 is thus a linear combination of the columns of U_1. In equation form, this means that $AU_1 = U_1 \tilde{A}_{11}$ for appropriate $\tilde{A}_{11}$.   ∎

Proposition 20.4.

$$A[U_1 \ \ U_2] = [U_1 \ \ U_2]\begin{bmatrix} \tilde{A}_{11} & \tilde{A}_{12} \\ 0 & \tilde{A}_{22} \end{bmatrix}$$

for appropriate $\tilde{A}_{ij}$.

Proof.   From proposition 20.3,

$$AU_1 = [U_1 \ \ U_2]\begin{bmatrix} \tilde{A}_{11} \\ 0 \end{bmatrix}$$

for an appropriate $\tilde{A}_{11}$. Now, the columns of $[U_1 U_2]$ represent a basis for $\mathbf{R}^n$. But each column of AU_2 is an element of $\mathbf{R}^n$. Hence, there must exist matrices $\tilde{A}_{12}$ and $\tilde{A}_{22}$ such that

$$AU_2 = [U_1 \ \ U_2]\begin{bmatrix} \tilde{A}_{12} \\ \tilde{A}_{22} \end{bmatrix}$$

which establishes the proposition.   ∎

Proposition 20.5.   The appropriate A_{ij} matrices of proposition 20.4 are given by

$$\begin{bmatrix} \tilde{A}_{11} & \tilde{A}_{12} \\ 0 & \tilde{A}_{22} \end{bmatrix} = [U_1 \ \ U_2]^{-1} A[U_1 \ \ U_2] \qquad (20.5)$$

Proof.   The proof follows from propositions 20.2 and 20.4.   ∎

Proposition 20.6.   For an appropriate $p \times m$ matrix B_1,

$$[U_1 \ \ U_2]^{-1}B = \begin{bmatrix} B_1 \\ 0 \end{bmatrix}$$

Proof.   By construction, col-sp$[Q]$ = col-sp$[B \ \ AB \ \ \dots \ \ A^{n-1}B]$ contains the column space of B. Hence, there must exist a $p \times m$ matrix B_1 such that $B = U_1 B_1$, or, equivalently,

$$B = [U_1 \ \ U_2]\begin{bmatrix} B_1 \\ 0 \end{bmatrix}$$

Inverting $[U_1 \ \ U_2]$ then produces the desired result.   ∎

Propositon 20.7.   Given $\dot{x} = Ax + Bu$, the state transformation

$$[U_1 \ \ U_2]z = x$$

yields the equivalent state dynamics

$$\begin{bmatrix} \dot{z}_1 \\ \dot{z}_2 \end{bmatrix} = \begin{bmatrix} \tilde{A}_{11} & \tilde{A}_{12} \\ 0 & \tilde{A}_{22} \end{bmatrix} \begin{bmatrix} z_1 \\ z_2 \end{bmatrix} + \begin{bmatrix} B_1 \\ 0 \end{bmatrix} u \qquad (20.6)$$

where $\text{rank}[B_1 \ \tilde{A}_{11}B_1 \ \dots \ \tilde{A}_{11}^{p-1}B_1] = p$.

Proof. The proof is a direct consequence of propositions 20.2–20.6, lemma 20.2 and problem 10 at the end of the chapter, which guarantees that $\text{rank}[B_1 \ \tilde{A}_{11}B_1 \ \dots \ \tilde{A}_{11}^{p-1}B_1] = p$. ∎

Equation 20.6 represents the *Kalman controllable form* of the original state dynamics; see [9–11] for a complete discussion of the particulars. The reason for the term "controllable form" stems from the fact that the z_1-part of the new state vector embodies the "controllable part" of the system in the z-coordinates. The last of the propositions dealing with $[U_1 \ U_2]$ sets up this interpretation.

Proposition 20.8. If $x \in \text{col-sp}[Q]$, then

$$[U_1 \ U_2]^{-1} x = \begin{bmatrix} z_1 \\ \theta \end{bmatrix}$$

where $z_1 \in \mathbf{R}^p$ and θ is the zero vector of $\mathbf{R}^{n-p}$, i.e., $[z_1^t \ \theta^t]^t$ represents the column space of Q.

Numerically speaking, the material of chapter 19 suggests that the singular value decomposition (equation 19.1) of the controllability matrix takes the form

$$Q = [U_1 \ U_2] \begin{bmatrix} S & 0 \\ 0 & 0 \end{bmatrix} \begin{bmatrix} V_1^t \\ V_2^t \end{bmatrix}$$

The matrix U_1 has orthonormal columns spanning col-sp[Q], and U_2 has columns which complete an orthonormal basis for $\mathbf{R}^n$. Thus, $[U_1 \ U_2]$ of propositions 20.2–20.8 can be taken to be the $[U_1 \ U_2]$ in the SVD of Q. Further, if $[U_1 \ U_2]$ is so taken, then $[U_1 \ U_2]^{-1} = [U_1 \ U_2]^t$, simplifying the computation of equation 20.6. As pointed out in [12], the SVD of $Q = [B \ AB \ \dots \ A^{n-1}B]$ is sensitive to the condition of A, i.e., the ratio of the largest to the smallest singular value of A. If the condition of A is close to unity (i.e., A is close to being orthogonal), then conclusions regarding the controllability of the system, bases for the controllable subspace, etc., are reliable. However, if the condition of A is much greater than unity, then erroneous conclusions can result (a forthcoming section has an illuminating example). Reference [13] describes a stable method called the *staircase algorithm* for determining the controllable subspace of the pair (A, B).

The next lemma completes our preparation for the statement and proof of the main result, theorem 20.2. It states that a certain matrix K, constructed in terms of the matrices $\tilde{A}_{11}$ and B_1 of the Kalman controllable form, is nonsingular. The inverse of this K-matrix is central to a control input which drives an initial state to a desired state.

Lemma 20.3. The *controllability Gramian matrix*

$$K = \int_0^{t_1} \exp[-\tilde{A}_{11}\tau]B_1 B_1^t \exp[-\tilde{A}_{11}^t\tau]d\tau \tag{20.7}$$

is nonsingular whenever $\text{rank}[\tilde{Q}] = \text{rank}[B_1 \quad \tilde{A}_{11}B_1 \quad \ldots \quad \tilde{A}_{11}^{p-1}B_1] = p$.

    Proof. Using the technique of proof by contradiction, suppose there exists a vector $v \neq \theta$, $v \in \mathbf{R}^p$, such that $v^t K = \theta_p^t$; i.e., suppose K is singular. The goal entails showing that this assumption contradicts the hypothesis that $\text{rank}[\tilde{Q}] = p$. Hence, K must be nonsingular. To demonstrate this, we prove that if $v^t K = \theta_p^t$, then $v^t \tilde{Q} = \theta_{pm}^t$.

Step 1. If $v^t K = \theta^t$, then $v^t Kv = 0$, i.e.,

$$0 = v^t Kv = \int_0^{t_1} v^t \exp[-\tilde{A}_{11}\tau]B_1 B_1^t \exp[-\tilde{A}_{11}^t\tau]v d\tau$$

Now let $c(\tau) = B_1^t \exp[-\tilde{A}_{11}^t\tau]v \triangleq [c_1(\tau),\ldots,c_p(\tau)]^t$. Observe that in terms of $c(\tau)$,

$$v^t Kv = \int_0^{t_1} c^t(\tau)c(\tau)d\tau = \int_0^{t_1} [c_1^2(\tau) + \ldots + c_p^2(\tau)]d\tau = 0$$

Since the integrand is always nonnegative, $v^t Kv$ equals zero if and only if each $c_i(\tau)$ is identically equal to zero over $[0,t_1]$; i.e., for all i, $c_i(\tau) \equiv 0$ for $\tau \in [0,t_1]$.

Step 2. Since $c_i(\tau) \equiv 0$, all its derivatives are identically equal to zero as well. That is,

$$\frac{d^j c^t}{d\tau^j}(\tau) \equiv 0$$

for all j. In particular, $c^t(\tau) \equiv \theta^t$ implies that $v^t \exp[-\tilde{A}_{11}\tau]B_1 \equiv \theta^t$ for all τ. Now, at $\tau = 0$,

$$v^t B_1 = \theta^t$$

Evaluating the first derivative of $c^t(\tau) = \theta$ at $\tau = 0$ produces

$$\left.\frac{dc^t}{d\tau}(\tau)\right|_{\tau=0} = \left.v^t \exp[-\tilde{A}_{11}\tau](-\tilde{A}_{11})B_1\right|_{\tau=0} = -v^t \tilde{A}_{11}B_1 = \theta^t$$

For $j \geqslant 1$,

$$\left.\frac{d^j c^t}{d\tau^j}(\tau)\right|_{\tau=0} = (-1)^j v^t \tilde{A}_{11}^j B_1 = \theta^t$$

Step 3. From the evaluation of the derivatives of $c^t(\tau)$ at $\tau = 0$, it follows that

$$v^t \tilde{Q} = v^t [B_1 \quad \tilde{A}_{11}B_1 \quad \ldots \quad \tilde{A}_{11}^{p-1}B_1] = \theta_{pm}^t$$

This contradicts the basic hypothesis that rank$[\tilde{Q}] = p$, and hence, K must be non-singular. ∎

Exercise. Suppose the controllability matrix for a particular system is

$$Q = \begin{bmatrix} 1 & 1 & 1 \\ 0 & 0 & 0 \\ 1 & -1 & 1 \end{bmatrix}$$

A state transformation which will produce a matrix A having the structure

$$\begin{bmatrix} A_{11} & A_{12} \\ 0 & A_{22} \end{bmatrix}$$

is

(a) $\begin{bmatrix} 1 & 1 & 1 \\ 0 & 0 & 0 \\ 1 & -1 & 1 \end{bmatrix}$ (b) $\begin{bmatrix} 1 & 1 & 2 \\ 0 & 0 & 0 \\ 1 & -1 & 0 \end{bmatrix}$ (c) $\begin{bmatrix} 1 & 1 & 0 \\ 0 & 0 & 1 \\ 1 & -1 & 0 \end{bmatrix}$ (d) $\begin{bmatrix} 1 & 1 & 1 \\ 0 & 0 & 1 \\ -1 & 1 & 0 \end{bmatrix}$

(e) two of the above (f) none of the above

True-false exercises

1. If A is 3×3, then A^4B is linearly dependent on the set $\{A^3B, A^2B, AB\}$.

2. If three states, x^1, x^2, and x^3, are controllable, then $x^4 = 2x^1 - 7x^2 + 3x^3$ is controllable.

3. A certain switched time-varying system is controllable over the interval $[0,2]$. It must therefore be controllable over $[0,1]$.

4. If A is 4×4, rank$[B \; AB \; ... \; A^4B] = $ rank$[B \; AB \; ... \; A^3B]$.

With the preliminary work done, the development is sufficiently advanced to prove the main theorem.

Theorem 20.2.   For the state dynamics $\dot{x} = Ax + Bu$, the state $\hat{x}$ is controllable if and only if $\hat{x} \epsilon$col-sp$[Q]$, where $Q = [B \; AB \; ... \; A^{n-1}B]$.

Proof.

Part 1. Suppose $\hat{x}$ is controllable; show that $\hat{x} \epsilon$col-sp$[Q]$.

Step 1. Since $\hat{x}$ is controllable, there exists a function $u(\cdot)$ defined over $[0,t_1]$ such that

$$\hat{x} = \int_0^{t_1} \exp[A(t_1 - \tau)]Bu(\tau)d\tau$$

Step 2. Expanding the exponential in a Taylor series produces

$$\hat{x} = \int_0^{t_1} [I + A(t_1 - \tau) + A^2\frac{(t_1 - \tau)^2}{2!} + ...]Bu(\tau)d\tau$$

Step 3. Distributing the integral over the infinite sum and factoring constant matrices out to the left produces

$$\hat{x} = B\int_0^{t_1} u(\tau)d\tau + AB\int_0^{t_1}(t_1-\tau)u(\tau)d\tau + A^2B\int_0^{t_1}\frac{(t_1-\tau)^2}{2!}u(\tau)d\tau + \dots$$

$$= \sum_{j=0}^{\infty} A^jBv_j(t_1) \tag{20.8}$$

where

$$v_j(t_1) = \int_0^{t_1}\frac{(t_1-\tau)^j}{j!}u(\tau)d\tau$$

Step 4. From lemma 20.2, for any $p > n-1$, A^pB has columns that depend on the columns of Q. Hence, the infinite summation in equation 20.8 is equivalent to the finite sum

$$\hat{x} = \sum_{j=0}^{n-1} A^jB\hat{v}_j = Q[\hat{v}_0^t, \hat{v}_1^t,\dots, \hat{v}_{n-1}^t]^t$$

for appropriate $\hat{v}_j$. This $\hat{x}$ is a linear combination of the columns of Q and lies in the column space of Q.

Part 2. Suppose $\hat{x} \in \text{col-sp}[Q]$, and show that $\hat{x} = x(t_1)$ is controllable. The crux of this proof is to actually construct an input which will drive θ to $\hat{x}$.

Step 1. If $\hat{x} \in \text{col-sp}[Q]$, then the state model $\dot{x} = Ax + Bu$ has the equivalent representation

$$\begin{bmatrix} \dot{z}_1 \\ \dot{z}_2 \end{bmatrix} = \begin{bmatrix} \tilde{A}_{11} & \tilde{A}_{12} \\ 0 & \tilde{A}_{22} \end{bmatrix}\begin{bmatrix} z_1 \\ z_2 \end{bmatrix} + \begin{bmatrix} B_1 \\ 0 \end{bmatrix}u$$

in a new coordinate system under the state transformation $[U_1 \ U_2]z = x$ as per proposition 20.7. Here, it is presumed that

$$\tilde{Q} = [B_1 \ \tilde{A}_{11}B_1 \ \dots \ \tilde{A}_{11}^{p-1}B_1]$$

has rank p, where $\tilde{A}_{11}$ is $p \times p$; i.e., rank $[Q] = \text{rank}[\tilde{Q}]$. Further by proposition 20.8, $\text{col}(\hat{z}_1, \theta)$ represents a vector $\hat{x}$ in the column space of Q in the new coordinates:

$$[U_1 \ U_2]^{-1}\hat{x} = \begin{bmatrix} \hat{z}_1 \\ \theta \end{bmatrix}$$

From lemma 20.3, the $p \times p$ matrix

$$K = \int_0^{t_1} \exp[-\tilde{A}_{11}\tau]B_1B_1^t\exp[-\tilde{A}_{11}^t\tau]d\tau \tag{20.9}$$

is nonsingular.

Step 2. Define the control input

$$u(q) = B_1' \exp[-\tilde{A}_{11}'q] K^{-1} \exp[-\tilde{A}_{11}t_1] \hat{z}_1 \tag{20.10}$$

for $0 \leqslant q \leqslant t_1$. This input explicitly depends on both K of equation 20.9 and the desired final state col$[\hat{z}_1, \theta]$. The remainder of the proof demonstrates that $u(q)$ will drive θ to col$[z_1, \theta]$. Toward that end, let $\Phi_z(t_1 - q)$ represent the system state transition matrix in the z-coordinates. Then observe that

$$\Phi_z(t_1 - q) = \exp \begin{bmatrix} \tilde{A}_{11}(t_1 - q) & \tilde{A}_{12}(t_1 - q) \\ 0 & \tilde{A}_{22}(t_1 - q) \end{bmatrix} = \begin{bmatrix} \Phi_{11}(t_1 - q) & \Phi_{12}(t_1 - q) \\ 0 & \Phi_{22}(t_1 - q) \end{bmatrix}$$

where $\Phi_{ii}(t_1 - q) = \exp[\tilde{A}_{ii}(t_1 - q)]$ (see problem 12 of chapter 12) and also that

$$\Phi_z(t_1 - q) \begin{bmatrix} B_1 \\ 0 \end{bmatrix} = \exp[\tilde{A}_{11}t_1] \begin{bmatrix} \exp[-\tilde{A}_{11}q]B_1 \\ 0 \end{bmatrix}$$

and

$$\int_0^{t_1} \Phi_z(t_1 - q) \begin{bmatrix} B_1 \\ 0 \end{bmatrix} u(q)dq = \exp[\tilde{A}_{11}t_1] \int_0^{t_1} \begin{bmatrix} \exp[-\tilde{A}_{11}q]B_1 u(q) \\ \hline 0 \end{bmatrix} dq$$

The problem then reduces to showing that

$$\hat{z}_1 = \exp[\tilde{A}_{11}t_1] \int_0^{t_1} \exp[-\tilde{A}_{11}q]B_1 u(q)dq \tag{20.11}$$

Step 3. The right-hand side of equation 20.11 becomes

$$\exp[\tilde{A}_{11}t_1] \int_0^{t_1} \exp(-\tilde{A}_{11}q)B_1[B_1'\exp(-\tilde{A}_{11}'q)K^{-1}\exp(-\tilde{A}_{11}t_1)z_1]dq$$

$$= \exp[\tilde{A}_{11}t_1]KK^{-1}\exp[-\tilde{A}_{11}t_1]\hat{z}_1 = \hat{z}_1$$

completing the proof. ∎

Corollary 20.2. The state model

$$\dot{x} = Ax + Bu$$

$$y = Cx + Du$$

is (completely) controllable if and only if rank$[Q] = n$.

The next theorem follows directly from theorem 20.2.

Theorem 20.3. The following are equivalent:

(*i*) If $\hat{x}$ ϵcol-sp$[Q]$, then $\hat{x}$ can be driven to θ by some input.

(*ii*) If $\hat{x}$ ϵcol-sp$[Q]$, then θ can be driven to $\hat{x}$ by some input.

(*iii*) If $\hat{x}$ and $\tilde{x}$ are both in col-sp$[Q]$, then $\hat{x}$ can be driven to $\tilde{x}$ by some input.

The proof of theorem 20.2 delivers an input which will drive $x_0 = x(0)$ to θ over the interval $[0, t_1]$. For the more general case, define

$$\hat{K} = \int_{t_0}^{t_1} \exp[\tilde{A}_{11}(t_1 - q)] B_1 B_1' \exp[\tilde{A}_{11}'(t_1 - q)] \, dq$$

An input which will drive $x_0 = x(t_0) \in \text{col-sp}[Q]$ to $x_1 = x(t_1) \in \text{col-sp}[Q]$ is

$$u(q) = -B_1' \exp[\tilde{A}_{11}'(t_1 - q)] \hat{K}^{-1} [\exp[\tilde{A}_{11}(t_1 - t_0)] \hat{z}_0 - \hat{z}_1]$$

where

$$\begin{bmatrix} \hat{z}_0 \\ 0 \end{bmatrix} = [U_1 \ U_2]^{-1} x_0 \quad \text{and} \quad \begin{bmatrix} \hat{z}_1 \\ 0 \end{bmatrix} = [U_1 \ U_2]^{-1} x_1$$

Exercise. Verify the above remarks for the general case.

EXAMPLE 20.1

Consider the system

$$\dot{x} = \begin{bmatrix} 1 & 0 \\ 0 & -1 \end{bmatrix} x + \begin{bmatrix} 0 \\ 1 \end{bmatrix} u$$

The controllability matrix is

$$Q = \begin{bmatrix} 0 & 0 \\ 1 & -1 \end{bmatrix}$$

which has rank 1. The column space of Q is therefore one-dimensional and is spanned by the vector $(0 \ 1)'$. Figure 20.1 portrays the controllable subspace in the state space.

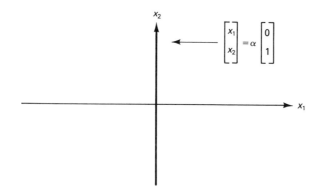

Figure 20.1 The controllable subspace in $\mathbf{R}^2$ when $x = (x_1, x_2)'$.

The conclusion is that any $x = \beta(0,1)'$ is controllable, whereas any $x = (\alpha, \beta)$ with $\alpha \neq 0$ is not controllable—i.e., one cannot drive θ to such an $(\alpha, \beta)'$ in finite time.

EXAMPLE 20.2

Let the state dynamics for a system be

$$\dot{x} = \begin{bmatrix} -1 & 0 \\ 0 & -1 \end{bmatrix} x + \begin{bmatrix} 1 \\ 1 \end{bmatrix} u$$

The resulting controllability matrix clearly has rank 1, and col-sp[Q] = span{(1 1)^t}. Hence, any vector of the form α(1 1)^t is controllable, whereas any vector of the form $(\alpha, \beta)^t$ with $\alpha \neq \beta$ is not controllable.

Exercise. Investigate the controllability of the pair (A,B) when

$$A = \begin{bmatrix} 1 & 2 \\ -1 & -2 \end{bmatrix}, \ B = \begin{bmatrix} 2 \\ -1 \end{bmatrix}$$

True-false exercise. If the Q matrix for a certain linear time-invariant state model is

$$Q = \begin{bmatrix} -1 & 1 & 2 & -2 \\ 1 & -1 & -2 & 2 \end{bmatrix}$$

then the state $[1 \ 1]^t$ is controllable.

EQUIVALENCES TO SYSTEM CONTROLLABILITY

The following theorem delineates a number of equivalences for complete controllability of a system. Note that a matrix M is *positive definite* if all its eigenvalues are real and strictly greater than zero.

Theorem 20.4. The following statements are equivalent:

(*i*) The pair (A,B) is controllable.
(*ii*) Rank$[(\lambda_i I - A) \mid B] = n$ for each eigenvalue λ_i of A.
(*iii*) Rank$[Q] = n$, where Q is the controllability matrix.
(*iv*) Rank$[\exp(-At)B] = n$; i.e., there are n linearly independent rows (row functions) of $\exp(-At)B$ over $[0, \infty)$.
(*v*) The matrix

$$\hat{K} = \int_{t_0}^{t_1} \Phi(t_1 - q) B \ B^t \Phi^t(t_1 - q) dq \tag{20.12}$$

is positive definite. Moreover, the input

$$u(t) = -B^t \Phi^t(t_1 - t) \hat{K}^{-1}[\Phi(t_1 - t_0)x_0 - x_1] \tag{20.13}$$

transfers $x_0 = x(t_0)$ to $x_1 = x(t_1)$.

The existence of the input $u(t)$ in theorem 20.4 follows directly from the special form given in equation 20.10 and can be easily shown to drive $x_0 = x(t_0)$ to $x_1 = x(t_1)$. An example best illustrates the equivalences mentioned in the theorem and serves as preparation for their proof.

EXAMPLE 20.3

Suppose the state dynamics of a particular state model are given by

$$\begin{bmatrix} \dot{x}_1 \\ \dot{x}_2 \end{bmatrix} = \begin{bmatrix} 0 & 1 \\ 0 & -1 \end{bmatrix} \begin{bmatrix} x_1 \\ x_2 \end{bmatrix} + \begin{bmatrix} 1 \\ 1 \end{bmatrix} u$$

Step 1. Determine the controllability matrix and its rank. The matrix is

$$Q = \begin{bmatrix} 1 & 1 \\ 1 & -1 \end{bmatrix}$$

which has rank 2.

Step 2. Check whether $\text{rank}[(\lambda_i I - A)|B]$ equals 2 for each eigenvalue, $\lambda_1 = 0$ and $\lambda_2 = -1$, of A. We have

$$\text{rank}[(\lambda_1 I - A)|B] = \text{rank} \begin{bmatrix} 0 & -1 & 1 \\ 0 & 1 & 1 \end{bmatrix} = 2$$

and

$$\text{rank}[(\lambda_2 I - A)|B] = \text{rank} \begin{bmatrix} -1 & -1 & 1 \\ 0 & 0 & 1 \end{bmatrix} = 2$$

as the theorem implies.

Step 3. Show that $[\exp(-At)B]$ has two linearly independent rows. Since

$$\exp[-At] = \begin{bmatrix} 1 & 1 \\ 0 & -1 \end{bmatrix} \begin{bmatrix} 1 & 0 \\ 0 & e^t \end{bmatrix} \begin{bmatrix} 1 & 1 \\ 0 & -1 \end{bmatrix} \triangleq Te^{Dt}T^{-1}$$

it follows that

$$\exp[-At]B = \begin{bmatrix} 2 - e^t \\ e^t \end{bmatrix}$$

For the rows to be linearly dependent over some interval $[t_0, t_1]$, we must have, for $t_0 \leq t \leq t_1$ and $t_0 < t_1$,

$$\alpha_1(2 - e^t) + \alpha_2 e^t \equiv 0$$

which requires that $\alpha_1 = \alpha_2 = 0$. Thus, the two rows are linearly independent over $[t_0, t_1]$ and in particular $[0, \infty]$.

Step 4. Show that $\hat{K}$ is positive definite. For simplicity, let $t_0 = 0$ and $t_1 = 1$. Then

$$\hat{K} = \int_0^1 \Phi(t_1 - q)B\,B^t\Phi^t(t_1 - q)dq = T\int_0^1 e^{D(1-q)}(T^{-1}B)\,(T^{-1}B)^t e^{D^t(1-q)}dqT^t$$

$$= T\int_0^1 \begin{bmatrix} 1 & 0 \\ 0 & e^{-(1-q)} \end{bmatrix}\begin{bmatrix} 4 & -2 \\ -2 & 1 \end{bmatrix}\begin{bmatrix} 1 & 0 \\ 0 & e^{-(1-q)} \end{bmatrix}dqT^t$$

$$= \begin{bmatrix} 1 & 1 \\ 0 & -1 \end{bmatrix}\begin{bmatrix} 4 & -2(1-e^{-1}) \\ -2(1-e^{-1}) & 0.5(1-e^{-1}) \end{bmatrix}\begin{bmatrix} 1 & 0 \\ 1 & -1 \end{bmatrix}$$

$$= \begin{bmatrix} 1.9039 & 0.8319 \\ 0.8319 & 0.4323 \end{bmatrix}$$

This matrix has eigenvalues at 2.228 and 0.0575 and is thus positive definite.

Proof of Theorem 20.4. The proof of the theorem proceeds as follows. First, the equivalence of conditions (*i*), (*ii*), and (*iii*) is proven. After this, condition (*iii*) is shown to imply condition (*iv*), condition (*iv*) to imply (*v*), and (*v*) to imply (i). The proof relies heavily on theorem 20.2 and the lemmas preceding it.

Step 1. The equivalence of (*i*) and (*iii*) follows directly from the corollary to theorem 20.2.

Step 2. The proof that condition (*iii*) implies condition (*ii*) is established by contradiction. Suppose there exists a λ_i such that $\text{rank}[(\lambda_i I - A)|B] < n$. Then there exists a $v \in \mathbf{R}^n, v \neq \theta$, such that

$$v^t[(\lambda_i I - A)|B] = \theta^t$$

or, equivalently, such that

$$v^t(\lambda_i I - A) = \theta^t \qquad\qquad (20.14a)$$

and

$$v^t B = \theta^t \qquad\qquad (20.14b)$$

where we have not distinguished among the various dimensions of θ. From equation 20.14a and the fact that $v \neq \theta$, it follows that $\bar{v} = w_i$, a left eigenvector of A. Hence,

$$v^t A^k = w_i^* A^k = \lambda_i^k w_i^*$$

From this equation, together with equation 20.14b, it follows that

$$v^t Q = v^t[B \quad AB \quad \ldots \quad A^{n-1}B] = \theta^t$$

contradicting the linear independence of the rows of Q. Therefore, condition (*iii*) implies condition (*ii*).

Step 3. The equivalence of conditions (*ii*) and (*iii*) is complete upon showing that condition (*ii*) implies condition (*iii*). Again, the verification relies on the technique of proof by contradiction: assume that $\text{rank}[Q] < n$, and show that

$\text{rank}[(\lambda_i I - A)|B] < n$ for some eigenvalue λ_i of A. Accordingly, if $\text{rank}[Q] < n$, then there exists at least one $v \neq \theta$ such that $v^t Q = \theta^t$. By lemmas 20.1 and 20.2, it follows that $v^t A^k B = \theta^t$ for all $k \geqslant 0$. It is now necessary to establish the existence of a $w_i \neq \theta$ such that $w_i^t[(\lambda_i I - A)|B] = \theta^t$ for some eigenvalue λ_i. Clearly, w_i must be a left eigenvector of A satisfying $w_i^t A = \lambda_i w_i^t$ or, equivalently, $A^t w_i = \lambda_i w_i$, i.e., w_i must be a right eigenvector of A^t.

In order to show that there exists such a w_i, consider the set

$$W = \{w \mid w^t A^k B = \theta^t, \; k \geqslant 0\}$$

This set is nonempty, since $v \in W$. Now, the set W is invariant with respect to multiplication by A^t: let $w \in W$; then $A^t w \in W$ since, $(A^t w)^t A^k B = w^t A^{k+1} B = \theta^t$. Since W is A^t-invariant, it must contain an eigenvector, say, w_i, of A^t, associated with λ_i. But since the eigenvalues of A and A^t coincide, it follows that $(\overline{w}_i)^t[(\lambda_i I - A)|B] = \theta^t$. This is the statement in contradiction with our assumption.

Step 4. We must prove that $\text{rank}[Q] = n$ implies $\text{rank}[\exp(-At)B] = n$. Again using the technique of contradiction (and not for the last time, either), suppose that $\text{rank}[\exp(-At)B] \neq n$—i.e., there exists $v \neq \theta$ such that $v^t \exp[-At]B \equiv \theta^t$, the zero function. Taking derivatives and evaluating at $t = 0$ yields the result that $v^t A^k B = \theta^t$ for all $k \geqslant 0$. Thus, $v^t Q = \theta^t$, contradicting the assumption that $\text{rank}[Q] = n$. Hence, $\text{rank}[Q] = n$ implies that $\text{rank}[\exp(-At)B] = n$.

Step 5. We seek to prove that if $\text{rank}[\exp(-At)B] = n$ over $[0,\infty)$, then $\hat{K}$ is positive definite. Again, the proof uses the method of contradiction. Suppose $\hat{K}$ is not positive definite. Then because equation 20.12 precludes the possibility of a negative definite $\hat{K}$, $\hat{K}$ must be singular. But then, a singular $\hat{K}$ implies the existence of a $v \neq \theta$ such that $v^t \hat{K} = \theta^t$, which in turn implies that $v^t \hat{K} v = 0$. In particular,

$$v^t \hat{K} v = \int_{t_0}^{t_1} v^t \Phi(t_1 - q) B\, B^t \Phi^t(t_1 - q) v \, dq = 0$$

Analogously to the proof of lemma 20.3, the integrand must be identically zero for all $t_1 \geqslant q \geqslant t_0$, so that

$$v^t \Phi(t_1 - q) B = v^t \exp[A(t_1 - q)]B = \theta^t$$

Thus, the rows of $\exp[At]B$ are linearly dependent, from which it follows that the rows of $\exp[-At]B$ are also linearly dependent.

Step 6. Here, we prove that a positive definite $\hat{K}$ implies the controllability of the pair (A,B). Since the input of equation 20.13 is well defined, whenever $\hat{K}$ is positive definite, "plugging and chugging" on

$$x(t_1) = \Phi(t_1 - t_0)x(t_0) + \int_{t_0}^{t_1} \Phi(t_1 - q)Bu(q)dq$$

demonstrates that $u(t)$ drives $x(t_0)$ to $x(t_1)$, as per the definition of a controllable pair (A,B). This completes the proof.   ∎

For a similar development, see [7]. Also, some interesting controllability aspects having a circuit-theoretic flavor can be found in [14].

CONTROLLABILITY AND DIRECTED MODES

Suppose a system matrix A has distinct eigenvalues $\{\lambda_1,...,\lambda_n\}$ and right eigenvectors $\{e_1,...,e_n\}$. Then if $T = [e_1,...,e_n]$, it is always possible to choose the associated left eigenvectors $\{w_1,...,w_n\}$ so that

$$T^{-1} = \begin{bmatrix} w_1^* \\ \cdot \\ \cdot \\ \cdot \\ w_n^* \end{bmatrix}$$

The aim is to demonstrate that a mode (λ_i,e_i) is controllable if and only if $w_i^*B \neq \theta^t$. The following lemma is the cornerstone of this result.

Lemma 20.4. Let A be as just mentioned. Then the column space of B has a projection onto e_i with respect to the right eigenvector basis for $\mathbf{R}^n$, if and only if e_i lies in the column space of Q. (*Note:* For the case of complex eigenvectors, e_i has a projection if and only if its real and/or imaginary parts do.)

Proof.
Step 1. Since A has distinct eigenvalues, the set of right eigenvectors is a basis for $\mathbf{R}^n$ (see theorem 17.1). Consequently, there exist matrices (row vectors) $v_1^t,...,v_n^t$ such that $B = e_1v_1^t + ... + e_nv_n^t$. Now, the column space of B has a projection onto e_i with respect to the right eigenvector basis if and only if $v_i^t \neq \theta^t$. From the properties of the right and left eigenvectors,

$$v_i^t = w_i^*B$$

Thus, the column space of B has a projection onto e_i with respect to the right eigenvector basis of $\mathbf{R}^n$ if and only if $w_i^*B \neq \theta^t$.

Step 2. Since $B = e_1v_1^t + ... + e_nv_n^t$ for appropriate row vectors v_1^t to v_n^t, then, as per the proof of theorem 17.2 or the spectral mapping theorem of problem 9, chapter 17,

$$A^jB = e_1(\lambda_1^j v_1^t) + ... + e_n(\lambda_n^j v_n^t)$$

Consequently,

$$Q = [B\ AB\ ...\ A^{n-1}B] = [e_1\ e_2 ... e_n] \begin{bmatrix} v_1^t & & & & \\ & v_2^t & & & \\ & & \cdot & & \\ & & & \cdot & \\ & & & & v_n^t \end{bmatrix} \begin{bmatrix} I & \lambda_1 I & ... & \lambda_1^{n-1}I \\ I & \lambda_2 I & ... & \lambda_2^{n-1}I \\ \vdots & \vdots & \vdots & \vdots \\ I & \lambda_n I & ... & \lambda_n^{n-1}I \end{bmatrix}$$

Since $v_i^t \neq \theta^t$ if and only if $w_i^*B \neq \theta^t$, each column of Q can be expressed as a linear combination of only those e_i on which the column space of B has a nonzero projection. Hence, $x \in \text{col-sp}[Q]$ implies that x lies in the space spanned by those e_i for which $w_i^*B \neq \theta^t$.

Step 3. It remains to show that the number of nonzero v_i^t's precisely equals p, in which case col-sp$[Q]$ equals the span of the eigenvectors for which $w_i^*B \neq \theta^t$. This follows directly from the preceding equation: since $[e_1,...,e_n]$ is nonsingular, and since the Vandermonde matrix

$$\begin{bmatrix} I & \lambda_1 I & ... & \lambda_1^{n-1}I \\ I & \lambda_2 I & ... & \lambda_2^{n-1}I \\ \vdots & \vdots & \vdots & \vdots \\ I & \lambda_n I & ... & \lambda_n^{n-1}I \end{bmatrix}$$

is always nonsingular for distinct λ_i,

$$\text{rank}[Q] = \text{rank} \begin{bmatrix} v_1^t & & & & \\ & v_2^t & & & \\ & & \cdot & & \\ & & & \cdot & \\ & & & & \cdot \\ & & & & v_n^t \end{bmatrix}$$

which equals p. Hence, there are only p nonzero v_i^t determined by $w_i^*B = v_i^t$, and a basis for the column space of Q is the set of associated right eigenvectors. ∎

EXAMPLE 20.4

Let

$$\dot{x} = \begin{bmatrix} 1 & -3 & -1 \\ 2 & -4 & -1 \\ 2 & -2 & -3 \end{bmatrix} x + \begin{bmatrix} 3 \\ 2 \\ 2 \end{bmatrix} u$$

Then the eigenvalues of the A-matrix are $\lambda_1 = -1, \lambda_2 = -2$, and $\lambda_3 = -3$. Now, observe that the matrix of right eigenvectors is

$$T = [e_1 \ e_2 \ e_3] = \begin{bmatrix} 2 & 1 & 1 \\ 1 & 1 & 1 \\ 1 & 0 & 1 \end{bmatrix}$$

and that

$$T^{-1} = \begin{bmatrix} w_1^* \\ w_2^* \\ w_3^* \end{bmatrix} = \begin{bmatrix} 1 & -1 & 0 \\ 0 & 1 & -1 \\ -1 & 1 & 1 \end{bmatrix}$$

With this information, it is straightforward to decompose the state transition matrix into a sum of matrix directed modes:

$$\Phi(t) = \exp[At] = e^{\lambda_1 t}R_1 + e^{\lambda_2 t}R_2 + e^{\lambda_3 t}R_3$$

where

$$R_1 = [e_1 \quad -e_1 \quad \theta] = \begin{bmatrix} 2 & -2 & 0 \\ 1 & -1 & 0 \\ 1 & -1 & 0 \end{bmatrix}$$

$$R_2 = [\theta \quad e_2 \quad -e_2] = \begin{bmatrix} 0 & 1 & -1 \\ 0 & 1 & -1 \\ 0 & 0 & 0 \end{bmatrix}$$

and

$$R_3 = [-e_3 \quad e_3 \quad e_3] = \begin{bmatrix} -1 & 1 & 1 \\ -1 & 1 & 1 \\ -1 & 1 & 1 \end{bmatrix}$$

Let us now consider the matrix products R_iB and then $\exp[A(t_1-q)]B$:

$$R_1B = e_1 = \begin{bmatrix} 2 \\ 1 \\ 1 \end{bmatrix}, \; R_2B = \theta = \begin{bmatrix} 0 \\ 0 \\ 0 \end{bmatrix}, \; R_3B = e_3 = \begin{bmatrix} 1 \\ 1 \\ 1 \end{bmatrix}$$

Observe that e_2 has no projection onto the column space of B and hence onto Q, since $R_2B = \theta$. Our conclusion is that all states which are scalar multiples of e_2 or which have components in the direction of e_2 are *not* controllable, whereas all states which are linear combinations of e_1 and e_3 are controllable, since they lie in the column space of Q.

This same conclusion can be drawn from another line of reasoning. Consider the range of the integral

$$\int_0^{t_1} \exp[A(t_1-\tau)]Bu(\tau)d\tau$$

for arbitrary inputs $u(\cdot)$. For each t_1, this integral produces a fixed vector. The range of the integral is the set of all possible vectors which result from an admissible input.

Based on the matrix products R_iB and the decomposition of the state transition matrix into directed modes, the integral takes the form

$$\int_0^{t_1} \exp[A(t_1-\tau)]Bu(\tau)d\tau = \xi_1(t_1,u(\cdot))e_1 + \xi_2(t_1,u(\cdot))e_3$$

where the (possibly complex) scalars $\xi_i(\cdot,\cdot)$ are given by

$$\xi_i(t_1,u(\tau)) = \int_0^{t_1} \exp[\lambda_i(t_1-\tau)]u(\tau)d\tau$$

Therefore, the range of the integral is simply span$\{e_1,e_3\}$.

As a final point of illustration, notice that col-sp[Q] equals span$\{e_1,e_3\}$, since

(i)

$$Q = [B\,|\,AB\,|\,A^2B] = \begin{bmatrix} 3 & -5 & 11 \\ 2 & -4 & 10 \\ 2 & -4 & 10 \end{bmatrix}$$

which clearly has rank 2 since the second and third rows are identical,

(ii)

$$\begin{bmatrix} 3 \\ 2 \\ 2 \end{bmatrix} = \begin{bmatrix} 2 \\ 1 \\ 1 \end{bmatrix} + \begin{bmatrix} 1 \\ 1 \\ 1 \end{bmatrix} = e_1 + e_3$$

and

(iii)

$$\begin{bmatrix} -5 \\ -4 \\ -4 \end{bmatrix} = \begin{bmatrix} -2 \\ -1 \\ -1 \end{bmatrix} + \begin{bmatrix} -3 \\ -3 \\ -3 \end{bmatrix} = -e_1 - 3e_3$$

Thus, each column of Q can be expressed as a linear combination of e_1 and e_3. Proof of the converse is easy and is left to the reader.

The following points summarize the discussion of directed modes and controllability:

(i) A directed mode $\exp[\lambda_i t]e_i$ is controllable if and only if $R_iB \neq [0]$.
(ii) A directed mode $\exp[\lambda_i t]e_i$ is uncontrollable if and only if $R_iB = [0]$.
(iii) $R_iB = [0]$ if and only if $w_i^*B = \theta^t$, and $R_iB \neq [0]$ if and only if $w_i^*B \neq \theta^t$.

CONTROLLABILITY OF DISCRETE–TIME SYSTEMS

Consider the discrete-time state dynamics

$$x(k+1) = Ax(k) + Bu(k)$$

where A is $n \times n$ and B is $n \times m$. Then the following definition applies.

Definition 20.4.   A state $\hat{x} \in \mathbf{R}^n$ is *controllable*, or more commonly, *reachable*, if and only if there exists a finite index N and an input sequence $\{u(0),u(1),...,u(N-1)\}$ such that if $x(0) = \theta$, then $x(N) = \hat{x}$.

This definition reverses the continuous-time statement of controllability, where one needs to find an input which drives $\hat{x}$ to θ. This is contrary to the literature where discrete-time controllability means the ability to drive $\hat{x}$ to θ and reachability is defined as per definition 20.3. It appears that such a situation exists due to an originally ill stated definition of discrete-time controllability. As justification, definition 20.4 provides us with the intuitive and physically expected equivalences and characterizations similar to those found in the continuous-time case. The following example shows a more concrete justification for definition 20.4.

EXAMPLE 20.5

As a point of reference, just because a discrete-time system can drive $\hat{x}$ to θ in a finite number of steps does not necessarily mean that it is controllable. For example, consider Figure 20.2,

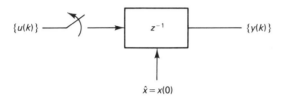

$\hat{x} = x(0)$

Figure 20.2   An uncontrollable discrete time system.

and let $x(0) = 1$. Then, for arbitrary input sequences, which are obviously decoupled from the system, $y(0) = 1$ and $y(k) = 0$ for all $k \geqslant 1$. In other words, the system is uncontrollable.

Using the definition, then, a controllable state is "reachable" from θ. On the other hand, θ is reachable from many uncontrollable states in a finite number of steps. For example, the discrete-time system

$$x(k + 1) = \begin{bmatrix} 1 & 0 & 0 \\ 1 & 1 & 0 \\ 0 & 0 & 0 \end{bmatrix} x(k) + \begin{bmatrix} 0 \\ 1 \\ 0 \end{bmatrix} u(k)$$

can send any vector of the form $x(0) = [0\ 0\ \alpha]^t$ to θ in one step by setting $u(0) = 0$, similarly to example 20.5. This follows because the null space of A is spanned by vectors of the form $[0\ 0\ \alpha]$. Here, however, the controllability matrix

$$Q = \begin{bmatrix} 0 & 0 & 0 \\ 1 & 1 & 1 \\ 0 & 0 & 0 \end{bmatrix}$$

has rank 1, so that the only states reachable from θ are vectors of the form $[0\ \alpha\ 0]^t$. Note that if $x(k) = \theta$, then $x(k + 1) = [0\ \alpha\ 0]^t$ is attainable in exactly one step by setting $u(k) = \alpha$.

This simple example sparks a question: what states are reachable from θ in exactly one step? Since $x(k + 1) = Ax(k) + Bu(k)$, if $x(k) = \theta$, then only those states contained in the image of B are attainable in one step. What about two steps? Again, if $x(k) = \theta$, then $x(k + 2) = ABu(k) + Bu(k + 1)$. Thus, $x(k + 2)$ is a linear combination of the columns of AB and B, so those states reachable in exactly two steps are those contained in the image of $[B\ AB]$. This, of course, can be extrapolated to k steps with the obvious characterization. These notions are consistent with those of chapters 5 and 9, where an input sequence $\{u(0),...,u(n-1)\}$ which drives $x(0)$ to $x(n)$ must satisfy

$$\left[x(n) - A^n x(0)\right] = \left[B \ AB \ \dots \ A^{n-1}B\right] \begin{bmatrix} u(n-1) \\ u(n-2) \\ \vdots \\ u(0) \end{bmatrix} \qquad (20.15)$$

From our continuous-time development (lemmas 20.1 and 20.2), observe that $Q = [B \ AB \ \dots \ A^{n-1}B]$ has maximal rank, and thus the index $N \leqslant n$ in equation 20.3. Equation 20.15, however, completely describes the controllability aspects of the pair (A,B). Now, if $x(0)$ is taken to be θ, then $x(n)$ can be reached from θ only when $x(n) \in \text{col-sp}[Q]$, i.e., only when the equations are consistent. As with the continuous-time case, the *controllable subspace* is defined as col-sp$[Q]$. It is straightforward to show that $\hat{x} \in \mathbf{R}^n$ if and only if $\hat{x} \in \text{col-sp}[Q]$.

Exercise. Prove that $\hat{x}$ is controllable if and only if $\hat{x} \in \text{col-sp}[Q]$. (The ideas involved here suggest the following theorem.)

Theorem 20.5.   The following are equivalent:

(i) There exists a finite index $N \leqslant n$ such that θ can be driven to $\hat{x} = x(N)$ by some input sequence $\{u(k), u(k+1),\dots,u(k+N-1)\}$.

(ii) $\hat{x} \in \text{col-sp}[Q]$ where $Q = [B \,|\, AB \,|\, \cdots \,|\, A^{n-1}B]$.

(iii) If $\hat{x}_1$ and $\hat{x}_2 \in \text{col-sp}[Q]$ then there exists a finite index N such that $\hat{x}_1(k)$ can be driven to $\hat{x}_2(k+N)$ by some input sequence $\{u(k), u(k+1),\dots,u(k+N-1)\}$.

Exercise. Construct the proof of this theorem.

As a final theoretical comment, we point out that controllability and directed modes for continuous-time systems have a direct analog in discrete-time systems. In particular,

$$\Phi(k) = A^k = R_1\lambda_1^k + \dots + R_n\lambda_n^k$$

whenever A has distinct eigenvalues. A directed mode $R_i\lambda_i^k$ is controllable if and only if $R_i B \neq [0]$. That is to say, a directed mode is uncontrollable if and only if $R_i B = [0]$.

Several features of the controllability of discrete-time systems are illustrated by the following two examples.

EXAMPLE 20.6

Consider a switched discrete-time system having the structure

$$x(k+1) = Ax(k) + Bu(k)$$

where $B = [1 \ 0]^t$ and

$$A_i = \begin{cases} A_1 = \begin{bmatrix} 0 & 1 \\ 0 & 1 \end{bmatrix} & \text{for } k = 0,1,4,5,6,\ldots \\[2em] A_2 = \begin{bmatrix} -1 & 0 \\ 1 & 0 \end{bmatrix} & \text{for } k = 2,3 \end{cases}$$

It is simple to show that the pair (A_1,B) is not completely controllable, whereas the pair (A_2,B) is. The question arises, does there exist an input sequence which will drive $x(0) = [1\ 1]^t$ to $x(5) = [1\ -1]^t$? Observe that neither $x(0)$ nor $x(5)$ live in the column space of

$$Q_1 = [B\ \ A_1B]$$

Intuition might suggest that no such input sequence exists. On the other hand, the system is controllable over the subinterval [2,3] of [0,5]. So maybe by proper adjustment of the inputs $u(2)$ and $u(3)$ something can be done.

Step 1. Since the system is uncontrollable over [0,1], set $u(0) = u(1) = 0$. Then

$$x(2) = A_1^2 x(0) = \begin{bmatrix} 0 & 1 \\ 0 & 1 \end{bmatrix}^2 \begin{bmatrix} 1 \\ 1 \end{bmatrix} = \begin{bmatrix} 1 \\ 1 \end{bmatrix}$$

The original problem now becomes, Find $\{u(2), u(3), u(4)\}$ which will drive $x(2) = [1\ 1]^t$ to $x(5) = [1\ -1]^t$.

Step 2. If A_1 were nonsingular and $u(4)$ were set to zero, then that $x(4)$ needed to produce $x(5) = [1\ -1]^t$ would satisfy $x(4) = A_1^{-1}[1\ -1]^t$. Unfortunately, A_1 is singular. So let us derive a modified equation 20.15. Clearly,

$$x(4) = A_2^2 x(2) + A_2 Bu(2) + Bu(3)$$

and

$$x(5) = A_1 A_2^2 x(2) + A_1 A_2 Bu(2) + A_1 Bu(3) + Bu(4)$$

Rearranging the latter equation and putting it in matrix form yields

$$[x(5) - A_1 A_2^2 x(2)] = [B\ \ A_1 B\ \ A_1 A_2 B] \begin{bmatrix} u(4) \\ u(3) \\ u(2) \end{bmatrix} \tag{20.16}$$

Step 3. Substituting the appropriate quantities into equation 20.16 yields

$$\begin{bmatrix} 2 \\ 0 \end{bmatrix} = \begin{bmatrix} 1 & 0 & 1 \\ 0 & 0 & 1 \end{bmatrix} \begin{bmatrix} u(4) \\ u(3) \\ u(2) \end{bmatrix}$$

Since the rank of $[B\ \ A_1 B\ \ A_1 A_2 B]$ is obviously 2, the equations are always consistent and always solvable using any of the many right inverses or, specifically, using the Moore-Penrose inverse discussed in chapter 19. In this case the "minimum energy" solution is

$$
\begin{bmatrix} u(4) \\ u(3) \\ u(2) \end{bmatrix} = \begin{bmatrix} 1 & -1 \\ 0 & 0 \\ 0 & 1 \end{bmatrix} \begin{bmatrix} 2 \\ 0 \end{bmatrix} = \begin{bmatrix} 2 \\ 0 \\ 0 \end{bmatrix}
$$

It is important to note that if A_1 were nonsingular, and if the pair (A_2,B) were controllable, then any $x(0)$ could be driven to any $x(k)$ for $k \geqslant 4$ by a suitable input sequence. In the example A_1 was singular, and it appears that $x(0)$ can still be driven to any $x(k)$, $k \geqslant 4$, by some input sequence. However, this is possible only because $\text{Null}[A_1] \subset \text{Im}(B)$. If this condition were not satisfied, then it would not be possible, as the following example illustrates.

EXAMPLE 20.7

Consider again a switched system

$$
x(k + 1) = A_i x(k) + Bu(k)
$$

where $B = [1 \ 0]^t$ and

$$
A_i = \begin{cases} A_1 = \begin{bmatrix} 1 & -1 \\ 0 & 0 \end{bmatrix} & \text{for } k = 0,1,4,5,6,... \\[2ex] A_2 = \begin{bmatrix} -1 & 0 \\ 1 & 0 \end{bmatrix} & \text{for } k = 2,3 \end{cases}
$$

Is it possible to drive $x(0) = \theta$ to $x(5) = (1 \ -1)^t$?

Step 1. Set $u(0) = u(1) = 0$, in which case

$$
x(2) = \begin{bmatrix} 1 & -1 \\ 0 & 0 \end{bmatrix}^2 \begin{bmatrix} 1 \\ 1 \end{bmatrix} = \begin{bmatrix} 0 \\ 0 \end{bmatrix}
$$

Step 2. Again, equation 20.16 must be satisfied for the given A_1 and A_2. Accordingly, we have

$$
\begin{bmatrix} 1 \\ -1 \end{bmatrix} = \begin{bmatrix} 1 & 1 & -2 \\ 0 & 0 & 0 \end{bmatrix} \begin{bmatrix} u(4) \\ u(3) \\ u(2) \end{bmatrix}
$$

The equation is clearly inconsistent and no solution exists. Thus, $x(5)$ is not reachable from the origin.

Note that in the preceding example $\text{Null}[A_1] \not\subset \text{Im}(B)$. Thus, those parts of the state eliminated by multiplication by A_1 in the state equation $x(k + 1) = A_1 x(k) + Bu(k)$ cannot be reinserted by action of the input. On the other hand, if it were the case that $\text{Null}[A_1] \subset \text{Im}(B)$, they could.

By contrast, the continuous-time case does not suffer from this anomaly. Let $\dot{x}(t) = A(t)x(t) + Bu(t)$, where (i) $A(t) = A_1$ for $t < t_0$ and $t > t_1$ and (ii) $A(t) = A_2$ for $t_0 \leqslant t \leqslant t_1$. Then whenever the pair (A_2, B) is controllable over $[t_0, t_1]$, the system is controllable over $[t_0', t_1']$, where $t_0' \leqslant t_0 \leqslant t_1 \leqslant t_1'$. In fact, by time invariance, any $x(t_0)$ can be driven to any $x(t_1)$ regardless of the choices of t_0 and t_1 (see problem 13). This follows because of the nonsingularity of $\exp[At]$.

POLE PLACEMENT BY STATE FEEDBACK

Preliminary Material

The state dynamical equation $\dot{x} = Ax + Bu$ models a physical process such as a power system. Often, such systems have poles bordering the imaginary axis but in the left half of the complex plane. Responses consequently have an oscillatory behavior similar to the bouncing motion of a car with a failed shock absorber. Such behavior calls for increased damping, which translates to the mathematical requirement that poles have negative real parts of sufficient magnitude. Accomplishment of this pole shifting typically occurs with state feedback. The input to the state model takes the form $u = u_c + u_e$, where u_c denotes a control input and u_e designates the usual external system input. In addition, $u_c = Fx$, where F is an $m \times n$ state feedback matrix designed so as to achieve desired eigenvalue locations of the controlled system. The controlled system thus has the new state dynamical equation

$$\dot{x} = (A + BF)x + Bu_e$$

Since it is tacitly assumed that u represents an external input, the subscript e on u_e will hereafter be dropped.

Mathematically, the goal is to fabricate F so that the eigenvalues of $A + BF$ coincide with a prespecified symmetric set of complex numbers denoted by $\sigma(A)$. To say that the set is symmetric means that all nonreal numbers appear as complex conjugate pairs.

Before solving the general problem of spectral assignability, we solve a special case: a single-input problem in which A and B have a well-known canonical form. With the technique for solving the single-input case in hand, we then develop a string of *transformations which convert a multi-input model into a canonical single-input model*. Using the spectral assignability technique that we employed for the special case, we construct a feedback matrix which assigns the appropriate spectrum to the canonical single-input model. The feedback for the multi-input problem results by taking the *inverse transformation*.

To begin, recall that

$$\pi_A(\lambda) = det[\lambda I - A] = (\lambda - \lambda_1)(\lambda - \lambda_2) \dots (\lambda - \lambda_n) \qquad (20.17)$$

$$= a_n + a_{n-1}\lambda + \dots + a_1\lambda^{n-1} + \lambda^n$$

is the characteristic polynomial of A. Clearly, if

$$A = \begin{bmatrix} 0 & 1 & 0 & \ldots & 0 & 0 \\ 0 & 0 & 1 & \ldots & 0 & 0 \\ . & . & . & \ldots & . & . \\ 0 & . & 0 & \ldots & 0 & 1 \\ -a_n & -a_{n-1} & -a_{n-2} \ldots & . & -a_1 \end{bmatrix} \qquad (20.18)$$

then equation 20.17 specifies the characteristic polynomial of A. The proof of this assertion is straightforward and is left as an exercise for the reader. (*Hint:* Expand $\det[\lambda I - A]$ along the bottom row.)

Suppose now that equation 20.18 specifies A and that $B = [0,\ldots,0,1]^t$ is $n \times 1$. This is the controllable canonical form described in chapter 4, also termed the *rational canonical form* by some authors [5]. With this special structure, the state feedback matrix F is $1 \times n$ and has the form

$$F \triangleq \begin{bmatrix} f_n & f_{n-1} & \ldots & f_1 \end{bmatrix}$$

Observe that $(A + BF)$ has the structure

$$\begin{bmatrix} 0 & 1 & \ldots & 0 \\ 0 & 0 & \ldots & 0 \\ . & . & \ldots & . \\ 0 & . & \ldots & 1 \\ (-a_n + f_n) & (-a_{n-1} + f_{n-1}) & \ldots & (-a_1 + f_1) \end{bmatrix} \qquad (20.19)$$

This observation is the key to the spectral assignability problem for a system that is in the controllable canonical form. Note that such a system is completely controllable. (Prove this!)

The objective, of course, is to choose $f_n,\ldots, f_1$ so that the characteristic polynomial of equation 20.19 has a desired set of coefficients, which is equivalent to assigning a desired set of eigenvalues. Specifically, if $\Lambda = \{\lambda_1,\ldots,\lambda_n\}$ represents a desired set of system eigenvalues, then $f_n,\ldots,f_1$ should satisfy

$$\pi_{A+BF}(\lambda) = (a_n - f_n) + (a_{n-1} - f_{n-1})\lambda + \ldots + (a_1 - f_1)\lambda^{n-1} + \lambda^n$$
$$= (\lambda - \lambda_1)(\lambda - \lambda_2) \ldots (\lambda - \lambda_n)$$
$$\triangleq \hat{a}_n + \hat{a}_{n-1}\lambda + \ldots + \hat{a}_1\lambda^{n-1} + \lambda^n$$

where the $\hat{a}_i$'s specify the coefficients of the desired characteristic polynomial.

Exercise. Suppose the A- and B-matrices of a state model are

$$A = \begin{bmatrix} 0 & 1 & 0 \\ 0 & 0 & 1 \\ 0 & 1 & 0 \end{bmatrix}$$

and

$$B = \begin{bmatrix} 0 \\ 0 \\ 1 \end{bmatrix}$$

(*i*) Is the pair (A,B) controllable?

(*ii*) Find the eigenvalues of A.

(*iii*) Find a feedback matrix F such that the spectrum of $A + BF$ is given by $\Lambda = \{-1, -2 \pm j\}$.

The General Single-Input Case

Spectral assignability of general single-input systems proceeds by converting the given pair (A,B) to its controllable canonical form via a nonsingular state transformation $z = Vx$. Following this, one assigns the spectrum to this equivalent canonical system and then takes the inverse transformation to obtain the needed state feedback matrix. In particular, suppose A is $n \times n$, B is $n \times 1$, and both have arbitrary structures and form a controllable pair. Then the controllability matrix $Q = [B \ AB \ ... \ A^{n-1}B]$ is $n \times n$ and rank$[Q] = n$, implying that Q^{-1} exists. Depending on the eigenvalues of A, it is sometimes more convenient, numerically speaking, to work with the matrix $[B \ AB \ (A^2B)/2! \ ... \ (A^{n-1}B)/(n-1)!]$. For example, if the eigenvalues of A are large in magnitude, a much more tenable Q matrix results. Let v denote the last row of Q^{-1}, and construct the state transformation V in $z = Vx$ as follows:

$$V = \begin{bmatrix} v \\ vA \\ vA^2 \\ \vdots \\ vA^{n-1} \end{bmatrix} \qquad (20.20)$$

Then, by construction, V^{-1} exists. (Prove this.) Executing this state transformation on $\dot{x} = Ax + Bu$ produces

$$V^{-1}\dot{z} = AV^{-1}z + Bu$$

which reduces to

$$\dot{z} = VAV^{-1}z + VBu$$

Now, define $\hat{A} = VAV^{-1}$ and $\hat{B} = VB$. Then the pair $(\hat{A}, \hat{B})$ has the controllable canonical form (see chapter 4).

To assign the spectrum of the given system, first assign the desired set of eigenvalues to the canonical system

$$\dot{z} = \hat{A}z + \hat{B}u \qquad (20.21)$$

by a suitable choice of state feedback $\hat{F}$, where $u = \hat{F}z$ is the obvious control law. Thus, $\sigma(\hat{A} + \hat{B}\hat{F}) = \Lambda$, where Λ represents the set of desired eigenvalues and $\sigma(M)$ denotes the spectrum or eigenvalues of M. Applying the inverse state transformation $z = Vx$ to equation 20.21 results in

$$V\dot{x} = (\hat{A} + \hat{B}\hat{F})Vx \qquad (20.22)$$

from which it follows that

$$\dot{x} = [V^{-1}\hat{A}V + V^{-1}\hat{B}\hat{F}V]x = [A + B(\hat{F}V)]x \qquad (20.23)$$

The required state feedback takes the form

$$F = \hat{F}V \qquad (20.24)$$

where

$$\sigma(\hat{A} + \hat{B}\hat{F}) = \sigma(A + BF) = \Lambda \qquad (20.25)$$

The latter follows because the similarity transformation

$$(A + BF) = V^{-1}(\hat{A} + \hat{B}\hat{F})V \qquad (20.26)$$

leaves all the eigenvalues unchanged.

EXAMPLE 20.8

Suppose the pair (A,B) is such that

$$A = \begin{bmatrix} 1 & 1 \\ 0 & -1 \end{bmatrix}$$

and

$$B = \begin{bmatrix} 0 \\ 1 \end{bmatrix}$$

Then the eigenvalues of A are ± 1. The characteristic polynomial of A is

$$\pi_A(\lambda) = \lambda^2 - 1$$

The controllability matrix

$$Q = [B \,|\, AB] = \begin{bmatrix} 0 & 1 \\ 1 & -1 \end{bmatrix} \qquad (20.27)$$

has rank 2; therefore, the pair (A,B) is controllable. Consequently, the spectrum of the system

$$\dot{x} = Ax + Bu$$

may be arbitrarily assigned through appropriate state feedback.

Suppose our goal is to find an F such that $\sigma(A + BF)$ is $\{0, -2\}$. The inverse of Q given in equation 20.27 is

$$Q^{-1} = \begin{bmatrix} 1 & 1 \\ 1 & 0 \end{bmatrix}$$

Since the last row of Q^{-1} is $v = (1,0)$, the transformation matrix V is

$$V = \begin{bmatrix} v \\ vA \end{bmatrix} = \begin{bmatrix} 1 & 0 \\ 1 & 1 \end{bmatrix}$$

The canonical form of A is

$$\hat{A} = VAV^{-1} = \begin{bmatrix} 0 & 1 \\ 1 & 0 \end{bmatrix}$$

Clearly, $\hat{B} = B$. The characteristic polynomial of the *desired* system is

$$\pi_{A+BF}(\lambda) = \lambda(\lambda + 2) = \lambda^2 + 2\lambda$$

Consequently, the intermediate feedback matrix is

$$\hat{F} = [0 - 1 \quad -2 - 0] = [-1 \quad -2]$$

The desired feedback matrix is

$$F = \hat{F}V = [-3 \quad -2]$$

Observe that

$$A + BF = \begin{bmatrix} 1 & 1 \\ -3 & -3 \end{bmatrix}$$

has the required spectrum.

Exercise. Write a computer program which checks whether the pair (A,B) is controllable. Assume A is at most 10×10. (*Hint*: The number of nonzero singular values in a singular value decomposition of Q is identical to the rank of Q.)

Exercise. Write a computer program which constructs the state feedback matrix F given a controllable pair (A,B) and a spectral assignment set $\Lambda = \{\lambda_1,...,\lambda_n\}$. Assume A is at most 10×10. Make sure your program includes a part which checks whether $\sigma(A + BF)$ is correct. Assume B is $n \times 1$.

The Multi-input Case

The conditions for arbitrary spectral assignability through state feedback require that the system be completely controllable. For the multi-input case, let B be $n \times m$. Then, assuming the pair (A,B) is controllable,

$$Q = [B \,|\, AB \,|\, ... \,|\, A^{n-1}B]$$

has rank n.

Again, the objective is to assign a symmetric set of n complex numbers Λ by constructing a state feedback F such that $\sigma(A + BF) = \Lambda$. To accomplish this, we convert the multi-input system to a single-input system, assign the appropriate spectrum to this single-input system, and convert back [7].

Before proceeding, a comment on the phrase "with probability 1." Intuitively, this phrase is equivalent to "almost surely," "most of the time," "except for a set of measure zero," etc. Algebraic geometry provides a rigorous foundation for these phrases. Of course, we unceremoniously omit further rigor. As a meaningful illustration, suppose one randomly generates entries of a square matrix. Then, *with probability 1,* that matrix is invertible.

Part of the spectral assignment algorithm requires that one take a random (linear) combination of the columns of the matrix B. Specifically, it is necessary to construct a vector

$$\underline{B}_0 = B\mu$$

where, in this case (keeping with the notation of [7]), μ is a column vector whose entries are randomly generated. Suppose the entries of μ are to be random integers between -3 and 3 inclusive. Suppose also that a computer program generates random numbers uniformly between 0 and 1. Then multiply the number by 7, subtract 3.5, and take the integer part. This scheme will randomly generate integers between -3 and 3.

THE MULTI-INPUT SPECTRAL ASSIGNABILITY ALGORITHM

1. Check whether the pair (A,B) is controllable. If so, record the dimensions of A and B.
2. Randomly pick an appropriately dimensioned F_0 and define $A_0 \triangleq A + BF_0$. With probability 1, A_0 has distinct eigenvalues. Check. If not, repeat step 2.
3. Take a random linear combination of the columns of B (as just discussed) to generate $B_0 = B\mu$. Record μ. With probability 1, the pair (A_0,B_0) is controllable. Check. If not, choose another B_0. (*Note:* The pair (A_0,B_0) represents an "equivalent" single-input system.)
4. Assign the desired spectrum Λ via the methods discussed earlier by constructing a feedback matrix F' such that $\sigma(A_0 + B_0F') = \Lambda$.
5. The desired feedback matrix for the system is $F = F_0 + \mu F'$. Check to make sure that $A + BF$ has the required spectrum.

Exercise. Modify the computer program of the earlier exercise so that it handles the multi-input case.

Exercise. Suppose the A- and B-matrices of a state model are

$$A = \begin{bmatrix} 1 & 0 & 0 & 0 \\ 0 & 0 & 1 & 0 \\ 0 & 0 & 0 & 0 \\ 1 & 0 & 0 & 0 \end{bmatrix}$$

and

$$B = \begin{bmatrix} 1 & 0 \\ 1 & 0 \\ 0 & 1 \\ 0 & 0 \end{bmatrix}$$

Find a state feedback F such that the eigenvalues of the controlled system are $\{-1,-1,-1\pm j\}$. (*Hint:* Choose

$$F_0 = \begin{bmatrix} 0 & 0 & 0 & 1 \\ 0 & 1 & 0 & 0 \end{bmatrix}$$

and $\mu = (1,-1)^t$. Your answer should be

$$F = \begin{bmatrix} -5 & -10 & -10 & -8 \\ 5 & 11 & 10 & 9 \end{bmatrix}$$

Exercise. Given that the spectrum of $A + BF = \Lambda$, completely describe the set of all $\hat{F}$ such that $\sigma(A + BF + B\hat{F}) = \Lambda$.

The multi-input spectral assignability algorithm *constructively* verifies the forward direction of the following theorem, which elegantly characterizes controllability in terms of spectral assignability by state feedback.

Theorem 20.6. The pair (A,B) is controllable if and only if, for every symmetric set Λ of n complex numbers, there exists a state feedback map F such that $\sigma(A + BF) = \Lambda$.

Proof. To prove the converse of this theorem assume that for any symmetric set Λ of n complex numbers, there exists a state feedback map F such that $\sigma(A + BF) = \Lambda$. We seek to show that the pair (A,B) is controllable.

Recall that (A,B) is controllable if and only if rank$(Q) = n$ and that a vector x in $\mathbf{R}^n$ is controllable if and only if x is in the image or range of Q, denoted Im(Q), where Im(Q) is simply the span of the columns of Q. But the latter is true if and only if the controllable subspace of the state space is the *whole* space $\mathbf{R}^n$.

Now, let $\lambda_i \in \sigma(A)$ $(i = 1,...,n)$ be real and distinct. By hypothesis, there exists an F such that $\sigma(A + BF) = \{\lambda_1,...,\lambda_n\}$. Let $\{e_1,...,e_n\}$ be the associated right eigenvectors. Then

$$(A + BF)e_i = \lambda_i e_i$$

Some algebraic manipulation yields

$$e_i = (\lambda_i I - A)^{-1}BFe_i \qquad (20.28)$$

The inverse $(\lambda_i I - A)^{-1}$ exists, since, as assumed, $\lambda_i \in \sigma(A)$.

Now it is possible to show that

$$(\lambda I - A)^{-1} = \sum_{j=1}^{n} \rho_j(\lambda)A^{j-1} \qquad (20.29)$$

for a suitable choice of rational functions $\rho_j(\lambda)$ that are analytic in $\mathbf{C}$ except at points in $\sigma(A)$. A straightforward analysis of Leverrier's algorithm (see chapter 18) indicates how to construct the $\rho_j(\lambda)$'s. Substituting equation 20.29 into equation 20.28 yields

$$e_i = \sum_{j=1}^{n} \rho_j(\lambda_i)A^{j-1}BFe_i$$

in which case $e_i \in \text{Im}(Q)$. This follows because

$$\rho_j(\lambda_i)A^{j-1}BFe_i = A^{j-1}B[\rho_j(\lambda_i)Fe_i] \in \text{Im}(A^{j-1}B) \subset \text{Im}(Q)$$

Since the set of vectors $\{e_1,...,e_n\}$ spans $\mathbf{R}^n$ (theorem 17.1), Im$(Q) = \mathbf{R}^n$, meaning that the controllable subspace of the pair (A,B) is the whole space. Hence, the pair (A,B) is controllable.

Exercise. Let A and B both be 2×2. Construct an example showing that step 2 of the multi-input assignability algorithm is necessary.

Spectral Assignability and Incomplete Controllability

It turns out that the controllable part of a system, of a pair (A,B) that is not completely controllable, can be assigned an arbitrary spectrum by appropriate state feedback. The key to the solution of this problem is a nonsingular transformation which converts the given state description into the Kalman controllable form which specifically identifies and extracts the completely controllable modes of the system. As per propositions 20.2–20.8, there exists a state transformation $[U_1 \ U_2]z = x$ such that

$$\dot{x} = Ax + Bu$$

has the equivalent Kalman canonical form

$$\begin{bmatrix} \dot{z}_1 \\ \dot{z}_2 \end{bmatrix} = \begin{bmatrix} \tilde{A}_{11} & \tilde{A}_{11} \\ 0 & \tilde{A}_{22} \end{bmatrix} \begin{bmatrix} z_1 \\ z_2 \end{bmatrix} + \begin{bmatrix} B_1 \\ 0 \end{bmatrix} u \qquad (20.30)$$

where (i) rank$[Q]$ = rank$[B \ AB \ \dots \ A^{n-1}B] = p$, (ii) U_1 is $n \times p$ and has columns which span col-sp$[Q]$, (iii) U_2 is $n \times (n-p)$ and has columns which together with those of U_1 form a basis for $\mathbf{R}^n$, (iv) $\tilde{A}_{11}$ is $p \times p$, (v) $\tilde{Q} = [B_1 \ \tilde{A}_{11}B_1 \ \dots \ \tilde{A}_{11}^{p-1}B_1]$ has rank p, and (vi) the z_1 part of $z = \text{col}[z_1, z_2]$ embodies the controllable part of the state space. A brief example illustrates this form for the multi-input case.

EXAMPLE 20.9

Suppose a system has the following dynamics:

$$\dot{x} = \begin{bmatrix} -3 & 2 & 2 \\ 0 & -1 & 0 \\ -6 & 6 & 4 \end{bmatrix} x + \begin{bmatrix} 1 & 1 & 2 \\ 0 & 1 & 1 \\ 2 & 0 & 2 \end{bmatrix} u$$

Then the controllability matrix is

$$Q = \begin{bmatrix} 1 & 1 & 2 & 1 & -1 & 0 & 1 & 1 & 2 \\ 0 & 1 & 1 & 0 & -1 & -1 & 0 & 1 & 1 \\ 2 & 0 & 2 & 2 & 0 & 2 & 2 & 0 & 2 \end{bmatrix}$$

which has rank 2. Observe that the vector $(2, -2, -1)^t$ is independent of the columns of Q; hence, we may define the transformation

$$[U_1 \ U_2] = \begin{bmatrix} 1 & 1 & 2 \\ 0 & 1 & -2 \\ 2 & 0 & -1 \end{bmatrix}$$

so that

$$[U_1 \ U_2]^{-1} = \frac{1}{9} \begin{bmatrix} 1 & -1 & 4 \\ 4 & 5 & -2 \\ 2 & -2 & -1 \end{bmatrix}$$

Applying this transformation to obtain an equivalent state description results in

$$
\begin{bmatrix} \tilde{A}_{11} & \tilde{A}_{12} \\ 0 & \tilde{A}_{22} \end{bmatrix} =
\left[\begin{array}{cc|c} 1 & 0 & -14 \\ 0 & -1 & 2 \\ \hline 0 & 0 & 0 \end{array} \right]
$$

and

$$
\begin{bmatrix} B_1 \\ \hline 0 \end{bmatrix} =
\left[\begin{array}{ccc} 1 & 0 & 1 \\ 0 & 1 & 1 \\ \hline 0 & 0 & 0 \end{array} \right]
$$

Obviously, the reduced controllability matrix

$$
\tilde{Q} = \begin{bmatrix} B_1 | \tilde{A}_{11} B_1 \end{bmatrix} =
\left[\begin{array}{ccc|ccc} 1 & 0 & 1 & 1 & 0 & 1 \\ 0 & 1 & 1 & 0 & -1 & -1 \end{array} \right]
$$

has rank 2, and the reduced component state model $(\tilde{A}_{11}, B_1)$ is completely controllable.

Exercise. Given the system

$$
\dot{x} = \begin{bmatrix} -1 & -2 & 2 \\ -1 & -3 & 2 \\ -2 & -4 & 3 \end{bmatrix} x + \begin{bmatrix} 0 & 0 & 0 \\ -1 & 1 & -1 \\ -1 & 1 & -1 \end{bmatrix} u
$$

find the Kalman canonical form clearly expressing each step in the procedure. Note that, as mentioned earlier, these computations proceed more reliably via the SVD of Q.

In order to assign the poles of the controllable part of the system in example 20.9, we postulate a control law $\tilde{F} = \begin{bmatrix} \tilde{F}_1 & \tilde{F}_2 \end{bmatrix}$ for which

$$
u = \begin{bmatrix} \tilde{F}_1 & \tilde{F}_2 \end{bmatrix} \begin{bmatrix} z_1 \\ z_2 \end{bmatrix} + u_e
$$

Equation 20.30 then becomes

$$
\begin{bmatrix} \dot{z}_1 \\ \dot{z}_2 \end{bmatrix} = \begin{bmatrix} \tilde{A}_{11} + B_1 \tilde{F}_1 & \tilde{A}_{12} + B_1 \tilde{F}_2 \\ 0 & \tilde{A}_{22} \end{bmatrix} \begin{bmatrix} z_1 \\ z_2 \end{bmatrix} + \begin{bmatrix} B_1 \\ 0 \end{bmatrix} u_e
$$

Clearly, the eigenvalues of the system are given by the eigenvalues of $\tilde{A}_{11} + B_1 \tilde{F}_1$ and the eigenvalues of $\tilde{A}_{22}$. To arbitrarily assign the spectrum of the controllable part of the original system, we use the methods developed earlier. The feedback matrix $\tilde{F}_2$ can be chosen so as to minimize coupling from the uncontrollable to the controllable modes if desired. Specifically, we use SVD techniques to find a least squares solution to the equation $B_1 \tilde{F}_2 = \tilde{A}_{12}$.

Once $\tilde{F} = [\tilde{F}_1 \; \tilde{F}_2]$ has been chosen, the feedback matrix F in the x-coordinates is given by $F = [\tilde{F}_1 \; \tilde{F}_2][U_1 \; U_2]^{-1}$. If $[U_1 \; U_2]$ is computed using the SVD of Q, then $F = [\tilde{F}_1 \; \tilde{F}_2][U_1 \; U_2]^t$.

Note that this kind of spectral assignability is extremely important, especially in the system stabilization context. In particular, if the unstable modes of a system are controllable, then the system may be internally stabilized by state feedback. Such systems are said to be *stabilizable*.

Exercise. Assign the spectrum $-1 \pm j$ to the controllable part of the system of example 20.9.

Exercise. Develop a computer program which will assign a given spectrum to the controllable part of a system. How would one numerically choose F_2 so that coupling between the controllable and uncontrollable states is minimized? Incorporate such a feature into your program.

Before closing this section, some conceptual tidying up is in order. Implicit throughout the development of the pole placement procedure is the assumption that feedback does not alter the controllability of the system. Step 2 of the multi-input assignment procedure clearly displays this implicit hypothesis: here one randomly generates a feedback F_0 and forms $A_0 = A + BF_0$, expecting that the pair (A_0, B) is controllable. Fortunately, this is indeed the case, as the proof of theorem 20.7 next shows. Then, in step 3 of the multi-input algorithm, one generates a vector, denoted B_0, as a random linear combination of the columns of B. Again, one expects the pair (A_0, B_0) to be controllable. Fortunately again, theorem 20.8 tells us that for each vector B_0 contained in the column space of Q, there exists an F_0 such that $(A + BF_0, B_0)$ is controllable. It then turns out that this property is generic in that randomly chosen pairs of F_0 and B_0 yield the property, legitimating step 3.

Theorem 20.7. Suppose that Q_A is the controllability matrix of the pair (A,B) and Q_{A+BF} the controllability matrix of the pair $(A + BF, B)$, where F is an arbitrary feedback matrix. Then $\text{Im}(Q_A) = \text{Im}(Q_{A+BF})$.

Proof.

Part 1. We show that $\text{Im}(Q_{A+BF}) \subset \text{Im}(Q_A)$. First, observe that $(A + BF)B = AB + BFB$; hence, $\text{Im}[(A + BF)B] \subset \text{Im}[B \; AB]$. Similarly, $(A + BF)^2 B = A^2B + AB(FB) + B(FBFB)$; hence, $\text{Im}[(A + BF)^2 B] \subset \text{Im}[B \; AB \; A^2B]$. By induction it then follows that $\text{Im}[(A + BF)^k B] \subset \text{Im}[B \; AB \; ... \; A^k B]$ for all k. Consequently, $\text{Im}(Q_{A+BF}) \subset \text{Im}(Q_A)$.

Part 2. We show that $\text{Im}(Q_A) \subset \text{Im}(Q_{A+BF})$. Let $A_0 = A + BF$. From before, we know that the pair (A_0, B) is controllable. Thus, from part 1, $\text{Im}(Q_{A_0-BF}) \subset \text{Im}(Q_{A_0})$. But $\text{Im}(Q_{A_0-BF}) = \text{Im}(Q_A)$, yielding the desired result.

We now show that given B_0 in the column space of B, there exists an F_0 such that the pair $(A + BF_0, B_0)$ is controllable. Formally stated, we have theorem 20.8.

Theorem 20.8. Let the pair (A,B) be controllable, and let b_0 be an arbitrary vector in the column space of B. Then there exists a state feedback F_0 such that the pair $(A + BF_0, b_0)$ is controllable.

Proof. To prove the theorem, we define a sequence of vectors v^i which span $\mathbf{R}^n$, the controllable space, in such a way that the choice of F_0 is obvious. Begin the sequence by setting $v^1 = b_0 = Bu_0$ for an appropriate m-vector u_0. Define

$$v^{i+1} = Av^i + b_0$$

Let p_0 be the largest integer for which the set $\{v^i\}$ is linearly independent. If $p_0 = n$, stop; otherwise choose a vector b_1 in the column space of B which is independent of $\{v^i\}$. Such a vector must exist, since the pair (A,B) is assumed controllable; i.e., if such a b_1 did not exist, then the pair (A,B) would not be controllable. In any event, $b_1 = Bu_1$ for an appropriate m-vector u_1. Let $v^{p_0+1} = b_1$, and for $i \geq p_0$, define

$$v^{i+1} = Av^i + b_1$$

Let p_1 be the largest integer for which the new set $\{v^i\}$ is linearly independent. If $p_1 = n$, stop; otherwise continue this process until the set spans $\mathbf{R}^n$. Thus, there are n-vectors satisfying

$$v^{i+1} = Av^i + b^i$$

where b^i equals the appropriate $b_0 = Bu_0, b_1 = Bu_1,\ldots$, depending on the exact nature of the preceding construction.

The idea now is to choose F_0 so that $(A + BF_0)v^i = v^{i+1}$, in which case the controllability matrix of the pair $(A + BF_0, b_0)$ has columns equal to the v^i which span $\mathbf{R}^n$, making the pair controllable. Toward this end, choose F_0 so that

$$F_0[v^1 \; v^2 \; \ldots \; v^n] = [u_0 \; u_1 \; \ldots \; u_n]$$

Since the set $\{v^i\}$ is a basis for $\mathbf{R}^n$,

$$F_0 = [u_0 \; u_1 \; \ldots \; u_n][v^1 \; v^2 \; \ldots \; v^n]^{-1}$$

which has the desired properties.

NUMERICAL CONSIDERATIONS

Up to this point, the theory of controllability appears rather elegant. However, a question arises regarding the numerical soundness of several of the tests for controllability. Let us examine three controllability tests, drawing from the exposition of [12].

The question alluded to is, are there problems with implementing the test for rank on $Q = [B \; AB \; \ldots \; A^{n-1}B]$? The answer is yes. The controllability of the pair (A,B) translates into a rank characterization on a matrix Q fabricated from terms of the form A^kB. Thus, a poorly conditioned A-matrix (one for which the ratio of the largest to smallest singular value is large) can lead to a more poorly conditioned Q. And a poorly conditioned Q can suggest a deficiency in rank and, therefore, a lack of controllability when in fact there is no such lack. The following example is illustrative. Suppose a pair (A,B) is given by

$$A = \begin{bmatrix} 1 & 0 & 0 & \cdots & 0 \\ 0 & 2^{-1} & 0 & \cdots & 0 \\ 0 & 0 & 2^{-2} & \cdots & 0 \\ \vdots & \vdots & \vdots & \ddots & \vdots \\ 0 & 0 & 0 & \cdots & 2^{-9} \end{bmatrix}, \quad B = \begin{bmatrix} 1 \\ 1 \\ 1 \\ \vdots \\ 1 \end{bmatrix}$$

Obviously, the pair is controllable. The controllability matrix $Q = [q_{ij}] = [B \ AB \ \cdots \ A^9 B]$ has ij entry, $q_{ij} = 2^{(i-1)(1-j)}$. The three smallest singular values of Q are 6.13×10^{-13}, 3.64×10^{-10}, and 7.12×10^{-8}. If the relative precision of the particular computer doing the calculations is no smaller than 10^{-12}, then the effective rank of Q will be less than 10, indicating a system that is not completely controllable. Similar, more general matrices would insert rounding errors into the calculation of Q, exacerbating the problem. On the other hand, if A had a condition number close to 1, i.e., if the columns/rows of A were relatively orthonormal, then an SVD of Q would produce a reliable test for rank.

In general, it is not advisable to discard the role of the matrix Q even if it is poorly conditioned. One striking reason is that an input sequence which will drive $x(k)$ to $x(k+n)$ is the solution to

$$\left[x(k+n) - A^n x(k) \right] = Q \begin{bmatrix} u(k+n-1) \\ u(k+n-2) \\ \vdots \\ u(k) \end{bmatrix}$$

The point is, of course, to be cautious in the use of Q.

By contrast, the condition that $\text{rank}[(\lambda_i I - A) \ B] = n$ for each eigenvalue λ_i of A does much better, provided that the eigenvalues are known exactly. Specifically, the smallest singular value of $[(\lambda_i I - A) \ B]$ is greater than $2^{-11} = 4.883 \times 10^{-4}$ [12]. One concludes that any computer having a relative precision of 10^{-5} or smaller would determine a completely controllable pair, as is in fact the case. The kicker, of course, is the assumption that the eigenvalues can be computed exactly.

As illustrated in chapter 15, there exist matrices whose eigenvalues are highly sensitive to noisy data. In [12], the uncontrollable system

$$A = Q \hat{A} Q^t, \ B = [1 \ 1 \ \cdots \ 1 \ 0]^t Q$$

where Q is a random orthogonal matrix and

$$\hat{A} = \begin{bmatrix} 20 & 20 & 0 & 0 & 0 & 0 & 0 \\ 0 & 19 & 20 & 0 & 0 & 0 & 0 \\ 0 & 0 & 18 & 20 & 0 & 0 & 0 \\ & & & & \cdot & & \\ & & & & & \cdot & \\ & & & & & & \cdot \\ 0 & 0 & 0 & 0 & 0 & 2 & 20 \\ 0 & 0 & 0 & 0 & 0 & 0 & 1 \end{bmatrix}$$

was checked for uncontrollability using the criterion that rank$[(\lambda_i I - A) \ B] < n$ for some λ_i. The test was carried out by first numerically computing the eigenvalues of A and then determining the inverse condition number (the ratio of the smallest singular value to the largest) of $[(\lambda_i I - A) \ B]$ for each calculated eigenvalue. Since the system is uncontrollable, the ratio of the smallest singular value to the largest should be zero. The calculations reported in [12] indicate that the smallest such number is 0.002, suggesting a completely controllable system. The problem arises from the computed eigenvalues for the matrix A, whose spectrum is highly sensitive to round-off and noisy data. If the correct eigenvalues are used, the test appears reliable.

Another method for computing a basis for the controllable subspace which seems to be reliable is the following [12]. Suppose B is $n \times 1$ and single input and A is $n \times n$. Let $\{v_1, \ldots, v_0\}$ be the orthonormal basis for $\text{Im}(Q)$ which must be computed. Choose v_1 to be B properly normalized to unit length. Let v_2 be the new basis vector obtained by finding an orthonormal basis for $[v_1 \ Av_1]$; i.e., the orthonormal basis for $[B \ AB]$ is given by $[v_1 \ v_2]$. The third basis vector, v_3, follows by orthogonalizing $[v_1 \ v_2 \ Av_2]$. Continuing this process yields a basis for $\text{Im}(Q)$. This procedure is an oversimplified version of the *staircase algorithm* [12].

To accomplish the orthogonalization, we can use the well-known Gram-Schmidt process, the SVD algorithm, or, for more efficiency, the QR algorithm, where Q is the matrix whose columns are the desired basis vectors and R an upper triangular matrix. In using the QR algorithm, the appropriate diagonal entry of R determines whether or not the new vector is added to the basis. Specifically, suppose the decision is whether or not the vector v_i is to be added to the set $\{v_1, \ldots, v_{i-1}\}$. If the contribution of the ii entry of R to the Euclidean length of the ith column of R is negligible compared with the other entries in the column, then $\{v_1, \ldots, v_{i-1}\}$ is the desired basis.

The QR algorithm was carried out on the completely controllable system

$$A = \begin{bmatrix} 1 & 0 & 0 & 0 \\ 0 & (0.5)^8 & 0 & 0 \\ 0 & 0 & (0.5)^{16} & 0 \\ 0 & 0 & 0 & (0.5)^{24} \end{bmatrix}, \quad B = \begin{bmatrix} 1 \\ 1 \\ 1 \\ 1 \end{bmatrix}$$

The basis for the controllable subspace obtained by a sequential application of the algorithm is given by the columns of the matrix

$$\begin{bmatrix} 0.5 & 0.866 & 0.00319 & -0.108 \times 10^{-7} \\ 0.5 & -0.286 & -0.8175 & 0.00277 \\ 0.5 & -0.29 & 0.4048 & -0.7085 \\ 0.5 & -0.29 & 0.4096 & 0.7057 \end{bmatrix}$$

CONCLUDING REMARKS

Besides the many ideas covered in this chapter, other points of view, extensions, and specialized investigations dot the literature on controllability. Notions of controllability and reachability for discrete-time periodic systems can be found in [15]. Extensions of these ideas to linear time varying systems [4] are briefly reviewed in chapter 24. Extensions to nonlinear systems have been and are currently an interesting area of research as well. Lastly, the area of pole placement has received much attention in the literature, including various numerically stable approaches [16–20].

PROBLEMS

1. Suppose

$$\dot{x} = \begin{bmatrix} 1 & 1 \\ 0 & 3 \end{bmatrix} x + \begin{bmatrix} b_1 \\ b_2 \end{bmatrix} u$$

$$y = [c_1 \ \ c_2] x$$

(a) What conditions on b_1 and b_2 make the system controllable?

(b) What conditions on c_1 and c_2 make the system observable? Recall from chapter 5 that the observability matrix is

$$R = \begin{bmatrix} C \\ CA \end{bmatrix}$$

for a two-dimensional system.

(c) If $[b_1 \ b_2]^t = [1 \ -1]$, find the Kalman canonical form of the state dynamics.

2. Consider the scalar model

$$\dot{x} = \lambda x + Bu$$

Find inputs $u_1(t)$ and $u_2(t)$ which will drive $x_0 = x(t_0)$ to $x_1 = x(t_1)$.

3. Consider the state dynamical equation

$$\dot{x} = Ax + Bu$$

where

$$A = \begin{bmatrix} 1 & 1 & 1 \\ 1 & 1 & 0 \\ -2 & 0 & 1 \end{bmatrix}, \qquad B = \begin{bmatrix} 1 & 1 \\ 0 & -1 \\ 0 & 2 \end{bmatrix}$$

(a) Find the eigenvalues of A.
(b) Find the right eigenvectors of A.
(c) Find the left eigenvectors of A.
(d) Find the controllability matrix Q.
(e) Find the rank of Q, and then find a basis for the controllability subspace.
(f) Find a state transformation $x = [U_1, U_2]z$ which will convert the system to the *Kalman controllable canonical* form.
(g) Find an input $u(t)$ which will drive the initial state $x(0) = [1 \ 0 \ 0]^t$ to the final state $x(1) = [0 \ 1 \ -2]^t$.
(h) Verify your answer to (g) by evaluating the formula

$$x_1(t_1) = \Phi_x(t_1 - t_0)x_0(t_0) + \int_{t_0}^{t_1} \Phi_x(t_1 - q)Bu(q)\,dq$$

when Φ_x is expressed in terms of Φ_z and $[U_1 \ U_2]$.
(i) Decompose $\Phi_z(t)$ into a sum of real matrix directed modes.

4. Consider the state model

$$\dot{x} = \begin{bmatrix} 0 & 1 & -1 \\ -2 & 0 & 0 \\ -1 & 0 & 0 \end{bmatrix} x + \begin{bmatrix} 0 \\ 1 \\ 1 \end{bmatrix} u$$

$$y = [1 \ 1 \ -1]\,x$$

(a) Compute the system eigenvalues.
(b) Compute the system eigenvectors.
(c) Express the B-matrix as a linear combination of the eigenvectors.
(d) Is the C-matrix orthogonal to any of the eigenvectors? Would such modes appear in the system response?
(e) Discuss all aspects of the controllability of this system.

5. Consider the switched discrete-time system

$$x(k+1) = A_1 x(k) + Bu(k), \quad k \leqslant 1$$

and

$$x(k+1) = A_2 x(k) + Bu(k), \quad k \geqslant 2$$

where

$$A_1 = \begin{bmatrix} 1 & 1 \\ 0 & 2 \end{bmatrix}, \ B = \begin{bmatrix} 0 \\ 1 \end{bmatrix}$$

and

$$A_2 = \begin{bmatrix} 1.6 & 0 \\ 1 & 1 \end{bmatrix} \begin{bmatrix} 0.8 & 0 \\ 0 & -0.8 \end{bmatrix} \begin{bmatrix} 0.625 & 0 \\ -0.625 & 1 \end{bmatrix}$$

(a) Find $x(2)$ and $x(4)$ when $u(k) = 0$ for all k and $x(0) = [1 \ 1]^t$.

(b) Find an input sequence $\{u(k)\}$ which will drive $x(0) = \theta$ to the $x(2)$ computed in (a).

(c) If possible, find an input which will drive $x(0) = \theta$ to the $x(4)$ computed in (a). Explain your reasoning.

(d) Is the system controllable over the set of points $\{2,3,4,5,...\}$? Justify your answer using the controllability matrix.

(e) Is the system controllable over the set of points $\{0,1,2,3,4,...\}$? Justify your answer.

6. Write a state model for the circuit of Figure P20.6. Discuss the controllability properties of this circuit.

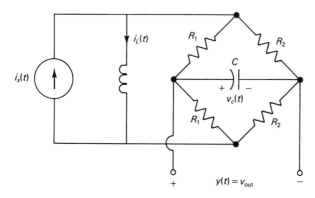

Figure P20.6   Circuit for problem 6.

7. Consider the system $\dot{x} = Ax + Bu$, where $B = [a \ \ b]^t$ and $A = \begin{bmatrix} 0 & -2 \\ 1 & -3 \end{bmatrix}$.

(a) What are the eigenvalues of A?

(b) What are the right eigenvectors of A?

(c) What are the left eigenvectors of A?

(d) Find the directed matrix mode $\exp(\lambda_1 t) \ R_1$.

(e) Find the directed matrix mode $\exp(\lambda_2 t) \ R_2$.

(f) What conditions on a and b of the B-matrix will make the directed mode $\exp(\lambda_1 t)$ *uncontrollable*?

8. Consider the continuous-time state dynamical equation $\dot{x} = Ax + Bu$, where

$$A = \begin{bmatrix} 1 & -3 & -1 \\ 2 & -4 & -1 \\ 2 & -2 & -3 \end{bmatrix}, \quad B = \begin{bmatrix} 3 \\ 2 \\ 2 \end{bmatrix}$$

Note that the eigenvalues of A are $\lambda_1 = -1$, $\lambda_2 = -2$, and $\lambda_3 = -3$. Answer the following questions:

(i) The rank of the controllability matrix Q is
(a) 0 (b) 1 (c) 2 (d) 3 (e) 4 (f) cannot be determined

(ii) A left eigenvector of A is
(a) $(1 \ -1 \ 0)^t$ (b) $(1 \ 0 \ -1)^t$ (c) $(1 \ 1 \ 1)^t$
(d) $(-2 \ 2 \ 0)^t$ (e) none of the above (f) two of the above

(iii) Which of the following directed modes (is) are controllable?
(a) $e^{-t}e_1$ (b) $e^{-2t}e_2$ (c) $e^{-3t}e_3$ (d) (a) and (b)
(e) (a) and (c) (f) (b) and (c) (g) all modes

(iv) Which of the following initial states are controllable?
(a) $(3 \ 2 \ 1)^t$ (b) $(1.5 \ 1 \ 1)^t$ (c) $(6 \ 2 \ 2)^t$ (d) $(-2 \ -2 \ -2)^t$
(e) none of the above (f) two of the above

(v) The height a of a step input $u(t) = a \ 1^+(t)$ which will drive $x(0) = \theta$ to $x(1) = e_1$ is _____.

9. Suppose a particular $n \times n$ matrix M has a p-dimensional null space $N\{M\}$ which is A-invariant—i.e., $v \in N\{M\}$ has the property that $Av \in N\{M\}$, where A is square. Show that there exists a nonsingular matrix Q such that

$$Q^{-1}AQ = \begin{bmatrix} \tilde{A}_{11} & \tilde{A}_{12} \\ 0 & \tilde{A}_{22} \end{bmatrix}$$

where A_{11} is $p \times p$.

10. Let $\dot{x} = Ax + Bu$ have controllability matrix $Q_x = [B \ AB \ ... \ A^{n-1} B]$. Let $Tz = x$ be a nonsingular state transformation. Show that rank $[Q_z] = \text{rank}[Q_x]$, where Q_z is the controllability matrix in the z-coordinates. (*Note:* A nonsingular state transformation thus leaves the rank of the controllability matrix unchanged, as intuitively expected.)

11. Consider the discrete-time state model

$$x(k + 1) = \begin{bmatrix} 0 & 1 \\ 0 & 0 \end{bmatrix} x(k) + \begin{bmatrix} 1 & -1 \\ -1 & -1 \end{bmatrix} u(k), \ x(0) = \begin{bmatrix} -1 \\ -1 \end{bmatrix}$$

(a) Find a general expression for $x(k)$ given that

$$u(k) = \begin{bmatrix} (-1)^k \\ -(-1)^k \end{bmatrix}$$

(b) Is the system completely controllable? If so, prove it. If not, find all uncontrollable modes.

12. Verify the equivalences of theorem 20.4 for the state dynamics

$$\begin{bmatrix} \dot{x}_1 \\ \dot{x}_2 \end{bmatrix} = \begin{bmatrix} 0 & 1 \\ 0 & 0 \end{bmatrix} \begin{bmatrix} x_1 \\ x_2 \end{bmatrix} + \begin{bmatrix} 1 \\ 1 \end{bmatrix} u$$

13. Use equation 20.13 to show that if the pair (A, B) is controllable over the interval $[t_0, t_1]$, then it is controllable over $[t_0', t_1']$ for arbitrary t_0', t_1'. (*Hint:* You will need to use the notion of time invariance.)

14. Let

$$Q = \begin{bmatrix} 1 & 1 & 1 & 0 & 1.05 & 1 \\ 1 & 1 & 1 & 0 & 1.05 & 1 \\ 1 & -1 & 0.05 & 1 & 1.0025 & -0.95 \end{bmatrix}$$

(a) Using a canned program from IMSL or EISPACK, compute the SVD of Q.

(b) Recall that the numerical algorithm for finding the SVD actually computes the SVD for the matrix $Q + \Delta Q$, where $Q + \Delta Q$ is very close to Q in that $\|\Delta Q\|/\|Q\|$ (the ratio of the spectral norms) is a modest multiple of the machine epsilon.

   (i) Find ΔQ and $\|\Delta Q\|$ (at least approximately).
   (ii) Find $\|Q\|$ (at least approximately).
   (iii) Find $\|\Delta Q\|/\|Q\|$.
   (iv) Find the machine epsilon (single precision) and compare with your answer for part (iii).

(c) What is the condition number of Q?

(d) Discuss the relative orthogonality of the columns of Q with each other. Be thorough. (*Hint:* Draw each column as a vector in three-space.)

(e) What angle does the vector $[1 \ 1 \ 0.05]^t$ make with the subspace spanned by the first two columns of Q?

(f) If Q represents the controllability matrix of some system, what is the "useful" dimension of the controllable subspace? Explain your answer in light of parts (c), (d), and (e).

(g) What is a basis for the "useful" controllable subspace?

(h) If Q^t represents the observability matrix of some system, what is a basis for the "useful" unobservable subspace, i.e., for a basis for $N[Q^t]$?

(i) If

$$A = 0.5 \begin{bmatrix} 1 & 0 & 1 \\ 1 & 0 & 1 \\ 1.05 & 0 & -0.95 \end{bmatrix}$$

then, using the results of the SVD on Q, find a representation of A which (1) is block upper triangular and (2) identifies the effectively controllable and effectively uncontrollable modes of A. (In other words, find the Kalman controllable form of A.)

15. The following pair (A,B) is not completely controllable:

$$A = \begin{bmatrix} 1 & 1 & 0 & 0 & 0 & 0 & 0 & 0 \\ 0 & 0 & 0 & 0 & 0 & 0 & 0 & 0 \\ 0 & 0 & 2 & 0 & 0 & 0 & 0 & 0 \\ 0 & 0 & 0 & 4 & 1 & 4 & 3 & 0 \\ 0 & 0 & 0 & 1 & -2 & 1 & 0 & -1 \\ 0 & 0 & 0 & -3 & 0 & -3 & -3 & 0 \\ 0 & 0 & 0 & -4 & -1 & -4 & -3 & 0 \\ 0 & 0 & 0 & 1 & 1 & 1 & 0 & 0 \end{bmatrix} \quad B = \begin{bmatrix} 0 & 0 & 0 \\ 1 & 0 & 0 \\ 0 & 1 & 0 \\ 0 & 0 & 0 \\ 0 & 0 & 1 \\ 0 & 0 & 0 \\ 0 & 0 & 0 \\ 0 & 0 & 0 \end{bmatrix}$$

The rank of Q is either 4, 5, or 6. Assign the following spectrum as indicated:

(a) If rank$[Q] = 4$,   assign spectrum $\{-1, -1, -2, -3\}$

(b) If rank$[Q] = 5$,   assign spectrum $\{-1 \pm j, -2 \pm j, -3\}$

(c) If rank$[Q] = 6$,   assign spectrum $\{-1 \pm j, -2 \pm j, -3 \pm j\}$

Try to minimize controllable and uncontrollable state interaction. In effect, you are to find the appropriate feedback matrix F. Make sure you turn in

1. Your computer program listing
2. A list of the eigenvalues of A
3. The matrices, in controllability canonical form
4. The desired feedback matrix when the system is in controllability canonical form
5. The desired system feedback matrix
6. The eigenvalues of $A + BF$ as computed by some canned routine

16. Show that the only poles of the transfer function matrix $H(s) = (sI - A)^{-1}B$ are those which are controllable. (*Hint:* Consider a state transformation relating A and B to the Kalman controllable canonical form.)

17. The state dynamical equations of two systems,

$$\dot{x} = A_1 x + B_1 u \tag{20.31}$$

and

$$\dot{z} = A_2 z + B_2 u \tag{20.32}$$

are zero-state equivalent, i.e., the zero-state state responses coincide for all admissible u.

(a) Show that the transfer function matrices are equal.

(b) If A_1 has some eigenvalues different from A_2, what can you say about the controllability of equations 20.31 and 20.32? Based on this information, can you obtain a "tight" upper bound on the dimension of the controllable subspace?

18. Suppose the nth order single-input discrete-time system

$$x(k + 1) = Ax(k) + Bu(k)$$

is completely controllable. From the controllability treatment, any state can be driven to the zero state within n steps. A feedback design problem centers on finding a state feedback control which will drive an arbitrary state to zero in the shortest possible number of steps.

(a) Since the system is completely controllable, Q^{-1} (the inverse of the controllability matrix) exists. Let v_i^t equal the ith row of Q^{-1}. Show that the input sequence

$$u(k) = - v_{n-k}^t A^n x(0) \quad \text{for} \quad k = 0, 1, 2,...,n-1$$

will drive $x(0)$ to the zero vector in the smallest number of steps.

(b) Let the state feedback $F = - v_n^t A^n$. Show that the state feedback law $u(k) = Fx(k)$ will also drive the state $x(0)$ to zero in the shortest number of steps.

(c) Show that the closed-loop system has all its eigenvalues at zero, and find the Jordan form of $(A + BF)$.

19. (a) Consider the system $x(k +1) = Ax(k) + Bu(k)$, where

$$A = \begin{bmatrix} -2 & 1 & 0 \\ 0 & -2 & 1 \\ 0 & 0 & -2 \end{bmatrix}, \quad B = \begin{bmatrix} 0 \\ 1 \\ 0 \end{bmatrix}$$

Find a state feedback law which will drive $x(0) = (-2 \ 1 \ 0)^t$ to zero in the shortest possible time. What is the minimum number of steps?

(b) Can you generalize the result of problem 18 to systems which are not completely controllable? How?

20. Consider the discrete-time system $x(k +1) = (A + BF)x(k) + Bu(k)$, where

$$A = \begin{bmatrix} -3 & 0 & 0 \\ 0 & -2 & 0 \\ 0 & 0 & -1 \end{bmatrix}, \quad B = \begin{bmatrix} 1 \\ 1 \\ 1 \end{bmatrix}$$

Find a state feedback such that the zero-input response aproaches zero as $k \rightarrow \infty$ for any arbitrary initial condition.

21. Let C_+ denote complex numbers z such that $\text{Re}(z) \geqslant 0$, and C_- the set z such that $\text{Re}(z) < 0$. The pair (A,B) is stabilizable if there exists state feedback F such that $\sigma(A +BF) \subset C_-$. Show that the pair (A,B) is stabilizable if and only if all unstable modes (real parts greater than or equal to zero) are controllable, i.e., the space generated by the eigenvector or generalized eigenvectors associated with an unstable eigenvalue in C_+ must lie in the controllable subspace.

REFERENCES

1. Charles A. Desoer, *Notes for a Second Course on Linear Systems* (New York: Van-Nostrand-Reinhold, 1970).

2. L. Zadeh and C. A. Desoer, *Linear System Theory: The State Space Approach* (New York: McGraw-Hill, 1963).

3. R. A. DeCarlo and R. Saeks, *Interconnected Dynamical Systems* (New York: Marcel Dekker, 1981).

4. David L. Russell, *Mathematics of Finite Dimensional Control Systems* (New York: Marcel Dekker, 1979).

5. W. M. Wonham, *Linear Multivariable Control: A Geometric Approach* (New York: Springer-Verlag, 1979).

6. Thomas Kailath, *Linear Systems* (Englewood Cliffs, NJ: Prentice-Hall, 1980).

7. W. A. Wolovich, *Linear Multivariable Systems* (New York: Springer-Verlag, 1974).

8. David Luenberger, *Introduction to Dynamical Systems: Theory, Models, and Applications* (New York: Wiley, 1979).

9. R. E. Kalman, "Mathematical Description of Linear Dynamical Systems," *SIAM Journal of Control*, Ser. A, Vol. 1, 1963, pp. 152–192.

10. R. E. Kalman, "On the Computation of the Reachable/Observable Canonical Form," *SIAM Journal of Control and Optimization*, Vol. 20, No. 2, 1982, pp. 258–260.

11. E. Bruce Lee, Stanislaw Zak, and Stephen Brierley, "Remarks on Kalman's Canonical Form," *Proceedings of the Fourth Polish-English Seminar on Real-Time Process Control*, June 1983, pp. 238–249.

12. Chris C. Paige, "Properties of Numerical Algorithms Related to Computing Controllability," *IEEE Transactions on Automatic Control*, Vol. AC-26, No. 1, February 1981, pp. 130–138.

13. Daniel Boley, "A Perturbation Result for Linear Control Problems," *SIAM J. Alg. Disc. Math.*, Vol. 6, No. 1, January 1985, pp. 66–72.

14. P. M. Lin and Y. L. Kuo, "Some New Results on the Controllability and Observability of Linear Systems and Networks," *Circuit Theory and Applications*, Vol. 7, 1979, pp. 413–422.

15. Sergio Bittanti and Pavlo Bolzern, "Reachability and Controllability of Discrete-Time Linear Periodic Systems," *IEEE Transactions on Automatic Control*, Vol. AC-30, No. 4, April 1985, pp. 399–400.

16. G. S. Miminis and Chris C. Paige, "An Algorithm for Pole Assignment of Time Invariant Multi-input Linear Systems," *Proceedings of the IEEE Conference on Decision and Control*, December 1982, pp. 62–67.

17. R. V. Patel and P. Misra, "A Numerical Algorithm for Eigenvalue Assignment by Output Feedback," *Proceedings of the Seventh International Symposium on Mathematical Theory of Networks and Systems, Stockholm, Sweden, June 1985* (Amsterdam: North Holland, 1985).

18. Thomas Harris, Raymond DeCarlo, and Stephen Richter, "A Continuation Approach to Eigenvalue Assignment," *Automatica*, Vol. 19, No. 5, 1983, pp. 551–555.

19. Rikus Eising, "Pole Assignment, A New Proof and Algorithm," *Systems and Control Letters*, Vol. 2, No. 1, July 1982, pp. 6–12.

20. P. Petkov, N. D. Christov, and M. Konstantinov, "A Computational Algorithm for Pole Assignment of Linear Multi-input Systems," *IEEE Transactions on Automatic Control*, Vol. AC-31, No. 11, November 1986, pp. 1044–1047.

21

OBSERVABILITY AND
THE BASICS OF
OBSERVER DESIGN

INTRODUCTION

Generally speaking, the problem of determining the system state vector from input-output measurements breaks down into two radically different approaches. In the one approach, the input $u(t)$ and the system output $y(t)$ are known in a neighborhood of some time, say t_1. The observability problem here is to determine $x(t_1)$ given the foregoing measurements and the time-invariant state model matrices A, B, C, and D. In effect, this is a capsule summary of generalized problem 1 of chapter 5. In the second approach, continuous measurements of $u(t)$ and $y(t)$ over $[t_0, t_1]$ drive a dynamical system known as a *dynamic observer* whose output $\hat{x}(t)$ approximates (i.e., asymptotically tracks) the plant state vector $x(t)$. The structure of the dynamic observer mirrors the usual state model equations and depends implicitly on the known plant matrices A, B, C, and D.

These two approaches of determining system states reflect the problem of reconstructability more than the philosophical notion of observability. Many authors simply equate the two in view of their intertwined nature and the obvious engineering importance of reconstructability. For the present, we will also identify observability with reconstructability.

BASIC DEFINITIONS AND EQUIVALENCES

What does it mean for a state to be observable? Parroting the controllability development, we begin with the following definition.

Definition 21.1. The state $x(t_0)$ in $\mathbf{R}^n$ is *observable* if and only if there exists $t_1 > t_0$ such that measurements of $y(t)$ and $u(t)$ over $[t_0, t_1]$ and knowledge of A, B, C, and D are sufficient to determine $x(t_0)$.

When is the system model observable or more concisely when is the pair (A,C) observable?

Definition 21.2. The pair (A,C) representing the system $\{A,B,C,D\}$ is *observable* (often called *completely observable*) if and only if every state in $\mathbf{R}^n$ is observable.

One approach for nondynamic state vector reconstruction stems from the development in chapter 5. Recall equation 5.17,

$$[Y(t) - TU(t)] = Rx(t) \tag{21.1}$$

where the *observability matrix* R has the form

$$R = \begin{bmatrix} C \\ CA \\ CA^2 \\ \cdot \\ \cdot \\ \cdot \\ CA^{n-1} \end{bmatrix} \tag{21.2}$$

Recall as well from lemmas 20.1 and 20.2 that the addition of further terms of the form CA^k, $k \geq n$, will fail to increase the rank of R.

Theorem 21.1. If equation 21.1 is consistent, then there exists a unique solution $x(t)$ if and only if rank$[R] = n$. Moreover, when it exists, the unique solution is given by

$$x(t) = R^{-L}[Y(t) - TU(t)] \tag{21.3}$$

where R^{-L} is any left inverse of R.

Proof. *Part 1.* Suppose there exists a unique solution $x*(t)$, to equation 21.1. Then if the rank of R were not n, R would have a nontrivial null space. Let v be any nonzero vector in this null space. Then a second solution would be $x*(t) + v$, contradicting the uniqueness of $x*(t)$. Therefore, the rank of R must be n.

Part 2. Now suppose rank$[R] = n$. This implies there are n-linearly independent equations (of the possible rn equations) available to solve for the n entries of $x*(t)$.

But since the *rn* equations are consistent, any set of *n* linearly independent equations will yield the same solution $x*(t)$. Therefore, a unique solution exists.

Part 3. Finally, since rank$[R] = n$, R has a left inverse, denoted R^{-L}, satisfying $R^{-L}R = I$. Thus,

$$R^{-L}[Y(t) - TU(t)] = R^{-L}Rx(t) = x*(t) \qquad \blacksquare$$

which completes the proof.

Recall that one possible left inverse is the Moore-Penrose pseudo inverse $R^{-L} = (R^tR)^{-1}R^t$. This, of course, should be computed using the SVD techniques described in chapter 19.

Corollary 21.1. The time-invariant state model $\{A,B,C,D\}$ is observable if and only if rank$[R] = n$.

Proof. Since the measurements needed in equation 21.1 arise from the solution to the time-invariant linear state model, equation 21.1 must be consistent, assuming perfect measurements. The result follows immediately. $\blacksquare$

Question 1: What is the form of the solution when R does not have full column rank — i.e., rank[R] = n - p, for 0 < p < n?

Answer: Let x_0 denote a particular solution to equation 21.1 at $t = t_0$. Since rank$[R]=n-p$, R has a p-dimensional null space denoted by N[R]. Let the set $\{v_1, v_2,..., v_p\}$ be a basis for the null space of R. Then every solution to $Rx(t_0) = [Y(t_0) - T U(t_0)]$ has the form

$$x(t_0) = x_0 + \sum_{i=1}^{p} \alpha_i v_i \qquad (21.4a)$$

or, in matrix form,

$$x = x_0 + [v_1\, v_2 ... v_p] \begin{bmatrix} \alpha_1 \\ \alpha_2 \\ . \\ . \\ . \\ \alpha_p \end{bmatrix} = x_0 + V\underline{\alpha} \qquad (21.4b)$$

Question 2: How does one compute x_0?

Answer: Using the Moore-Penrose pseudo inverse R^+, given by equation 19.3, one computes a minimum norm x_0 according to

$$x_0 = R^+[Y(t_0) - TU(t_0)]$$

From the foregoing development, uncertainty will characterize the reconstruction of x_0 from input-output measurements whenever the null space of R is nontrivial. Interestingly enough, elements of $N[R]$ have a special physical interpretation: any x_0 in $N[R]$ produces a zero-input response identically equal to zero. To verify this, recall the Caley-Hamilton theorem: let A be an $n \times n$ matrix with characteristic polynomial $\pi_A(\lambda) = \lambda^n + a_1 \lambda^{n-1} + ... + a_n$; then

$$A^n + a_1 A^{n-1} + ... + a_n I = [0]$$

This theorem directly implies that A^n is a linear combination of lower powers of A—i.e.,

$$A^n = -a_1 A^{n-1} - ... - a_n I$$

Indirectly speaking,

$$A^p = \beta_1 A^{n-1} + ... + \beta_n I$$

for appropriate scalars β_i and all p. With these facts in mind consider the zero-input state response when the initial condition x_0 lies in $N[R]$. For $t_0 = 0$, the zero-input response is given by

$$y(t) = C \exp[At]x_0$$

Expanding the right side in a Taylor series produces

$$y(t) = C \exp[At] x_0 = C[I + At + A^2 t^2/2! + ...] x_0$$

$$= Cx_0 + (CAx_0)t + (CA^2 x_0) t^2/2! + ...$$

Since x_0 is in $N[R]$, $Rx_0 = \theta$. Hence, by the Caley-Hamilton theorem or by lemmas 20.1 and 20.2, $CA^p x_0 = \theta$ for all $p \geq n$. Therefore,

$$y(t) = C \exp[At] x_0 = [0]$$

This physical interpretation suggests the following definition.

Definition 21.3. $N[R]$ is the *unobservable subspace* of the state space.

Suppose now that x_0 is in $N[R]$. *Can one reconstruct this state from input-output measurements?* Categorically, no. For, from the preceding derivation, zero inputs can produce zero outputs with nonzero initial states. Hence, equation 21.1 would become $\theta = Rx(t_0)$. The only possible conclusion is that $x(t_0)$ is in $N[R]$.

Another question now comes to mind: *Can a state that is in $N[R]$ escape from $N[R]$ by some kind of system action?* For this development, "system action" means multiplication by A.

Proposition 21.1.   Let x_0 be the unobservable subspace. Then Ax_0 lies in the unobservable subspace.

The proof of proposition 21.1 follows directly from the previous discussion. Framed differently, the proposition states that *the unobservable subspace is A-invariant.* In fact, one can show that the unobservable subspace is the largest A-invariant subspace contained in the kernel (null space) of C [1].

An interesting question is, *If states in the unobservable subspace evade reconstruction, are any states reconstructable with certainty?* The answer is yes: if one knows that a state lives in $\text{Im}[R^t]$, then one can reconstruct that state with certainty. This can be shown via duality, where the state x is observable with regard to the pair (A,C) if and only if it is controllable with respect to the pair $(-A^t, C^t)$ or (A^t, C). Hence, one can view $\text{Im}[R^t]$ as the observable subspace of the state space. Note that each vector in $\text{Im}[R^t]$ is orthogonal to every vector in $N[R]$ suggesting the notation $N[R]^\perp = \text{Im}[R^t]$.

Proposition 21.2.   The state space $X = \mathbf{R}^n$ has the direct sum decomposition

$$X = \text{Im}[R^t] \oplus N[R] = N[R]^\perp \oplus N[R]$$

which decomposes the state space into its observable and unobservable parts.

The proof of proposition 2 is obvious in light of the SVD of R. Indeed, orthonormal bases for $N[R]$ and $N[R]^\perp$ are supplied by the SVD discussed in chapter 19.

Corollary 21.2a.   Each x in $\mathbf{R}^n$ has the decomposition

$$x = x_{\text{ob}} + x_{\text{unob}}$$

where x_{ob} is in $\text{Im}[R^t]$ and x_{unob} is in $N[R]$.

Corollary 21.2b. $x \in \mathbf{R}^n$ is observable if and only if $x_{\text{unob}} = \theta$.

A state coordinate change proves helpful in concretizing the preceding ideas. In the new coordinates, say, the z-coordinates, the upper part of z, designated z_1, embodies the unobservable part of the state space and the lower part of z, denoted z_2, embodies the observable or reconstructible part. This separation into observable and unobservable parts parallels the development in propositions 20.2 through 20.8, which partition the state space into controllable and uncontrollable segments.

To accomplish the change of coordinates, let the observability matrix R of equation 21.2 have the singular value decomposition (see chapter 19)

$$R = [U_1^R \ U_2^R] \begin{bmatrix} S^R & 0 \\ 0 & 0 \end{bmatrix} \begin{bmatrix} (V_1^R)^t \\ (V_2^R)^t \end{bmatrix} \tag{21.5}$$

where the superscript R denotes dependence on the observability matrix. From this decomposition, an orthonormal basis for the unobservable subspace $N[R]$ is given by the columns of V_2^R, i.e., $N[R] = \text{Im}(V_2^R)$. Similarly, $N[R]^\perp = \text{Im}(V_1^R)$. From proposition 21.1, $N[R] = \text{Im}(V_2^R)$ is A-invariant, and hence,

$$A[V_2^R \ V_1^R] = [V_2^R \ V_1^R] \begin{bmatrix} \bar{A}_{11} & \bar{A}_{12} \\ 0 & \bar{A}_{22} \end{bmatrix}$$

by the same arguments used to prove proposition 20.4. Further, if $x \in N[R]$, then $Cx = \theta$, or, equivalently, $CV_2^R = [0]$. Consequently,

$$C = [0 \ \bar{C}_2][V_2^R \ V_1^R]^t$$

Finally, since R is independent of B,

$$B = [V_2^R \ V_1^R] \begin{bmatrix} \bar{B}_1 \\ \bar{B}_2 \end{bmatrix}$$

These relationships suggest the coordinate transformation $[V_2^R \; V_1^R]z = x$, leading to the equivalent state model called the *Kalman observable form,* viz.,

$$\begin{bmatrix} \dot{z}_1 \\ \dot{z}_2 \end{bmatrix} = \begin{bmatrix} \bar{A}_{11} & \bar{A}_{12} \\ 0 & \bar{A}_{22} \end{bmatrix} \begin{bmatrix} z_1 \\ z_2 \end{bmatrix} + \begin{bmatrix} \bar{B}_1 \\ \bar{B}_2 \end{bmatrix} u \qquad (21.6a)$$

$$y = \begin{bmatrix} 0 & \bar{C}_2 \end{bmatrix} \begin{bmatrix} z_1 \\ z_2 \end{bmatrix} \qquad (21.6b)$$

By construction, the pair $(\bar{C}_2, \bar{A}_{22})$ is observable. To see this, suppose the contrary, i.e., that $(\bar{C}_2, \bar{A}_{22})$ is not observable. Then, there is a state of the form $[0 \; z_2']^t$ in the null space of the new observability matrix. But then the state $x' = [V_2^R \; V_1^R]$ $[0 \; z_2']^t \in \text{Im}(V_2^R)$, contradicting the partitioned structure of the equivalent state model. Hence, z_2 embodies the observable part of the state space. Further, from the structure of $\bar{C} = [0 \; \bar{C}_2]$ and the observability of the pair $(\bar{C}_2, \bar{A}_{22})$, state reconstruction from input-output measurements can arrive only at vectors of the form $[0 \; \bar{z}_2]^t$ with certainty. Clearly, then, the observable part of the state space is orthogonal to the unobservable part.

These ideas combine with those of chapter 20 to form the Kalman canonical form of the state model matrix triple (C, A, B). Here, the A-matrix transforms to a partitioned matrix $T^{-1}AT$, whose partitions represent the controllable and observable parts of the state space, the controllable and unobservable parts, the uncontrollable and observable parts, etc. A full discussion of this partition as well as its construction appears later.

EQUIVALENCES FOR SYSTEM OBSERVABILITY

The duality of the notions of controllability and observability grows clearer with theorem 21.2 to follow. This theorem, the twin of theorem 20.4, establishes parallel statements dealing with system observability.

Theorem 21.2. For the usual linear, time-invariant state model $\dot{x}(t) = Ax(t) + Bu(t), y(t) = Cx(t) + Du(t)$, the following are equivalent:

(i) The pair (C, A) is observable.

(ii)

$$\text{Rank} \begin{bmatrix} C \\ \lambda_i I - A \end{bmatrix} = n$$

for each eigenvalue λ_i of A.

(iii) Rank$[R] = n$, where R is the observability matrix (equation 21.2).

(iv) Rank$[C\exp(At)] = n$, i.e., there are n linearly independent columns each of which is a vector-valued function of time defined over $[0, \infty]$.

(v) The matrix called the *observability Gramian,* given by

$$W_O(t_0, t_1) = \int_{t_0}^{t_1} e^{A^t q} C^t C e^{Aq} dq \qquad (21.7)$$

is nonsingular for all $t_1 > t_0$. Moreover, if $W_O(t_0,t_1)$ is nonsingular, then

$$x(t_0) = e^{At_0} W_O^{-1}(t_0,t_1) e^{A't_0} \int_{t_0}^{t_1} e^{A'(q-t_0)} C'y^M(q)\,dq \qquad (21.8)$$

where $y^M(t)$ plays the role of an output measurement and is defined as

$$y^M(t) = y(t) - C \int_{t_0}^{t} e^{A(t-q)} Bu(q)\,dq - Du(t) \qquad (21.9)$$

Observe that if some device measures $y(t)$ and $u(t)$, then $y^M(t)$ can be generated numerically or by an analog integrator-summer type of circuit. After computing $x(t_0)$, evaluation by the well-known formula

$$x(t) = e^{A(t-t_0)} x(t_0) + \int_{t_0}^{t} e^{A(t-q)} Bu(q)\,dq$$

will generate the entire state trajectory $x(t)$ over $[t_0,t_1]$. This method in which $x(t_0)$ is computed now offers an alternative to the impractical method developed in chapter 5 and given in equation 21.1.

Before formulating a proof of theorem 21.2 based on a precise statement of duality, let us consider an example which will instill some familiarity with the equivalences stated in the theorem.

EXAMPLE 21.1

The aim of this example is to show that if a pair (C,A) is not observable, then the tests (ii) through (v) of theorem 21.1 fail. Accordingly, consider the system model

$$\begin{bmatrix} \dot{x}_1 \\ \dot{x}_2 \\ \dot{x}_3 \end{bmatrix} = \begin{bmatrix} 2 & 0 & 1 \\ 0 & -2 & 0 \\ 1 & 0 & 2 \end{bmatrix} \begin{bmatrix} x_1 \\ x_2 \\ x_3 \end{bmatrix} \qquad (21.10)$$

$$= \begin{bmatrix} 0 & 1 & 1 \\ 1 & 0 & 0 \\ 0 & -1 & 1 \end{bmatrix} \begin{bmatrix} -2 & 0 & 0 \\ 0 & 1 & 0 \\ 0 & 0 & 3 \end{bmatrix} \begin{bmatrix} 0 & 1 & 0 \\ 0.5 & 0 & -0.5 \\ 0.5 & 0 & 0.5 \end{bmatrix} \begin{bmatrix} x_1 \\ x_2 \\ x_3 \end{bmatrix}$$

and

$$\begin{bmatrix} y_1 \\ y_2 \end{bmatrix} = \begin{bmatrix} 1 & 0 & 1 \\ 1 & 0 & -1 \end{bmatrix} \begin{bmatrix} x_1 \\ x_2 \\ x_3 \end{bmatrix}$$

Part 1. Applying condition (ii) to each eigenvalue λ_i of A yields

$$\text{rank}\begin{bmatrix} C \\ ---- \\ (\lambda_i I - A) \end{bmatrix} = \text{rank}\begin{bmatrix} 1 & 0 & 1 \\ 1 & 0 & -1 \\ \hline \lambda_i - 2 & 0 & -1 \\ 0 & \lambda_i + 2 & 0 \\ -1 & 0 & \lambda_i - 2 \end{bmatrix}$$

Inspection of the lower portion of the right-hand matrix clearly shows that the rank is 3 whenever $\lambda_i \neq -2$. Thus, since A has an eigenvalue at -2, the system is not completely observable.

Part 2. The observability matrix of the system, given by

$$R = \begin{bmatrix} C \\ CA \\ CA^2 \end{bmatrix} = \begin{bmatrix} 1 & 0 & 1 \\ 1 & 0 & -1 \\ \hline 3 & 0 & 3 \\ 1 & 0 & -1 \\ \hline 9 & 0 & 9 \\ 1 & 0 & -1 \end{bmatrix}$$

has rank 2 because of the central column of zeros. This, of course, is consistent with the previous test.

Part 3. Condition (iv) of the theorem requires looking at $\text{rank}[C\exp(-At)]$. Using equation 21.10, we have

$$Ce^{-At} = \begin{bmatrix} 0 & 0 & 2 \\ 0 & 2 & 0 \end{bmatrix}\begin{bmatrix} e^{2t} & 0 & 0 \\ 0 & e^{-t} & 0 \\ 0 & 0 & e^{-3t} \end{bmatrix}\begin{bmatrix} 0 & 1 & 0 \\ 0.5 & 0 & -0.5 \\ 0.5 & 0 & 0.5 \end{bmatrix}$$

$$= \begin{bmatrix} e^{-3t} & 0 & e^{-3t} \\ e^{-t} & 0 & -e^{-t} \end{bmatrix} \tag{21.11}$$

As vector functions of time, the first and third columns are linearly independent. Consequently, $\text{rank}[C\exp(-At)] = 2 \neq 3$, and the system is not observable.

Part 4. Using equation 21.11 with -3 replaced by 3 and -1 by 1 to produce Ce^{At} versus Ce^{-At}, one obtains

$$W_O(0,1) = \int_0^1 \begin{bmatrix} e^{3t} & e^t \\ 0 & 0 \\ e^{3t} & -e^t \end{bmatrix}\begin{bmatrix} e^{3t} & 0 & e^{3t} \\ e^t & 0 & -e^t \end{bmatrix} dt$$

$$= \int_0^1 \begin{bmatrix} e^{6t} + e^{2t} & 0 & e^{6t} - e^{2t} \\ 0 & 0 & 0 \\ e^{6t} - e^{2t} & 0 & e^{6t} + e^{2t} \end{bmatrix} dt = \begin{bmatrix} 70.27 & 0 & 63.88 \\ 0 & 0 & 0 \\ 63.88 & 0 & 70.27 \end{bmatrix}$$

Thus, $W_O(0,1)$ is clearly singular, as expected.

Numerical evaluation of conditions $(ii)-(iv)$ involving rank in the foregoing tests must progress with the cautions and warnings described in chapter 20. For example, the test of condition (ii) in part 1 of the example requires accurate computation of eigenvalues as well as of rank. The potential for an erroneous conclusion based on a determination of the rank of the observability matrix follows the same route as for the controllability matrix.

As mentioned earlier, the proof of theorem 21.2 depends on a duality between the state models (A,B,C) and the associated *adjoint* system model $(-A^t,C^t,B^t)$ (see problem 8 of chapter 12). The choice of $-A^t$ instead of A^t grows out of parameter optimization investigations and the design of optimal controllers for the model (A,B,C) [2].

Theorem 21.3 (Duality Theorem).   The state model (A,B,C) is completely controllable if and only if the state model $(-A^t,C^t,B^t)$ is completely observable. Moreover, (A,B,C) is completely observable if and only $(-A^t,C^t,B^t)$ is completely controllable.

Proof.   If (A,B,C) is completely controllable, then

$$\text{rank}[B \ AB \ \cdots \ A^{n-1}B] = n$$

Taking the transpose and multiplying appropriate rows by -1 yields

$$\text{rank} \begin{bmatrix} B^t \\ -B^t A^t \\ \vdots \\ B^t(-A^t)^{n-1} \end{bmatrix} = n$$

so that $(-A^t,C^t,B^t)$ is completely observable. The converse follows by a symmetric argument.

Next, (A,B,C) is completely observable if and only if

$$\text{rank} \begin{bmatrix} C \\ CA \\ \vdots \\ CA^{n-1} \end{bmatrix} = n$$

if and only if

$$\text{rank}[C^t | -A^t C^t | \ \cdots \ | (-A^t)^{n-1}C^t] = n$$

if and only if $(-A^t,C^t,B^t)$ is completely observable.   ■

Corollary 21.3. The model (A,B,C) is completely controllable (observable) if and only if (A^t, C^t, B^t) is completely observable (controllable).

Proof of Theorem 21.2. Combining the results of the corollary to theorem 21.3 with theorem 20.4, we see that the equivalences of $(i)-(v)$ follow immediately. ∎

OBSERVABILITY AND DIRECTED MODES

The objective in this section is to describe the observability properties of the system in a vein analogous to the controllability of directed modes. Assuming distinct eigenvalues, consider the zero-input system response decomposed into matrix directed modes, i.e.,

$$y(t) = C \exp(At) \, x(0) = C[R_1 \exp(\lambda_1 t) + \dots + R_n \exp(\lambda_n t)] \, x(0)$$
$$= [CR_1 \exp(\lambda_1 t) + \dots + CR_n \exp(\lambda_n t)] \, x(0) \tag{21.12}$$

where $R_i = e_i w_i^*$ and e_i (w_i) is a right (left) eigenvector of A. (Cf. the development of equation 17.1.)

Theorem 21.4. The mode $\exp(\lambda_i t) R_i$ (or equivalently, $\exp(\lambda_i t) e_i$) is unobservable if and only if $CR_i = [0]$.

Proof. *Part 1.* Suppose $CR_i = [0]$ for some i. Then for any $x(0)$, the contribution of the mode $\exp(\lambda_i t) R_i x(0)$ to the zero-input system response is identically zero. In other words, the mode is unobservable.

Part 2. For the converse, suppose that the mode $\exp(\lambda_i t) e_i$ is an unobservable mode—i.e., that e_i lies in $N[R]$. Then the contribution of this mode to the zero-input response is identically zero. Now, since e_i is in $N[R]$, it directly follows that $Ce_i = 0$, and since $R_i = e_i w_i^*$, $CR_i = [0]$, as was to be shown. ∎

Clearly, the observability of a directed mode (λ_i, e_i) depends on the relationship between C and the right eigenvectors of A, whereas the controllability of the mode depends on the kinship with the left eigenvectors and B.

Corollary 21.4. The set of unobservable eigenvectors spans the unobservable subspace.

This corollary implies that the unobservable subspace of the pair (C,A) is the *largest* A-invariant subspace contained in the null space of C. This property proves useful in various applications [1].

EXAMPLE 21.2

Consider the state model

$$\dot{x} = Ax + Bu$$
$$y = Cx$$

where

$$A = \begin{bmatrix} 0 & 0 & -2 \\ 0 & 0 & 0 \\ -2 & 0 & 0 \end{bmatrix}, \; B = \begin{bmatrix} 1 & 1 \\ 0 & 0 \\ -1 & 1 \end{bmatrix}, \; C = [1 \; 1 \; 1]$$

Some straightforward arithmetic leads to the usual eigenvalue-eigenvector factorization of $A = TDT^{-1}$:

$$A = \begin{bmatrix} 0 & 0 & -2 \\ 0 & 0 & 0 \\ -2 & 0 & 0 \end{bmatrix} = \begin{bmatrix} 0 & 1 & 1 \\ 1 & 0 & 0 \\ 0 & -1 & 1 \end{bmatrix} \begin{bmatrix} 0 & 0 & 0 \\ 0 & 2 & 0 \\ 0 & 0 & -2 \end{bmatrix} \begin{bmatrix} 0 & 1 & 0 \\ 0.5 & 0 & -0.5 \\ 0.5 & 0 & 0.5 \end{bmatrix}$$

$$B = \begin{bmatrix} 1 & 1 \\ 0 & 0 \\ -1 & 1 \end{bmatrix} = \begin{bmatrix} 0 & 1 & 1 \\ 1 & 0 & 0 \\ 0 & -1 & 1 \end{bmatrix} \begin{bmatrix} 0 & 0 \\ 1 & 0 \\ 0 & 1 \end{bmatrix}$$

$$C = [1 \; 1 \; 1] = [1 \; 0 \; 2] \begin{bmatrix} 0 & 1 & 0 \\ 0.5 & 0 & -0.5 \\ 0.5 & 0 & 0.5 \end{bmatrix}$$

It follows that the state transition matrix $\Phi(t) = e^{At}$ has the factorization

$$e^{At} = \begin{bmatrix} 0 & 1 & 1 \\ 1 & 0 & 0 \\ 0 & -1 & 1 \end{bmatrix} \begin{bmatrix} 1 & 0 & 0 \\ 0 & e^{2t} & 0 \\ 0 & 0 & e^{-2t} \end{bmatrix} \begin{bmatrix} 0 & 1 & 0 \\ 0.5 & 0 & -0.5 \\ 0.5 & 0 & 0.5 \end{bmatrix}$$

$$= \begin{bmatrix} 0 & 0 & 0 \\ 0 & 1 & 0 \\ 0 & 0 & 0 \end{bmatrix} + \begin{bmatrix} 0.5 & 0 & -0.5 \\ 0 & 0 & 0 \\ -0.5 & 0 & 0.5 \end{bmatrix} e^{2t} + \begin{bmatrix} 0.5 & 0 & 0.5 \\ 0 & 0 & 0 \\ 0.5 & 0 & 0.5 \end{bmatrix} e^{-2t}$$

If $\lambda_1 = 0$, then since $CR_1 \neq [0]$, the mode is observable. However, since $R_1 B = [0]$, $\lambda_1 = 0$ is uncontrollable. Similarly, if $\lambda = 2$, $CR_2 = [0]$ and $R_2 B \neq [0]$, from which it follows that the mode is not observable but is controllable. Finally, since $CR_3 B \neq [0]$, the mode $\lambda_3 = -2$ is both controllable and observable. These results yield certain advantages in regard to computing the system response. For suppose $x(0) = \theta$ and

$$u(t) = \begin{bmatrix} 0 \\ 0.5 \end{bmatrix} e^{-2t} 1^+(t)$$

Then the resulting zero-state system response is

$$y(1) = C \int_0^1 e^{A(1-\tau)} Bu(\tau) d\tau = CR_3 B \begin{bmatrix} 0 \\ 0.5 \end{bmatrix} \int_0^1 e^{-2(1-\tau)} e^{-2\tau} d\tau$$

$$= \int_0^1 e^{-2} e^{2\tau} e^{-2\tau} d\tau = e^{-2}$$

Thus, knowledge of the controllability and observability of the various modes can lead to a straightforward evaluation of the system response.

Directed modes have certain other important and peculiar characteristics not evidenced in the preceding example. *Do they remain in the unobservable subspace with state feedback? Is the unobservable subspace invariant under state feedback as is the controllable subspace? Given that observability and controllability are duals of each other, shouldn't there be similar properties for directed modes?* The answers to these questions, in reverse order, are (*i*) to some extent, (*ii*) no, and (*iii*) no. The following example demonstrates how a completely observable system can become unobservable by an appropriate state feedback.

EXAMPLE 21.3

Consider the usual state model $\dot{x} = Ax + Bu$, $y = Cx$, where

$$A = \begin{bmatrix} -1 & 0 & 0 \\ 0 & -2 & 0 \\ 0 & 0 & -3 \end{bmatrix}, B = \begin{bmatrix} 0 \\ 0 \\ 1 \end{bmatrix}, C = [1 \ 1 \ 1] \qquad (21.13)$$

The observability matrix

$$R = \begin{bmatrix} 1 & 1 & 1 \\ -1 & -2 & -3 \\ 1 & 4 & 9 \end{bmatrix}$$

has rank 3, confirming the complete observability of the model. The state feedback $F = [f_1 \ f_2 \ f_3]$ is claimed to affect and possibly make the system unobservable. To see this, let $F = [0 \ 0 \ 1]$. Then the new A-matrix is

$$\hat{A} = A + BF = \begin{bmatrix} -1 & 0 & 0 \\ 0 & -2 & 0 \\ 0 & 0 & -2 \end{bmatrix}$$

The new observability matrix is

$$\hat{R} = \begin{bmatrix} C \\ C\hat{A} \\ C(\hat{A})^2 \end{bmatrix} = \begin{bmatrix} 1 & 1 & 1 \\ -1 & -2 & -2 \\ 1 & 4 & 4 \end{bmatrix}$$

Since columns 2 and 3 coincide, $\text{rank}[\hat{R}] = 2$. But then the pair $(C,\hat{A}) = (C,A + BF)$ is unobservable. Hence, the structure of the unobservable subspace may in fact change under state feedback.

The seemingly undesirable property that the structure of the unobservable subspace may change under state feedback in fact greatly aids the solution of various practical problems. One such problem is the *disturbance decoupling problem* (DDP) [1], which the following example considers.

EXAMPLE 21.4

Consider a modified state model of the form

$$\dot{x} = Ax + Bu + Ed$$

$$y = Cx$$

where $d(t) \epsilon \mathbf{R}^{\hat{m}}$ is an input disturbance. Typically, $d(t)$ is assumed to be the response of a properly initialized, possibly unknown state model. Let A, B, and C be as given in equation 21.13. Let

$$E = \begin{bmatrix} 1 \\ -2 \\ 1 \end{bmatrix}$$

The objective of the example is to find a state feedback $u = Fx$ such that the response y is free of the influence of the disturbance d which enters the state space through the matrix E. The system with control $u = Fx$ has the form

$$\dot{x} = (A + BF)x + Ed$$

$$y = Cx$$

The response $y(t)$ is given by the usual formula,

$$y(t) = Ce^{(A + BF)t}x(0) + C \int_0^t e^{(A + BF)(t-q)} Ed(q)dq$$

For $y(t)$ to be free of the disturbance $d(t)$, it must happen that

$$C \int_0^t e^{(A + BF)(t-q)} Ed(q)dq = [0]$$

From the development of chapter 20, the range of the integral

$$x(t) = \int_0^t e^{(A + BF)(t-q)} Ed(q)dq$$

is precisely the controllable subspace of the pair $(A + BF, E)$. Hence $y(t)$ is free of the effects of $d(t)$ only when the controllable subspace of the pair $(A + BF, E)$ is contained in the null space of C, i.e., the pair is unobservable. Before delving into a brute-force solution to this problem, note that the controllable subspace of the pair (A, B) is invariant with respect to state feedback. Since the feedback F affects the state space through B (not E), the controllable subspace of the pair $(A + BF, E)$ depends intimately on F, unless of course $\mathrm{Im}(E) \subset \mathrm{Im}(B)$.

Now, by direct computation,

$$A + BF = \begin{bmatrix} -1 & 0 & 0 \\ 0 & -2 & 0 \\ f_1 & f_2 & -3 + f_3 \end{bmatrix}$$

The controllable subspace of the pair $(A + BF, E)$ is contained in $N[C]$ if and only if $CE = [0]$, $C(A + BF)E = [0]$, and $C(A + BF)^2E = [0]$. $CE = [0]$ is easily shown.

At present, the controllable subspace of the pair (A,E) is $\mathbf{R}^3$. A solution requires the existence of an appropriate nonzero feedback. The condition $C(A + BF)E = [0]$ reduces to

$$f_1 - 2f_2 + f_3 = 0$$

and $C(A + BF)^2E = [0]$ becomes

$$-4f_1 + 10f_2 - 6f_3 + f_1f_3 - 2f_2f_3 + f_3^2 = -2$$

A solution to these equations is

$$F = [-2 \ -1 \ 0]$$

Thus, if $u = Fx = [-2 -1 \ 0]x$, the disturbance $d(t)$ will be completely unobservable at $y(t)$; i.e., the disturbance is decoupled from the output. Note finally that F can be constructed by linear means as described in chapter 4 of [1]. Such techniques, however, require careful construction far beyond the brute-force approach illustrated in this example. A complete explanation, however, is given in chapter 25.

BASIC DESIGN OF DYNAMIC OBSERVERS

Identity Observers

In this chapter, as well as in chapter 5, we have derived some results on state reconstruction from input-output measurements. These techniques developed out of an algebraic perspective which required derivatives of the output and input measurements as entries in a set of linear algebraic equations. An alternative algebraic approach was given by equation 21.8; although this approach requires no derivatives, it depends heavily on a series of tedious calculations. By contrast, a dynamic observer builds around a replica of the given system to provide an on-line, continuous estimate of the system state. For example, suppose a particular plant has a scalar state model

$$\dot{x} = \lambda x + \beta u$$
$$y = \xi x \tag{21.14}$$

for nonzero scalars λ, β, and ξ. The following replica

$$\dot{\hat{x}} = \lambda \hat{x} + \beta u \tag{21.15}$$

of the plant could serve as a dynamic observer. *How so?* Intuitively, for the replica to be a dynamic observer, the state $\hat{x}(t)$ must asymptotically approach $x(t)$ for large t. In other words, it must happen that $[x(t) - \hat{x}(t)] \to 0$ for large t. The impending question is, *under what conditions will this occur?* The answer becomes clear after looking at the error $[x(t) - \hat{x}(t)]$, i.e.,

$$[\dot{x}(t) - \dot{\hat{x}}(t)] = \lambda[x(t) - \hat{x}(t)]$$

which has the solution

$$[x(t) - \hat{x}(t)] = e^{\lambda t}[x(0) - \hat{x}(0)]$$

If $x(0)$ is known *a priori* and $\hat{x}(0)$ is chosen equal to $x(0)$, then theoretically (certainly not realistically) speaking $\hat{x}(t) = x(t)$ for all $t \geq 0$. On the other hand, $x(0)$ generally remains unknown. Thus, if $\lambda < 0$, $[x(t) - \hat{x}(t)] \rightarrow 0$ for large t. The rate at which $\hat{x}(t)$ converges to $x(t)$ depends squarely on the magnitude of λ. This restricts the flexibility of the observer, since λ equals the pole of the plant. A more desirable observer would have its own dynamics independent of the plant dynamics. A designer could then choose the rate at which the observer output tracks the state.

What would be the structure of a more flexible observer? A perusal of the information available to an observer, as well as that utilized by equation 21.15, indicates that the observer of equation 21.15 ignores the output measurement y. If $\hat{x}(t)$ is close to $x(t)$, then $\xi\hat{x}(t)$ should be close to $y(t)$. Any difference should provide direction to $\hat{x}(t)$ in its task of tracking $x(t)$. Specifically, the derivative of $\hat{x}(t)$ should be proportional to $[y(t) - \xi\hat{x}(t)]$, i.e., $\dot{\hat{x}}(t) \approx \kappa[y(t) - \xi\hat{x}(t)]$. For example, assuming $\xi > 0$, if $\hat{x}$ lags behind x, then $\xi\hat{x}$ lags behind y and $\hat{x}$ should grow proportionately in order to "quickly" close up the distance. This intuitive reasoning suggests the following scalar dynamic observer structure:

$$\dot{\hat{x}} = \lambda\hat{x} + \kappa(y - \xi\hat{x}) + \beta u$$

The error dynamical equation, $x(t) - \hat{x}(t)$, has the form

$$(\dot{x} - \dot{\hat{x}}) = \lambda(x - \hat{x}) + \kappa(y - \xi\hat{x}) = (\lambda - \kappa\xi)(x - \hat{x}) \qquad (21.16)$$

The solution of the error dynamics equation 21.16 is

$$[x(t) - \hat{x}(t)] = e^{(\lambda - \kappa\xi)t}[x(0) - \hat{x}(0)]$$

Clearly, proper choice of κ establishes convergence, with the rate of convergence independent of the plant dynamics.

In a more general context, suppose

$$\dot{x} = Ax + Bu$$
$$y = Cx$$

is the usual multi-input, multi-output (MIMO) state model. Parroting the preceding discussion, the observer dynamical equation will have the form

$$\dot{\hat{x}} = A\hat{x} + K[y - C\hat{x}] + Bu \qquad (21.17)$$

where, again, the term $K[y - C\hat{x}]$ determines the degree to which $\dot{\hat{x}}$, the derivative of the state estimate, is proportional to the difference between the system output measurement y and the estimated output $C\hat{x}$. The observer of equation 21.17 is commonly called an *identity observer* [3].

Substituting Cx for y in equation 21.17 and looking at the error vector $(x - \hat{x})$ leads to the error dynamical equation

$$(\dot{x} - \dot{\hat{x}}) = (A - KC)(x - \hat{x}) \qquad (21.18)$$

which has the solution

$$[x(t) - \hat{x}(t)] = e^{(A - KC)t}[x(0) - \hat{x}(0)]$$

The two requirements imposed on the dynamics are: (*i*) that $\hat{x}(t) \to x(t)$ for large t (i.e., all eigenvalues of $(A - KC)$ lie in the open left-half complex plane) and (*ii*) the rate at which $\hat{x}(t)$ approaches $x(t)$ is determined by the design engineer by suitable choice of K. The first condition is easier to meet than the second, which requires the eigenvalues of $(A - KC)$ to be chosen arbitrarily in the left-half plane. In other words, one must be able to assign the eigenvalues of $(A - KC)$ by proper choice of K.

Theorem 21.5. If the pair (A,C) is completely observable, then the spectrum of $(A - KC)$ can be arbitrarily assigned by proper choice of K.

Proof. If (C,A) is completely observable, then the pair (A^t,C^t) is completely controllable, by duality. Hence, by the spectral assignability property developed in chapter 20, the spectrum (eigenvalues) of $(A^t + C^t\hat{K})$ can be arbitrarily placed by suitable choice of $\hat{K}$. Choosing $K = -\hat{K}^t$, the spectrum of $(A - KC)$ can be arbitrarily placed since $(A^t + C^t\hat{K})$ and $(A - KC)$ have the same characteristic polynomial.   ∎

EXAMPLE 21.5

Consider the second-order state model

$$\dot{x} = \begin{bmatrix} 0 & -2 \\ 1 & -2 \end{bmatrix} x + \begin{bmatrix} 0 \\ 1 \end{bmatrix} u$$

$$y = \begin{bmatrix} 0 & 1 \end{bmatrix} x$$

which must have a dynamic observer structure

$$\dot{\hat{x}} = \begin{bmatrix} 0 & -2 \\ 1 & -2 \end{bmatrix} \hat{x} + \begin{bmatrix} k_1 \\ k_2 \end{bmatrix} [y - \begin{bmatrix} 0 & 1 \end{bmatrix} \hat{x}] + \begin{bmatrix} 0 \\ 1 \end{bmatrix} u$$

Suppose the observer error dynamics

$$(\dot{x} - \dot{\hat{x}}) = \left[\begin{bmatrix} 0 & -2 \\ 1 & -2 \end{bmatrix} - \begin{bmatrix} k_1 \\ k_2 \end{bmatrix} \begin{bmatrix} 0 & 1 \end{bmatrix} \right] (x - \hat{x})$$

$$= \begin{bmatrix} 0 & -2-k_1 \\ 1 & -2-k_2 \end{bmatrix} (x - \hat{x})$$

is to have spectrum $\{-5, -6\}$. Then the desired characteristic polynomial is $\pi_{A-KC}(\lambda) = \lambda^2 + 11\lambda + 30$. Since the characteristic polynomial of the error dynamics is $\pi_{A-KC}(\lambda) = \lambda^2 + (2+k_2)\lambda + (2+k_1)$, the desired feedback must satisfy

$$K = \begin{bmatrix} k_1 \\ k_2 \end{bmatrix} = \begin{bmatrix} 28 \\ 9 \end{bmatrix}$$

A straightforward calculation shows that the error vector has the form

$$[x(t) - \hat{x}(t)] = \begin{bmatrix} 1 & 1 \\ 6 & 5 \end{bmatrix} \begin{bmatrix} e^{-5t} & 0 \\ 0 & e^{-6t} \end{bmatrix} \begin{bmatrix} -5 & 1 \\ 6 & -1 \end{bmatrix} [x(0) - \hat{x}(0)]$$

If $[x(0) - \hat{x}(0)] = [50 \quad 75]^t$, a large initial deviation, then after 5 seconds,

$$[x(5) - \hat{x}(5)] = \begin{bmatrix} 1 & 1 \\ 6 & 5 \end{bmatrix} \begin{bmatrix} -175e^{-25} \\ 225e^{-30} \end{bmatrix} \approx \begin{bmatrix} -2 \times 10^{-9} \\ -10^{-8} \end{bmatrix}$$

a rather small error.

Observers for Unobservable Systems

What if (C,A) is not observable and only part of the state space is reconstructible? What modifications in the just-developed observer structure permit partial state reconstruction, i.e., reconstruction of that part of the state space orthogonal to the unobservable subspace? The Kalman observable form [4,5] breaks open the solution package. Recall the Kalman observable form of equations 21.6, viz.,

$$\begin{bmatrix} \dot{z}_1 \\ \dot{z}_2 \end{bmatrix} = \begin{bmatrix} \bar{A}_{11} & \bar{A}_{12} \\ 0 & \bar{A}_{22} \end{bmatrix} \begin{bmatrix} z_1 \\ z_2 \end{bmatrix} + \begin{bmatrix} \bar{B}_1 \\ \bar{B}_2 \end{bmatrix} u$$

$$y = [0 \quad \bar{C}_2] \begin{bmatrix} z_1 \\ z_2 \end{bmatrix}$$

obtained from the usual state model under the state transformation $[T_1 \quad T_2]z = x$, where T_1 is any matrix whose columns span the unobservable subspace and T_2 is any matrix whose columns combined with those of T_1 form a basis for $\mathbf{R}^n$. A good choice is $[T_1 \quad T_2] = [V_2^R \quad V_1^R]$, where the V_i^R are given by the SVD of R, equation 21.5. The dynamic observer must use measurements $y(t)$ and $u(t)$ to reconstruct $z_2(t)$ in the new coordinate frame. The reconstructable part of the state space orthogonal to the unobservable subspace is given by

$$\hat{x} = [T_1 \quad T_2] \begin{bmatrix} 0 \\ z_2 \end{bmatrix}$$

Since $\hat{x}$ has no projection onto the unobservable subspace of (C,A), $\hat{x}$ is a least squares approximation. Figure 21.1 illustrates the block diagram structure of the observer. If (C,A) is observable, the structure reduces to that of equation 21.17.

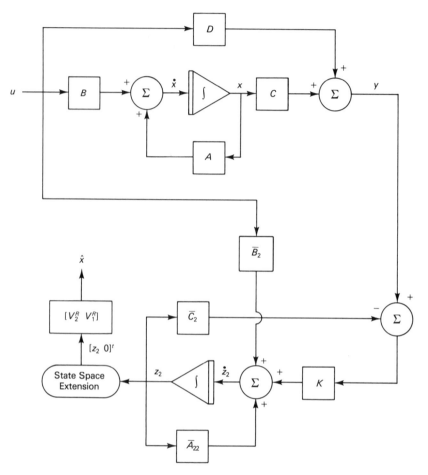

Figure 21.1 Block diagram of plant observer when pair (C,A) is not completely observable.

Reduced-Order Observers

The perspicatious reader might have begun to ponder the relationship between two observer design facts: (*i*) observers require the *p*-output measurements and the *m*-input measurements and (*ii*) the order of the observer equals the dimension of the observable part of the state space. If the *p*-output measurements are independent (equivalent to rank$[C] = p$) then there exists a one-to-one relationship between the measurements $y(t)$ and a *p*-dimensional subspace of the state space. Of course, this *p*-dimensional subspace must be observable. But then one might ask, *if (C,A) is observable, why does one need an n-dimensional observer?* Intuitively, an $(n - p)$-dimensional observer would seem sufficient. In fact it is, and the constructions that follow fabricate the structure of this so-called *reduced-order observer* [3,5].

Suppose (C,A) is an observable pair. Whatever their structure, reduced-order observers must utilize the measurement $y(t)$ in some direct fashion. A coordinate (state) transformation in which the measurements $y(t)$ form part of the new state vector provides the key to the direct use of measurements. In particular, define a new set of state variables as

$$\begin{bmatrix} w(t) \\ y(t) \end{bmatrix} = \begin{bmatrix} V \\ C \end{bmatrix} x(t) \qquad (21.19)$$

where C is $p \times n$, for $p < n$, and rank$[C] = p$, and V is any maximal-rank matrix such that $CV^t = [0]$. Clearly the columns of V^t span $N[C]$. An optimal choice for V^t would be $V^t = V_2^C$, where V_2^C arises from a singular value decomposition of C according to

$$C = [U_1^C \quad U_2^C] \begin{bmatrix} S^C & 0 \\ 0 & 0 \end{bmatrix} \begin{bmatrix} (V_1^C)^t \\ (V_2^C)^t \end{bmatrix}$$

Of course, the values of the variables $w(t)$ in equation 21.19 are unknown.

Equation 21.19 specifies a nonsingular state transformation. The equivalent state model has the form

$$\begin{bmatrix} \dot{w} \\ \dot{y} \end{bmatrix} = \begin{bmatrix} V \\ C \end{bmatrix} A \begin{bmatrix} V \\ C \end{bmatrix}^{-1} \begin{bmatrix} w \\ y \end{bmatrix} + \begin{bmatrix} V \\ C \end{bmatrix} Bu(t) \qquad (21.20a)$$

$$\begin{bmatrix} \dot{w} \\ \dot{y} \end{bmatrix} \triangleq \begin{bmatrix} \tilde{A}_{11} & \tilde{A}_{12} \\ \tilde{A}_{21} & \tilde{A}_{22} \end{bmatrix} \begin{bmatrix} w \\ y \end{bmatrix} + \begin{bmatrix} \tilde{B}_1 \\ \tilde{B}_2 \end{bmatrix} u(t) \qquad (21.20b)$$

The objective at this point is to subtract Ky from w, where K is an unspecified observer gain matrix. By defining a new vector variable $v = w - Ky$, one may determine a dynamical state model in v driven by the measurable variables y and u. By building a dynamic (identity) observer for this new state model one can then generate the estimate $\hat{v}$ of v. Hence, the estimate $\hat{w}$ of w becomes $\hat{w} = \hat{v} + Ky$. In consequence, the estimate $\hat{x}$ of the plant state x will satisfy

$$\hat{x} = \begin{bmatrix} V \\ C \end{bmatrix}^{-1} \begin{bmatrix} \hat{w} \\ y \end{bmatrix} = \begin{bmatrix} V \\ C \end{bmatrix}^{-1} \begin{bmatrix} \hat{v} + Ky \\ y \end{bmatrix} \qquad (21.21)$$

It remains only to fill in the particulars. As mentioned, we subtract Ky from w in equation 21.20b to produce

$$(\dot{w} - K\dot{y}) = (\tilde{A}_{11} - K\tilde{A}_{21})w + (\tilde{A}_{12} - K\tilde{A}_{22})y + (\tilde{B}_1 - K\tilde{B}_2)u \qquad (21.22)$$

Defining $v = w - Ky$ leads to the following dynamical state model in v:

$$\dot{v} = (\tilde{A}_{11} - K\tilde{A}_{21})v + [\tilde{A}_{12} - K\tilde{A}_{22} + \tilde{A}_{11}K - K\tilde{A}_{21}K]y + (\tilde{B}_1 - K\tilde{B}_2)u \qquad (21.23)$$

Equation 21.23 has the form of an (identity) observer. To see this, suppose $\hat{v}$ designates an estimate of v and replaces v in the equation. Then the resulting error dynamical equation is

$$(\dot{v} - \dot{\hat{v}}) = (\tilde{A}_{11} - K\tilde{A}_{21})(v - \hat{v}) \qquad (21.24)$$

Hence, equation 21.23 has an identity observer structure, and if $(\tilde{A}_{21}(=C),$ $\tilde{A}_{11}(=A))$ is an observable pair, by proper choice of K equation 21.24 will converge to zero at any desired rate of convergence. Note that the observability of the pair $(\bar{A}_{21}, \bar{A}_{11})$ follows from the observability of the pair (C, A) as hypothesized (see problem 19).

For further background and other results on dynamic observer design, see [1, 3, 6–9].

The Separation Theorem

The eigenvalue separation theorem paves the way for implementing state feedback based on an estimate of the state, obtained from a dynamic observer. Oftentimes only the output variables or some subset thereof are available for measurement. Limited sensor technology may preclude direct measurement of most state variables. Yet, many feedback design schemes require complete state information. For example, the eigenvalue assignment scheme detailed in chapter 20, or other schemes as in [10,11], utilize a full state feedback law. To alleviate the difficulty, control engineers insert dynamic observer structures in the feedback loop, and use the state estimate in the feedback control law. Insertion of a dynamic observer in the feedback path creates additional system dynamics—additional eigenvalues or natural frequencies. The question then arises, *Do the additional dynamics interfere with the desired system behavior in some unforseen way?* Fortunately, they do not. The eigenvalue separation theorem states that the characteristic polynomial of the feedback system with a dynamic observer (see Figure 21.2) equals the product of the characteristic polynomials of the observer and that of the state feedback control without the observer. Hence, the dynamic behavior of the observer does not interfere with the desired eigenstructure of the controlled plant.

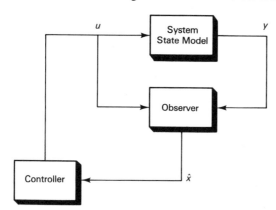

Figure 21.2 System, observer, and controller structure suggesting need for eigen value separation theorem.

Theorem 21.6 (The Eigenvalue Separation Theorem).   Let a plant have the usual state model

$$\dot{x} = Ax + Bu$$

$$y = Cx$$

and the identity observer given by

$$\dot{\hat{x}} = A\hat{x} + K[y - C\hat{x}] + Bu \qquad (21.25)$$

Suppose the control law for the plant is

$$u = F\hat{x}$$

Then the characteristic polynomial of the interconnected system is the product of the characteristic polynomials of the observer dynamics $A - KC$ and the feedback dynamics $A + BF$.

Proof. From the hypothesis of the theorem,

$$\dot{x} = Ax + Bu = Ax + BF\hat{x} = (A + BF)x - BF(x - \hat{x}) \qquad (21.26)$$

The dynamics of $\hat{x}$ follow from equation 21.25 and can be written as

$$\dot{\hat{x}} = (A - KC)\hat{x} + KCx + BF\hat{x}$$

Forming the error dynamics implies that

$$(\dot{x} - \dot{\hat{x}}) = (A - KC)(x - \hat{x}) \qquad (21.27)$$

Putting equations 21.26 and 21.27 in matrix form yields

$$\begin{bmatrix} \dot{x} \\ \dot{x} - \dot{\hat{x}} \end{bmatrix} = \begin{bmatrix} A + BF & -BF \\ 0 & A - KC \end{bmatrix} \begin{bmatrix} x \\ x - \hat{x} \end{bmatrix} \qquad (21.28)$$

whose characteristic polynomial is

$$\pi(\lambda) = \pi_{A+BF}(\lambda)\pi_{A-KC}(\lambda)$$

which was to be proven.   ∎

THE KALMAN CANONICAL FORM

As mentioned earlier, the ideas of proposition 20.2–20.8 which developed the Kalman controllable form, interconnect with the Kalman observable form given by equation 21.6 to form the *Kalman canonical form* [4,5]. The logical progression to this canonical form begins with the usual linear, time-invariant state model

$$\dot{x} = Ax + Bu$$
$$y = Cx$$

having controllability matrix $Q = [B \ AB \ \dots \ A^{n-1}B]$ with rank$[Q] = p$. Now suppose Q has the singular value decomposition (see chapter 19)

$$Q = [U_1^Q \ U_2^Q] \begin{bmatrix} S^Q & 0 \\ 0 & 0 \end{bmatrix} \begin{bmatrix} (V_1^Q)^t \\ (V_2^Q)^t \end{bmatrix} \qquad (21.29)$$

where (*i*) the superscript Q refers the SVD back to the matrix Q, (*ii*) U_1^Q has orthonormal columns spanning the Im(Q), and (*iii*) U_2^Q has orthonormal columns spanning $N[Q^t] = [\text{Im}(Q)]^\perp$, where the superscript $\perp$ denotes the orthogonal complement of Im(Q) in $\mathbf{R}^n$. (For example, if Im$(Q) = \text{span}\{v_1, v_2\}$, where $v_1 = (1 \ 0 \ 1)^t$ and $v_2 = (0 \ 1 \ 0)^t$, then Im$(Q)^\perp = \text{span}\{v_3\}$, where $v_3 = (1 \ 0 \ -1)^t$.) Then

$$[1 \; 0 \; -1] \begin{bmatrix} 1 & 0 \\ 0 & 1 \\ 1 & 0 \end{bmatrix} = [0 \; 0]$$

confirming the orthogonality of v_3 with v_1 and v_2.

This analysis suggests a decomposition of the state space into a controllable part and a part perpendicular to the controllable part, i.e.,

$$\mathbf{R}^n = X = X_C \oplus X_{\bar{C}} \tag{21.30}$$

where X_C denotes the controllable subspace and $X_{\bar{C}}$ denotes the orthogonal complement of the controllable subspace. From the foregoing SVD of Q, an orthonormal basis for X_C is given by the columns of U_1^Q, and one for $X_{\bar{C}}$ is given by the columns of U_2^Q. The representation of this decomposition (see equation 20.6) is very elegantly given by the equivalent state model

$$\begin{bmatrix} \dot{z}_1 \\ \dot{z}_2 \end{bmatrix} = \begin{bmatrix} \tilde{A}_{11} & \tilde{A}_{12} \\ 0 & \tilde{A}_{22} \end{bmatrix} \begin{bmatrix} z_1 \\ z_2 \end{bmatrix} + \begin{bmatrix} \tilde{B}_1 \\ 0 \end{bmatrix} u \tag{21.31}$$
$$y = \tilde{C}z$$

where

$$\begin{bmatrix} \tilde{A}_{11} & \tilde{A}_{12} \\ 0 & \tilde{A}_{22} \end{bmatrix} = [U_1^Q \; U_2^Q]^t A [U_1^Q \; U_2^Q] \tag{21.32}$$

$$\begin{bmatrix} \tilde{B}_1 \\ 0 \end{bmatrix} = [U_1^Q \; U_2^Q]^t B \tag{21.33}$$

and

$$\tilde{C} = C[U_1^Q \; U_2^Q]. \tag{21.34}$$

As mentioned, this equivalent system extracts the controllable part of the state space and embodies it in the state vector $(z_1 \; 0)^t$. Consistent with the decomposition described, each vector z has the form

$$z = \begin{bmatrix} z_1 \\ 0 \end{bmatrix} + \begin{bmatrix} 0 \\ z_2 \end{bmatrix}$$

where $(0 \; z_2)^t$ is the orthogonal complement of $(z_1 \; 0)^t$. Thus, $(0, z_2)^t$ is uncontrollable.

The goals of this chapter are (i) to further decompose the state space into observable and unobservable partitions and (ii) to represent the complete partitioning by an equivalent state representation which extracts the controllable and unobservable part, the projection of the observable part onto the controllable space, the projection of the uncontrollable subspace onto the unobservable part, and finally the last partition, the complement of these spaces in $\mathbf{R}^n$, the so-called observable and uncontrollable part. In other words, the objective is to generate bases for the decomposition of the state space given by

$$\mathbf{R}^n = X = X_{\bar{C}\bar{O}} \oplus X_{CO} \oplus X_{C\bar{O}} \oplus X_{\bar{C}O} \tag{21.35}$$

where

(i) $X_{C\bar{O}}$ is the intersection of the controllable subspace with the unobservable subspace, i.e., those states that are controllable but unobservable.

(ii) X_{CO} is the (orthogonal) complement of $X_{C\bar{O}}$ contained in X_C, i.e., the space of reconstructable states projected onto the controllable subspace.

(iii) $X_{\bar{C}\bar{O}}$ is the (orthogonal) complement of $X_{C\bar{O}}$ contained in $N(R)$; i.e., the projection of the uncontrollable subspace (the (orthogonal) complement of X_C) onto the unobservable subspace.

(iv) $X_{\bar{C}O}$ is that space which is the (orthogonal) complement of $X_{C\bar{O}} \oplus X_{CO} \oplus X_{\bar{C}\bar{O}}$, i.e., those states which are observable (reconstructable) but have no projection onto the controllable subspace and are thus uncontrollable.

If $x \in X_{C\bar{O}} \oplus X_{CO}$, then x is controllable; if $x \in X_{C\bar{O}} \oplus X_{\bar{C}\bar{O}}$, then x is unobservable. Note also that the decomposition must begin with $X_{C\bar{O}} = X_C \cap X_{\bar{O}}$ because only $X_{C\bar{O}}$ is A-invariant and is thus the only space definable in a coordinate-free fashion [5]. Finally, the use of the word *orthogonal* in the preceding explanation is more restrictive than necessary, but maintains consistency with the SVD method of constructing the decomposition. For alternative approaches, see [12–13].

The SVD techniques provide a straightforward means for obtaining bases for these spaces, which in turn allows us to develop a state transformation. This state transformation then allows us to convert the original state matrices to a set whose partitioned structure will clearly pick off each of the spaces. Since there are four spaces, the transformation must have the structure $T = [T_1, T_2, T_3, T_4]$ where the columns of T_1 are a basis for $X_{C\bar{O}}$, those of T_2 a basis for X_{CO}, those of T_3 a basis for $X_{\bar{C}\bar{O}}$, etc.

Clearly, the SVD of Q given by equation 21.29, will play a key role in the construction. Also, the SVD of the observability matrix R will play a dual role. This SVD is given by

$$R = [U_1^R \ U_2^R] \begin{bmatrix} S^R & 0 \\ 0 & 0 \end{bmatrix} \begin{bmatrix} (V_1^R)^t \\ (V_2^R)^t \end{bmatrix} \tag{21.36}$$

A pictorial view of the decomposition appears in Figure 21.3.

The columns of T_1 must span the intersection of $\text{Im}(Q) \cap N[R] = \text{Im}(U_1^Q) \cap \text{Im}(V_2^R)$. This intersection property is equivalent to the columns of T_1 being orthogonal to both $\text{Im}(Q)^\perp = \text{Im}(U_2^Q)$ and $N[R]^\perp = \text{Im}(V_1^R)$. Thus, T_1 is a maximal-rank matrix satisfying

$$\begin{bmatrix} (U_2^Q)^t \\ (V_1^R)^t \end{bmatrix} T_1 \triangleq M_1 T_1 = [0] \tag{21.37}$$

Computation of T_1 occurs by performing an SVD on M_1 and setting $T_1 = V_2^{M_1}$.

The matrix T_2 must have columns that serve as a basis for X_{CO}, the orthogonal complement of $X_{C\bar{O}} = \text{Im}(T_1)$ contained within $X_C = \text{Im}(U_1^Q)$, i.e.,

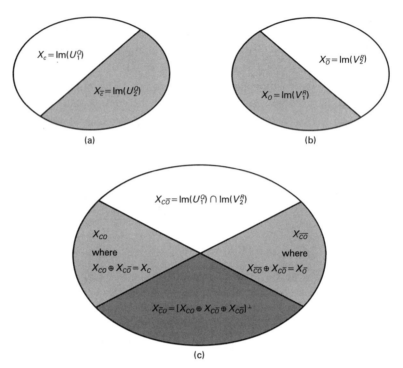

Figure 21.3 Decomposition of the state space into (a) controllable/uncontrollable partition, (b) observable/unobservable partition, and (c) controllable/unobservable, controllable/observable, etc., partitions.

$X_C = X_{C\bar{O}} \oplus X_{CO}$. Therefore, the columns of T_2 are orthogonal to those of T_1 and to those of U_2^Q. Hence, T_2 is a maximal-rank matrix satisfying

$$\begin{bmatrix} (U_2^Q)^t \\ T_1^t \end{bmatrix} T_2 \triangleq M_2 T_2 = [0] \tag{21.38}$$

The computation of T_2 is similar to that of T_1, proceeding by executing an SVD on M_2 and setting $T_2 = V_2^{M_2}$.

$\text{Im}(T_3) = X_{\bar{C}\bar{O}}$ represents the orthogonal complement of $X_{C\bar{O}}$ contained within $X_{\bar{O}} = \text{Im}(V_2^R)$. In other words, $X_{\bar{C}\bar{O}} \oplus X_{C\bar{O}} = X_{\bar{O}}$, the unobservable subspace. Clearly, then, T_3 is a maximal-rank matrix satisfying

$$\begin{bmatrix} (V_1^R)^t \\ T_1^t \end{bmatrix} T_3 \triangleq M_3 T_3 = [0] \tag{21.39}$$

So we set $T_3 = V_2^{M_3}$, where $V_2^{M_3}$ has columns spanning the null space of M_3.

Finally, T_4 has columns spanning the orthogonal complement of the previous three spaces. Thus, T_4 is a maximal-rank matrix satisfying

$$\begin{bmatrix} T_1^t \\ T_2^t \\ T_3^t \end{bmatrix} T_4 \triangleq M_4 T_4 = [0] \tag{21.40}$$

Again, we execute an SVD of M_4 and set $T_4 = V_2^{M_4}$.

Note that the matrix $T = [T_1 \ T_2 \ T_3 \ T_4]$ is the optimal choice for the state transformation $Tz = x$ converting the original model to the Kalman canonical form [14]. But optimal in what sense? Let σ_1 be the largest and σ_n the smallest singular values of T. Then T is optimal in that the condition number of T, $\kappa(T) = \sigma_1/\sigma_n$, is minimal with respect to all other state transformations leading to a Kalman canonical form. This makes T the best conditioned state transformation with respect to inversion. See [14] for a proof. In general, T is not orthogonal, since X_{CO} and $X_{C\bar{O}}$ are not orthogonal spaces. Thus, $T_2^t T_3 \neq [0]$.

In any event, executing the state transformation $Tz = x$ on the usual state model produces the equivalent model

$$\dot{z} = \tilde{A}z + \tilde{B}u$$
$$y = \tilde{C}z$$

where

$$\tilde{A} = T^{-1}AT = \begin{bmatrix} \tilde{A}_{11} & \tilde{A}_{12} & \tilde{A}_{13} & \tilde{A}_{14} \\ 0 & \tilde{A}_{22} & 0 & \tilde{A}_{24} \\ 0 & 0 & \tilde{A}_{33} & \tilde{A}_{34} \\ 0 & 0 & 0 & \tilde{A}_{44} \end{bmatrix} \qquad (21.41)$$

$$\tilde{B} = T^{-1}B = \begin{bmatrix} \tilde{B}_1 \\ \tilde{B}_2 \\ 0 \\ 0 \end{bmatrix}$$

and

$$\tilde{C} = CT = [0 \ \tilde{C}_2 \ 0 \ \tilde{C}_4]$$

As should be clear from the decomposition, each vector z in the new coordinates has the form $z = [z_1, z_2, z_3, z_4]^t$. Moreover, every vector of the form $z = [z_1, z_2, 0, 0]^t$ is controllable, and every vector of the form $z = [0, z_2, 0, z_4]^t$ is observable, etc.

EXAMPLE 21.6

This example is due to Boley [13,14] and was also investigated by Lee et al. in [12]. Suppose the matrices A, B, and C of the usual state model are

$$A = \begin{bmatrix} -2 & -1 & \dfrac{3}{\sqrt{2}} & \dfrac{1}{\sqrt{2}} \\ 0 & -1 & \dfrac{8}{\sqrt{2}} & -\dfrac{2}{\sqrt{2}} \\ 0 & 0 & 3.5 & -1.5 \\ 0 & 0 & 2.5 & -0.5 \end{bmatrix}, \quad B = \begin{bmatrix} 1 \\ 1 \\ 0 \\ 0 \end{bmatrix}, \quad C^t = \begin{bmatrix} 0 \\ 1 \\ 0 \\ -\sqrt{2} \end{bmatrix}$$

These are almost in the desired canonical form. Some straightforward calculations show that the controllability matrix is

$$Q = \begin{bmatrix} 1 & -3 & 7 & -15 \\ 1 & -1 & 1 & -1 \\ 0 & 0 & 0 & 0 \\ 0 & 0 & 0 & 0 \end{bmatrix}$$

clearly indicating that only the first two states are controllable. Orthogonal bases for the controllable and uncontrollable subspaces, computed by executing an SVD on Q, are given by

$$U_1^Q = \begin{bmatrix} -0.9958 & -0.0917 \\ -0.0917 & 0.9958 \\ 0 & 0 \\ 0 & 0 \end{bmatrix}, \quad U_2^Q = \begin{bmatrix} 0 & 0 \\ 0 & 0 \\ 1 & 0 \\ 0 & 1 \end{bmatrix}$$

The observability matrix takes the form

$$R = \begin{bmatrix} 0 & 1 & 0 & -1.4142 \\ 0 & -1 & 2.1213 & -0.7071 \\ 0 & 1 & -0.209 \times 10^{-6} & -1.4142 \\ 0 & -1 & 2.1213 & -0.7071 \end{bmatrix}$$

Executing an SVD on R leads to the following orthogonal bases for the unobservable subspace $X_{\bar{O}}$ and the observable subspace $X_O = X_{\bar{O}}^{\perp}$:

$$V_2^R = \begin{bmatrix} 0 & 1 \\ 0.7071 & 0 \\ 0.5 & 0 \\ 0.5 & 0 \end{bmatrix}, \quad V_1^R = \begin{bmatrix} 0 & 0 \\ -0.4082 & 0.5774 \\ 0.866 & 0.314 \times 10^{-6} \\ -0.2887 & -0.8165 \end{bmatrix}$$

Each of the spaces spanned by the columns of T_i is one-dimensional, so, following the algorithm previously described, T can be computed as

$$T = [T_1 \; T_2 \; T_3 \; T_4] = \begin{bmatrix} 1 & 0 & 0 & 0 \\ 0 & 1 & 0.7071 & -0.88 \times 10^{-8} \\ 0 & 0 & 0.5 & -0.7071 \\ 0 & 0 & 0.5 & 0.7071 \end{bmatrix}$$

The resulting inverse matrix is

$$T^{-1} = \begin{bmatrix} 1 & 0 & 0 & 0 \\ 0 & 1 & -0.7071 & -0.7071 \\ 0 & 0 & 1.0 & 1.0 \\ 0 & 0 & -0.7071 & 0.7071 \end{bmatrix}$$

And the resulting canonical state matrices are

$$\tilde{A} = T^{-1}AT = \begin{bmatrix} -2.0 & -1 & 0.7071 & -1 \\ 0 & -1 & 0.999 \times 10^{-6} & -1 \\ 0 & 0 & 2.0 & -5.6568 \\ 0 & 0 & 0.247 \times 10^{-7} & 1 \end{bmatrix}$$

where we consider 0.999×10^{-6} and 0.247×10^{-7} to be zero and

$$\tilde{B} = T^{-1}B = \begin{bmatrix} 1 \\ 1 \\ 0 \\ 0 \end{bmatrix}, \quad \tilde{C} = CT = [0 \quad 1 \quad -0.37 \times 10^{-6} \quad -1]$$

As an exercise, the reader might try to show that transforming the system

$$A = \begin{bmatrix} 2 & 0 & -3 \\ 0 & 0 & 0 \\ 1.5 & 0 & 2.5 \end{bmatrix}, \quad B = \begin{bmatrix} -1 \\ 0 \\ -2 \end{bmatrix}, \quad C = [1 \quad 1 \quad -1]$$

to the Kalman canonical form leads, among others, to one transformation matrix

$$T = [T_1 \ T_2 \ T_4] = \begin{bmatrix} 0.7071 & -0.7071 & 0 \\ 0 & 0 & 1 \\ 0.7071 & 0.7071 & 0 \end{bmatrix}$$

The resulting canonical form is

$$\tilde{A} = \begin{bmatrix} -1.0 & -4.5 & 0 \\ 0 & 0.5 & 0 \\ 0 & 0 & 0 \end{bmatrix}, \quad \tilde{B} = \begin{bmatrix} -2.121 \\ -0.7071 \\ 0 \end{bmatrix} \quad \tilde{C} = [0 \quad -1.414 \quad 1]$$

As a final remark, note that the algorithm presented in this section depends critically on the bases for X_C, $X_{\bar{C}}$, $X_{\bar{O}}$ and X_O. Although computations using the intermediary matrices Q and R can often lead to numerical errors, there are certain advantages to using these intermediaries anyway. For example, the SVD of Q orders the basis vectors for the controllable space; one difficulty arises in choosing the number of columns of U assigned to U_1, i.e., in determining the useful dimension of X_C, and another in the accurate construction of Q and R to begin with.

OBSERVABILITY AND DISCRETE-TIME SYSTEMS

The ideas of observability carry over in the obvious way modulo a finite number of measurements versus a finite time interval.

Definition 21.4. The discrete-time state model is *completely observable* if and only if there exists a finite index such that knowledge of

$\{y(0), y(1),...,y(N-1)\}$ and $\{u(0), u(1),...,u(N-1)\}$, as well as of the pair (A,C), is sufficient to determine $x(0)$ for arbitrary $x(0)$ in $\mathbf{R}^n$.

Theorem 21.7.   The discrete-time system

$$x(k+1) = Ax(k) + Bu(k)$$

$$y(k) = Cx(k) + Du(k)$$

is completely observable if and only if the $rn \times n$ *observability matrix*

$$R = \begin{bmatrix} C \\ CA \\ CA^2 \\ \vdots \\ CA^{n-1} \end{bmatrix}$$

has rank n.

Proof.   Without loss of generality, suppose the input observations are all zero. Let $\text{rank}[R] = n$, and suppose measurements $\{y(0),...,y(N-1)\}$ are known, as is the pair (A,C). Then from the discrete-time counterpart to equation 21.1 (see chapter 9),

$$Y(n) = \begin{bmatrix} y(0) \\ y(1) \\ \vdots \\ y(n-1) \end{bmatrix} = \begin{bmatrix} C \\ CA \\ \vdots \\ CA^{n-1} \end{bmatrix} x(0)$$

Since $\text{rank}[R] = n$, R^{-L} exists and $x(0) = R^{-L}Y(n)$ is uniquely determined.

For the converse, suppose an arbitrary initial state $x(0)$ can be uniquely determined by a finite number of observations $\{y(0),...,y(N-1)\}$ for some finite integer N. Define the matrix

$$R_N = \begin{bmatrix} C \\ CA \\ CA^2 \\ \vdots \\ CA^{N-1} \end{bmatrix}$$

By a unique determination, it is meant that the solution to

$$Y(N) = R_N x(0)$$

is unique. This requires that $\text{rank}(R_N)$ be maximal—i.e., $\text{rank}[R_N] = n$. But from lemmas 20.1 and 20.2 of the previous chapter, if $\text{rank}[R_N] = n$, then $\text{rank}[R] = \text{rank}[R_N] = n$, as was to be shown.

As with the continuous-time case, a state x_0 is said to be *completely unobservable* if its zero-input system response is identically zero. It is then a simple matter to prove the following proposition.

Proposition 21.3. x_0 is completely unobservable if and only if x_0 is in $N[R]$.

Proof.   Recall again the discrete-time equivalent of equation 21.1, viz.,

$$Y - TU = Rx_0$$

Part 1. Suppose x_0 is completely unobservable. Then

$$y(0) = y(1) = y(2) = \ldots = y(n-1) = 0$$

Hence $\theta = Rx_0$ implies $x_0 \in N[R]$.

Part 2. Now suppose $x_0 \in N[R]$. Then $Rx_0 = \theta$, which implies that

$$y(0) = y(1) = y(2) = \ldots = y(n-1) = \theta$$

The problem is to show that $y(N) = \theta$ for all $N \geqslant n$. To see this again, define

$$R_N = \begin{bmatrix} C \\ CA \\ CA^2 \\ \vdots \\ CA^{N-1} \end{bmatrix}$$

Then by the Caley-Hamilton theorem or lemmas 20.1 and 20.2, CA^{N-1} has rows which are dependent on the rows of R. Therefore $N[R_N] = N[R]$, and

$$y(k) = CA^k x(0) = 0$$

for all k, i.e., $x(0)$ is completely unobservable.   ∎

Proposition 21.3 continues the parallelism with the continuous time case by implying that *$N[R]$ indeed represents the unobservable subspace* of the system. It then follows immediately from proposition 21.1 that $N[R]$ is invariant under multiplication by A. In particular, if $x_0 \in N[R]$, then

$$RA^k x_0 = \begin{bmatrix} CA^k x_0 \\ CA^{k+1} x_0 \\ \vdots \\ CA^{k+n-1} x_0 \end{bmatrix}$$

So by the Caley-Hamilton theorem or by lemmas 20.1 and 20.2,

$$R(A^k x_0) = CA^k x_0 = CA^{k+1} x_0 = \ldots = CA^{k+n-1} x_0 = \theta$$

as was to be shown.   ∎

Theorem 21.8.   Let A have distinct λ_i. The discrete-time directed mode (λ_i, e_i) is unobservable if and only if $Ce_i = \theta$.

The proof of this theorem is left as problem 20.   ∎

Of course, the state space of the discrete state model decomposes into observable and unobservable parts. This partition is representable as a Kalman observable form computed in the same manner as the continuous-time partition given in equations 21.6. The counterpart of those equations is

$$\begin{bmatrix} z_1(k+1) \\ z_2(k+1) \end{bmatrix} = \begin{bmatrix} \bar{A}_{11} & \bar{A}_{12} \\ 0 & \bar{A}_{22} \end{bmatrix} \begin{bmatrix} z_1(k) \\ z_2(k) \end{bmatrix} + \begin{bmatrix} \bar{B}_1 \\ \bar{B}_2 \end{bmatrix} u(k)$$

$$y(k) = [0 \quad \bar{C}_2] \begin{bmatrix} z_1(k) \\ z_2(k) \end{bmatrix}$$

computed via the state transformation $\begin{bmatrix} V_2^R & V_1^R \end{bmatrix} z(k) = x(k)$, with V_1^R and V_2^R as given in equation 21.5.

Naturally, it follows that for a discrete-time system the pair (C,A) is observable if and only if

$$\text{rank} \begin{bmatrix} C \\ \lambda_i I - A \end{bmatrix} = n \qquad (21.42)$$

for each eigenvalue λ_i of A, or, in turn, if and only if

$$\text{rank} [CA^k] = n \qquad (21.43)$$

where each column of CA^k is viewed as a vector-valued sequence dependent on k. A set of n sequences $\{s_k^1, s_k^2, ..., s_k^n\}$ is linearly independent if

$$\alpha_1 s_k^1 + \alpha_2 s_k^2 + ... + \alpha_n s_k^n \equiv 0$$

(i.e., the equality holds for every term in the sequence indexed by k), which in turn implies $\alpha_1 = \alpha_2 = ... = \alpha_n = 0$. Thus, equations 21.42 and 21.43 are equivalent conditions for complete observability of the pair (C,A) along with the condition rank$[R] = n$.

Continuing the symmetry between continuous- and discrete-time state models, a discrete-time state model identity observer will have the structure

$$\hat{x}(k+1) = A\hat{x}(k) + K[y(k) - C\hat{x}(k)] + Bu(k) \qquad (21.44)$$

with error dynamics

$$[x(k+1) - \hat{x}(k+1)] = (A - KC)[x(k) - \hat{x}(k)] \qquad (21.45)$$

Thus, the observer gain K must be chosen so that each eigenvalue λ_i of $(A - KC)$ satisfies $|\lambda_i| < 1$. This insures that the error approaches zero for large k. Finally, the concept of minimal-order observers, the eigenvalue separation theorem, and the Kalman canonical form have virtually identical developments as their continuous-time counterparts.

EXAMPLE 21.7

This example points out an often overlooked subtlety in the state reconstruction of discrete-time state models having a singular A-matrix. The problem looks into state reconstruction using a switched discrete-time state model

$$x(k+1) = \begin{bmatrix} 0 & 0 \\ 0 & 1 \end{bmatrix} x(k) + \begin{bmatrix} 2 \\ 1 \end{bmatrix} u(k)$$

$$y(k) = \begin{cases} [0 \ 1] \, x(k), & 0 \leqslant k \leqslant 2 \\ [1 \ 1] \, x(k), & 3 \leqslant k \end{cases}$$

Suppose $u(k) = 1^{+}(k)$ and the following output measurements are taken:

k	0	1	2	3	4
y(k)	1	1	2	6	7

We reconstruct $x(3)$ and, if possible, $x(2)$, $x(1)$, and $x(0)$. Notice that the model is observable over $[3,4]$, but unobservable over $[0,2]$. Now if A were nonsingular, knowing $x(3)$ would lead to a quick computation of $x(k)$, $k = 0,1,2$. It turns out that $x(1)$ and $x(2)$ are computable, but $x(0)$ is not.

To compute $x(3)$, observe that

$$\begin{bmatrix} y(3) \\ y(4) \end{bmatrix} - \begin{bmatrix} 0 & 0 \\ CB & 0 \end{bmatrix} \begin{bmatrix} u(3) \\ u(4) \end{bmatrix} = \begin{bmatrix} C \\ CA \end{bmatrix} x(3)$$

Thus,

$$\begin{bmatrix} 6 \\ 4 \end{bmatrix} = \begin{bmatrix} 1 & 1 \\ 0 & 1 \end{bmatrix} x(3)$$

so that $x(3) = [2 \ 4]^{t}$.

In order to compute $x(2)$, carefully consider the pair of equations

$$x(3) = Ax(2) + Bu(2) \tag{21.46a}$$

and

$$x(2) = Ax(1) + Bu(1) \tag{21.46b}$$

First, observe that $Ax(1)$ in equation 21.46b lies in the image of A, with no component arising from $N(A)$. Also, if $x(2)$ has any component that is not in the image of A, then only $Bu(1)$ contributes to that component. Second, rewrite equation 21.46a as

$$\begin{bmatrix} 0 & 0 \\ 0 & 1 \end{bmatrix} x(2) = x(3) - \begin{bmatrix} 2 \\ 1 \end{bmatrix} u(2) = \begin{bmatrix} 0 \\ 3 \end{bmatrix}$$

A least squares solution (use, for example, the Moore-Penrose pseudo inverse described in chapter 19) yields $\hat{x}(2) = [0 \ 3]^{t}$. This solution is orthogonal to the null space of A. But as was just mentioned, any component of $x(2)$ in the null space of A must arise from $Bu(1) = [2 \ 1]^{t}$. Now, the projection of $Bu(1)$ onto $N[A]$ is $[2 \ 0]^{t}$. Hence,

$$x(2) = \begin{bmatrix} 0 \\ 3 \end{bmatrix} + \begin{bmatrix} 2 \\ 0 \end{bmatrix} = \begin{bmatrix} 2 \\ 3 \end{bmatrix}$$

A similar argument leads to the reconstruction of $x(1)$ $[2 \ 2]^{t}$. Unfortunately, $x(0)$ is the system initial condition, and there is no information as to whether or not it has a component in the direction of $N[A]$. Hence, $x(0)$ is not reconstructable.

The above example suggests a slight enhancement of our understanding of reconstructability of an initial state $x(k)$ when A is singular. Suppose for example that $x(k + n)$ is known. Then

$$x(k+n) - [B\ AB\ \cdots\ A^{n-1}B]\begin{bmatrix} u(k+n-1) \\ u(k+n-2) \\ \vdots \\ u(k) \end{bmatrix} = A^n x(k)$$

A least squares solution computed via the singular value decomposition of A^n would yield a vector $x_1(k)$ having no projection onto $N[A^n]$. Since

$$x(k) = A^n x(k-n) + Q\begin{bmatrix} u(k-1) \\ \vdots \\ u(k-n) \end{bmatrix} \triangleq A^n x(k-n) + QU(k)$$

where Q is the usual controllability matrix, only $QU(k)$ can contribute a component in $N[A^n]$ to $x(k)$. Hence

$$x(k) = x_1(k) + x_2(k)$$

where $x_2(k)$ is the projection of $QU(k)$ onto $N[A^n]$. Extending this analysis to output measurements leads to the conclusion that a state $x(k)$ is reconstructable from output measurements over $[k, k + n - 1]$ and input measurements over $[k - n, k + n - 1]$ provided that $N[A^n] \subset N[R]$ where R is the observability matrix.

REALIZATION THEORY

Realization theory grows out of a need to fabricate useful time-domain system models from differential equation representations, frequency response, and/or impulse response measurements. At this juncture, the reader may remember various references to such ideas. For example, in chapter 4, various canonical state models were derived in terms of the parameters of scalar nth-order differential equations. The responses of the differential equation and the state model coincided for zero initial conditions. These notions lead to the concept of the zero-state equivalence of state models whose state vectors are related by a state transformation. Here, identical inputs lead to identical outputs for relaxed initial state vectors. Since differential equations have a one-to-one relationship with transfer functions, the ideas concerning the realization of differential equations would naturally carry over to the realization of transfer functions (transfer function matrices). The difficulty - and the advantage, of course is — in choosing a state model representation from among the infinite variety of equivalent ones.

Several other points pertinent to realization theory occur in the text. For example, problem 23 of chapter 18 asks for the proof that two state models with the property that $CA^iB = \tilde{C}\tilde{A}^i\tilde{B}$ for all $i \geq 0$ are zero-state equivalent. Thus, if two state models have impulse responses whose Taylor series expansions have matching terms, then the responses coincide and the models must be zero-state equivalent. Another interesting property arose in chapter 20 in the context of the Kalman controllable form. Using this form, one concludes that the transfer function matrix of a state model embodies at most the controllable portion of the state model (see also problem 20 of chapter 18). This thread lengthened in chapter 21 relative to the Kalman canonical form, in which one can show that the transfer function matrix of a system consists only of the controllable and observable poles— i.e., the poles of the transfer function matrix are the eigenvalues of $\tilde{A}_{22}$ in equation 21.41.

These properties further complicate the process of constructing an (A,B,C,D) state model from either $H(t)$ or $H(s)$. Specifically, since $H(t)$ and $H(s)$ are input-output descriptions, the choice and number of state variables are not apparent. Often the choice is determined by the physical system being modeled. Coordinate transformations from these physical variables can often make certain numerical calculations easier or certain physical properties such as controllability and observability more transparent. However, the transformation distorts the physical meaning of the plant state variables, which typically have an intuitive behavior understood by the design engineer. The choice and number of state variables from $H(t)$ and $H(s)$, however, remain unclear. This requires a broadening of the notion of zero-state equivalence defined earlier in the text.

Although perhaps not patently clear, zero-state equivalence does not presume that state models have identical dimensional state spaces.

Definition 21.5.   Two state models (A,B,C,D) and $(\tilde{A},\tilde{B},\tilde{C},\tilde{D})$, where A is $n \times n$, B is $n \times m$, C is $r \times n$, D is $r \times m$, $\tilde{A}$ is $\tilde{n} \times \tilde{n}$, $\tilde{B}$ is $\tilde{n} \times \tilde{m}$, $\tilde{C}$ is $\tilde{r} \times \tilde{n}$, and $\tilde{D}$ is $\tilde{r} \times \tilde{m}$, are *zero-state equivalent* if (*i*) $r = \tilde{r}$, (*ii*) $m = \tilde{m}$, and (*iii*) for any given input, the zero-state responses of the two models coincide.

Since, according to this definition, zero-state equivalence does not presume identical state spaces, it is necessary to sharpen our usual notion of equivalent state models. State models which are zero-state equivalent and have the same state spaces are said to be *algebraically equivalent* [15]. This is also true of time-varying systems and nonsingular time-varying state transformations $T(t)z(t) = x(t)$. Clearly, then, algebraic equivalence implies zero-state equivalence, but the converse is not true in general.

Proposition 21.4.   Let two state models (A,B,C,D) and $(\tilde{A},\tilde{B},\tilde{C},\tilde{D})$ be algebraically equivalent—i.e., there exists a nonsingular T such that $Tx(t) = \tilde{x}(t)$. Then the associated state transition matrices are related according to $\Phi(t) = T^{-1}\tilde{\Phi}(t)T$.

Proof.   Clearly, the left- and right-hand sides of the latter equation equal the identity matrix at $t = 0$. Further, since $\Phi(t)$ and $\tilde{\Phi}(t)$ are fundamental matrices, $\dot{\Phi}(t) = A\,\Phi(t)$ and $\dot{\tilde{\Phi}}(t) = \tilde{A}\,\tilde{\Phi}(t)$. Now, observe that

$$T^{-1}\dot{\tilde{\Phi}}(t)T = T^{-1}\tilde{A}\,\tilde{\Phi}(t)T = (T^{-1}\tilde{A}T)\,(T^{-1}\tilde{\Phi}(t)T) = AT^{-1}\tilde{\Phi}(t)T$$

Therefore, the left- and right-hand sides satisfy the same differential equation. So by the uniqueness theorem (theorem 10.4), they are equal.   ∎

Corollary 21.4. The impulse response matrices of the state models of proposition 21.4 are equal, i.e. $H(t) = \tilde{H}(t)$.

Proof. The proof follows directly from the relation between the state transition matrices given in proposition 21.4. ∎

Utilizing the preceding basic relationships, it is possible to precisely define the notion of a realization.

Definition 21.6. A *realization* of a time-invariant impulse response matrix $H(t)$ (transfer function matrix $H(s)$) is any state model (A,B,C,D), whose impulse response matrix is $H(t) = C\exp(At)B + D\delta(t)$ (whose transfer function matrix is $H(s) = C(sI - A)^{-1}B + D)$.
The following definition categorizes realizations.

Definition 21.7. A *controllable* (*observable*) *realization* is a realization (A,B,C,D) that is controllable (observable). A realization is *complete* if (A,B) is controllable and (C,A) is observable.

Efficiency or economy of variables often shapes the kind of realization one constructs. To measure such desirable properties, one defines the notion of a minimal realization.

Definition 21.8. A *minimal realization* of an impulse response matrix $H(t)$ (transfer function matrix $H(s)$) is a realization whose state space is of minimal dimension among all possible realizations.
The following example puts these definitions into a more meaningful perspective.

EXAMPLE 21.8

The transfer function $H(s) = (s + 1)/[(s + 1)(s + 3)] = 1/(s + 3)$ helps illustrate the ambiguity in a state model realization. Specifically, each of the following state models (A,B,C,D) realizes $H(s)$ with different controllability and observability properties as well as different dimensional state spaces:

(i) A controllable canonical realization which is not completely observable is

$$A = \begin{bmatrix} 0 & 1 \\ -3 & -4 \end{bmatrix}, \quad B = \begin{bmatrix} 0 \\ 1 \end{bmatrix}, \quad C = [1 \ 1]$$

(ii) An observable canonical realization which is not completely controllable is

$$A = \begin{bmatrix} 0 & -3 \\ 1 & -4 \end{bmatrix}, \quad B = \begin{bmatrix} 1 \\ 1 \end{bmatrix}, \quad C = [0 \ 1]$$

(iii) A diagonal realization which is neither controllable nor observable is

$$A = \begin{bmatrix} -1 & 0 \\ 0 & -3 \end{bmatrix}, \ B = \begin{bmatrix} 0 \\ 1 \end{bmatrix}, \ C = [0 \ 1]$$

(iv) A minimal realization which is both controllable and observable is

$$A = [-3], \ B = [1], \ C = [1]$$

(v) A three-dimensional diagonal realization which is observable but not controllable is

$$A = \begin{bmatrix} -1 & 0 & 0 \\ 0 & -3 & 0 \\ 0 & 0 & -6 \end{bmatrix}, \ B = \begin{bmatrix} 0 \\ 1 \\ 0 \end{bmatrix}, \ C = [1 \ 1 \ 1]$$

In the example, the controllable and observable realization (*iv*) has a minimal-dimensional state space (one dimension). The following theorem verifies the generality of this property.

Theorem 21.8. The realization (A,B,C,D) is a minimal realization if and only if it is complete.

Proof. Suppose that (A,B,C,D) is a minimal realization, but that (A,B) is not controllable and (C,A) not observable. Then, using the Kalman canonical form (equation 21.41) of (A,B,C), it is possible to extract the complete (controllable and observable) triple $(\tilde{A}_{22},\tilde{B}_2,\tilde{C}_2)$ which is zero-state equivalent to (A,B,C) but has a strictly lower dimensional state space. But this contradicts the minimality of (A,B,C,D). Hence, (A,B,C,D) must be complete.

Now suppose (A,B,C,D) is complete but not minimal. Since completeness depends only on (A,B,C), without loss of generality set $D = [0]$.

Step 1. In this step, we generate an equivalence needed for the next step. Note that the impulse response is $H(t) = C\exp(At)B$. If (A,B,C) is not minimal, there exists a zero-state equivalent realization $(\tilde{A},\tilde{B},\tilde{C})$ with $\tilde{A} \ \tilde{n} \times \tilde{n}$ and $\tilde{n} < n$ such that

$$C\exp(At)B = \tilde{C}\exp(\tilde{A}t)\tilde{B}$$

Expanding the exponentials into their Taylor series and equating terms with like powers of t yields

$$CA^kB = \tilde{C}\tilde{A}^k\tilde{B}$$

for all $k \geqslant 0$.

Step 2. Now consider the product RQ of the observability matrix R and the controllability matrix Q. Each entry of RQ has the form CA^kB for some $0 \leqslant k \leqslant 2(n-1)$. Using the foregoing equivalence, replace each of these terms by $\tilde{C}\tilde{A}^k\tilde{B}$, in which case $RQ = \tilde{R}\tilde{Q}$. Now, since rank(R) = rank(Q) = n, rank(RQ) = n. On the other hand, rank$(\tilde{R}\tilde{Q})$ = $\tilde{n} < n$, a contradiction. Hence, if (A,B,C,D) is complete, it is minimal.

Consequently, (A,B,C,D) is minimal if and only if it is complete.   ∎

Corollary 21.8a. $\text{Rank}(RQ) = n$ if and only if (A,B,C,D) is complete.

Proof.   Using   Sylvester's   inequality   [15,16], $n = \text{rank}(RQ) \leqslant$ $\min\{\text{rank}(R),\text{rank}(Q)\}$. That is, the ranks of both R and Q have n as an upper bound. But then, $\text{rank}[R] = \text{rank}[Q] = n$, and (A,B,C,D) is both observable and controllable.   ∎

Corollary 21.8b.   Two realizations have the same impulse response matrix if and only if (i) $CA^kB = \tilde{C}\tilde{A}^k\tilde{B}$, $k = 0,1,2,\ldots$, and (ii) $D = \tilde{D}$.

The following theorem ties various minimal realizations together through an intuitively expected nonsingular state transformation.

Theorem 21.9.   Let (A,B,C,D) be a minimal realization and $(\tilde{A},\tilde{B},\tilde{C},\tilde{D})$ be another realization of the $r \times m$ impulse response matrix $H(t)$. Then $(\tilde{A},\tilde{B},\tilde{C},\tilde{D})$ is minimal if and only if there exists a nonsingular (square) coordinate transformation T such that $Tx = \tilde{x}$. Furthermore,

$$T = (\tilde{R}^t\tilde{R})^{-1}\tilde{R}^tR$$

and

$$T^{-1} = Q\tilde{Q}^t(\tilde{Q}\tilde{Q}^t)^{-1}$$

Proof.   *Part 1.*   Suppose there exists a nonsingular square T such that $\tilde{x} = Tx$. Then $\text{dimension}(\tilde{x}) = \text{dimension}(x) = n$. Thus, since (A,B,C,D) and the minimal $(\tilde{A},\tilde{B},\tilde{C},\tilde{D})$ both realize $H(t)$, $(\tilde{A},\tilde{B},\tilde{C},\tilde{D})$ is also minimal.

Part 2.   Now suppose both (A,B,C,D) and $(\tilde{A},\tilde{B},\tilde{C},\tilde{D})$ are minimal realizations of $H(t)$. We seek to show the existence of a nonsingular state transformation T such that $Tx = \tilde{x}$. The procedure is to show that (i) the matrices $(\tilde{R}^t\tilde{R})$ and $(\tilde{Q}\tilde{Q}^t)$ are nonsingular, (ii) $T_1 = T_2 \triangleq T$, where $T_1 = (\tilde{R}^t\tilde{R})^{-1}\tilde{R}^tR$ and $T_2^{-1} = Q\tilde{Q}^t$ $(\tilde{Q}\tilde{Q}^t)^{-1}$, (iii) $\tilde{B} = TB$ and $\tilde{C} = CT^{-1}$, and (iv) $A = T^{-1}\tilde{A}T$.

Step 1.   The minimality of both realizations leads to the conclusion:

$$\text{rank}(Q) = \text{rank}(R) = \text{rank}(\tilde{Q}) = \text{rank}(\tilde{R}) = n$$

Consequently, $(\tilde{R}^t\tilde{R})$ and $(\tilde{Q}\tilde{Q}^t)$ are nonsingular and thus invertible.

Step 2.   As developed in the proof of theorem 21.8, since (A,B,C,D) and $(\tilde{A},\tilde{B},\tilde{C},\tilde{D})$ are both minimal and realize $H(t)$,

$$RQ = \tilde{R}\tilde{Q} \tag{21.47}$$

Multiplying both sides of equation 21.47 by $\tilde{R}^t$, and then multiplying by $(\tilde{R}^t\tilde{R})^{-1}$, yields

$$\tilde{Q} = (\tilde{R}^t\tilde{R})^{-1}\tilde{R}^tRQ \triangleq T_1Q \tag{21.48}$$

On the other hand, multiplying equation 21.47 on the right by $\tilde{Q}^t$ and then by $(\tilde{Q}\tilde{Q}^t)^{-1}$ produces

$$\tilde{R} = RQ\tilde{Q}^t(\tilde{Q}\tilde{Q}^t)^{-1} = RT_2^{-1} \qquad (21.49)$$

Implicit in equation 21.49 is the invertibility of T_2. Since, from equation 21.47, $RQ = \tilde{R}\tilde{Q}$, it follows that $RQQ^t = \tilde{R}\tilde{Q}Q^t$. But, from Sylvester's inequality [2], $n = \text{rank}(RQQ^t) = (\tilde{R}\tilde{Q}Q^t)$; hence, $\text{rank}(\tilde{Q}Q^t) = n$.

Step 3. From equations 21.48 and 21.49, $RQ = \tilde{R}\tilde{Q} = RT_2^{-1}T_1Q$. Now, since $\text{rank}(R) = n$, R has a left inverse, and since $\text{rank}(Q) = n$, Q has a right inverse in which case $T_2^{-1}T_1 = I$. Thus $T_2 = T_1 = T$.

Step 4. Combining the results of steps 2 and 3, we have

$$\tilde{R} = RT_2^{-1}, \quad \tilde{Q} = T_1Q \qquad (21.50)$$

From the structures of the observability matrices R and $\tilde{R}$ and the controllability matrices Q and $\tilde{Q}$, it follows immediately that $\tilde{B} = TB$ and $\tilde{C} = CT^{-1}$. Thus, T is the state transformation with the desired structure, provided that $TAT^{-1} = \tilde{A}$. We seek to demonstrate this equality in the next step.

Step 5. Once again, from corollary 21.8b, it follows that $CA^kB = \tilde{C}\tilde{A}^k\tilde{B}$. Also, since the matrices have conformable dimensions, $RAQ = \tilde{R}\tilde{A}\tilde{Q}$. So using equation 21.50, we obtain $RAQ = RT^{-1}\tilde{A}TQ$. And again, from the left invertibility of both R and Q, it follows that $A = T^{-1}\tilde{A}T$, or, equivalently, $\tilde{A} = TAT^{-1}$. ∎

The preceding material sets the stage for a delineation of an algorithm to construct a state model (A,B,C,D) from a transfer function matrix $H(s)$. One presumes that after numerous frequency response tests on the physical system of interest, the design engineer has concluded that $H(s)$ models the system. From equation 18.9, if $H(s)$ has a state representation (A,B,C,D), then

$$H(s) = C(sI-A)^{-1}B + D$$

and from equation 18.10,

$$(sI-A)^{-1} = \frac{R(s)}{\pi_A(s)}$$

where the degrees of the polynomial entries of $R(s)$ are strictly less than n and $\pi_A(s)$ is the characteristic polynomial of the A-matrix.

This structure suggests the following assumption.

Assumption 1. Each entry of $H(s)$ is a proper rational function, i.e., the degree of the numerator does not exceed the degree of the denominator.

Under this assumption,

$$H(\infty) \triangleq \lim_{s \to \infty} H(s) = \lim_{s \to \infty} [C(sI-A)^{-1}B + D] = D \qquad (21.51)$$

Thus, realization of the D-matrix arises from a very straightforward computation.

Now, let $\pi(s)$ be a polynomial equation equal to the least common multiple of the denominators of the entries in $H(s)$. Suppose the degree of $\pi(s)$ is n. Let $N(s)$ be that polynomial matrix such that

$$H(s) = \frac{N(s)}{\pi(s)} + H(\infty) \tag{21.52}$$

which is a valid decomposition for every proper rational transfer function matrix. The realization problem reduces to finding A, B, and C such that

$$\frac{N(s)}{\pi(s)} = C(sI - A)^{-1}B$$

From the decomposition of equation 21.52, $N(s)$ has entries whose degree is strictly less than n. Hence, $N(s)/\pi(s)$ has a Laurent expansion that is valid for $|s| > |\lambda_{max}|$, where λ_{max} is equal to largest root of $\pi(s)$. This expansion is

$$\frac{N(s)}{\pi(s)} = \frac{M_1}{s} + \frac{M_2}{s^2} + \frac{M_3}{s^3} + \ldots = \sum_{k=1}^{\infty} \frac{M_k}{s^k} \tag{21.53}$$

The matrices M_k are termed *Markov parameters*. Synthetic division provides one means of computing the series.

The right-hand side of the series representation above, equation 21.53 is a very general system representation which includes rational transfer function matrices as a subclass. Hence, one might suspect the Markov parameters to satisfy some interdependency similar to the dependency A^k has on A^j for $j = 0, \ldots, n$ in accordance with the Caley-Hamilton theorem. Lemma 21.1 sets forth this relationship.

Lemma 21.1. Let the Markov parameters $M_j, j \geqslant 1$, $N(s)$, and $\pi(s)$ be as given in equation 21.53, with the nth order polynomial

$$\pi(s) \triangleq a_n s^n + a_{n-1} s^{n-1} + \ldots + a_0 \tag{21.54}$$

where $a_n \neq 0$. Then the Markov parameters satisfy the recursion

$$M_k = \sum_{j=1}^{n} - \frac{a_{n-j}}{a_n} M_{k-j} \tag{21.55}$$

Proof. Multiplying both sides of equation 21.53 by $\pi(s)$ and grouping like powers of s, one obtains

$$N(s) = \pi(s) \sum_{k=1}^{\infty} M_k s^{-k} = \left[\sum_{j=0}^{n} a_j s^j \right] \left[\sum_{k=1}^{\infty} M_k s^{-k} \right]$$

$$= \sum_{k=1}^{\infty} \left[\sum_{j=0}^{n} a_{n-j} M_{k-j} \right] s^{n-k}$$

where $M_i \triangleq [0]$ for all $i \leqslant 0$. The left-hand side of this equation, $N(s)$, is a matrix whose entries are polynomials, i.e., only positive powers of s are present. The right-hand side contains, in general, both positive and negative powers of s. Equality requires that all coefficients of the negative powers of s be zero, i.e., for $k > n$,

$$\sum_{j=0}^{n} a_{n-j} M_{k-j} = 0$$

This is equivalent to equation 21.55, the desired result.   ∎

Armed with the recursion equation 21.55, define the A-matrix of the desired realization to be a block matrix in controllable canonical form dependent on the coefficients of $\pi(s)$:

$$A = \begin{bmatrix} 0 & I & 0 & & 0 \\ 0 & 0 & I & & 0 \\ & & & & \\ 0 & 0 & 0 & & I \\ -\dfrac{a_0}{a_n} I & -\dfrac{a_1}{a_n} I & -\dfrac{a_2}{a_n} I & & -\dfrac{a_{n-1}}{a_n} I \end{bmatrix} \tag{21.56}$$

Also, we define the set of *Hankel matrices* H_i, dependent on the Markov parameters:

$$H_i = \begin{bmatrix} M_i & M_{i+1} & \dots & M_{i+n-1} \\ M_{i+1} & M_{i+2} & & M_{i+n} \\ \vdots & \vdots & & \vdots \\ M_{i+n-1} & M_{i+n} & \dots & M_{i+2(n-1)} \end{bmatrix} \tag{21.57}$$

We then have the following proposition.

Proposition 21.5. With A given by equation 21.56 and H_i defined by equation 21.57, it follows that

$$AH_i = H_{i+1} \tag{21.58a}$$

and more generally,

$$A^k H_i = H_{i+k} \tag{21.58b}$$

Proof. The proof follows by direct computation coupled with the result of lemma 21.1.   ∎

With D and A defined in equations 21.51 and 21.56, respectively, the following theorem specifies the complete realization of $H(s)$.

Theorem 21.10. If $H(s)$ is a proper rational transfer function matrix, then a realization (A,B,C,D) is given by (*i*) $D = H(\infty)$, (*ii*) A, as specified in equation 21.56, (*iii*) $B = \mathrm{col}(M_1, M_2, \dots, M_n)$, and (*iv*) $C = [I, 0, \dots, 0]$.

Proof. From the discussion prior to the statement of the theorem,

$$H(s) = H(\infty) + \sum_{k=1}^{\infty} M_k s^{-k} \tag{21.59}$$

On the other side of the coin, any realization (A,B,C,D) must have a transfer function matrix

$$C(sI - A)^{-1}B + D = D + \sum_{k=1}^{\infty} CA^{k-1}Bs^{-k} \qquad (21.60)$$

which is valid for $|s| > |\lambda_{max}|$, where λ_{max} is that eigenvalue of A having the largest magnitude.

Comparing the coefficients of equations 21.59 and 21.60 at the zeroth power of s yields $H(\infty) = D$. It remains to show that $CA^{k-1}B = M_k$ for all $k \geqslant 1$. Accordingly, let M_k be $r \times m$ and define a matrix $P = \text{col}[I,0,0,...,0]$, where I is an $m \times m$ identity and the zeros represent $m \times m$ zero blocks. Observe that, as defined by the theorem,

$$B = H_1P$$

With $C = [I,0,...,0]$, consider the terms $CA^{k-1}B$. We obtain

$$CA^{k-1}B = CA^{k-1}H_1P = CH_kP = M_k$$

where $A^{k-1}H_1 = H_k$ results from proposition 21.5 and equation 21.58b. But this is what was left to be proven.   ∎

EXAMPLE 21.9

Compute a realization (A,B,C,D) of

$$H(s) = \frac{1}{2s^2 + 2s + 1} = \frac{N(s)}{\pi(s)} + H(\infty)$$

Using synthetic division,

$$H(s) = \frac{0.5}{s^2} - \frac{0.5}{s^3} + \frac{0.25}{s^4} + ...$$

Since the degree of $\pi(s)$ is 2, only the first two Markov parameters are necessary for the realization. In particular, $D = H(\infty) = [0]$, $B = \text{col}[M_1,M_2] = [0 \ 0.5]^t$, $C = [1 \ 0]$, and

$$A = \begin{bmatrix} 0 & 1 \\ -0.5 & -1 \end{bmatrix}$$

Of course, the Hankel matrix method produces only one of any number of realizations. Moreover, the realization it does produce is not necessarily of minimal dimension. However, using the transformation to the Kalman canonical form, one can always extract a minimal realization.

An alternative realization can be developed as follows. Consider again equation 21.52, which has the expansion

$$H(s) = H(\infty) + \frac{N(s)}{\pi(s)} = H(\infty) + \frac{N_1s^{n-1} + N_2s^{n-2} + ... + N_n}{a_ns^n + a_{n-1}s^{n-2} + ... + a_0} \qquad (21.61)$$

With $D \triangleq H(\infty)$ and A again defined as in equation 21.56, a realization (A,B,C,D) results upon setting

$$C = [N_n \ N_{n-1} \ldots N_1], \ B = \text{col}[0,\ldots,0,I] \qquad (21.62)$$

For other realization techniques, consult [8,9,17].

A few brief statements with regard to the degree of the transfer function matrix ($\deg_{TFM}$) and the degree of the state model realization ($\deg_{SM}$) will conclude this section.

The following are corollaries to theorem 21.10.

Corollary 21.10a.   Let $H(s)$ be a proper rational (i.e., single-input, single output) transfer function whose numerator and denominator polynomials are relatively prime. Then $\deg_{TFM} = \deg_{SM}$ for the realization given by theorem 21.10.

Proof.   The proof is left as exercise 33.   ∎

Corollary 21.10b.   Under the same hypotheses as corollary 21.10a, the realization given by theorem 21.10 is of minimal dimension.

The degree of a multi-input, multi-output transfer function matrix is naturally more complex than that of a single-input, single-output function. The pertinent concept here is the so-called *McMillan degree* [17]. This degree depends on the orders of various minors of the transfer function matrix. The following theorem, stated without proof, links $\deg_{SM}$ and $\deg_{TFM}$.

Theorem 21.11.   If $\deg_{TFM}$ is taken to be the McMillan degree, [17] then if $H(s)$ is a proper rational transfer function matrix, $\deg_{TFM} = \deg_{SM} = \text{rank}[H_1]$, where H_1 is the Hankel matrix defined in equation 21.57.

As a final point, note that all the preceding realization results and properties have their obvious discrete-time counterparts.

PROBLEMS

1. Suppose, for a state model, that

$$A = \begin{bmatrix} 0 & 0 & 0 \\ 0 & -1 & 0 \\ 0 & 0 & -2 \end{bmatrix}$$

and

$$C = \begin{bmatrix} 1 & 0 & 1 \\ 1 & 0 & 0 \end{bmatrix}$$

   (a) Determine the observability matrix R and its rank.
   (b) Determine a basis for the unobservable subspace.
   (c) Determine a basis for the observable subspace.

(d) If the zero-input response $y(t) = Ce^{At}x(0)$ satisfies $y(1) = [5.135 \quad 5]'$, determine a least squares solution for $x(1)$.

(e) Given your answer to (d), characterize the set of all solutions for $x(1)$.

(f) Given your answer to (d), determine $x(0)$.

(g) Compute the zero-input responses to $x(0) = [0 \quad -17 \quad 0]'$ and $x(0) = [1 \quad 0 \quad 1]'$. Discuss your results.

2. Compute the Kalman observable form of the matrices A and C of problem 1. Verify that the resulting pair $(\bar{C}_2, \bar{A}_{22})$ is completely observable (see equations 21.6).

3. Consider again the matrices A and C of problem 1.

(a) Compute the eigenvalues and eigenvectors of A.

(b) Determine which modes are observable and which modes are unobservable.

(c) Verify your answer to (b) by checking against condition (ii) of Theorem 21.2.

4. Suppose, for a state model, that

$$A = \begin{bmatrix} 0 & 0 & 0 \\ 0 & -1 & 0 \\ 0 & 0 & -2 \end{bmatrix}$$

and

$$C = \begin{bmatrix} 1 & 0 & 1 \\ 0 & -1 & 0 \end{bmatrix}$$

(a) Show that the rank of the observability matrix is 3.

(b) Verify (a) using conditions (ii) and (iv) of Theorem 21.2.

(c) Compute the state transition matrix $\Phi(t) = e^{At}$. Now verify that the observability Gramian of equation 21.7 is nonsingular over $[0,1]$.

(d) If $y(t) = [5 + e^{-2t} \quad -3e^{-t}]'$, use equation 21.8 to determine $x(1)$.

(e) Using your answer to (d), determine $x(t)$ over $[0,1]$.

5. For the state model

$$\begin{bmatrix} \dot{x}_1 \\ \dot{x}_2 \end{bmatrix} = \begin{bmatrix} 0 & 0 \\ 1 & 0 \end{bmatrix} \begin{bmatrix} x_1 \\ x_2 \end{bmatrix}$$

$$y = [1 \quad 1] \begin{bmatrix} x_1 \\ x_2 \end{bmatrix}$$

(a) Show that the observability matrix has rank 2.

(b) Compute the state transition matrix $\Phi(t) = \exp(At)$, and show that

$$W_o^{-1}(0,1) = \begin{bmatrix} 12 & -18 \\ -18 & 28 \end{bmatrix}$$

(c) Show that if $y(t) = 1^+(t), 0 \leqslant t \leqslant 1$, then $x(0) = [0 \quad 1]'$.

6. For the matrix A of problem 1, find a linear combination of A^k, $k = 0,1,2$, which equals A^3.

7. *True or False.* Let (A,B,C) specify the usual state model equations, with B and C both nonzero matrices. If (A,B) is a controllable pair, then there exists a state feedback F such that $(C,A + BF)$ is observable. If the statement is false, give a counterexample. If it is true, provide a proof.

8. Suppose a state model has matrices

$$A = \begin{bmatrix} 0 & 0 & 0 \\ 1 & -1 & -1 \\ 0 & 0 & 0 \end{bmatrix}, \quad B = C^t = \begin{bmatrix} 1 & 0 \\ 1 & 0 \\ 0 & 1 \end{bmatrix}$$

(a) Construct a dynamic identity observer whose error dynamics has all eigenvalues at -7. (*Hint:* Verify that

$$K = \begin{bmatrix} 24.5 & 0 \\ -11.5 & 0 \\ 0 & 7 \end{bmatrix}$$

does the job. Can you determine how this feedback was computed?)

(b) Use the methods of chapter 18 to determine an expression for the error. If the initial error is $e(0) = [10 \quad -10 \quad 20]^t$, determine the error after 3 seconds.

9. Consider the system

$$\dot{x} = \begin{bmatrix} -1 & 0 \\ 0 & 0 \end{bmatrix} x + \begin{bmatrix} 1 \\ 1 \end{bmatrix} u$$
$$y = [1 \quad -1] x$$

(a) Determine the structure of a dynamic (identity) observer for this system.

(b) Design a dynamic observer with eigenvalues at -4 and -5 for the system.

(c) Consider the observer error dynamics given by $A - KC$ with K computed in part (b). Compute the associated state transition matrix, $\exp[(A - KC)t]$, by the eigenvalue-eigenvector method.

(d) Suppose $u(t) = 0$ and $y(t) = 3 e^{-t}$ for $t \geqslant 0$. Use your dynamic observer to generate an estimate, denoted $\hat{x}(2)$, of $x(2)$, assuming that $\hat{x}(0) = [0 \quad 0]^t$, the initial condition on the observer.

(e) Repeat the preceding for the design of a minimal-order dynamic observer.

10. Suppose a state model of a plant has the structure

$$\dot{x} = \begin{bmatrix} 0 & 0 & -1 \\ 1 & 0 & -1 \\ 0 & 1 & -1 \end{bmatrix} x + \begin{bmatrix} 0 \\ 0 \\ 0 \end{bmatrix} u, \quad x(0) = \begin{bmatrix} 1 \\ 1 \\ 1 \end{bmatrix}$$
$$y = [0 \quad 0 \quad 1] x$$

(a) Find a minimal-order observer whose error dynamics has a double pole at -6.

(b) Verify your structure by computing an expression for the error, $x(t) - \hat{x}(t)$, in terms of $x(0) - \hat{x}(0)$, assuming that $\hat{x}(0) = \theta$.

11. Consider problem 10 again, except suppose that $y = [1 \quad 0 \quad -1]x$. The system is *not* completely observable. Design an observer which reconstructs the observable subspace. The error dynamics should have a double pole at -6.

12. Suppose

$$\dot{x} = \begin{bmatrix} 1 & 1 \\ 0 & 3 \end{bmatrix} x + \begin{bmatrix} b_1 \\ b_2 \end{bmatrix} u$$
$$y = [c_1 \quad c_2] x$$

(a) What conditions on b_1 and b_2 make the system controllable?

(b) What conditions on c_1 and c_2 make the system observable?

(c) Suppose $[b_1 \quad b_2] = [0 \quad 1]$ and $[c_1 \quad c_2] = [0 \quad 1]$. Find a state feedback which makes the system completely observable.

13. Consider the time-varying system

$$\dot{x} = A(t)x + Bu, \qquad x(0) = \begin{bmatrix} 1 \\ 0 \\ 0 \end{bmatrix}$$
$$y = Cx$$

where

$$A(t) = \begin{bmatrix} 0 & 0 & 0 \\ 1 & 0 & 0 \\ 1 & 1 & 0 \end{bmatrix} \text{ for } 0 \leqslant t < 2, \quad A(t) = \begin{bmatrix} -1.5 & 0.5 & 0 \\ 0.5 & -1.5 & 0 \\ 0 & 0 & 1 \end{bmatrix} \text{ for } 2 \leqslant t,$$

$$B = \begin{bmatrix} 0 \\ 1 \\ 1 \end{bmatrix}, \quad C = [0 \quad 0 \quad 1]$$

(a) Discuss the observability of this system over the intervals
  (i) [0, 2)
  (ii) [2, ∞)
  (iii) [0, ∞)
(b) Suppose $u(t)$ and $y(t)$ are known over the interval [0, 10]. Is it possible to determine $x(9)$? If so, describe an analytic procedure for computing $x(9)$.

14. Answer the following questions about the following $\{A,B,C\}$ state model by *carefully* justifying and/or deriving each answer:

$$A = \begin{bmatrix} 0 & -1 \\ -1 & 0 \end{bmatrix}, \quad B = \begin{bmatrix} 1 \\ -1 \end{bmatrix}, \quad C = [1 \quad 1]$$

(a) What are the eigenvalues of A?
(b) What are the eigenvectors of A?
(c) Decompose e^{At} as

$$e^{At} = R_1 e^{\lambda_1 t} + R_2 e^{\lambda_2 t}$$

(d) Is the system controllable? If not, find a basis for the controllable subspace.
(e) Is the system observable? If not, find a basis for the observable and unobservable subspaces.
(f) Do the controllable and observable subspaces overlap? Explain.
(g) Given that the initial condition $x(0) = x_0 = [0 \quad 0]'$, find an input, constant over the time interval [0, 1], which will drive the system from the zero state to $x(1) = [2 \quad -2]'$.
(h) Given that at $t = 1$ $y(1) = 3$, find an initial condition $x(0)$ (which is in the *observable subspace*) such that the zero-input response produces $y(1) = 3$.
(i) What is the transfer function of the system?

15. (Multiple choice) Consider the continuous-time state model

$$\dot{x} = Ax + Bu$$
$$y = Cx$$

where

$$A = \begin{bmatrix} -1 & 0 & 0 \\ -1 & -4 & 4 \\ -1 & -2 & 2 \end{bmatrix}, \quad B = \begin{bmatrix} 1 & 0 \\ 1 & 1 \\ 1 & 1 \end{bmatrix}, \quad C = [1 \quad 1 \quad -1]$$

(i) What is the sum of the ranks of the controllability and observability matrices? That is, what does rank$[Q]$ + rank$[R]$ equal?
(a) 2   (b) 3   (c) 4   (d) 5   (e) 6
(f) cannot be determined   (g) none of the above

(ii) Which state(s) are *not* controllable?
(a) $[0 \ 1 \ -1]^t$ (b) $[1 \ 0 \ 0]^t$ (c) $[-1 \ -1 \ -1]^t$
(d) $[2 \ 1 \ 1]^t$ (e) two of the above   (f) none of the above

(iii) Which state(s) are *not* observable?
(a) $[1 \ 2 \ 0]^t$ (b) $[0 \ 1 \ 1]^t$ (c) $[1 \ 1 \ -1]^t$
(d) $[0 \ 2 \ 6]^t$ (e) two of the above   (f) none of the above

(iv) The zero-input system response is identically zero for which initial state(s)?
(a) $[1 \ 2 \ 0]^t$ (b) $[0 \ 1 \ 1]^t$ (c) $[1 \ 1 \ -1]^t$
(d) $[0 \ 2 \ 6]^t$ (e) two of the above   (f) none of the above

(v) A basis for the controllable subspace is

(a) $\begin{bmatrix} 1 & 0 & 0 \\ 0 & 1 & 0 \\ 0 & 0 & 1 \end{bmatrix}$ (b) $\begin{bmatrix} 1 & 0 \\ 0 & 1 \\ 1 & 0 \end{bmatrix}$ (c) $\begin{bmatrix} 1 \\ 1 \\ 1 \end{bmatrix}$

(d) $\begin{bmatrix} 1 & 0 \\ 0 & 1 \\ 0 & 1 \end{bmatrix}$ (e) $\begin{bmatrix} 0 \\ 1 \\ 1 \end{bmatrix}$ (f) none of the above

16. (Multiple choice.) Consider the system

$$\dot{x}(t) = Ax(t) + B(t)u(t)$$
$$y(t) = C(t)x(t)$$

Suppose the eigenvalue-eigenvector triples of A are (λ_1, e_1, w_1), (λ_2, e_2, w_2), (λ_3, e_3, w_3), and (λ_4, e_4, w_4).

(i) Suppose $C(t) = C_1$ over $[0,1)$ and $C(t) = C_2$ over $[1,\infty)$. Let R_1 be the observability matrix for the pair (C_1, A) and R_2 be that for (C_2, A). If rank$[R_1] = 4$ and rank$[R_2] < 4$, which of the following is (are) true?
(a) If $u(t)$ and $y(t)$ are known over $[1,10]$, $x(2)$ can be computed.
(b) If $u(t)$ and $y(t)$ are known over $[0,9]$, $x(0.5)$ can be computed.
(c) If $u(t)$ and $y(t)$ are known over $[0,10]$, $x(0.5)$ can be computed.
(d) two of the above   (e) all of the above   (f) none of the above.

(ii) Same as (i), except rank$[R_1] < 4$ and rank$[R_2] = 4$. Which of the following (is) are true?
(a) If $u(t)$ and $y(t)$ are known over $[0,9]$, $x(0.5)$ can be computed.
(b) If $u(t)$ and $y(t)$ are known over $[1,10]$, $x(0.5)$ can be computed.
(c) If $u(t)$ and $y(t)$ are known over $[1,10]$, but not over $[0,1)$, $x(1)$ can be computed.
(d) a and b   (e) a and c   (f) b and c
(g) all of the above   (h) none of the above

(iii) Suppose that $C(t) = C_1$ over $[0,1)$ and $C(t) = C_2$ over $[1,\infty)$, with R_1 and R_2 the respective observability matrices. Suppose also that (λ_1, e_1) is an uncontrollable mode and $e_1 \in$ Null$[R_1]$ but $e_1 \in$ Null$[R_2]$. If $x(0) = e_1$, which of the following is (are) *not* true?
(a) The zero-input response is identically zero over $[0,1]$.

(b) The step response is identically zero over $[0,1]$.

(c) The zero-input response is identically zero over $[0,2]$.

(d) The step response is identically zero over $[0,2]$.

(e) two of the above   (f) all of above   (g) none of the above

(iv) Suppose $C(t) = C$ is a constant 3×4 matrix for all t. Suppose the rows of C are w_1^*, $Re(w_2^*)$, and $Im(w_2^*)$, respectively. Which modes are observable?

(a) (λ_1, e_1) (b) (λ_2, e_2) (c) (λ_3, e_3) (d) (λ_4, e_4)

(e) two of the above   (f) none of the above

17. Consider the discrete-time system

$$x(k+1) = Ax(k) + Bu(k)$$

$$y(k) = Cx(k)$$

where

$$A = \begin{bmatrix} \lambda_1 & 0 & 0 \\ 0 & \lambda_2 & 0 \\ 0 & 0 & \lambda_3 \end{bmatrix}, \quad B = \begin{bmatrix} b_1 \\ b_2 \\ b_3 \end{bmatrix}, \quad C^t = \begin{bmatrix} c_1 \\ c_2 \\ c_3 \end{bmatrix}$$

(a) If $c_1 = c_2 = c_3$, what condition(s) on the λ_i will guarantee complete observability of the system?

(b) Suppose A has distinct eigenvalues. What conditions on C guarantee the observability of the modes (λ_1, e_1) and (λ_2, e_2) and the nonobservability of (λ_3, e_3)?

(c) If $\lambda_1 = \lambda_2 = \lambda_3$, what conditions on C guarantee the complete observability of the system?

(d) Suppose $[c_1 \ c_2 \ c_3] = [-2 \ 0 \ 0]$. What conditions on the eigenvalues are necessary and sufficient for

$$\lim_{k\to\infty} |y(k)|_2 < \infty$$

(*Note*: The Vandermonde matrix

$$\begin{bmatrix} 1 & a & a^2 \\ 1 & b & b^2 \\ 1 & c & c^2 \end{bmatrix}$$

always has a nonzero determinant, provided that a, b, and c are distinct.)

18. Consider the discrete-time state model

$$\begin{bmatrix} x_1(k+1) \\ x_2(k+1) \end{bmatrix} = \begin{bmatrix} 0.5 & 1 \\ 0 & -1 \end{bmatrix} \begin{bmatrix} x_1(k) \\ x_2(k) \end{bmatrix} + \begin{bmatrix} b_1 \\ b_2 \end{bmatrix} u(k)$$

$$y(k) = [c_1 \ c_2] \begin{bmatrix} x_1(k) \\ x_2(k) \end{bmatrix}$$

The eigenvalue-eigenvector pairs of the A-matrix of the system are

$$(\lambda_1, e_1) = (0.5, [1 \ 0]^t)$$

and

$$(\lambda_2, e_2) = (-1, [2 \ -3]^t)$$

Fill in the blank in each of the following:

(a) The directed mode (λ_1, e_1) is *not* controllable, but (λ_2, e_2) is controllable, when $[b_1 \ b_2]^t =$ _____.

(b) The mode (λ_2, e_2) is *not* observable, but the mode (λ_1, e_1) is observable, when $[c_1 \ c_2] =$ _____.

(c) If $[b_1 \ b_2]^t = [0 \ 1]^t$ for $k = 0,1$, and $[b_1 \ b_2] = [1 \ 0]^t$ for $k \geqslant 2$, an input sequence which will drive $x(0) = \theta$ to $x(3) = [1 \ 1]^t$ is $\{u(0), u(1), u(2)\} =$ _____.

(d) If the input sequence $u(k) = 0$ for all k, if $[c_1 \ c_2] = [0 \ 1]$ for $k = 0,1,2$ and $[c_1 \ c_2] = [2 \ 0]$ for $k \geqslant 3$, and if $y(1) = -1$, $y(2) = 1$, $y(3) = 2$, and $y(4) = -1$, then $x(2) =$ _____.

19. Show that the pair $(\tilde{A}_{21}, \tilde{A}_{11})$ from equations 21.20 and 21.24 is observable, where A_{21} plays the role of the C-matrix and $\tilde{A}_{11}$ the role of the A-matrix of the system. (*Hint:* From the construction of equations 21.20, the observability of (C,A) implies the observability of 21.20b with output $y(t)$. Verify this—i.e., verify the observability of the pair

$$\left([0 \ I], \begin{bmatrix} \tilde{A}_{11} & \tilde{A}_{12} \\ \tilde{A}_{21} & \tilde{A}_{22} \end{bmatrix} \right)$$

and then prove that the pair $(\tilde{A}_{21}, \tilde{A}_{22})$ is observable by proving the dual result, i.e., by showing that $(\tilde{A}_{22}^t, \tilde{A}_{21}^t)$ is controllable.

20. Prove theorem 21.8.

21. Consider the discrete-time system

$$x(k+1) = \begin{bmatrix} 1 & 1 \\ 0 & -1 \end{bmatrix} x(k) + \begin{bmatrix} 0 \\ 1 \end{bmatrix} u(k)$$

$$y(k) = [1 \ 1] x(k) + [1] u(k)$$

Construct a discrete-time dynamic observer for this system whose error dynamics approach zero with convergence rate $(0.1)^k$.

22. Consider the discrete-time state model

$$x(k+1) = \begin{bmatrix} 1 & 1 \\ 0 & 0 \end{bmatrix} x(k) + \begin{bmatrix} 0 \\ 1 \end{bmatrix} u(k)$$

$$y(k) = \begin{cases} [0 \ 1] x(k), & 0 \leqslant k \leqslant 2 \\ [1 \ 0] x(k), & 3 \leqslant k \end{cases}$$

Let $u(k) = (-1)^k 1^+(k)$, and suppose the following measurements are taken:

k	0	1	2	3	4
$y(k)$	2	1	-1	1	2

(a) Determine $x(3)$.

(b) If possible, determine $x(2)$, $x(1)$, and $x(0)$.

23. Consider the Kalman canonical form of equation 21.41.

(a) Show that the transfer function of this system is $H(s) = \tilde{C}_2(sI - \tilde{A}_{22})^{-1} \tilde{B}_2$.

(b) Determine the form of the impulse response matrix.

(c) Given this decomposition, are there any advantages to using the state model over input-output models such as the transfer function matrix or the impulse response matrix?

24. Show that the impulse responses $H(t)$ and $\tilde{H}(t)$ of two time-invariant state models coincide if and only if the models are zero-state equivalent.

25. Let two state models $(A(t),B(t),C(t),D(t))$ and $(\hat{A}(t),\hat{B}(t),\hat{C}(t),\hat{D}(t))$ be algebraically equivalent; i.e., there exists a nonsingular state transformation $T(t)\hat{x}(t) = x(t)$ over the interval $[0,\infty)$.

     (a) Determine the relationship between the state model matrices in terms of the coordinate transformation $T(t)$.

     (b) Show that the associated state transition matrices are related according to $\Phi(t,\tau) = T^{-1}(t)\Phi(t,\tau)T(t)$. (*Hint:* Follow the pattern of the proof of proposition 21.4.)

     (c) Show that the associated impulse responses are identical.

26. Show that the impulse response matrix $H(t,\tau)$ of a linear, lumped, causal, time-varying system is realizable by a state model $\{A(t),B(t),C(t),[0]\}$ if and only if there exists matrices $G(t)$ and $P(t)$ such that $H(t,\tau) = P(t)G(\tau)$. (*Hint:* First suppose that there exists a desired realization; then write the impulse response in terms of the state transition matrix and express the state transition matrix in terms of a product of fundamental matrices. For the converse, let $A(t) = [0]$, define $B(t)$ and $C(t)$ appropriately, and draw a block diagram of the realization.)

27. Find five different two-dimensional realizations of the transfer function $H(s) = (s+3)/[(s+3)(s+2)]$. Identify the controllability and observability properties of each.

28. Find five different three-dimensional realizations of the impulse response matrix $H(t) = -[2\exp(-t) + 0.5\exp(-1.5t)]1^+(t)$.

29. Show that (A,B,C,D) is a minimal realization if and only if $\text{rank}[QR] = n$.

30. Let (A,B,C) be a single-input, single-output time-invariant state model. Show that $H(s) = C(sI-A)^{-1}B = n(s)/d(s)$, where $d(s) = (sI-A)^{-1}$ is minimal if and only if $n(s)$ and $d(s)$ have no common factors. (*Hint:* Use the method of contradiction for each direction.)

31. Prove that the decomposition given in equation 21.61 leads to a realization (A,B,C,D), where A is given by equation 21.56, $D = H(\infty)$, and B and C are given by equation 21.62.

32. Find realizations (A,B,C,D) for the following transfer functions:

     (a)
$$H(s) = \frac{s^3}{s^3 + 2s^2 - s + 2}$$

     (b)
$$\frac{\begin{bmatrix} 2s^2 + 5s + 4 & -(s+2) \\ s+2 & s^2 + 3s + 2 \end{bmatrix}}{(s+2)(s^2 + 2s + 2)}$$

33. Prove corollary 21.10a.

34. Prove that if A is singular such that $N[A] \subset N[R]$, then $x(k)$ is reconstructable from output measurements over $[k,k+n-1]$ and input measurements over $[k-n,k+n-1]$.

35. Prove that $N[R] = N[W_O]$ where W_O is the observability Gramian of equation 21.7.

36. Show that if A is a stability matrix (all eigenvalues in open left half complex plane) then the observability Gramian W_O is the solution to the Lyapunov equation

$$AW_O + A'W_O + C'C = -\frac{\partial}{\partial t}W_O$$

REFERENCES

1. W. M. Wonham, *Linear Multivariable Control: A Geometric Approach* (New York: Springer-Verlag, 1979).

2. George Leitmann, *The Calculus of Variations and Optimal Control: An Introduction* (New York: Plenum Press, 1981).

3. David G. Luenberger, "An Introduction to Observers," *IEEE Transactions on Automatic Control*, Vol. AC-16, No. 6, December 1971, pp. 596–602.

4. R. E. Kalman, "Mathematical Description of Linear Dynamical Systems," *SIAM Journal of Control*, Ser. A, Vol. 1, 1963, pp. 152–192.

5. R. E. Kalman, "On the Computation of the Reachable/Observable Canonical Form," *SIAM Journal of Control and Optimization*, Vol. 20, No. 2, March 1982, pp. 258–260.

6. David Luenberger, *Introduction to Dynamic Systems: Theory, Models, and Applications* (New York: Wiley, 1979).

7. David Russell, *Mathematics of Finite-Dimensional Control Systems: Theory and Design* (New York: Marcel Dekker, 1979).

8. Thomas Kailath, *Linear Systems* (Englewood Cliffs, NJ: Prentice Hall, 1980).

9. W. A. Wolovich, *Linear Multivariable Systems* (New York: Springer-Verlag, 1974).

10. Thomas Harris, Steve Richter, and Raymond DeCarlo, "A Continuation Approach to Eigenvalue Assignment," *Automatica*, Vol. 19, No. 5, 1983, pp. 551–555.

11. Dale Sebok and Raymond DeCarlo, "Feedback Gain Optimization in Decentralized Eigenvalue Assignment," *Automatica*, Vol. 22, No. 4, 1986, pp. 433–447.

12. E. Bruce Lee, Stanislaw Żak, and Stephen Brierley, "Remarks on Kalman's Canonical Form," Proceedings of the Fourth Polish-English Seminar on Real-Time Process Control, Jablonna, Poland, June 1983, pp. 238–249.

13. Daniel L. Boley, "On Kalman's Procedure for the Computation of the Controllable/Observable Canonical Form," *SIAM Journal of Control and Optimization*, Vol. 18, No. 6, November 1980, pp. 624–626.

14. Daniel Boley, "Computing the Kalman Decomposition: An Optimal Method," *IEEE Transactions on Automatic Control*, Vol. AC-29, No. 1, January 1984, pp. 51–53.

15. C. A. Desoer, *Notes for a Second Course in Linear Systems* (New York: Van Nostrand Reinhold, 1970).

16. F. R. Gantmacher, *The Theory of Matrices*, Vol. 1 (New York: Chelsea, 1960).

17. C. T. Chen, *Linear System Theory* (New York: Holt, Rinehart, and Winston, 1984).

18. H. H. Rosenbrock, *State-Space and Multivariable Theory* (New York: Wiley, 1970).

22

STABILITY OF LUMPED
TIME-INVARIANT SYSTEMS

INTRODUCTION

What is stability? What is instability?

Imagine the experience of having the cruise control of your car automatically and continuously accelerate your car without kicking out. The car's speed would climb until an accident occurred or the engine died. This exemplifies instability. By contrast, the cruise control operates stably when it kicks out at the appropriate set-point speed.

From a system-theoretic viewpoint, what is stability/instability?

System stability means that well-behaved inputs produce well-behaved outputs. By contrast, if any well-behaved input produces an undesirable response (monotonically increasing energy, for example) the system is unstable.

To render these concepts concrete, consider Figure 22.1, which depicts a system whose output is always identically zero for arbitrary inputs. Inside the system, however, things are quite different: even for inputs as well behaved as a step function, the energy in $x_1(t)$ and $x_2(t)$ grows arbitrarily large as time moves forward.

Hence, such a system is *internally unstable and externally stable*. This behavior occurs because the system is completely unobservable, i.e., nothing penetrates to the outside world from the inside of the system.

In Figure 22.2 information clearly penetrates to the system output. In particular, the effects of the internal states $x_1(t)$ and $x_2(t)$ appear in the response $y(t)$. If the input is again a simple step function, the state $x_2(t)$ is a ramp function whose energy grows to infinity as time moves to infinity. This scenario illustrates an *internally and externally unstable system*.

Figure 22.3 is the same system as in Figure 22.2, except that the system input is decoupled from the $1/s$ block. Since the input cannot affect this block, the mode it represents is uncontrollable. For any arbitrary initial condition, the response $x_2(t)$ will always be constant and will not grow to infinity. Also, the response $x_1(t)$ will always be well behaved whenever the input is well behaved. Hence, this system is *internally and externally stable*.

Finally, Figure 22.4 represents a device which can predict nonzero input values prior to observing the moment the input turns nonzero. Intuitively speaking, this requires infinite energy and thus the system is unstable. By definition, all non-causal systems are unstable.

These examples bring to mind a number of conjectures: (*i*) internal stability implies external stability; (*ii*) internally unstable modes which are not observable do not make the system response unstable; and (*iii*) if a system has a first-order pole on the imaginary axis, this pole must be uncontrollable for the system to be

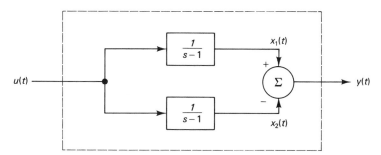

Figure 22.1 An internally unstable, externally stable system.

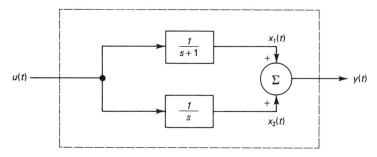

Figure 22.2 An internally and externally unstable system.

internally stable. The verification of these conjectures as well as others pervades the forthcoming development.

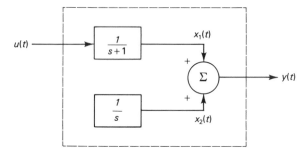

Figure 22.3 An internally and externally stable system.

Figure 22.4 Ideal predictor is noncausal, and hence is an unstable system.

MATHEMATICAL PRELIMINARIES

We begin by developing further the notion of norm introduced in chapter 10 with regard to the existence and uniqueness of solutions to $\dot{x} = f(x,t)$. Essentially, the concept of a norm arises out of the need to measure the sizes of functions in various function classes and thereby to distinguish among them. The norm or size of the difference between two functions provides a handle on the distance between them. Intuitively speaking, the norm of a function measures size relative to a given yardstick. Height, width, and bulk are various measures of object size; by analogy, one expects various measures of function size, such as maximum value and total absolute area. Our purpose here is to explore various notions of norm and to mathematically pin down the ideas of "well behaved" inputs and responses. Of particular interest is the equivalence of "well behaved" with "bounded." This is important because two types of stability discussed in this chapter are *bounded input bounded output* (BIBO) stability and *bounded input bounded state* (BIBS) stability.

The norm of a function $u(\cdot)$, denoted $||u(\cdot)||$, is *always* a nonnegative real number. That is, norms always map functions in a particular space to nonnegative real numbers. (See [1,2] for a treatment of norms; also, [3] treats norms in the context of state space theory.) Throughout the development, all signals are assumed admissible as outlined in chapter 1.

Definition 22.1. An admissible function $u(\cdot):\mathbf{R}\to\mathbf{R}$ has a *sup-norm* or L_∞-*norm* given by

$$||u(\cdot)||_\infty = \sup_t |u(t)|$$

In accordance with this definition, if $u(t) = (1 - e^{-t})1^+(t)$, then $||u(\cdot)||_\infty = 1$. However, if $u(t) = (1 - e^t)$, then, since $|u(t)| \to \infty$ as $t \to +\infty$, the sup-norm technically does not exist. By convention, in this case we denote the sup-norm by $||u(\cdot)||_\infty = \infty$. The extension of the sup-norm to vector-valued functions is straightforward.

Definition 22.2.   Let $u(\cdot):\mathbf{R}\rightarrow\mathbf{R}^m$, i.e., for each t in $\mathbf{R}$, $u(t)\in\mathbf{R}^m$. Then the *sup-norm* or L_∞-*norm* is given by

$$||u(\cdot)||_\infty = \max_i ||u_i(\cdot)||_\infty \qquad (22.1)$$

where $u(\cdot) = \mathrm{col}[u_1(\cdot),...,u_m(\cdot)]$.

Of course, one can also define the sup-norm of a matrix as in the following definition.

Definition 22.3.   Let $H(\cdot) = [h_{ij}(\cdot)]$ be an $n\times m$ matrix whose entries $h_{ij}(\cdot)$ map real numbers into real numbers. Then the *sup-norm* is given by

$$||H(\cdot)||_\infty = \max_i || \sum_{j=1}^m | h_{ij}(\cdot)| ||_\infty$$

The sup-norm of a vector-valued function is the largest sup-norm of its entries. The sup-norm of a matrix is the largest sup-norm of the sup-norms of the sum of the absolute value of the entries in each row. This leads to the following definition of a bounded function.

Definition 22.4.   A (scalar-, vector-, or matrix-valued) function is *bounded* if and only if its L_∞-norm (sup-norm) is finite. Equivalently, the function is bounded if and only if there exists a finite constant K such that

$$||u(\cdot)||_\infty \leqslant K$$

Note that for any function $u(\cdot):D\rightarrow\mathbf{R}$ where D is some subset of $\mathbf{R}$,

$$||u(\cdot)||_{\infty,D} = \sup_{t\in D} |u(t)|$$

and similarly for vector- and matrix-valued functions. This merely indicates that one can restrict the domain of a function and consider its sup-norm over the domain D.

The last notion of a norm germane to the characterizations of stability in this text is the L_1-norm, which measures the total absolute area under a function.

Definition 22.5.   Let $H(\cdot):\mathbf{R}\rightarrow\mathbf{R}$. Then the L_1-*norm* of $H(\cdot)$ is

$$||H(\cdot)||_1 = \int_{-\infty}^{+\infty} |H(q)|\,dq$$

For our purposes, $H(\cdot)$ will usually be matrix-valued—i.e., $H(\cdot) = [h_{ij}(\cdot)]$ is $r\times m$, where $h_{ij}(\cdot):\mathbf{R}\rightarrow\mathbf{R}$. In this case the L_1-norm of $H(\cdot)$ is given by

$$||H(\cdot)||_1 = \max_i || \sum_{j=1}^m | h_{ij}(\cdot)| ||_1 \qquad (22.2)$$

In other words, the L_1-norm of the matrix-valued function $H(\cdot)$ is the largest of the L_1-norms of the sum of the absolute values of the entries of each row.

With the preceding notions of norm, it is possible to characterize input-output stability in terms of the L_1-norm of the impulse response.

STABILITY VIA THE IMPULSE RESPONSE MATRIX

Recall that the time-invariant state model $\{A,B,C,D\}$ has an external, or input-output, convolutional representation

$$y(t) = \int_{-\infty}^{+\infty} H(t-q)u(q)dq \tag{22.3}$$

where the impulse response matrix has the form

$$H(t) = C \exp[At] B \ 1^{+}(t) + D\delta(t) \tag{22.4}$$

Although this relationship is the one that is most germane to our development, the results that follow are applicable to the impulse response of any time-invariant, lumped, causal linear system. Generalizations to the time-varying case are straightforward and follow in a later chapter. Since the impulse response matrix depicts the input-output properties of the system for admissible input-output pairs defined over $(-\infty,\infty)$, no initial conditions are presumed, e.g., at $t = -\infty$. Relative to the state model, this determines a zero-state stability criterion.

The overall objective of the section is to characterize the input-output stability of the system in terms of $H(\cdot)$. By adapting these ideas to the state response, it is possible to characterize the internal stability of a system having a state model.

Definition 22.6.   A system is said to be *BIBO stable* if, for each admissible bounded input $u(\cdot)$ (i.e., for $||u(\cdot)||_\infty < \infty$), the response $y(\cdot)$ is bounded (i.e., $||y(\cdot)||_\infty < \infty$).

Of course, if a bounded input were to give rise to an unbounded response, then the energy in the response signal, on a per-unit basis, would eventually exceed and continue at a rate of increase beyond the energy of the input signal. Such a scenario indicates an infinite gain and, hence, system *instability*.

The following theorem characterizes BIBO stability in terms of the L_1-norm of the impulse response.

Theorem 22.1.   A causal system modeled by the convolution equation 22.3 is BIBO stable if and only if there exists a finite constant K such that

$$||H(\cdot)||_1 \leqslant K \tag{22.5}$$

Proof.   The proof will consider only the single-input single-output case. The multi-input, multi-output case is left as an exercise (with hints) at the end of the chapter.

Part 1. Suppose that the system is causal and that equation 22.5 is true. The goal is to show that the system is BIBO stable, according to definition 22.6. Let

$$K = ||H(\cdot)||_1 \triangleq \int_{-\infty}^{+\infty} |H(q)| \, dq = \int_0^{\infty} |H(q)| \, dq < \infty$$

Then the sup-norm of the response $y(\cdot)$ of the system is, as per equation 22.3;

$$||y(\cdot)||_\infty = \sup_t \left| \int_{-\infty}^t H(t-q)u(q)dq \right| \leqslant \sup_t \int_0^{\infty} |H(q)| \, |u(t-q)| \, dq$$

where the magnitudes have been brought inside the integral and a change of variable has been executed. Now, since the "sup" is taken over t, it can be placed inside the integral to yield

$$||y(\cdot)||_\infty \leqslant \int_0^{\infty} |H(q)| \left[\sup_t |u(t-q)| \right] dq \leqslant ||u(\cdot)||_\infty \int_0^{\infty} |H(q)| \, dq$$

$$\leqslant K ||u(\cdot)||_\infty$$

Therefore, if $||u(\cdot)||_\infty$ is finite (tantamount to well behaved), then $||y(\cdot)||_\infty$ is finite and the system is BIBO stable.

Part 2 [4]. Our method of proof is by contradiction: suppose that the system is causal and that bounded inputs map to bounded outputs, but that the L_1-norm of $H(\cdot)$ is unbounded—i.e., for each finite constant K, there exists a time t_K (the particular time depends on the value of K) such that

$$\int_{-\infty}^{t_K} |H(q)| \, dq > K \tag{22.6}$$

By properly defining a bounded convergent sequence of inputs, say, $\{u_k(\cdot)\}_{k=0}^{\infty}$ in conjunction with equation 22.6, it is possible to construct a convergent output sequence, $\{y_k(\cdot)\}_{k=0}^{\infty}$ whose L_∞ norms blow up—i.e., a bounded input leads to an unbounded response.

Step 1. From equation 22.6 and the causality of the system, there exists a t_1 such that

$$\int_0^{t_1} |H(q)| \, dq > 1$$

Define the bounded input

$$u_1(t) = \begin{cases} \text{sgn}[H(t_1-t)], & 0 \leqslant t \leqslant t_1 \\ 0, & \text{otherwise} \end{cases}$$

where

$$\text{sgn}(\beta) = \begin{cases} \dfrac{\beta}{|\beta|}, & \beta \neq 0 \\ 0, & \beta = 0 \end{cases} \tag{22.7}$$

(If β is complex, a slight modification is needed; see the remarks at the end of the proof.) Thus, the response to $u_1(\cdot)$ at time t_1 is

$$y_1(t_1) = \int_0^{t_1} H(t_1-q)u_1(q)dq = \int_0^{t_1} H(q)u_1(t_1-q)dq = \int_0^{t_1} |H(q)|dq > 1$$

However, from the BIBO stability assumption, there exists a finite K_1 such that for all $t \geqslant t_1$,

$$\left| \int_0^t H(t_1-q)u_1(q)dq \right| < K_1$$

Step 2. From equation 22.6, given the K_1 of step 1, there exists a $t_2 > t_1$ such that

$$\int_0^{t_2-t_1} |H(q)|dq > 2 + K_1$$

With this in mind, define the input $u_2(\cdot)$ as

$$u_2(t) = \begin{cases} u_1(t), & 0 \leqslant t \leqslant t_1 \\ \text{sgn}[H(t_2-t)], & t_1 \leqslant t \leqslant t_2 \\ 0, & \text{otherwise} \end{cases}$$

The response to $u_2(\cdot)$, denoted $y_2(\cdot)$, must satisfy

$$y_2(t_2) = \int_0^{t_2} H(t_2-q)u_2(q)dq = \int_0^{t_2} H(q)u_2(t_2-q)dq$$

$$= \int_0^{t_2-t_1} |H(q)|dq + \int_{t_2-t_1}^{t_2} H(q)u_1(t_2-q)dq$$

$$= \int_0^{t_2-t_1} |H(q)|dq + \int_0^{t_2} H(q)u_1(t_2-q)dq$$

$$\geqslant 2 + K_1 - K_1 = 2$$

where $u_1(t_2-q) = 0$ for $t_1 < t_2 - q$. Again, from the BIBO stability assumption, for all $t > t_2$, there exists a finite constant K_2 such that

$$\left| \int_0^t H(q)u_2(t-q)dq \right| < K_2$$

Step 3. Repeating the foregoing argument a countably infinite number of times, it follows that there exists a sequence of times $t_3,t_4,\ldots,t_n,\ldots$, with $t_j > t_{j-1}$, and a sequence of finite constants $K_3,K_4,\ldots,K_n,\ldots$ such that

(i) $\displaystyle\int_0^{t_n - t_{n-1}} |H(q)|\, dq > n + K_{n-1}$ by the unboundedness of the L_1-norm of $H(\cdot)$.

(ii) For all $t > t_{n-1}$, there exists K_{n-1} such that

$$\left| \int_0^t H(q) u_{n-1}(t-q)\, dq \right| < K_{n-1}$$

from the BIBO stability assumption. With the nth input then defined as

$$u_n(t) = \begin{cases} u_{n-1}(t), & 0 \leqslant t \leqslant t_{n-1} \\ \mathrm{sgn}[H(t_n - t)], & t_{n-1} < t \leqslant t_n \\ 0, & \text{otherwise} \end{cases} \qquad (22.8)$$

It follows that

$$y_n(t_n) = \int_0^{t_n} H(q) u_n(t_n - q)\, dq = \int_0^{t_n - t_{n-1}} |H(q)|\, dq + \int_{t_n - t_{n-1}}^{t_n} H(q) u_{n-1}(t_n - q)\, dq$$

$$= \int_0^{t_n - t_{n-1}} |H(q)|\, dq + \int_0^{t_n} H(q) u_{n-1}(t_n - q)\, dq > n + K_{n-1} - K_{n-1} = n$$

Step 4. Define the bounded input

$$u(t) = \lim_{n \to \infty} u_n(t)$$

Clearly, $u(t)$ exists, since $\{u_n(t)\}_{n=1}^\infty$ is ultimately constant. Evaluate $u(t)$ by observing that $u_{n+1}(t)$, $u_{n+2}(t)$,... are all identical to $u_n(t)$ on the interval $(-\infty, t_n]$, i.e.,

$$u(t) = u_n(t) \quad \text{for all } t < t_n .$$

Now observe that the response $y(\cdot)$ to $u(\cdot)$ must satisfy

$$y(t_n) = \int_0^{t_n} H(q) u(t_n - q)\, dq = \int_0^{t_n} H(q) u_n(t_n - q)\, dq > n$$

Hence, $u(\cdot)$ is a bounded-input, satisfying $u(t) = 0$ for $t \leqslant 0$, which produces an unbounded output, i.e., $y(t_n) \to \infty$ as $n \to \infty$. But then, this contradicts the BIBO hypothesis, and $\displaystyle\int_0^\infty |H(t)|\, dt$ is finite, as was to be shown.

Theorem 22.1 presumes $H(t)$ is real. Some of the forthcoming analysis, however, requires a complex $H(t)$, say, of the form $H(t) = e^{\lambda t} 1^+(t)$, where λ is a complex number. This exacts a slight modification in equation 22.8 in order to achieve the equality

$$|H(t_1 - q) u(q)| = |H(t_1 - q)|$$

for complex $H(t)$. If we define

$$\text{sgn}(\beta) = \begin{cases} \dfrac{\bar{\beta}}{|\beta|}, & \beta \neq 0 \\ 0, & \text{otherwise} \end{cases}$$

where $\bar{\beta}$ is the complex conjugate of β, the sequence of inputs defined in equation 22.8 becomes amenable to a complex impulse response.

The generalization of the above proof to the MIMO case is straightforward except for some notational contrivances. The key is to define the vector input $u_n(t)$ as follows: (i) determine what row of $H(t)$ has the largest L_1-norm as defined in equation 22.2, (ii) if this is the kth row, define each entry in the input so that, over the proper interval, the kth row multiplied by the input vector will yield the sum of the absolute values of the entries of the row.

Corollary 22.1. A causal system with impulse response $H(t)$ is BIBO stable if and only if there exists a single finite constant K such that

$$\|y(\cdot)\|_\infty \leqslant K \|u(\cdot)\|_\infty \tag{22.9}$$

for all input-output pairs related by the convolution integral of equation 22.3.

Proof. If the system is BIBO stable, then the L_1-norm of the impulse response is finite. Then, with $K = \|H(\cdot)\|_1$ equation 22.7, follows. (This equation 22.9, of course, is true in general but not with finite K.) Conversely, if equation 22.9 is true, then BIBO stability is guaranteed.   ∎

Corollary 22.1 connects BIBO stability with the finite gain of the system. Since K is a universal constant, it represents an upper bound on the system gain. Indeed, K can be taken to be the L_1-norm of the impulse response matrix.

STABILITY VIA THE TRANSFER FUNCTION MATRIX

Throughout this chapter, the transfer function matrix $H(s)$ is presumed to represent a linear, lumped, time-invariant, causal system having an underlying (zero-state) state model, although this is not a critical assumption. What is critical is that each entry of $H(s)$ be a proper rational function in s. Since $H(s)$ equals, in the present circumstances, the one-sided Laplace transform of the impulse response matrix $H(t)$, BIBO stability of the system should have some special characterization in terms of the attributes of $H(s)$. The attributes of $H(s)$ crucial to the characterization are the pole locations of $H(s)$. Specifically, $H(s)$ represents the input-output behavior of a BIBO stable system if and only if all poles of $H(s)$ lie in the open left-half complex plane.

To develop the case for this claim, assume that each entry of $H(s)$ is a proper rational function. Then

$$H(s) = H(\infty) + \frac{P(s)}{\psi(s)}$$

where $P(s)$ is a polynomial matrix and $\psi(s)$ is a polynomial whose zeros are the poles of $H(s)$. One can think of $\psi(s)$ as the least common multiple of the denominators of the entries of $H(s)$, although this is not true in general. Now, let $u(s)$ be the Laplace transform of any bounded input $u(t)$. Note that $\mathcal{L}^{-1}\{H(\infty)u(s)\} = H(\infty)u(t)$ which is bounded since $u(t)$ is bounded. Hence, without loss of generality, our stability considerations may focus on the term $P(s)/\psi(s)$, which is a *strictly* proper rational matrix.

In trying to interpret the boundedness of the L_1-norm of $H(t)$ with regard to $P(s)/\psi(s)$, suppose that

$$\psi(s) = (s - \lambda_1)^{\hat{m}_1}(s - \lambda_2)^{\hat{m}_2} \dots (s - \lambda_{\hat{\sigma}})^{\hat{m}_{\hat{\sigma}}}$$

where $\lambda_1, \lambda_2, ..., \lambda_{\hat{\sigma}}$ are the distinct poles of $H(s)$ and the $\hat{m}_i$ are the appropriate multiplicities. Adapting the partial fraction expansion spelled out in theorem 18.4 to our immediate needs implies

$$\frac{P(s)}{\psi(s)} = \sum_{i=1}^{\hat{\sigma}} \sum_{j=1}^{\hat{m}_i} \frac{P_i^j}{(s - \lambda_i)^j} \qquad (22.10)$$

where the constant matrices

$$P_i^j = \frac{1}{(\hat{m}_i - j)} \lim_{s \to \lambda_i} \frac{d^{\hat{m}_i - j}}{ds^{\hat{m}_i - j}} \left[(s - \lambda_i)^{\hat{m}_i} \frac{P(s)}{\psi(s)} \right]$$

Without loss of generality, assume that the matrices $P_i^j \neq [0]$; if any is zero, the corresponding term has no effect on the stability of the system. Since the L_1-norm condition on $H(t)$ deals with time functions, it becomes necessary to take the inverse transform of equation 22.10 to represent $H(t)$ as a linear combination of independent time functions (exponentials) parameterized by the poles of $H(s)$, i.e.,

$$H(t) = \sum_{i=1}^{\hat{\sigma}} \sum_{j=1}^{\hat{m}_i} P_i^j \frac{t^{j-1}}{(j-1)!} \exp(\lambda_i t) 1^+(t)$$

From this decomposition, the L_1-norm of $H(t)$ is clearly bounded whenever $\text{Re}(\lambda_i) < 0$; in which case

$$\|H(\cdot)\|_1 = \left\| \sum_{i=1}^{\hat{\sigma}} \sum_{j=1}^{m_i} P_i^j \frac{t^{j-1}}{(j-1)!} \exp(\lambda_i \cdot) 1^+(\cdot) \right\|_1$$

$$\leqslant \sum_{i=1}^{\hat{\sigma}} \sum_{j=1}^{m_i} \left\| P_i^j \frac{t^{j-1}}{(j-1)!} \exp(\lambda_i \cdot) 1^+(\cdot) \right\|_1$$

Since each of the norms

$$\left\| \frac{t^{j-1}}{(j-1)!} \exp(\lambda_i \cdot) 1^+(\cdot) \right\|_1 \leqslant K_i^j < \infty$$

for each i and j, it follows that

$$\|H(\cdot)\|_1 \leqslant \sum_{i=1}^{\hat{\sigma}} \sum_{j=1}^{m_i} K_i^j < \infty$$

Hence, a sufficient condition for BIBO stability is that all the poles of $H(s)$ (which have nonzero residues) lie in the open left-half complex plane.

To see the necessity of this condition, suppose $\text{Re}(\lambda_i) \geqslant 0$. If this property is sufficient for instability, then $\text{Re}(\lambda_i) < 0$ is necessary for BIBO stability. To verify this, proceed as follows. Without loss of generality, suppose $\text{Re}(\lambda_1) \geqslant \text{Re}(\lambda_i)$ for all i; it is always possible to reorder the poles of $H(s)$ to achieve this. Define the scalar β as the minimum of the nonzero absolute values of the entries of P_1^1, i.e.,

$$\beta = \min_{k,m} [\,|P_1^1(k,m)\,| \neq 0]$$

Further, let E^q be a projection operator defined by

$$E^q H(t) = \begin{cases} H(t) & \text{for } t \leqslant q \\ 0 & \text{for } t > q \end{cases} \tag{22.11}$$

Now consider the L_1 norm of $E^q H(\cdot)$:

$$||E^q H(t)||_1 = ||\sum_{i=1}^{\hat{\sigma}} \sum_{j=1}^{m_i} P_i^j E^q \frac{t^{j-1}}{(j-1)!} \exp(\lambda_i t) 1^+(t)||_1$$

$$\geqslant \beta\,||E^q \exp(\lambda_1 t) 1^+(t)||_1$$

The inequality obtains because the L_1 norm of a matrix is the maximum over the L_1 row-norms of the sum of the absolute values of the entries of each row. Certainly, the L_1 norm of the matrix is larger than (or equal to) the L_1 norm of one of its entries. However, as $q \to \infty$, $||E^q \exp(\lambda_1 t) 1^+(t)||_1 \to \infty$, indicating the unboundedness of $||H(\cdot)||_1$. Hence, for a transfer function matrix $H(s)$, if $\text{Re}(\lambda_i) \geqslant 0$ for some i where λ_i is a pole of $H(s)$, then the system is not BIBO stable. Consequently, $\text{Re}(\lambda_i) < 0$ is a necessary condition for the BIBO stability of the transfer function matrix model.

The preceeding assertions are formalized in the following theorem.

Theorem 22.2. A necessary and sufficient condition for the BIBO stability of the transfer function matrix model $H(s)$ having poles $\{\lambda_1, \lambda_2, ..., \lambda_{\hat{\sigma}}\}$ is that $\text{Re}(\lambda_i) < 0$ for all i.

Again, the transfer function matrix model is an input-output model lacking any cognizance of initial conditions present in the system. The state model, on the other hand, sets forth both an external and an internal system description with explicit provision for information regarding initial conditions. Molding the foregoing conditions to the framework of the usual state model follows in the next section.

BIBS AND BIBO STABILITY VIA THE STATE MODEL

The conditions for BIBO stability in the context of the state model are founded on an internal system equivalent of BIBO stability: BIBS, or bounded input, bounded state stability.

Definition 22.7. A state model $\{A,B,C,D\}$ having arbitrary initial conditions is said to be bounded input, bounded state (BIBS) stable if, for any bounded input $u(t)$, the state response

$$x(t) = \exp(At)x(0) + \int_0^t \exp[A(t-q)]Bu(q)dq \qquad (22.12)$$

is bounded.

For the internal system representation, this definition almost duplicates the BIBO stability definition (22.6). The interesting twist stems from the proviso of an arbitrary initial condition, say, $x(0)$. This means that our stability considerations must account for the excitations of all system modes, i.e., the zero-input state response as well as the zero-state state response. The arbitrariness of the initial condition and the requirement of bounded responses over the entire class of bounded inputs precludes the zero-input and zero-state responses canceling each other out in any generic way. Hence, the state model is BIBS stable if and only if both the zero-input response and the zero-state state response have finite L_∞ norms. The ultimate objective is to characterize the finiteness of these norms in terms of the eigenvalues of A and the controllability (or lack thereof) of certain directed modes.

Let us initiate our discussion by considering the zero-state response part of equation 22.12: when does

$$x(t) = \int_0^t \exp[A(t-q)]Bu(q)dq$$

have a finite L_∞ norm? By observing that $\exp[At]B1^+(t)$ is the impulse state response, a necessary and sufficient condition becomes $\|\exp[A \cdot]B1^+(\cdot)\|_1 < \infty$ by theorem 22.1. We state this as the following proposition.

Proposition 22.1. A necessary and sufficient condition for the zero-state state response to have a finite L_∞ norm is that $\|\exp[A \cdot]B1^+(\cdot)\|_1 < \infty$.

Taking the Laplace transform of $\exp[At]B1^+(t)$ and utilizing the partial fraction expansion results of chapter 18 yields

$$(sI-A)^{-1}B = \sum_{i=1}^{\sigma} \sum_{j=1}^{m_i} \frac{R_i^j B}{(s-\lambda_i)^j} \qquad (22.13)$$

where $\{\lambda_1, \lambda_2,...,\lambda_\sigma\}$ are the distinct eigenvalues of A, the m_i are their respective multiplicities in the minimal polynomial $\psi_A(s)$ of A, and the R_i^j are the residue matrices defined in the usual way, viz.,

$$R_i^j = \frac{1}{(m_i-j)!} \lim_{s \to \lambda_i} \frac{d^{m_i-j}}{ds^{m_i-j}} [(s-\lambda_i)^{m_i} (sI-A)^{-1}]$$

From equation 22.13, we obtain the following proposition as a direct corollary to theorem 22.2.

Proposition 22.2. Necessary and sufficient conditions for the zero-state state response to be bounded are that (*i*) $\text{Re}(\lambda_i) < 0$ for all i for which $R_i^j B \neq [0]$ and (*ii*) $R_i^j B = [0]$ for all j if $\text{Re}(\lambda_i) \geq 0$.

Condition (*ii*) means that any mode in the closed right-half complex plane must be uncontrollable.

We next consider the zero-input state response $\exp[At]x(0)$ for arbitrary $x(0)$. When is $\|\exp[A \cdot]x(0)\|_\infty < \infty$? Taking the Laplace transform of $\exp[At]x(0)$ and executing a partial fraction expansion produces

$$(sI - A)^{-1}x(0) = \sum_{i=1}^{\sigma} \sum_{j=1}^{m_i} \frac{R_i^j x(0)}{(s - \lambda_i)^j} \tag{22.14}$$

Since stability is a class condition— i.e., the property is maintained over the entire class of bounded inputs and the entire class of initial conditions— we assume in the forthcoming analysis that the quantities $R_i^j x(0) \neq \theta$. To determine necessary and sufficient conditions for $\|\exp[A \cdot]x(0)\|_\infty < \infty$, take the inverse transform of equation 22.14 and analyze the associated time function. For sufficiency, consider

$$\|\exp[At]x(0)\|_\infty = \left\| \sum_{i=1}^{\sigma} \sum_{j=1}^{m_i} R_i^j x(0) \frac{t^{j-1}}{(j-1)!} \exp(\lambda_i t) 1^+(t) \right\|_\infty$$

$$\leq \sum_{i=1}^{\sigma} \sum_{j=1}^{m_i} \left\| R_i^j x(0) \frac{t^{j-1}}{(j-1)!} \exp(\lambda_i t) 1^+(t) \right\|_\infty \tag{22.15}$$

$$\leq \sum_{i=1}^{\sigma} \sum_{j=1}^{m_i} \| R_i^j x(0) \| \left\| \frac{t^{j-1}}{(j-1)!} \exp(\lambda_i t) 1^+(t) \right\|_\infty$$

where $\| R_i^j x(0) \|$ is just the usual Euclidean vector norm in $\mathbf{R}^n$. It is clear that $\|\exp(\lambda_i t) 1^+(t)\|_\infty < \infty$ whenever $\text{Re}(\lambda_i) \leq 0$: if $\text{Re}(\lambda_i) < 0$, then the functions are decaying exponentials or exponentially damped sinusoids; if $\text{Re}(\lambda_i) = 0$, then the functions are constant or undamped sinusoids which are bounded. Thus, sufficient conditions for boundedness are that (*i*) $\text{Re}(\lambda_i) \leq 0$, and (*ii*) if $\text{Re}(\lambda_i) = 0$, then m_i (the multiplicity of λ_i in the minimal polynomial) must be unity, i.e., $m_i = 1$; otherwise there will be a ramp-like function response proportional to $t^j 1^+(t)$, $j \geq 1$, or a linearly increasing sinusoidal response proportional to $t^j \sin(\omega t + \phi) 1^+(t)$.

The necessity of these conditions follows by considering sufficient conditions for instability. There are two cases: (*i*) $\text{Re}(\lambda_i) > 0$ and (*ii*) $\text{Re}(\lambda_i) = 0$. If $\text{Re}(\lambda_i) > 0$ for some i, then

$$\|E^q \exp[At]x(0)\|_\infty = \left\| \sum_{i=1}^{\sigma} \sum_{j=1}^{m_i} R_i^j x(0) \frac{t^{j-1}}{(j-1)!} E^q \exp(\lambda_i t) 1^+(t) \right\|_\infty$$

$$\geq \beta \| E^q \exp(\lambda_i t) 1^+(t) \|_\infty \tag{22.16}$$

where the projection operators E^q are defined as in equation 22.11 and where β is the minimum of the absolute values of the nonzero entries of $R_i^1 x(0)$. As $q \to \infty$, $\| E^q \exp(\lambda_i t) 1^+(t) \|_\infty \to \infty$, establishing instability.

Now suppose $\text{Re}(\lambda_i) = 0$ and the R_i^2-matrix exists and $R_i^2 x(0) \neq \theta$. In this case equation 22.16 reduces to

$$\| E^q \exp[At] x(0) \|_\infty \geq \beta \| E^q t \exp(\lambda_i t) 1^+(t) \|_\infty$$

where β is the minimum of the absolute values of the nonzero entries of $R_i^2 x(0)$ and $t\exp(\lambda_i t) 1^+(t)$ is proportional to either $t 1^+(t)$ or $t\exp[\text{Re}(\lambda_i)] 1^+(t)$. Again, as $q \to \infty$, $\| E^q \exp(\lambda_i t) 1^+(t) \|_\infty \to \infty$, establishing instability.

These conditions, yield the following.

Proposition 22.3. Necessary and sufficient conditions for the boundedness of the zero-input state response are that (i) $\text{Re}(\lambda_i) \leqslant 0$ for all eigenvalues λ_i of A and (ii) if $\text{Re}(\lambda_i) = 0$, then the order of the associated factor in the minimal polynomial of A must be 1.

Combining propositions 22.2 and 22.3 produces the following theorem.

Theorem 22.3. Necessary and sufficient conditions for BIBS stability of the usual time-invariant state model $\{A,B,C,D\}$ are that (i) $\text{Re}(\lambda_i) \leqslant 0$ for all eigenvalues λ_i of A, (ii) if $\text{Re}(\lambda_i) = 0$, then the order of the associated factor in the minimal polynomial of A must be 1, and (iii) if $\text{Re}(\lambda_i) = 0$, then the mode must be uncontrollable.

With this characterization of BIBS stability in terms of the eigenvalues of A and the controllability of the directed modes, theorem 22.3 permits a characterization of BIBO stability over all possible initial conditions in addition to the class of bounded inputs.

Corollary 22.3a. If the usual time-invariant state model is BIBS stable, then it is BIBO stable.

Corollary 22.3b. If the usual time-invariant state model is not BIBS stable, then it is BIBO stable if and only if all unstable modes lie in the unobservable subspace, i.e., if λ_i represents an unstable mode, then the model is BIBO stable if and only if $CR_i^j = [0]$ for all j.

Exercise (Circle all possibilities). The impulse response matrix of a time-invariant state model $\{A,B,C,D\}$ has a finite L_1-norm. If an eigenvalue $\lambda_i = 0$ has multiplicity 2 in the minimal polynomial of A, then the mode (λ_i, e_i) must be

  (a) controllable and observable   (b) uncontrollable and observable
  (c) controllable and unobservable (d) uncontrollable and unobservable

Exercise (Circle all possibilities). Suppose all eigenvalues of a particular time-invariant state model are in the open left-half plane except $\lambda_i = 0$, which has multiplicity 1 in the minimal polynomial of A. Then the L_1-norm of the impulse response matrix is finite if the mode is:

  (a) controllable and observable   (b) uncontrollable and observable
  (c) controllable and unobservable (d) uncontrollable and unobservable

EXAMPLE 22.1

Consider the linear switched state model

$$\dot{x} = Ax + B_i u, \qquad x(0) = x_0$$
$$y = C_i x$$

where

$$A = \begin{bmatrix} 0 & 1 & -1 \\ -1 & -1 & 2 \\ -1 & 1 & 0 \end{bmatrix} \qquad \text{for all } t$$

$$B_1 = \begin{bmatrix} 2 \\ 1 \\ 3 \end{bmatrix} \quad \text{for } t < 1, \; B_2 = \begin{bmatrix} 1 \\ -1 \\ 1 \end{bmatrix} \quad \text{for } t \geq 1$$

$$C_1 = [0 \;\; -1 \;\; 1] \quad \text{for } t < 1$$

and

$$C_2 = [1 \;\; 0 \;\; 0] \quad \text{for } t \geq 1$$

(a) *Find the eigenvalues and eigenvectors of A.*
By some straightforward calculations, $\lambda_1 = 0, \lambda_2 = -2, \lambda_3 = 1$, and a matrix of right eigenvectors is

$$T = \begin{bmatrix} 1 & 1 & 0 \\ 1 & -1 & 1 \\ 1 & 1 & 1 \end{bmatrix} = [e_1 \;\; e_2 \;\; e_3]$$

The matrix of left eigenvectors is

$$T^{-1} = \begin{bmatrix} 1 & 0.5 & -0.5 \\ 0 & -0.5 & 0.5 \\ -1 & 0 & 1 \end{bmatrix} = \begin{bmatrix} w_1^* \\ w_2^* \\ w_3^* \end{bmatrix}$$

(b) *Compute the state transition matrix.*
From the results of part (a),

$$\Phi(t) = \exp(At) = T \begin{bmatrix} 1 & 0 & 0 \\ 0 & e^{-2t} & 0 \\ 0 & 0 & e^t \end{bmatrix} T^{-1}$$

(c) *Investigate controllability over [0,1). Specifically determine the controllability of each mode.*
The controllability of each mode depends on the product $w_i^* B_1$. If $w_{i*} B_1$ is zero, the mode is uncontrollable; if $w_{i*} B_1$ is nonzero, the mode is controllable. An efficient check is to compute

$$T^{-1}B_1 = \begin{bmatrix} w_1^* \\ w_2^* \\ w_3^* \end{bmatrix} \quad B_1 = \begin{bmatrix} 1 & 0.5 & -0.5 \\ 0 & -0.5 & 0.5 \\ -1 & 0 & 1 \end{bmatrix} \begin{bmatrix} 2 \\ 1 \\ 3 \end{bmatrix} = \begin{bmatrix} 1 \\ 1 \\ 1 \end{bmatrix}$$

Clearly, all modes are controllable. The rank of

$$Q_1 = [B_1 \ AB_1 \ldots A^{n-1}B_1] = \begin{bmatrix} 2 & -2 & 4 \\ 1 & 3 & -3 \\ 3 & -1 & 5 \end{bmatrix}$$

should then be 3. Some straightforward computation shows this to be true.

(d) *Repeat part c for the interval $[1,\infty)$.*
The analysis here mimics that of part (c). Specifically, consider

$$T^{-1}B_2 = \begin{bmatrix} w_1^* \\ w_2^* \\ w_3^* \end{bmatrix}, \quad B_2 = \begin{bmatrix} 1 & 0.5 & -0.5 \\ 0 & -0.5 & 0.5 \\ -1 & 0 & 1 \end{bmatrix} \begin{bmatrix} 1 \\ -1 \\ 1 \end{bmatrix} = \begin{bmatrix} 0 \\ 1 \\ 0 \end{bmatrix}$$

Thus, mode 1, $(\lambda_1, e_1) = (0, e_1)$, and mode 3, $(\lambda_3, e_3) = (1, e_3)$, are uncontrollable, whereas mode 2, $(\lambda_2, e_2) = (-2, e_2)$, is controllable. A simple calculation shows that $\text{rank}[Q_2] = \text{rank}[B_2 \ AB_2 \ldots A^{n-1}B_2]$.

(e) *Investigate the system's observability over $[0,1)$. Specifically determine what modes are observable.*
The solution to this task parrots that of parts c and d. Specifically, compute

$$C_1 T = C_1[e_1, e_2, e_3] = [0 \ -1 \ 1] \begin{bmatrix} 1 & 1 & 0 \\ 1 & -1 & 1 \\ 1 & 1 & 1 \end{bmatrix} = [0 \ 2 \ 0]$$

Since the product is $[0, 2, 0]$, modes 1 and 3 are completely unobservable, i.e., the zero-input response is identically zero for initial conditions of the form $x(0) = \beta_1 e_1 + \beta_2 e_3$. A simple construction shows that $\text{rank}[R] = 1$, corroborating this fact. The remaining question concerns the observability of mode 2. To determine this, recall that

$$\mathbf{R}^n = \text{Im}[R^t] \oplus N[R]$$

where $\mathbf{R}^n$ is the usual Euclidean n-space and R is the observability matrix over $[0,1)$. From before, $\{e_1, e_3\}$ spans $N[R]$ which is orthogonal to $\text{Im}[R^t]$. Hence, $\text{Im}[R^t]$ is spanned by w_2, the second left eigenvector. It follows that only initial conditions of the form $x(0) = \beta w_2$ are reconstructible and hence observable. Moreover, $x(0) = e_2$ is not observable, yet the natural frequency $e^{\lambda_2 t}$ is present in the response. This comes about because e_2 has a nonzero projection onto $\text{Im}[R^t] = \text{span}\{w_2\}$.

The natural frequency (mode) λ_2 ($e^{\lambda_2 t}$) is thus considered observable, although the directed mode (λ_2, e_2) is not.

(f) *Repeat part e for the interval* $[1,\infty)$.
Again, the focus zooms in on $C_2 T$:

$$C_2 T = C_2[e_1, e_2, e_3] = [1 \ \ 0 \ \ 0] \begin{bmatrix} 1 & 1 & 0 \\ 1 & -1 & 1 \\ 1 & 1 & 1 \end{bmatrix} = [1 \ \ 1 \ \ 0]$$

so that mode 3 is completely unobservable. The observable subspace is spanned by w_1 and w_2. The modes λ_1 or λ_2 are observable only if they lie in this span. Although e_1 does not, observe that $e_2 = w_1 + 3w_2$, making the directed mode (λ_2, e_2) observable. However, the frequency $e^{\lambda_1 t}$ is still present in a typical zero-input response, since e_1 has a nonzero projection onto span$\{w_1, w_2\}$.

(g) *How would one drive* $x(0) \in \text{Im}[Q_1]$ *to an arbitrary* $x(2)$ *where* $Q_1 = [B_1, AB_1, \dots, A^{n-1}B_1]$?
The problem is that the system is uncontrollable over the interval $[1,\infty)$. However, the solution to the state dynamics implies that

$$x(2) = \exp[A(2-1)]x(1) = \exp[A]$$

whenever the input is zero over $[1,2]$. Thus, if $u(t) = 0$ for $t \in [1,2]$, then

$$\hat{x}(1) = \exp[-A]x(2)$$

Since the system is controllable over $[0,1)$, there exists a $u_{[0,1)}(t)$ which will drive any $x(0)$ to $\hat{x}(1)$ when both live in the controllable subspace. Hence, a control which will drive $x(0)$ to $x(2)$ is

$$u(t) = \begin{cases} u_{[0,1)}(t), & 0 \leqslant t < 1 \\ 0, & 1 \leqslant t \leqslant 2 \end{cases}$$

Exercise. Construct the said control explicitly.

(h) *Given measurements over* $[1,\infty)$, *describe the space of states that are reconstructible over* $[0,1)$. *How would one go about computing these states?*
For $1 \leqslant t_0 < \infty$, measurements over $[1,\infty)$ permit reconstruction of all $x(t_0)$ of the form

$$x(t_0) = \beta_1 w_1 + \beta_2 w_2$$

for arbitrary scalars β_1 and β_2. Once $x(t_0)$ is found, it is then possible to determine $x(1)$ according to

$$x(1) = e^{A(1-t_0)} \left[x(t_0) - \int_1^{t_0} e^{A(t_0-q)} B_2 u(q) \, dq \right]$$

Thus, for any t_0', where $0 \leqslant t_0' < 1$,

$$x(t_0') = e^{A(t_0'-1)} \left[x(1) - \int_{t_0'}^{1} e^{A(1-q)} B_1 u(q) dq \right]$$

where it is assumed that the system input $u(t)$ is known over $[0,1)$ but that the response $y(t)$ may not be known.

(i) *Investigate the BIBS stability of the system.*
Because the A-matrix has an eigenvalue $\lambda_3 = 1$, the initial condition $x(0) = e_3$ will always produce an unbounded zero-input state response. Thus, the system is not **BIBS** stable.

(j) *Investigate the BIBO stability of the system.*
For arbitrary initial conditions and bounded inputs, the state trajectory cannot blow up over a finite interval, such as the interval $[0,1)$. Thus, BIBO stability depends on the behavior of the system over the interval $[1,\infty)$. From part (f), the mode $(\lambda_3 = 1, e_3)$ is completely unobservable. Since the other two modes are observable, BIBO stability rests on the boundedness of those parts of the state trajectory that are linked to (λ_1, e_1) and (λ_2, e_2). Since $\mathrm{Re}[\lambda_2] < 0$, that part of the state trajectory associated with (λ_2, e_2) remains bounded for arbitrary initial conditions and all bounded inputs. From part (d), the mode $(\lambda_1 = 0, e_1)$ is uncontrollable. Hence, for arbitrary initial conditions and all bounded inputs, that part of the state trajectory associated with the mode (λ_1, e_1) remains bounded. The system is **BIBO** stable.

Multiple-Choice Exercises: Consider the continuous-time system, $\dot{x} = Ax + Bu$, $y = Cx$, where

$$A = [e_1 \ e_2 \ e_3] \begin{bmatrix} \lambda_1 & 0 & 0 \\ 0 & \lambda_2 & 0 \\ 0 & 0 & \lambda_3 \end{bmatrix} \begin{bmatrix} w_1^* \\ w_2^* \\ w_3^* \end{bmatrix}$$

1. A necessary and sufficient condition for λ_2 to be an uncontrollable mode is that

    (a) $e_1^* B = \theta^t$ (b) $w_1^* B \neq \theta^t$ (c) $w_2^* B = \theta^t$
    (d) $w_3^* B \neq \theta^t$ (e) $e_2^* B = \theta^t$ (f) two of the above

2. A necessary and sufficient condition for λ_1 to be an unobservable mode is that

    (a) $Ce_1 = \theta$ (b) $Cw_1 = \theta$ (c) $Ce_2 \neq \theta$
    (d) $Cw_2 = \theta$ (e) $Ce_3 \neq \theta$ (f) two of the above

3. If $\lambda_1 = -1, \lambda_2 = 0$, and $\lambda_3 = 1$, a necessary and sufficient condition for **BIBS** stability is:

(a) $Ce_2 = \theta$ and $Ce_3 = \theta$ (b) $Ce_3 = \theta$ and $w_2^* B = \theta$
(c) $w_2^* B = \theta$ and $w_3^* B = \theta$ (d) $w_2^* B = \theta$, $w_3^* B = \theta$, and $Ce_3 = \theta$
(e) system is not BIBS stable

4. If $\lambda_1 = -1$, $\lambda_2 = 0$, and $\lambda_3 = 1$, a sufficient condition for BIBO stability is that

(a) $Ce_2 = \theta$ and $Ce_3 = \theta$ (b) $Ce_3 = \theta$ and $w_3^* B = \theta$
(c) $w_2^* B = \theta$ and $Ce_3 = \theta$ (d) $w_2^* B = \theta$ and $w_3^* B = \theta$
(e) two of the above   (f) none of the above

ASYMPTOTIC STABILITY OF THE STATE TRAJECTORY

Another important stability notion pertinent to linear systems is that of asymptotic stability of the zero-input state response. Asymptotic stability requires not only that the state trajectory $x(t)$ remain bounded, but that $\lim x(t) = 0$ as $t \to \infty$.

Definition 22.8. The linear time-invariant (zero-input) state dynamical equation $\dot{x} = Ax$, is *asymptotically stable* if $\|x(\cdot)\|_\infty < \infty$ and $\lim_{t \to \infty} x(t) = 0$ for arbitrary initial conditions.

It is straightforward to show that a sufficient condition for $\lim_{t \to \infty} x(t) \neq 0$ is that $\mathrm{Re}[\lambda_i] \geq 0$ for some eigenvalue λ_i of A. Ergo the following theorem.

Theorem 22.4. The state dynamical equation $\dot{x} = Ax$ is asymptotically stable if and only if all eigenvalues of A have a real part strictly less than zero.

Exercise: Fill in the details of the proof of this theorem.

These results have a much more interesting extension to the time-varying state dynamics (see Chapter 24).

DISCRETE-TIME SYSTEM ANALOGS

The discrete-time system theorems come about by defining the relationship between the s-plane of the continuous-time world and the z-plane of the discrete-time world. The imaginary axis of the s-plane maps into the unit circle of the z-plane under the transformation $z = e^s$. The open left-half plane maps into the inside of the unit circle, and the open right-half plane folds into the exterior of the unit disc. The theorems about continuous-time system stability build on the location of various eigenvalues in the complex s-plane. Direct extensions to the discrete-time world result simply by translating the locations of these poles.

Theorem 22.5. $x(k + 1) = Ax(k) + Bu(k)$ is BIBS stable if and only if (*i*) all eigenvalues of A lie in the closed unit disc, (ii) eigenvalues on the unit circle have multiplicity 1 in the minimal polynomial of A, and (iii) unit circle modes are uncontrollable.

Given that A^k has the usual matrix directed mode decomposition, condition (iii) requires that $R_i B = [0]$ for such modes.

Theorem 22.6. $x(k+1) = Ax(k) + Bu(k)$ is BIBO stable if it is BIBS stable.

Theorem 22.7.   If $x(k+1) = Ax(k) + Bu(k)$ is not BIBS stable, then it is BIBO stable if and only if all unstable modes lie in the unobservable subspace of the pair (C,A).

Theorem 22.8.   The model $x(k+1) = Ax(k) + Bu(k)$ is asymptotically stable if and only if all eigenvalues of A have magnitude strictly less than unity.

Exercise. Consider the discrete-time system
$$x(k+1) = Ax(k) + Bu(k)$$
$$y(k) = Cx(k)$$
where $C = [1 \ 1 \ -1]$, $B = [-1 \ 0 \ -2]^t$, and
$$A = \begin{bmatrix} 1 & 0 & 2 \\ 0 & 1 & 0 \\ 1 & 0 & 1 \end{bmatrix} \begin{bmatrix} -1 & 0 & 0 \\ 0 & 0 & 0 \\ 0 & 0 & 0.5 \end{bmatrix} \begin{bmatrix} -1 & 0 & 2 \\ 0 & 1 & 0 \\ 1 & 0 & -1 \end{bmatrix}$$

True or False ?

1. The system is completely controllable. _____
2. The system is completely observable. _____
3. The system is BIBS stable. _____
4. The system is BIBO stable. _____

Exercise. Investigate the controllability, observability, and stability of the state model
$$\begin{bmatrix} x_1(k+1) \\ x_2(k+1) \\ x_3(k+1) \end{bmatrix} = \begin{bmatrix} 1 & 0 & 0 \\ 0 & 0 & 0 \\ 0 & 0 & -1 \end{bmatrix} \begin{bmatrix} x_1(k) \\ x_2(k) \\ x_3(k) \end{bmatrix} + \begin{bmatrix} 0 \\ 1 \\ 1 \end{bmatrix} u(k)$$
$$y(k) = [1 \ 1 \ 0] \begin{bmatrix} x_1(k) \\ x_2(k) \\ x_3(k) \end{bmatrix}$$

INTRODUCTION TO LYAPUNOV STABILITY FOR LINEAR SYSTEMS

This section expands on the notion of asymptotic stability for the zero-input state dynamical equation $\dot{x} = Ax$ from the point of view of Lyapunov [5] as presented in [6]. Recall that $\dot{x} = Ax$ is asymptotically stable if and only if all eigenvalues of A are in the open left-half complex plane. The term *stability matrix* will refer to an A-matrix, all of whose eigenvalues lie in the open left-half complex plane. The paramount goal of this section is to develop, interpret, and prove the following theorem: *a real matrix A is a stability matrix if and only if, for any given real, symmetric, positive definite matrix Q, the solution P of the continuous Lyapunov matrix equation $A^t P + PA = -Q$ is positive definite.* The key to establishing this result is the notion of a *quadratic form*. Quadratic forms determine closed surfaces— hyperspheres and hyperellipsoids— in $\mathbf{R}^n$. These closed surfaces permit one to mark the movement of a state trajectory from some initial point $x(t_0)$ to the origin.

By definition, asymptotic stability of $\dot{x} = Ax$ means that the state trajectory $x(t)$ remains bounded for all t and that $x(t) \to \theta$ as $t \to \infty$. To characterize this situation, imagine a nested set of elliptical shells in the state space (say, in $\mathbf{R}^3$, for convenience) through which the state trajectory must pass on its way to θ. Quadratic forms, yet to be defined, provide a characterization of these elliptical shells, and the asymptotic stability of $\dot{x} = Ax$ can be characterized in terms of the state trajectory having to pass through a nested set of these surfaces whose major and minor axes are progressively smaller and ultimately shrink to zero.

To formally set up this scenario, let T be a nonsingular state transformation of the form

$$z = Tx$$

Clearly, the nonsingularity of T guarantees that every x-coordinate is mapped uniquely to a z-coordinate, and conversely. Thus, as $x \to \theta$, so must $z \to \theta$. Define a set of closed surfaces in the z-coordinates as

$$z^t z = z_1^2 + z_2^2 + \dots + z_n^2 = K^2, K > 0$$

If $n = 2$, the surfaces are circles of radius K; if $n = 3$, they define spheres of radius K; and for $n \geqslant 4$, they are hyperspheres of radius K. From the relationship $z = Tx$,

$$z^t z = (Tx)^t Tx = x^t T^t Tx = K^2$$

Plainly, in the x-coordinates the surfaces are hyperellipsoids. The quantity $x^t T^t Tx$ is said to be a *positive definite quadratic form*. It is positive definite because $x^t T^t Tx$ is always positive and nonzero whenever $x \neq \theta$. And it is a quadratic form because $x^t T^t Tx$ equals a weighted sum of products of the entries of x. To see this observe that $T^t T = (T^t T)^t$ implying that $T^t T$ is a symmetric matrix, say, $M = T^t T = [m_{ij}]$, with the property that $m_{ij} = m_{ji}$ for all i and j. Suppose now that $n = 3$. Then

$$x^t T^t Tx = [x_1 \ x_2 \ x_3] \begin{bmatrix} m_{11} & m_{12} & m_{13} \\ m_{21} & m_{22} & m_{23} \\ m_{31} & m_{32} & m_{33} \end{bmatrix} \begin{bmatrix} x_1 \\ x_2 \\ x_3 \end{bmatrix}$$

$$= m_{11}x_1^2 + 2m_{12}x_1x_2 + 2m_{13}x_1x_3 + m_{22}x_2^2 + 2m_{23}x_2x_3 + m_{33}x_3^2$$

This sum consists of terms of the form $x_i x_j$ which are quadratic, and the structure of the equation is clearly that of an ellipsoid (hyperellipsoid for $n \geqslant 4$).

Now recall the development of the singular value decomposition of a matrix in chapter 19. It was pointed out there that a two-by-two matrix maps circles into ellipses and, more generally hyperspheres into hyperellipsoids determined by the singular values and singular vectors of the transformation. Hence since T is nonsingular, each closed hypersphere in the z-coordinates maps into a closed hyperellipsoid in the x-coordinates. In addition, if the hyperspheres are nested with progressively smaller radii, then the hyperellipsoids are nested with progressively smaller major and minor axes. The following example illustrates the property.

EXAMPLE 22.2

Suppose $z = Tx$, where

$$T^{-1} = \begin{bmatrix} 1 & 1 \\ 1 & 0 \end{bmatrix}$$

The singular values of T^{-1} are $\sigma_1 = 1.618$ and $\sigma_2 = 0.618$, with right singular vectors $e_1 = (0.8537 \ 0.5208)^t$ and $e_2 = (0.5257 \ -0.8507)^t$. Now, e_1 specifies the direction of the major axis and e_2 the direction of the minor axis in the x-coordinates. The circles of radius K in the z-coordinates become ellipses with lengths in the e_1-direction equal to $K\sigma_1$ and widths in the e_2-direction equal to $K\sigma_2$, as illustrated in Figure 22.5.

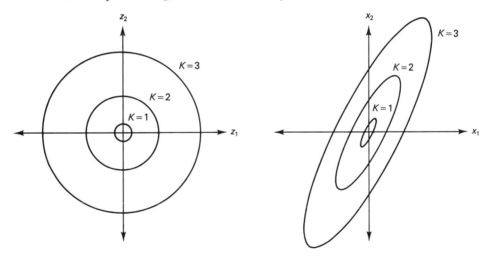

Figure 22.5 Relationship between nested circles and nested ellipses under a state transformation $z = Tx$.

It turns out that quadratic forms are sufficient for characterizing Lyapunov stability for linear time-invariant systems. However, for nonlinear systems this is not the case. So as not to restrict later developments, suppose that each nested set of surfaces is characterized by the positive definite function

$$V(x) = V(x_1, x_2, \ldots, x_n) = K, K \geqslant 0 \qquad (22.17)$$

where $V(x)$ is *positive definite* (*negative definite*) if (*i*) $V(x)$ has continuous partial derivatives with respect to the entries of x, (*ii*) $V(\theta) = 0$, and (*iii*) $V(x) > 0$ ($V(x) < 0$) for all $x \neq \theta$. If condition (*ii*) reads $V(x) \geqslant 0(V(x) \leqslant 0)$, then $V(x)$ is said to be *positive semidefinite (negative semidefinite)*. Also, it is possible to show that condition (*iii*) is equivalent to the statement, "If $r = \|x\|$, then there exists a function $\psi(r)$ such that $V(x) \geqslant \psi(r)$ for all x, provided that the function $\psi(r)$ is continuous, $\psi(0) = 0$, and $\psi(r)$ is strictly increasing" [7]. Observe that if $V(x)$ is positive definite, then $-V(x)$ is negative definite.

The preceding terminology helps to intuitively describe the derivation of the theorem regarding stability matrices mentioned at the beginning of the section. To begin the discussion, note that if $V(x) = K_i^2$ is a positive definite quadratic form defining a set of nested closed surfaces surrounding the origin, depicted in Figure 22.6 for two dimensions, then an asymptotically stable state trajectory must successively pierce these surfaces for all possible choices of $K_1^2 < K_2^2 < K_3^2 < ... < K_N^2$. With specific regard to Figure 22.6(a), if $\|x(t_0)\| < r_1$, then $V(x(t_0)) < K_2^2$. Further, if $\dot{V}(x) \leqslant 0$, then $V(x(t,t_0,x_0)) < K_2^2$ for all $t \geqslant t_0$, which means that $\|x(t,t_0,x_0)\| < R_2$. Finally, if $\dot{V}(x) < 0$, then as $t \rightarrow \infty$, $x(t) \rightarrow \theta$, since as $t \rightarrow \infty$ $V(x(t)) \rightarrow 0$ (see problem 16 at the end of the chapter). This property implies that asymptotic stability for linear systems requires that the derivative of $V(x(t))$ be negative for all $t \geqslant t_0$.

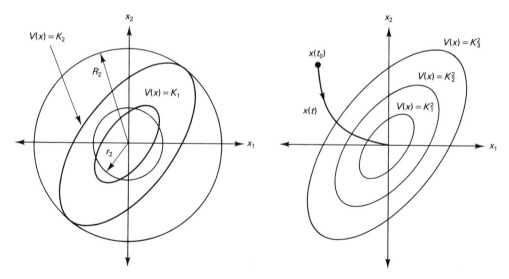

Figure 22.6 State trajectory motion from the Lyapunov perspective.

Fortunately, if $V(x) = x'Px$ for any real symmetric matrix P, it is possible to characterize the derivative of $V(x)$ for the system $\dot{x} = Ax$, viz.,

$$\frac{dV(x)}{dt}(x) = \dot{V} = \dot{x}'Px + x'P\dot{x} = x'A'Px + x'PAx$$

$$= x'(A'P + PA)x = -x'Qx$$

(22.18)

Now, for $\dot{V}$ to be negative, Q must be positive definite, and for $V(x) = x^t P x$ to define hyperellipsoids containing the origin P must be positive definite. Also, from equation 22.18 the relationship between P and Q is given by $A^t P + P A = -Q$, which is called the *matrix Lyapunov equation*. This brings us to the main result of the section.

Theorem 22.9. Let $V(x) = x^t P x$ where P is a solution to the matrix Lyapunov equation $A^t P + P A = -Q$. Then a real matrix A is a stability matrix if and only if, for any given real symmetric positive definite matrix Q, the solution P is symmetric positive definite.

Proof
Part 1. The proof is by contradiction. Suppose that A is a stability matrix and that $-x^t Q x < 0$, i.e., Q is positive definite. Suppose further that the solution P to the matrix Lyapunov equation is not positive definite, i.e., for some $x(0) = x_0, x_0^t P x_0 \leqslant 0$. Then since $Q > 0$ (Q is positive definite), $\dot{V} = -x^t Q x < 0$ for all $x \neq \theta$. This means that $x^t(t) P x(t)$ decreases as $t \to \infty$. But since $x_0^t P x_0 \leqslant 0$, and $\dot{V} = -x^t Q x < 0$ for all $x \neq 0$, then

$$x^t(t) P x(t) = x_0^t \exp(A^t t) P \exp(A t) x_0 \to -\infty$$

as $t \to \infty$. This implies that

$$x(t) = \exp(A t) x_o \to \infty$$

as $t \to \infty$, contradicting the assumption that A is a stability matrix. Hence, P must be positive definite.
Part 2. Suppose Q and P are positive definite, i.e., $-x^t Q x < 0$ and $x^t P x > 0$ for all $x \neq \theta$. Assume, for contradiction, that A is not a stability matrix. Then $x(t)$ does not approach zero as t approaches infinity. We show that this contradicts the positive definiteness of P and therefore, A must be a stability matrix.
Since $V[x(t)] = x^t(t) P x(t)$ and $\dot{V}(t) = d/dt [x^t(t) P x(t)] = -x^t(t) Q x(t)$, it follows that

$$x^t(t) P x(t) = x^t(0) P x(0) - \int_0^t x^t(q) Q x(q) dq \qquad (22.19)$$

Since $x(t)$ does not approach zero as t approaches infinity, there exists a constant $K > 0$ such that $-x^t(t) Q x(t) < -K < 0$ for all $x(t)$. Hence, from equation 22.19,

$$x^t(t) P x(t) \leqslant x^t(0) P x(0) - K t$$

It follows from this that for sufficiently large t, $x^t(t) P x(t) < 0$, contradicting the positive definiteness of P. Hence A must be a stability matrix. ∎

From another perspective, if P is positive definite and $\dot{V}(t) < 0$, then $V[x(t)] = x^t(t) P x(t)$ must decrease to zero as $t \to \infty$. Thus, $x(t) = \exp(A t) x(0) \to \theta$ as $t \to \infty$.

In order to determine the positive definiteness of P, it is necessary to solve the matrix Lyapunov equation for P. This can be accomplished in a straightforward manner by using the mathematics of Kronecker products. To this end, define the Kronecker product of two matrices $A = [a_{ij}]$ which is $m \times n$ and $B = [b_{ij}]$ which is $p \times q$ as the $mp \times nq$ matrix

$$A \otimes B = \begin{bmatrix} a_{11}B & a_{12}B & \cdots & a_{1n}B \\ a_{21}B & a_{22}B & \cdots & a_{2n}B \\ \vdots & \vdots & & \vdots \\ a_{m1}B & a_{m2}B & \cdots & a_{mn}B \end{bmatrix} \tag{22.20}$$

It is straightforward to show that the Kronecker product satisfies the properties (i) $(A \otimes B)(C \otimes D) = AC \otimes BD$ and (ii) $(A \otimes B)^t = A^t \otimes B^t$. Also, given a matrix $A = [a_1, a_2, \ldots, a_n]$ whose columns are a_1 through a_n, we define $\text{vec}(A) = \text{col}[a_1, a_2, \ldots, a_n]$, i.e., the vec operator makes a column vector out of the matrix A by stacking the columns in order— a_1 at the top and a_n at the bottom. The interesting and very useful property of the vec operator is that

$$\text{vec}(ABC) = C^t \otimes A \, \text{vec}(B) . \tag{22.21}$$

It is this property which pertains to the solution of the matrix Lyapunov equation. In particular, observe that $A^t P + PA = A^t PI + IPA = -Q$. Taking the vec of both sides produces

$$[I \otimes A^t + A^t \otimes I] \, \text{vec}(P) = - \text{vec}(Q) \tag{22.22}$$

which can be solved for $\text{vec}(P)$ and thus for P by Gaussian elimination, provided that $[I \otimes A^t + A^t \otimes I]$ is nonsingular. It turns out that this matrix is nonsingular if and only if $\lambda_i + \lambda_j \neq 0$ for all eigenvalues λ_i of A [7,8]. Clearly, if A is a stability matrix, this condition is satisfied. Hence, a unique solution exists whenever A is a stability matrix. This result can also be obtained by reasoning along a different path, as in the following theorem.

Theorem 22.10. If A is a stability matrix, then a solution to the matrix Lyapunov equation is

$$P = \int_0^\infty \exp(A^t q) Q \exp(A q) \, dq \tag{22.23}$$

Proof. The integral must exist, since A is a stability matrix. Direct substitution then yields

$$A^t P + PA = \int_0^\infty A^t \exp(A^t q) Q \exp(A q) \, dq + \int_0^\infty \exp(A^t q) Q \exp(A q) A \, dq$$

$$= \int_0^\infty \frac{d}{dt} [\exp(A^t q) Q \exp(A q)] \, dq = \exp(A^t q) Q \exp(A q) \big|_0^\infty = - Q$$

which means that P is indeed a solution to the matrix Lyapunov equation. ∎

Corollary 22.10.   If A is a stability matrix, then the matrix Lyapunov equation has a unique solution for every Q.

Proof.   Let P_1 and P_2 be two solutions satisfying the matrix Lyapunov equation $A^t P + PA = -Q$. Then, subtracting the two equations, one obtains

$$A^t(P_1 - P_2) + (P_1 - P_2)A = [0]$$

Multiplying on the left by $\exp(A^t t)$ and on the right by $\exp(At)$ produces

$$\exp(A^t t)[A^t(P_1 - P_2) + (P_1 - P_2)A]\exp(At)$$

$$= \frac{d}{dt}\exp(A^t t)[P_1 - P_2]\exp(At) = [0]$$

Thus, $\exp(A^t t)[P_1 - P_2]\exp(At)$ is a constant matrix for all t. Evaluating this expression at $t = 0$ and $t = T$ implies that

$$\exp(A^t T)[P_1 - P_2]\exp(AT) = P_1 - P_2$$

Now, since A is a stability matrix, the limit of the left hand side as $T \to \infty$ is $[0]$. Therefore, $P_1 - P_2 = [0]$ or $P_1 = P_2$.   ∎

For particulars in the discrete-time case, consult [9].

CONCLUDING REMARKS

This chapter has covered various aspects of the stability of linear time-invariant systems. Extensions of BIBS, BIBO, and Lyapunov stability are discussed in chapter 24. Unfortunately, stability characterizations for time-varying linear systems in terms of possibly time-dependent pole locations do not generalize from the time-invariant case. In fact, the literature contains examples for which a linear time-varying $A(t)$-matrix has constant eigenvalues in the left-half complex plane but for which the state trajectory is unbounded. Again chapter 24 takes up this topic.

Two interesting topics not covered in this chapter are the Routh-Hurwitz criterion and multivariable Nyquist stability theory for feedback systems. A Soviet researcher developed a generalization of the Routh-Hurwitz criterion known as Kharitonov's theorem in 1978 [10]. A simple proof of this new soon-to-be-classic result is published in [11,12]. Several works [13–17] deal with the Nyquist stability theorem. Finally, extensions of Lyapunov stability ideas to large-scale systems can be found in [18].

PROBLEMS

1. Let the impulse response matrix of a causal time-invariant linear system be

$$H(t) = 1^+(t)\begin{bmatrix} te^{-t}\cos(2t) \\ (t+1)^{-2} \end{bmatrix}$$

(a) Compute $\|H(\cdot)\|_1$. If you cannot compute it, can you obtain a bound (upper bound) for it?

(b) What can you say about the system's *external* stability?

2. Prove theorem 22.1 for the multi-input, multi-output case.

3. Let a single-input, single-output system have the state model

$$\dot{x} = \begin{bmatrix} a & b & 0 \\ -b & a & 0 \\ 0 & 0 & c \end{bmatrix} + Bu$$

$$y = Cx$$

(a) What conditions, if any, on B and C make the system BIBO stable when
   (i) $a = c = 0, b \neq 0$.   (iv) $a < 0, c = 0$.
   (ii) $a = b = c = 0$.   (v) $a < 0, c > 0$.
   (iii) $a > 0, c < 0$.

(b) Repeat part (a) for BIBS stability. (*Hint*: Compute the decomposition of $\exp[At]$.)

(c) Discuss the preceding results by viewing the matrix B as a linear combination of right eigenvectors and the matrix C as a linear combination of the conjugate transpose of the left eigenvectors.

4. Consider the system

$$\dot{x} = \begin{bmatrix} 0 & -1 & -2 \\ 0 & 1 & 2 \\ 0 & -1 & -1 \end{bmatrix} x + \begin{bmatrix} b_1 \\ b_2 \\ b_3 \end{bmatrix} u$$

$$y = [c_1 \ c_2 \ c_3] x$$

(a) What are the eigenvalues of A?

(b) What are the eigenvectors of A?

(c) Compute an expression for the state transition matrix.

(d) What conditions on $[b_1 \ b_2 \ b_3]^t$ make the system BIBS stable?

(e) Discuss the controllability of the resultant system.

(f) If $[b_1 \ b_2 \ b_3]^t = [1 \ 0 \ 0]^t$, find conditions on the vector $[c_1 \ c_2 \ c_3]^t$ which guarantee that the system is always BIBO stable.

5. Consider the time-varying system $\dot{x} = A(t)x + Bu$ with $x(0) = [1 \ 0 \ 0]^t$, where

$$A(t) = \begin{bmatrix} 0 & 0 & 0 \\ 1 & 0 & 0 \\ 1 & 1 & 0 \end{bmatrix}, 0 \leqslant t \leqslant 2$$

and

$$A(t) = \begin{bmatrix} -1.5 & 0.5 & 0 \\ 0.5 & -1.5 & 0 \\ 0 & 0 & 1 \end{bmatrix}, 2 \leqslant t$$

and

$$B = \begin{bmatrix} 0 \\ 1 \\ 1 \end{bmatrix}$$

and

$$C = \begin{bmatrix} 1 & 0 & 0 \\ 1 & 1 & 0 \end{bmatrix}$$

(a) Find the state transition matrix and the zero-input response for the system.
(b) Discuss the controllability of the system.
(c) Is the system BIBS stable? Justify your answer.
(d) Discuss the observability of the system.
(e) Is the system BIBO stable? Justify your answer.

6. Consider the system

$$\dot{x} = \begin{bmatrix} -1 & -2 & 2 \\ -1 & -3 & 2 \\ -2 & -4 & 3 \end{bmatrix} x + \begin{bmatrix} 0 & 0 \\ -1 & 1 \\ -1 & 1 \end{bmatrix} u$$

$$y = \begin{bmatrix} 1 & 1 & 0 \\ 0 & 1 & 1 \end{bmatrix} x$$

(a) Find a basis for the unobservable subspace in terms of the right eigenvectors. What states are observable?
(b) In terms of right eigenvectors, find a basis for the controllable subspace.
(c) Can θ be driven to $x_f = [0 \ 0 \ 1]^t$ in finite time? Explain.
(d) Can $x_0 = [0 \ 0 \ 1]^t$ be driven to θ in finite time? In infinite time?
(e) Repeat (d) for $x_0 = [1 \ 1 \ 1]^t$.
(f) Is the system BIBS stable? Justify your answer in terms of the R_i-matrices and their interaction with B.
(g) Is the system BIBO stable? Justify your answer in terms of the R_i-matrices and their interaction with C.

7. Answer the following questions about the state model $\{A,B,C\}$ by *carefully* justifying and/or deriving each answer. The model is

$$A = \begin{bmatrix} 0 & -1 \\ -1 & 0 \end{bmatrix}, \ B = \begin{bmatrix} 1 \\ -1 \end{bmatrix}, \ C = [1 \ 1]$$

(a) What are the eigenvalues of A?
(b) What are the eigenvectors of A?
(c) Decompose e^{At} into

$$e^{At} = R_1 e^{\lambda_1 t} + R_2 e^{\lambda_2 t}$$

(d) Is the system controllable? If not, find a basis for the controllable subspace.
(e) Is the system observable? If not, find a basis for the observable subspace.
(f) Do the controllable and observable subspaces overlap? Explain.
(g) Given that the initial condition $x(0) = x_0 = [0 \ 0]^t$, find an input, constant over the time interval $[0 \ 1]$, which will drive the system from the zero state to $x(1) = [2 \ -2]^t$.
(h) Given that at $t = 1$, $y(1) = 3$, find an initial condition $x(0)$ (which is *in the observable subspace*) such that the zero-input response produces $y(1) = 3$.
(i) What is the transfer function of the system?
(j) Is the system BIBS stable? Explain.
(k) Is the system BIBO stable? Explain.

8. A linear discrete-time state model is given by

$$\begin{bmatrix} x_1(k+1) \\ x_2(k+1) \end{bmatrix} = \begin{bmatrix} -2.5 & -4.5 \\ 3 & 5 \end{bmatrix} \begin{bmatrix} x_1(k) \\ x_2(k) \end{bmatrix} + \begin{bmatrix} 3 \\ -2 \end{bmatrix} u(k)$$

$$y(k) = [3 \quad 4] \begin{bmatrix} x_1(k) \\ x_2(k) \end{bmatrix}$$

   (a) Find the discrete-time transfer function $H(z)$.
   (b) Investigate the controllability of the system.
   (c) Investigate the observability of the system.
   (d) Investigate the stability of the system. Be sure to consider the directed modes.
   (e) If $u(k) = \sqrt{3} \cos(k\pi/3)$ and $x(0) = \theta$, find $y(k)$ as $k \to \infty$.
   (f) Repeat part (e) if $x(0) = [1 \ 0]^t$.

9. Let

$$x(k+1) = Ax(k) + B(k)u(k)$$

$$y(k) = C(k)x(k)$$

be a discrete-time system with distinct eigenvalues. Let the eigenvalue, right eigenvector, and left eigenvector triples of A be (λ_1,e_1,w_1), (λ_2,e_2,w_2), (λ_3,e_3,w_3), and (λ_4,e_4,w_4). Finally, suppose $\lambda_1 = 0$, $\lambda_2 = \lambda_3 \neq \lambda_4$, and $|\lambda_2| = 1$. Answer the following questions with specific reference to these eigenvalue-eigenvector triples. Adequately explain and/or justify your answers.

   (a) If $B(k)$ is *constant*, discuss the BIBS stability of the system if $|\lambda_4| = 1$. Describe and explain any conditions which will guarantee BIBS stability in terms of the (λ_i,e_i,w_i).
   (b) Suppose again that $B(k)$ is *constant* and the system is completely controllable. Suppose $|\lambda_4| > 1$ and suppose $C(k)$ is also constant. What conditions on the modes (λ_i,e_i,w_i) will guarantee the BIBO stability of the system?
   (c) Suppose $C(k) = C_1$ for $k \leqslant 9$ and the pair (C_1,A) is *not* completely observable. Suppose also $C(k) = C_2$ for $k \geqslant 10$ and the pair (C_2,A) is completely observable. Finally, suppose the input and output sequences $\{u(k)\}$ and $\{y(k)\}$ are known. Is it possible to determine $x(5)$? If so, how?
   (d) Suppose $B(k) = B_1$ for $k \leqslant 9$ and the pair (A,B_1) is controllable. Suppose $B(k) = B_2$ for $k \geqslant 10$ and the pair (A,B_2) is *not* completely controllable. Is it possible to drive $x(0) = \theta$ (the zero vector) to some arbitrarily given $x(15)$ by some input sequence? If so, describe how?

10. Consider the linear time-invariant state model

$$\dot{x} = \begin{bmatrix} 1 & 2 & 0 \\ 2 & 1 & 0 \\ 0 & 0 & 0 \end{bmatrix} x + \begin{bmatrix} 0 \\ 0 \\ 1 \end{bmatrix} u$$

$$y = \begin{bmatrix} c_{11} & c_{12} & c_{13} \\ c_{21} & c_{22} & c_{23} \end{bmatrix} x$$

   (a) What are the system eigenvalues?
   (b) What are the system eigenvectors?
   (c) Using the eigenvalue-eigenvector method, compute an expression for the state transition matrix of the system.
   (d) Decompose the state transition matrix into a sum of matrix directed modes.

(e) If $x(0) = [1\ 1\ 0]^t$, find $x(1)$ when $u(t) = \sin(t)\ 1^+(t)$.

(f) If possible, find an input which will drive $x(0) = [0\ 0\ 0]^t$ to $x(1) = [0\ 0\ 2]^t$. If it is not possible, explain why.

(g) Is the system BIBS stable? Justify your answer.

(h) What conditions on the entries of the C-matrix make the system BIBO stable?

(i) What is the transfer function matrix of the system if all unstable modes are unobservable? Justify your answer.

(j) Compute the transfer function matrix of the zero input state response.

11. Consider the discrete-time system model

$$x(k + 1) = \begin{bmatrix} 0 & 0.5 \\ -0.5 & 0 \end{bmatrix} x(k) + \begin{bmatrix} 1 \\ 1 \end{bmatrix} u(k)$$

$$y(k) = [1\ -1] x(k) + [1] u(k)$$

This model has the form

$$x(k + 1) = Ax(k) + Bu(k)$$

$$y(k) = Cx(k) + Du(k)$$

(a) Compute $(zI - A)^{-1}$.

(b) Construct a partial fraction expansion of $(zI - A)^{-1}$.

(c) Using the inverse z-transform, construct a decomposition of the discrete-time state transition matrix.

(d) Use the result of part (b) to construct the transfer function matrix.

(e) Determine the BIBS stability of the system.

(f) Determine the BIBO stability of the system.

(g) Prove or disprove the statement, "If $u(k) = 0$ for all k, then $y(k)$ approaches zero as k approaches ∞."

(h) Find a difference equation in terms of $u(k)$ and $y(k)$ which represents the input-output characteristics of the given state model.

12. For a discrete-time system

$$x(k + 1) = Ax(k) + Bu(k)$$

(a) Give a definition of when the state trajectory is *asymptotically* stable.

(b) Assuming A has distinct eigenvalues, what condition(s) must these eigenvalues satisfy to guarantee the asymptotic stability of the state trajectory?

(c) What is a sufficient condition to guarantee that the state trajectory is unbounded? Derive this condition, assuming A has distinct eigenvalues.

13. Consider the continuous-time state model

$$\dot{x} = Ax + Bu$$

$$y = Cx$$

having three states. The three eigenvalues-eigenvector triples (λ_i, e_i, w_i) of A are $(\lambda_1 = -1, (1, 1, 0)^t, (0, 1, -2)^t)$, $(\lambda_2 = 0, (1, 0, 0)^t, (1, -1, 3)^t)$, and $(\lambda_3 = 1, (-1, 2, 1)^t, (0, 0, 1)^t)$.

(a) Determine a 1×3 C-matrix for which modes 2 and 3 are completely unobservable.

(b) Characterize the class of B-matrices for which the system is BIBS stable.

(c) If mode 2 is uncontrollable, characterize the class of C-matrices for which the system is BIBO stable.

(d) If the system is single-input, single-output, mode 2 is uncontrollable, and mode 3 is unobservable, find $H(s)$ if $H(0) = 2$.

14. The linear differential equation $\dot{x}(t) = Ax(t) + Bu(t)$, where

$$A = \begin{bmatrix} 0 & 1 & 0 & 0 \\ 3\omega^2 & 0 & 0 & 2\omega \\ 0 & 0 & 0 & 1 \\ 0 & -2\omega & 0 & 0 \end{bmatrix}, \quad B = \begin{bmatrix} 0 & 0 \\ 1 & 0 \\ 0 & 0 \\ 0 & 1 \end{bmatrix}$$

is obtained as the result of linearizing the nonlinear equations of motion of an orbiting satellite about a steady-state solution. The state $x_1(t)$ is the differential radius $r_\Delta(t)$, while $x_3(t)$ is the differential angle $\theta_\Delta(t)$. The input $u_1(t)$ is the radial thrust and $u_2(t)$ is the tangential thrust.

(a) Is the system controllable? Consider directed modes.

(b) Is the system stable? Consider directed modes.

(c) If the radial thruster fails, does the system remain controllable?

(d) If the tangential thruster fails, does the system remain controllable?

(e) If

$$y(t) = \begin{bmatrix} x_1(t) \\ x_3(t) \end{bmatrix}$$

determine the transfer matrix of the system.

15. (a) Prove theorem 22.5.

(b) Prove theorem 22.6.

(c) Prove theorem 22.7.

(d) Prove theorem 22.8.

16. The discussion preceding theorem 22.9 says that if $V(x)$ determines a positive definite quadratic form and $\dot{V}(x) < 0$, then $x(t) \to \theta$ as $t \to \infty$. To prove this result, let $V(x) = K_1^2$ determine a closed surface and $V(x) = K_2^2$ a second such surface, where $K_2 < K_1$. Show that if $x(t_0) \in \{x \mid V(x) = K_1^2\}$, then $x(t)$ must pierce the surface $\{x \mid V(x) = K_2^2\}$ in a finite time. Continue this argument to show that $x(t) \to \theta$ as $t \to \infty$.

REFERENCES

1. H. L. Royden, *Real Analysis* 2d ed. (New York: Macmillan, 1983).

2. Frigyes Riesz and Bela Sz.-Nagy, *Functional Analysis* (New York: Frederick Ungar, 1955).

3. C. A. Desoer, *Notes for a Second Course on Linear Systems* (New York: Van Nostrand Reinhold, 1970).

4. Paul M. Lapsa, "A Proof of Stability Condition," *Signal Processing*, forthcoming.

5. M. A. Lyapunov, "Problème général de la stabilité du mouvement," in *Annals of Mathematical Studies,* 17, (Princeton, NJ: Princeton University Press, 1949, reprinted 1983).

6. Stan Żak, *Stability of Linear Time-Invariant Systems,* course notes, Purdue University, 1987.

7. Jacques L. Willems, *Stability Theory of Dynamical Systems* (New York: Wiley Interscience, 1970).

8. Thomas Kailath, *Linear Systems* (Englewood Cliffs, NJ: Prentice Hall, 1980).

9. J. P. LaSalle, *The Stability and Control of Discrete Processes* (New York: Springer-Verlag, 1986).

10. V. L. Kharitonov, "Asymptotic Stability of an Equilibrium Position of a Family of Linear Differential Equations," *Differential Uravnenia*, Vol. 14, No. 11, 1978, pp. 2086–2088.

11. K. S. Yeung and S. S. Wang, "A Simple Proof of Karitonov's Theorem," *IEEE Transactions on Automatic Control*, Vol. AC-32, No. 9, 1987, pp. 822–823.

12. N. K. Bose and Y. Q. Shi, "A Simple General Proof of Kharitonov's Generalized Stability Criterion," *IEEE Transactions of Circuits and Systems*, Vol. CAS-34, No. 8, October 1987.

13. John F. Barman and Jacob Katznelson, "A Generalized Nyquist-type Stability Condition for Multivariable Feedback Systems," *International Journal of Control*, Vol. 20, No. 4, 1974, pp. 593–622.

14. R. DeCarlo and R. Saeks, "The Encirclement Condition: An Approach Using Algebraic Topology," *International Journal of Control*, Vol. 26, No. 2, 1977, pp. 279–287.

15. Ian Postlethwaite and A. G. J. MacFarlane, *A Complex Variable Approach to the Analysis of Linear Multivariable Feedback Systems* (New York: Springer-Verlag, 1979).

16. C. A. Desoer and Y.-T. Wang, "On the Generalized Nyquist Stability Criterion," *IEEE Transactions on Automatic Control*, Vol. AC-25, No. 2, April 1980, pp. 187–196.

17. R. A. DeCarlo, J. Murray, and R. Saeks, "Multivariable Nyquist Theory," *International Journal of Control*, Vol. 25, No. 5, 1977, pp. 657–675.

18. A. N. Michel and R. K. Miller, *Qualitative Analysis of Large Scale Dynamical Systems* (New York: Academic Press, 1977).

23

CONTROLLABILITY AND OBSERVABILITY: THE TIME-VARYING CASE

INTRODUCTION

Beginning at Chapter 14, our developmental trek veered from the general linear time-varying setting to an in-depth coverage of the time-invariant case. A good deal of conceptual growth has taken place since then. With this maturity in place, our journey returns to the linear time-varying state model and re-investigates the notions of observability and controllability.

Recall the time-varying linear state model

$$\dot{x}(t) = A(t)x(t) + B(t)u(t) \qquad (23.1a)$$

$$y(t) = C(t)x(t) + D(t)u(t) \qquad (23.1b)$$

where each matrix has piecewise continuous entries over $[0,\infty)$. The state dynamics of this model thus satisfy a Lipschitz condition, resulting in the existence of a unique solution given by

$$x(t) = \Phi(t,t_0)x(t_0) + \int_{t_0}^{t} \Phi(t,\tau)B(\tau)u(\tau)d\tau \qquad (23.2)$$

and

$$y(t) = C(t)\Phi(t,t_0)x(t_0) + C(t)\int_{t_0}^{t} \Phi(t,\tau)B(\tau)u(\tau)d\tau + D(t)u(t) \qquad (23.3)$$

where $\Phi(t,t_0)$ is the state transition matrix. One general objective of this chapter is to define the notion of controllability for equations 23.1 and use the solution structure of equation 23.2 to characterize the concept. The dual goal is to use the solution structure of equation 23.3 to formulate a theory of observability.

Rather than start with controllability, as was done in the time-invariant case, our investigation reverses the direction by beginning with observability. Before proceeding recall some essential mathematical concepts.

MATHEMATICAL PRELIMINARIES

Recall definition 11.1: the set of piecewise continuous functions $\{\phi_1(\cdot),\phi_2(\cdot),...,\phi_n(\cdot)\}$ is *linearly independent over the interval* $[t_0,t_1]$ if, whenever

$$\alpha_1\phi_1(\cdot) + \alpha_2\phi_2(\cdot) + ... + \alpha_n\phi_n(\cdot) \equiv 0$$

over $[t_0,t_1]$ (i.e., the linear combination equals the identically zero function) implies that $\alpha_1 = \alpha_2 = ... = \alpha_n = 0$. This definition, as well as the following, are fundamental to our development.

Definition 23.1.   A complex $n \times n$ matrix M is *Hermitian* if $M^* = M$ where * denotes the complex conjugate transpose; if M is real and Hermitian, then M is *symmetric*.

Exercise. Let M be a Hermitian matrix. Prove that for all $z \in \mathbf{C}^n$ (complex n-space), the quantity z^*Mz is real.

The notion of a Hermitian matrix is central to the definition of a (complex) positive definite matrix defined informally in the previous chapter and formally next.

Definition 23.2.   An $n \times n$ Hermitian matrix M is *positive definite* (*positive semi-definite*) if and only if for all nonzero $z \in \mathbf{C}^n$, $z^*Mz > 0$ ($z^*Mz \geq 0$).

Exercise. Show that if $M = T^*T$ where T is an $m \times n$ complex matrix with $n \leq m$, then M is positive semi-definite.

Problems 1–3 at the end of the chapter deal with other properties of positive definite and semidefinite matrices.

Note that the quantity z^*Mz is a quadratic (sometimes called Hermitian) form. The quadratic form allows one to order $n \times n$ symmetric matrices in terms of the associated quadratic form: $M_1 \geq M_2$ if $x^*M_1x \geq x^*M_2x$ for all $x \in \mathbf{R}^n$. An analogous ordering holds for $M_1 \leq M_2$.

Exercise. Show that $M_1 \geqslant M_2$ if and only if $M_1 - M_2$ is positive semidefinite.

BASICS OF TIME-VARYING OBSERVABILITY

Essentially, the notion of observability involves the reconstructibility of an initial state $x(t_0)$ given input-output measurements $(y(t), u(t))$ over some interval $[t_0, t_1]$. Given a knowledge of the system matrices, the observations $(y(t), u(t))$, the fact that $\Phi(t, t_0)$ is known (at least theoretically), and the relationship defined by equation 23.3, it is possible, by doing the necessary calculations, to define a modified observation $\hat{y}(t)$ satisfying

$$\hat{y}(t) = y(t) - C(t) \int_{t_0}^{t} \Phi(t, \tau) B(\tau) u(\tau) d\tau - D(t) u(t)$$

$$= C(t) \Phi(t, t_0) x(t_0) \tag{23.4}$$

Thus the problem of determining $x(t_0)$ from observations $(y(t), u(t))$ is equivalent, at least theoretically, to the problem of determining $x(t_0)$ from observations $\hat{y}(t)$. The theory is thus essentially independent of the system input. The structure of equation 23.4 then justifies (at least for the present) consideration of only the zero-input state model,

$$\dot{x}(t) = A(t) x(t), \qquad x(t_0) = x_0 \tag{23.5a}$$

$$y(t) = C(t) x(t) \tag{23.5b}$$

That is to say, the state model of equation 23.1 is observable if and only if the state model of equations 23.5 is.

The notion of observability is best pinned down by first defining the notion of an unobservable state. See [1] for a similar development.

Definition 23.3. A state $x(t_0)$ is *completely unobservable over the interval* $[t_0, t_1]$ if the zero-input response $y(t) = C(t) \Phi(t, t_0) x(t_0) \equiv 0$ over $[t_0, t_1]$.

It is easy to show that the set of states defined in definition 23.3 is a linear subspace, referred to as the *unobservable subspace*. Thus, *a system is completely observable if no state is unobservable.* Following Russell [2], it is then possible to define system observability.

Definition 23.4. The time-varying linear state model, or, equivalently, the pair $(C(t), A(t))$ is *completely observable* (often, just observable) *over the interval* $[t_0, t_1]$ if and only if the zero-input response $y(t) = C(t) x(t) \equiv 0$ over $[t_0, t_1]$ implies that $x(t) \equiv 0$ over $[t_0, t_1]$, or, equivalently, $x(t_0) = 0$.

This definition differs from our earlier development in chapter 21 where observability was defined as the ability to determine $x(t_0)$ from observations $y(t)$ over $[t_0, t_1]$. In the continuous time case definition 23.4 is equivalent to this statement of reconstructability. In the discrete time case it is not.

EXAMPLE 23.1

Consider the system

$$\begin{bmatrix} \dot{x}_1 \\ \dot{x}_2 \end{bmatrix} = \begin{bmatrix} 0 & 0 \\ 0 & -2t \end{bmatrix} \begin{bmatrix} x_1 \\ x_2 \end{bmatrix}, \qquad x(0) = \begin{bmatrix} 1 \\ -1 \end{bmatrix}$$

$$y = \begin{bmatrix} 1 & e^{t^2} \end{bmatrix} \begin{bmatrix} x_1 \\ x_2 \end{bmatrix}$$

A simple calculation shows that

$$y(t) = C(t) x(t) = \begin{bmatrix} 1 & e^{t^2} \end{bmatrix} \begin{bmatrix} 1 & 0 \\ 0 & e^{-t^2} \end{bmatrix} \begin{bmatrix} 1 \\ -1 \end{bmatrix} = [0]$$

i.e., $y(t) = 0$ for all $t \geqslant 0$. Hence, the system is not observable over any interval $[t_0, t_1]$. In fact, any state of the form $(1 \ -1)^t$ is completely unobservable, and any state having a projection onto the unobservable subspace $\{x \mid x = \beta(1 \ -1)^t\}$ cannot be reconstructed from observations and is thus not observable.

Recall the notation from chapters 10 and 11 for a solution to 23.5a: a solution $\phi(t, t_0, x_0)$ to equation 23.5a satisfies that equation in such a manner that $\phi(t_0, t_0, x_0) = x_0$.

Theorem 23.1. Let $x_0^1, \ldots, x_0^n$ be vectors in $\mathbf{R}^n$ and let associated solutions to $\dot{x}(t) = A(t)x(t)$ on $[t_0, t_1]$ be denoted by $\phi_i(t, t_0, x_0^i)$. Define the associated responses (observations) by

$$y^i(t) = C(t)\phi_i(t, t_0, x_0^i) \qquad (23.6)$$

for t in $[t_0, t_1]$. Then the pair $(C(t), A(t))$ is observable on $[t_0, t_1]$ if and only if the $y^i(\cdot)$, $i = 1, \ldots, n$, are linearly independent vector functions on $[t_0, t_1]$ whenever the $\phi_i(\cdot, t_0, x_0^i)$, $i = 1, \ldots, n$, are.

Proof. From the development of chapters 11 and 12, the functions $\phi_i(t, t_0, x_0^i)$, $i = 1, \ldots, n$, are linearly independent over $[t_0, t_1]$ if and only if the vectors $x_0^1, \ldots, x_0^n$ are. We define an arbitrary solution to the state dynamical equations 23.5 by

$$x(t) = \sum_{i=1}^{n} \alpha_i \phi_i(t, t_0, x_0^i)$$

and the corresponding system response (observation) as

$$y(t) = C(t)x(t) = \sum_{i=1}^{n} \alpha_i \, C(t)\phi_i(t, t_0, x_0^i) = \sum_{i=1}^{n} \alpha_i \, y^i(t)$$

Now suppose the pair $(C(t),A(t))$ is observable. Then from definition 23.4, if

$$y(t) = \sum_{i=1}^{n} \alpha_i\, y^i(t) = 0$$

over $[t_0,t_1]$, then

$$x(t) = \sum_{i=1}^{n} \alpha_i\, \phi_i(t,t_0,x_0^i) = 0$$

over $[t_0,t_1]$. Now, if the x_0^i's $(i = 1,...,n)$ are independent, then the solutions $\phi_i(t,t_0,x_0^2)$ $(i = 1,...,n)$ are independent, and hence $\alpha_i = 0$ $(i = 1,...,n)$. But by definition, this means that the solutions $y^i(t)$ $(i = 1,...,n)$ are linearly independent vector functions over $[t_0,t_1]$.

To prove the converse, use the method of contradiction. Suppose the pair $(C(t),A(t))$ is not observable, but the vectors x_0^i $(i = 1,...,n)$ are linearly independent. Then there exists a set of (possibly complex) scalars α_i $(i = 1,...,n)$, not all zero, such that

$$y(t) = \sum_{i=1}^{n} \alpha_i\, y^i(t) = C(t) \sum_{i=1}^{n}\alpha_i\phi_i(t,t_0,x_0^i) = 0$$

However, the independence of the initial conditions x_0^i $(i = 1,...,n)$ implies independence of the functions $\phi_i(t,t_0,x_0^i)$ over $[t_0,t_1]$; i.e.,

$$x(t) = \sum_{i=1}^{n} \alpha_i\, \phi_i(t,t_0,x_0^i) \not\equiv 0$$

This represents the necessary contradiction proving the converse.   ∎

Note that the solutions $\phi_i(t,t_0,x_0^i)$ $(i = 1,...,n)$ constitute a fundamental set of solutions whenever the x_0^i $(i = 1,...,n)$ are independent. Recalling that a fundamental matrix $\Psi(t)$ (see chapter 12) is any square matrix whose columns are a fundamental set of solutions, one obtains the following corollary.

Corollary 23.1a.   The pair $(C(t),A(t))$ is observable over $[t_0,t_1]$ if and only if the columns of $C(t)\Psi(t)$ are a linearly independent set of functions over $[t_0,t_1]$.

Since the state transition matrix is a fundamental matrix, we obtain the following corollary as well.

Corollary 23.1b.   The pair $(C(t),A(t))$ is observable over $[t_0,t_1]$ if and only if the columns of $C(t)\Phi(t,t_0)$ are a linearly independent set of functions over $[t_0,t_1]$.

Testing the linear independence of functions over an interval— for example, the columns of $C(t)\Phi(t,t_0)$, is much more difficult than testing for the linear independence of vectors in $\mathbf{R}^n$. Fortunately, it is possible to define an $n \times n$ constant matrix in terms of $C(t)\Phi(t,t_0)$ called the *observability Gramian*, whose columns are independent if and only if the pair $(C(t),A(t))$ is observable. The following result generalizes equivalence (v) of theorem 21.2.

Theorem 23.2.   The pair $(C(t),A(t))$ is observable on $[t_0,t_1]$ if and only if the $n \times n$ observability Gramian

$$W_O(t_0,t_1) = \int_{t_0}^{t_1} \Phi^*(q,t_0)C^*(q)C(q)\Phi(q,t_0)dq \qquad (23.7)$$

is positive definite.

Proof.   Suppose the pair $(C(t),A(t))$ is observable on $[t_0,t_1]$. We prove that $W_O(t_0,t_1)$ is positive definite, by the method of contradiction. Suppose that $W_O(t_0,t_1)$ is not positive definite; we seek to show that the pair $(C(t),A(t))$ is not observable on $[t_0,t_1]$.

If the $n \times n$ matrix $W_O(t_0,t_1)$ is not positive definite, then, from problem 1 at the end of the chapter, its rank is less than n. Hence, there exists an n-vector $x_0 \neq 0$ such that

$$x_0^* W_O(t_0,t_1)x_0 = \int_{t_0}^{t_1} x_0^* \Phi^*(q,t_0)C^*(q)C(q)\Phi(q,t_0)x_0 dq = \int_{t_0}^{t_1} y^*(q)y(q)dq = 0$$

Now, since the integrand is a quadratic form whose value is always nonnegative, the integral equals zero if and only if the integrand is identically zero. Hence, if the observability Gramian is not positive definite, there exists a nonzero initial state x_0 whose zero-input response $y(t) = C(t)\Phi(t,t_0)x_0 = 0$ for all t in $[t_0,t_1]$. But then, the initial condition x_0 is an unobservable state, and hence the pair $(C(t),A(t))$ is not observable, as was to be shown.

For the converse, suppose $W_O(t_0,t_1)$ is positive definite and $y(t) = C(t)x(t) = 0$ for all t in $[t_0,t_1]$. Then by definition, if $x(t) = 0$ for all t in $[t_0,t_1]$, the pair $(C(t),A(t))$ is observable. Now, since $y(t) = C(t)\Phi(t,t_0)x_0 = 0$, it follows that

$$\int_{t_0}^{t_1} y^*(q)y(q)dq = \int_{t_0}^{t_1} x_0^* \Phi^*(q,t_0)C^*(q)C(q)\Phi(q,t_0)x_0 dq = x_0^* W_O(t_0,t_1)x_0 = 0$$

Since $W_O(t_0,t_1)$ is positive definite, (i.e., rank$[W_O(t_0,t_1)] = n$), $x_0 = 0$. But then it follows that $x(t) = 0$ on $[t_0,t_1]$.   ∎

EXAMPLE 23.2

Consider the state model

$$\dot{x}(t) = \begin{bmatrix} 0 & \cos(t) \\ 0 & 0 \end{bmatrix} x(t) \qquad (23.8a)$$

$$y(t) = [1 \quad -\cos(t)]x(t) \qquad (23.8b)$$

From some simple calculations, we obtain

$$C(t)\Phi(t,t_0) = [1 \quad -\cos(t)]\begin{bmatrix} 1 & \sin(t) - \sin(t_0) \\ 0 & 1 \end{bmatrix}$$

$$= [1 \quad \sin(t) - \cos(t) - \sin(t_0)] \qquad (23.9)$$

Let us use the Gramian test given in theorem 23.2 to determine the observability of the system over the interval $[0,\pi]$. To compute the Gramian over $[0,\pi]$, observe that

$$\Phi^*(q,t_0)C^*(q)C(q)\Phi(q,t_0) = \begin{bmatrix} 1 \\ \sin(q) - \cos(q) \end{bmatrix} [1 \quad \sin(q) - \cos(q)]$$

$$= \begin{bmatrix} 1 & \sin(q) - \cos(q) \\ \sin(q) - \cos(q) & 1 - \sin(2q) \end{bmatrix}$$

Hence,

$$W_O(t_0,t_1) = \begin{bmatrix} \pi & 2 \\ 2 & \pi \end{bmatrix}$$

which is easily shown to be positive definite, so that 23.8 is observable.

As intuitively developed in chapter 21, a system that is observable on the interval $[t_0,t_1]$ is observable on any interval containing $[t_0,t_1]$. The foregoing Gramian test leads to a straightforward proof of this well-known and useful result.

Corollary 23.2.   If the pair $(C(t),A(t))$ is observable over $[t_0,t_1]$, then the pair is observable over $[t_0',t_1']$, where $t_0' \leqslant t_0 < t_1 \leqslant t_1'$.

Proof.   Using the semigroup property of the state transition matrix (equation 23.10), we obtain $\Phi(t_1',t_0') = \Phi(t_1',t_0)\Phi(t_0,t_0')$. Thus, the Gramian over the interval $[t_0',t_1']$ is

$$W_O(t_0',t_1') = \Phi^*(t_0,t_0') \int_{t_0'}^{t_1'} \Phi^*(q,t_0)C^*(q)C(q)\Phi(q,t_0)dq \; \Phi(t_0,t_0') \qquad (23.10)$$

The integral on the right side of equation 23.10 has an integrand which is nonnegative for all q. Hence, by changing the interval of integration from $[t_0',t_1']$ to $[t_0,t_1]$, it follows that

$$W_O(t_0',t_1') \geqslant \Phi^*(t_0,t_0')W_O(t_0,t_1)\Phi(t_0,t_0')$$

using the ordering for symmetric matrices defined in the mathematical preliminaries. Now, by assumption, $W_O(t_0,t_1)$ is positive definite; consequently, $\Phi^*(t_0,t_0')W_O(t_0,t_1)\Phi(t_0,t_0')$ is positive definite, so that $W_O(t_0',t_1')$ is also positive definite. Hence, $(C(t),A(t))$ is observable over $[t_0,t_1]$.   ∎

The computation of the observability Gramian is numerically difficult since it first requires calculaton of $\Phi(t,t_0)$. Hence, one searches for other tests which are more convenient. The following two propositions afford such tests.

Proposition 23.1.   The pair $(C(t),A(t))$ is observable over $[t_0,t_1]$ if there are distinct points $t_1',...,t_p'$ such that

$$\text{rank} \begin{bmatrix} C(t'_1)\Phi(t'_1,t_0) \\ C(t'_2)\Phi(t'_2,t_0) \\ \vdots \\ C(t'_p)\Phi(t'_p,t_0) \end{bmatrix} \triangleq \text{rank}[\boldsymbol{O}(t'_1,t'_2,...,t'_p,t_0)] = n$$

where $\boldsymbol{O}$ denotes the matrix on the left-hand side of the equation.

Proof. The key to this proof is to relate the observability Gramian to the matrix $\boldsymbol{O}(t'_1,t'_2,...,t'_p,t_0)$ and show that if the rank of $\boldsymbol{O}$ is n, then the Gramian is positive definite. To accomplish this, note that if $\text{rank}[\boldsymbol{O}(t'_1,t'_2,...,t'_p,t_0)] = n$, then

$$\boldsymbol{O}^*(t'_1,t'_2,...,t'_p,t_0)\boldsymbol{O}(t'_1,t'_2,...,t'_q,t_0) = \sum_{i=1}^{p} \Phi^*(t'_i,t_0)C^*(t'_i)C(t'_i)\Phi(t'_i,t_0) \quad (23.11)$$

is positive definite (see problem 1 at the end of the chapter). To link equation 23.11 with the observability Gramian, choose δ such that $0 < \delta < 0.5\min|t'_j - t'_i|$ for $i \neq j$. Further, choose numbers $t'_i \pm \delta$ such that $t'_i \pm \delta$ always lies in the interval $[t_0,t_1]$—i.e., one is free to choose either "+" or "−" unless the choice puts the resulting number outside the interval. Then, given the integral relationship of the observability Gramian as defined in equation 23.7 to the ordering of symmetric matrices introduced in the mathematical preliminaries, we have the inequality

$$W_O(t'_0,t'_1) \geqslant \sum_{i=1}^{n} \left| \int_{t'_i}^{t'_i \pm \delta} \Phi^*(q,t_0)C^*(q)C(q)\Phi(q,t_0)dq \right| \quad (23.12a)$$

$$= \delta\,\boldsymbol{O}^*(t'_1,t'_2,...,t'_p,t_0)\boldsymbol{O}(t'_1,t'_2,...,t'_q,t_0) + E(\delta) \quad (23.12b)$$

where $E(\delta)$ represents an error matrix whose entries are proportional to δ^2 with the property that $[E(\delta)/\delta] \to 0$ as $\delta \to 0$. Further, the first term on the right-hand side of equation 23.12b is a forward Euler approximation to the sum of integrals on the right-hand side of equation 23.12a while $E(\delta)$ represents the actual error in the approximation. Clearly, $E(\delta)$ is symmetric, and, given the positive definiteness of $\boldsymbol{O}^*(t'_1,t'_2,...,t'_p,t_0)\boldsymbol{O}(t'_1,t'_2,...,t'_q,t_0)$, the right-hand side of equation 23.12b is positive definite for sufficiently small δ. But then, $W_O(t'_0,t'_1)$ is positive definite, and observability is assured. ∎

To illustrate the applicability of proposition 23.1, consider again example 23.2, which deals with the state model given by equations 23.8. From equation 23.9 we have

$$C(t)\Phi(t,t_0) = [1 \quad \sin(t) - \cos(t) - \sin(t_0)]$$

in which case

$$\boldsymbol{O}(t'_1,t'_2,t_0) = \begin{bmatrix} 1 & \sin(t'_1) - \cos(t'_1) - \sin(t_0) \\ 1 & \sin(t'_2) - \cos(t'_2) - \sin(t_0) \end{bmatrix} = \begin{bmatrix} 1 & -1 \\ 1 & 1 \end{bmatrix}$$

for the choices $t_0 = 0$, $t'_1 = 0$, and $t'_2 = 0.5\pi$. The rank is clearly 2.

The next proposition is similar to the rank test on the observability matrix in the time-invariant case. In fact, the time-invariant observability matrix rank test is a corollary to the proposition.

Proposition 23.2. The pair $(C(t),A(t))$ is observable over $[t_0,t_1]$ if there is some positive integer q and some point t' in $[t_0,t_1]$ such that

$$\text{rank}\begin{bmatrix} C(t')\Phi(t',t_0) \\ D[C(t)\Phi(t,t_0)]|_{t=t'} \\ \vdots \\ D^q[C(t)\Phi(t,t_0)]|_{t=t'} \end{bmatrix} \triangleq \text{rank}[R_q(t')] = n \qquad (23.13)$$

where $D^q = d^k/dt^k$ is the notation from chapter 4 for the derivative operator and $C(t)$ and $\Phi(t',t_0)$ are q times differentiable.

Proof. Suppose that $y(t) = C(t)x(t) = C(t)\Phi(t,t_0)x_0 = 0$ for all t in $[t_0,t_1]$. The crux of the proof is to show that the condition of equation 23.13 guarantees that $x_0 = 0$. Now, since $y(t)$ is identically zero on $[t_0,t_1]$, all its derivatives are also, i.e.,

$$D^k[C(t)\Phi(t,t_0)]x_0 = 0$$

for all t in $[t_0,t_1]$ and for $k = 1,...,q$. But the set of these equations is equivalent to the equation

$$R_q(t)x_0 = 0 \qquad (23.14)$$

At $t = t'$, $R_q(t')$ has rank n. But this means that for equation 23.14 to be satisfied, we must have $x_0 = 0$.   ∎

As an illustration of proposition 23.2, consider again the system of example 23.2 as given in equation 23.8. For that system, the matrix

$$R_q(t') = \begin{bmatrix} 1 & \sin(t') - \cos(t') - \sin(t_0) \\ 0 & \cos(t') + \sin(t') \end{bmatrix}$$

Given any interval $[t_0,t_1]$ with $t_1 > t_0$, any choice of t' for which $\cos(t') + \sin(t') \neq 0$ will work. Hence, this test is consistent with the results of the earlier examples.

The linear time-invariant test for complete observability follows as a natural corollary to this general result.

Corollary 23.2. The time-invariant state model is completely observable over $[t_0,t_1]$ if and only if the observability matrix

$$R = \begin{bmatrix} C \\ CA \\ \vdots \\ CA^{n-1} \end{bmatrix}$$

has rank n.

Proof. The "if part" of the corollary follows directly from proposition 23.2. The "only if part" is left as an exercise; it requires the use of either the Caley-Hamilton theorem or lemmas 20.1 and 20.2. ∎

Exercise. Fill in the details of the proof to corollary 23.2.

RECONSTRUCTION OF THE INITIAL STATE

Chapter 21 (theorem 21.2) presented a set of equivalences for complete observability in the time-invariant case. There, the observability Gramian (equation 21.7) was first mentioned, as was the formula for reconstructing the initial state in terms of the observability Gramian, (equation 21.8) viz.,

$$x(t_0) = \int_{t_0}^{t_1} W_O^{-1}(t_0,t_1) e^{A^t(q - t_0)} C^t y^M(q) \, dq \qquad (23.15)$$

where $y^M(t)$ is the measurement (or observation) function over $[t_0,t_1]$. In the present context $y^M(t) = C(t)\Phi(t,t_0)x(t_0)$. Following Russell [2], our goal here is to fit this formula into a more general setting, that of the *reconstruction kernel*. Specifically, equation 23.15 serves to motivate the following definition.

Definition 23.5. The $n \times r$ matrix function $\mathcal{R}(t)$ defined on $[t_0,t_1]$ is an initial state *reconstruction kernel* if and only if

$$\int_{t_0}^{t_1} \mathcal{R}(q) \, C(q)\Phi(q,t_0) \, dq = I \qquad (23.16)$$

where I is the $n \times n$ identity matrix.

Clearly if $\mathcal{R}(t)$ is a reconstruction kernel, then multiplying equation 23.16 on the right by $x_0 = x(t_0)$ indicates that if $y^M(t)$ is a measurement over $[t_0,t_1]$, then

$$\int_{t_0}^{t_1} \mathcal{R}(q) \, y^M(t) = x_0 \qquad (23.17)$$

since $y^M(t) = C(t)\Phi(t,t_0)x_0$. As one would expect such a reconstruction kernel exists if and only if the system is observable.

Proposition 23.3. The reconstruction kernel $\mathcal{R}(t)$, as defined in equation 23.16, exists on $[t_0,t_1]$ if and only if the system 23.5 is observable on $[t_0,t_1]$.

Proof. Suppose a reconstruction kernel $\mathcal{R}(t)$ exists. To show observability apply definition 23.4, i.e., assume $y^M(t) = C(t)\Phi(t,t_0)x_0 = 0$ on $[t_0,t_1]$ and show that $x_0 = 0$. This however follows immediately from equation 23.17.

To prove the converse suppose the system is observable. Then from theorem 23.2 it follows that the observability Gramian $W_O(t_0,t_1)$ is positive definite. Problem 1 at the end of the chapter then guarantees that $W_O(t_0,t_1)$ is nonsingular. Next, define the obvious candidate reconstruction kernel

$$\mathcal{R}(t) = W_O(t_0,t_1)^{-1}\Phi^*(t,t_0)C^*(t) \qquad (23.18)$$

To verify that this is in fact a valid reconstruction kernel, observe that

$$\int_{t_0}^{t_1} W_O(t_0,t_1)^{-1}\Phi^*(q,t_0)C^*(q)C(q)\Phi(q,t_0)dq = W_O(t_0,t_1)^{-1}W_O(t_0,t_1) = I$$

and thus $\mathcal{R}(t)$ satisfies definition 23.5.   ∎

EXAMPLE 23.3

Consider again example 23.2. With the interval $[0,\pi]$, the obvious reconstruction kernel has the form

$$\mathcal{R}(t) = (\pi^2 - 4)^{-1} \begin{bmatrix} \pi & -2 \\ -2 & \pi \end{bmatrix} \begin{bmatrix} 1 \\ \sin(t) - \cos(t) \end{bmatrix}$$

A straightforward substitution and subsequent calculation shows that $\mathcal{R}(t)$ satisfies definition 23.5.

The last phase of the development of observability takes up the case of determining the initial state given the driven state model of equations 23.1.

Proposition 23.4.   If $x(t)$ satisfies equation 23.1a and the pair $(C(t),A(t))$ is observable on $[t_0,t_1]$, then it is possible to compute

$$x_0 = \int_{t_0}^{t_1} [\mathcal{R}(q)y(q) - Q(q)B(q)u(q)]dq \qquad (23.19)$$

where $\mathcal{R}(q)$ is a reconstruction kernel and

$$Q(t) = \int_t^{t_1} \mathcal{R}(q)C(q)\Phi(q,t)dq \qquad (23.20)$$

Proof.   Consider again equation 23.4:

$$\hat{y}(t) = y(t) - C(t)\int_{t_0}^t \Phi(t,\tau)B(\tau)u(\tau)d\tau - D(t)u(t) = C(t)\Phi(t,t_0)x(t_0)$$

This equation permitted consideration of the zero-input state model given by equation 23.5 for the development of the observability results so far obtained. Implicitly, the $y(t)$ in equation 23.5 is in reality $\hat{y}(t)$, the hat notation dispensed with out of convenience. However by resurrecting that notation, we obtain, from the reconstruction kernel formula 23.17

$$\int_{t_0}^{t_1} \mathcal{R}(q)\hat{y}(q)dq = x_0$$

where $\hat{y}(t)$ is defined by equation 23.4 and $\mathcal{R}(q)$ by equation 23.18. Substituting for $\hat{y}(t)$ and distributing the integral over the sum yields

$$\int_{t_0}^{t_1} \mathcal{R}(q)[y(q) - D(q)u(q)]dq - \int_{t_0}^{t_1} \mathcal{R}(q)C(q) \int_{t_0}^{q} \Phi(q,\tau)B(\tau)u(\tau)d\tau \, dq$$

Noting that (1) the state transition matrix is separable, i.e., $\Phi(q,\tau) = \Psi(q)\Psi(\tau)^{-1}$ for any fundamental matrix $\Psi(q)$, (2) the inside integral results in a function of q, and (3) $t_0 \leqslant \tau \leqslant q \leqslant t_1$, one obtains the relationship

$$x_0 = \int_{t_0}^{t_1} \mathcal{R}(q)[y(q) - D(q)u(q)]dq - \int_{t_0}^{t_1}\int_{\tau}^{t_1} \mathcal{R}(q)C(q)\Phi(q,\tau)dq \, B(\tau)u(\tau)d\tau$$

Since q and τ are dummy variables of integration, a name change is valid. The result is

$$x_0 = \int_{t_0}^{t_1} [\mathcal{R}(q)[y(q) - D(q)u(q)] - Q(q)B(q)u(q)]dq$$

where

$$Q(\tau) = \int_{\tau}^{t_1} \mathcal{R}(q)C(q)\Phi(q,\tau)dq$$

as was to be proven.   ∎

CONTROLLABILITY DEFINITIONS AND DUALITY

In this section we adapt the controllability ideas of chapter 20 to the time-varying state dynamical equation 23.1a. However, a slight twist in definition 20.1 occurs: a controllable state is one which is reachable from the origin. This causes no difficulty because for the continuous time case the notions of reachability and controllability coincide. Furthermore this nonstandard but equivalent definition is more in accord with the physical/engineering motivation for the mathematical formulation.

Definition 23.6. A state $x_1 = x(t_1) \in \mathbf{R}^n$ is *controllable (reachable) over the interval* $[t_0,t_1]$ if and only if there exists a piecewise continuous input $u(\cdot)$ (often called a steering function) defined on $[t_0,t_1]$ such that

$$x(t_1) = \int_{t_0}^{t_1} \Phi(t_1,q)B(q)u(q)dq$$

The input $u(\cdot)$ is said to drive or steer θ to $x(t_1)$. In contrast to the time-invariant case, the notion is interval-dependent. Observe that, once again, the set of controllable states forms a linear subspace.

Exercise. Prove that the set of states that are controllable over $[t_0,t_1]$ forms a linear subspace, i.e., any linear combination of states in the space is controllable. The notation $Q_{[t_0,t_1]}$ will be used to identify this space.

Exercise. Show that if a state is controllable over $[t_0, t_1]$, then it is controllable over any interval containing $[t_0, t_1]$. This notion was utilized several times in the extensions of the time-invariant case in the context of switched time-invariant systems.

Definition 23.7. The system given by equations 23.1 is *controllable* if and only if every state in $\mathbf{R}^n$ is controllable.

As usual, "the system is controllable" and "the pair $(A(t), B(t))$ are controllable" are synonymous.

It is now possible to state and prove the time-varying analog of theorem 20.1.

Theorem 23.3. The following are equivalent:

(i) The pair $(A(t), B(t))$ is controllable on $[t_0, t_1]$.

(ii) For each x_1 in $\mathbf{R}^n$, there exists a $u(\cdot)$ defined over $[t_0, t_1]$ which will drive θ to $x(t_1) = x_1$.

(iii) For each x_0 in $\mathbf{R}^n$, there exists a $u(\cdot)$ defined over $[t_0, t_1]$ which will drive $x_0 = x(t_0)$ to $\theta = x(t_1)$.

(iv) For each pair of vectors $x_0 = x(t_0)$ and $x_1 = x(t_1)$, there exists a $u(\cdot)$ defined over $[t_0, t_1]$ which will drive $x_0 = x(t_0)$ to $x_1 = x(t_1)$.

Proof. The proof of these equivalences duplicates that of theorem 20.4, with the necessary modifications to the time-varying framework.   ∎

As in the continuous-time time-invariant case, the concepts of controllability and observability are duals of each other. Hence, the characterizations previously developed for time-varying observability suggest equivalent kinships for controllability. Proofs, of course, are immediate given a proper statement of duality. In examining the relations between observability and controllability, consider again problem 8 of chapter 12. This problem first mentioned the notation used for the adjoint system of a state model: for a given set of state dynamics

$$\dot{x}(t) = A(t)x(t) + B(t)u(t) \tag{23.21}$$

the associated adjoint system is

$$\dot{z}(t) = -A^*(t)z(t)$$
$$y(t) = B^*(t)z(t) \tag{23.22}$$

Observe that in equation 23.21 there is no output and in equation 23.22 no input, an obvious duality. As per problem 8 of chapter 12, some straightforward calculations produce the relationship

$$\Phi_x(t, t_0) = \Phi^*_z(t_0, t) \tag{23.23}$$

where the subscript x flags the state transition matrix of equation 23.21 and the subscript z that of equation 23.22. The following duality theorem establishes that the system of equation 23.21 is controllable whenever that of equation 23.22 is observable. The proof makes use of the relationship between the transition matrices given in equation 23.23.

Theorem 23.4.   The pair $(A(t),B(t))$ of equation 23.21 is controllable over $[t_0,t_1]$ if and only if the pair $(B^*(t), -A^*(t))$ of the adjoint system given by equation 23.22 is observable over $[t_0,t_1]$.

Proof.   The proof of the theorem rests upon the interval-dependent controllable space of the pair $(A(t),B(t))$ defined as

$$Q_{[t_0,t_1]} = \{x_1 \in \mathbf{R}^n \mid x_1 = \int_{t_0}^{t_1} \Phi_x(t_1,q)B(q)u(q)dq\} \qquad (23.24)$$

This space is often called the *reachable subspace*. The main thrust of the proof is to show that the system observability inherent in equation 23.22 implies that $Q_{[t_0,t_1]} = \mathbf{R}^n$ and that if the system of that equation is not observable, then $Q_{[t_0,t_1]} \neq \mathbf{R}^n$, thereby ensuring that the controllability of the system of equation 23.21 implies the observability of the system of equation 23.22.

To determine the extent or size of $Q_{[t_0,t_1]}$ in $\mathbf{R}^n$, let $z_1 \in \mathbf{R}^n$ have the property that

$$z_1^* x_1 = 0 \qquad (23.25)$$

for all $x_1 \in Q_{[t_0,t_1]}$. That is, if $z_1 = \theta$, then $Q_{[t_0,t_1]} = \mathbf{R}^n$; otherwise $Q_{[t_0,t_1]} \neq \mathbf{R}^n$. If z_1 satisfies equation 23.25, then

$$z_1^* \int_{t_0}^{t_1} \Phi_x(t_1,q)B(q)u(q)dq = 0$$

for all piecewise continuous functions $u(t)$ over $[t_0,t_1]$. Consequently,

$$z_1^* \Phi_x(t_1,t)B(t) \equiv \theta^t \qquad (23.26)$$

over $[t_0,t_1]$. Taking the conjugate transpose of this equation, one may define

$$y(t) \equiv B^*(t)\,\Phi_z(t,t_1)z_1 = B^*(t)\Phi_x^*(t_1,t)z_1 \qquad (23.27)$$

as a valid trajectory for equation 23.22 over $[t_0,t_1]$ with final condition z_1.

If the adjoint system is observable, then given that

$$y(t) = B^*(t)\,\Phi_z(t_1,t)z_1 \equiv \theta$$

it follows that $z_1 = \theta$. Thus, the only z_1 that satisfies equation 23.25 is the zero vector. But then, $Q_{[t_0,t_1]} = \mathbf{R}^n$.

If, however, there exists a $z_1 \neq \theta$ such that $y(t) = B^*(t)\,\Phi_z(t,t_1)z_1 = \theta$ — i.e., if the system of equation 23.22 is not observable—then, retracing our steps back to equation 23.25, it follows that $Q_{[t_0,t_1]} \neq \mathbf{R}^n$, disproving the controllability of equation 23.21 and establishing the forward direction of the statement of the theorem.   ∎

CHARACTERIZATIONS OF CONTROLLABILITY

This section undertakes the description and proof of various tests for controllability quite similar to those demonstrated for the testing of observability.

Theorem 23.5. The state model given by equations 23.1 is controllable on $[t_0,t_1]$ theta and only if the controllability Gramian

$$W_C(t_0,t_1) = \int_{t_0}^{t_1} \Phi(t_1,q)B(q)B^*(q)\Phi^*(t_1,q)dq \tag{23.28}$$

is positive definite.

Proof. Suppose $W_C(t_0,t_1)$ is positive definite, i.e., nonsingular. Define the control input

$$u(t) = B^*(t)\Phi^*(t_1,t)W_C(t_0,t_1)^{-1}x_1 \tag{23.29}$$

With this input, the terminal state is given by

$$\int_{t_0}^{t_1}\Phi(t_1,q)B(q)u(q)dq = \int_{t_0}^{t_1}\Phi(t_1,q)B(q)B^*(q)\Phi^*(t_1,q)dq\ W_C(t_0,t_1)^{-1}x_1$$

$$= W_C(t_0,t_1)W_C(t_0,t_1)^{-1}x_1 = x_1$$

To prove the converse, suppose the pair $(A(t),B(t))$ is controllable on $[t_0,t_1]$. Then the associated adjoint system of equation 23.22 is observable. Denote the state transition matrix of that system by $\Phi_z(t,q)$. Since the adjoint system is observable, its observability Gramian

$$W_0^{ad}(t_0,t_1) = \int_{t_0}^{t_1}[PHI_z^*(q,t_0)B^*(q)]^*B^*(q)\Phi_z(q,t_0)dq$$

$$= \int_{t_0}^{t_1}\Phi_z(q,t_0)B(q)B^*(q)\Phi_z(q,t_0)dq \tag{23.30}$$

is positive definite. The remainder of the proof generates a nonsingular relationship between this Gramian and the controllability Gramian thereby verifying the positive definiteness of 23.28. Now, from equation 23.23, the state transition matrix of the system of equations 23.1 and that of the adjoint system must satisfy

$$\Phi(t_0,q) = \Phi_z^*(q,t_0)$$

From the semigroup property of the state transition matrix, we then obtain

$$\Phi(t_1,q) = \Phi(t_1,t_0)\Phi(t_0,q) = \Phi(t_1,t_0)\Phi_z^*(q,t_0)$$

From this relationship, together with those defined in equations 23.28 and 23.29, it follows that

$$W_C(t_0,t_1) = \Phi(t_1,t_0)W_0^{ad}(t_0,t_1)\Phi^*(t_1,t_0)$$

But $\Phi(t_1,t_0)$ is nonsingular. Therefore, the positive definiteness of $W_0^{ad}(t_0,t_1)$ of the controllability Gramian implies that of $W_C(t_0,t_1)$, which is what was to be proven. ∎

The following two propositions are an immediate consequence of the duality theorem 23.4 and propositions 23.1 and 23.2.

Proposition 23.5.   The state model given by equations 23.1 is controllable on $[t_0,t_1]$ if there exist points $(p_1,...,p_q)$ such that

$$\text{rank}[\Phi(t_1,p_1)B(p_1),...,\Phi(t_1,p_q)B(p_q)] = n \qquad (23.31)$$

Proposition 23.6.   The state model given by equations 23.1 is controllable on $[t_0,t_1]$ if and only if there exists a point p_1 such that

$$\text{rank}[\Phi(t_1,p_1)B(p_1),D\Phi(t_1,t)B(t)\big|_{t=p_1},...,D^{q-1}\Phi(t_1,t)B(t)\big|_{t=p_1}] = n \qquad (23.32)$$

where D^j denotes j-th derivative.

This completes the characterizations of controllability for the time-varying state model. To complete this section, a few words are in order with respect to the finding of controls which will drive θ to x_1. First equation 23.29 provides a control structure which will always do the job. Transferring a state from x_0 to x_1 requires a different input which the interested reader is asked to generate in problem 8.

ISSUES OF REALIZATION

We close the chapter with some basic ideas regarding the realization of time-varying linear state models. No attempt is made to be complete or up to date. The interested reader can consult [3, 4] for a modern treatment.

Definition 23.8.   Two linear time-varying state models

$$\dot{x}(t) = A_x(t)x(t) + B_x(t)u(t)$$
$$y(t) = C_x(t)x(t) \qquad (23.33)$$

and

$$\dot{z}(t) = A_z(t)z(t) + B_z(t)u(t)$$
$$y(t) = C_z(t)z(t) \qquad (23.34)$$

are *algebraically equivalent* on $[t_0,t_1]$ if and only if there exists an $n \times n$ nonsingular state transformation $z(t) = T(t)x(t)$ with a piecewise continuous derivative such that

$$A_z(t) = [T(t)A_x(t) + \dot{T}(t)]T(t)^{-1}$$
$$B_z(t) = T(t)B_x(t) \qquad (23.35)$$
$$C_z(t) = C_x(t)T(t)^{-1}$$

It is simple to show that algebraic equivalence is a reflexive, symmetric, and transitive relation.

Exercise. Show that algebraic equivalence is a symmetric relation, i.e., if system I is algebraically equivalent to system II, then system II is algebraically equivalent to system I. If $T(t)$ determines the first equivalence, what is the state transformation for the second?

Exercise. Show that algebraic equivalence is a transitive relation, i.e., if system I is algebraically equivalent to system II and system II is algebraically equivalent to system III, then system I is algebraically equivalent to system III.

Intuitively, if two systems are algebraically equivalent, there ought to be a kinship between their respective state transition matrices as well as between their respective impulse response matrices. The following theorem delineates these relations.

Theorem 23.6. If the systems given by equations 23.33 and 23.34 are algebraically equivalent, then

(i) their state transition matrices are related by

$$\Phi_z(t,q) = T(t)\Phi_x(t,q)T(t)^{-1}$$

(ii) their impulse response matrices are related by

$$H_z(t,q) = H_x(t,q)$$

where subscript z refers back to equation 23.34 and the subscript x to equation 23.33.

Proof. The proof of the theorem is left as problem 16, part (a), at the end of the chapter.   ∎

The crux of the realization issue is, When does a time-varying impulse response matrix have a realization as a time-varying linear state model? The next theorem answers this question.

Theorem 23.7. The impulse response $H(t,\tau)$ has the state model realization

$$\dot{x}(t) = A(t)x(t) + B(t)u(t)$$
$$y(t) = C(t)x(t)$$

if and only if $H(t,\tau)$ has a factorization of the form

$$H(t,\tau) = M_1(t)M_2(\tau)$$

where $H(t,\tau)$ is $r \times m$, $M_1(t)$ is $r \times n$, and $M_2(\tau)$ is $n \times m$.

Proof. The proof of the theorem is left as problem 18 at the end of the chapter. However, if $H(t,\tau)$ does have a state space realization, then by writing down the form of the impulse response and by factoring the state transition matrix as a product of fundamental matrices, the desired factorization follows immediately. The key to the converse is to define the obvious realization in terms of $M_1(t), M_2(t)$, and an integrator (see [1]).   ∎

As a final result, we note that if one member of a set of algebraically equivalent systems is completely controllable (observable), then every member of the set is completely controllable (observable).

Proposition 23.5. Let $T(t)x(t) = z(t)$ represent a nonsingular state transformation over $[t_0, t_1]$, then the pair $(A_x(t), B_x(t))$ is completely controllable if and only if the pair $(A_z(t), B_z(t))$ is completely controllable. The obvious dual result holds for observability.

CONCLUDING REMARKS

This chapter has covered the rudiments of time varying state space controllability and observability, a very challenging topic, and a sprinkling of realization theory. By no means are the expositions complete. The literature contains abundant abstract formulations of these ideas [3, 4, 6]. Further numerical issues for time-varying systems have not had nearly the interest that the time-invariant case has had. The reader is encouraged to read the literature for various generalizations of the ideas expressed here. The next chapter caps off the time-varying development with a brief overview of stability for time-varying state models.

PROBLEMS

1. Prove that if $M = T^*T$, where T is an $m \times n$ complex matrix with $n \leqslant m$, then the following are equivalent: (i) M is positive definite; (ii) rank$[T] = n$; (iii) M is nonsingular.

2. Show that if M is positive definite, then all the eigenvalues of M are real, positive nonzero numbers.

3. A common criterion for positive semidefiniteness of a real symmetric matrix is Sylvester's criterion: a quadratic form $x^t M x$ is positive semidefinite if and only if all principal minors of M are nonnegative. Show by counterexample that the leading principal minors of M can be nonnegative but that the associated quadratic form is not. (*Hint*: See [5].)

4. **(a)** Show that the time-varying linear homogeneous differential equation

$$t^2 D^2 x + t Dx - x = 0$$

with output response $y = x$ has the zero-input state model

$$\begin{bmatrix} \dot{x}_1 \\ \dot{x}_2 \end{bmatrix} = \begin{bmatrix} 0 & 1 \\ t^{-2} & -t^{-1} \end{bmatrix} \begin{bmatrix} x_1 \\ x_2 \end{bmatrix}$$

$$y = \begin{bmatrix} 0 & 1 \end{bmatrix} \begin{bmatrix} x_1 \\ x_2 \end{bmatrix}$$

(b) Show that the state transition matrix for $t \geqslant 1$ is

$$\Phi(t, 1) = 0.5 \begin{bmatrix} t + t^{-1} & t - t^{-1} \\ 1 - t^{-2} & 1 + t^{-2} \end{bmatrix}$$

(c) Verify that the system is observable on the internal [1,2] using the corollary to theorem 23.2. (*Hint*: See [2].)

5. Consider the time-varying state model

$$\dot{x}(t) = \begin{bmatrix} 0 & \cos(\pi t) \\ 0 & 0 \end{bmatrix} x(t) + \begin{bmatrix} 0 \\ b(t) \end{bmatrix} u(t)$$

$$y(t) = [1 \quad -\sin(\pi t)] x(t)$$

where

$$b(t) = \begin{cases} 1 & 2k \leqslant t < 2k + 1 \\ 0 & \text{otherwise} \end{cases}$$

for $k = 0, 1, 2, \dots$. Determine the observability and controllability of the system over the intervals

    (a) [0,1].   **(c)** [1,3].

    (b) [1,2].

6. Show that the set of unobservable states is a linear subspace over $[t_0, t_1]$.

7. Prove the "only if" part of the corollary to proposition 23.2.

8. Prove the equivalences given in Theorem 23.3.

9. Develop a control along the lines of equation 23.29 which will drive the initial state $x_0 = x(t_0)$ to $x_1 = x(t_1)$ assuming controllability over the interval $[t_0, t_1]$.

10. Construct an example which shows that a system that is controllable over $[t_0, t_1]$ may not be controllable over a smaller subinterval.

11. The crux of the observability and controllability tests presented in the text is the positive definiteness of the associated Gramians. The Gramians depend on the structure of the state model. However, the form of the Gramian and its positive definite character have a context-free formulation. Accordingly, prove the following result. Let $M(t)$ be an $n \times m$ matrix defined over $[t_0, t_1]$ with piecewise-continuous, possibly complex-valued entries. Then the n columns of $M(t)$ are linearly independent vector functions over $[t_0, t_1]$ if and only if

$$W(t_0, t_1) = \int_{t_0}^{t_1} M(q) M^*(q) \, dq$$

is positive definite.

12. The use of the Gramian in developing controls and in reconstructing initial states also has a context-free interpretation. Here one considers the linear operation on the input $u(t)$ defined as

$$x(t_1) = \int_{t_0}^{t_1} M(q) u(q) \, dq$$

This more or less duplicates the formula for determining the zero-state state response. For each t_1, $x(t_1) \in \mathbf{R}^n$, and hence the range of this linear operation for each fixed t is a subspace of $\mathbf{R}^n$. Prove that the column space of the Gramian characterizes this range space—i.e., prove that for each t_1, $x(t_1)$ is in the range of the foregoing linear operation if and only if $x(t_1)$ lies in the column space of the constant matrix

$$W(t_0, t_1) = \int_{t_0}^{t_1} M(q) M^*(q) \, dq$$

13. Use problems 11 and 12 to construct a new proof of theorem 23.5. The proof must not rely on the principle of duality.

14. Again use problems 11 and 12 to construct new proofs of propositions 23.5 and 23.6.

15. Investigate the controllability of the following time-varying state model over the intervals **(a)** [0,1], **(b)** [1,2], and **(c)** [0,2]:

$$\dot{x}(t) = \begin{bmatrix} 0 & (t+1)^{-1} \\ (t+1)^{-1} & 0 \end{bmatrix} x(t) + B(t)u(t)$$

$$y(t) = B^*(t)x(t)$$

where

$$B(t) = \begin{cases} (1 \quad 1)^t, & k \leqslant t < k+1, k \text{ even} \\ (1 \quad -1)^t, & k \leqslant t < k+1, k \text{ odd} \end{cases}$$

16. **(a)** Prove part (i) of theorem 23.6.   **(b)** Prove part (ii) of theorem 23.6.

17. Prove theorem 23.7.

18. Prove that the impulse response $H(t,q)$ of a linear time-varying finite-dimensional system can be realized by a time-varying linear state model of the form of equation 23.1 with $D(t) = [0]$ if and only if it has a factorization of the form $H(t,q) = M_1(t)M_2(q)$ for appropriate matrices $M_1(t)$ and $M_2(q)$. (*Hint:* Think of the relationship between the state transition matrix and a fundamental matrix.)

REFERENCES

1. C. A. Desoer, *Notes for a Second Course on Linear Systems* (New York: Van Nostrand Reinhold, 1970).

2. David L. Russell, *Mathematics of Finite-Dimensional Control Systems: Theory and Design* (New York: Marcel Dekker, 1979).

3. E. W. Kamen, "New Results in Realization Theory for Linear Time-Varying Analytic Systems," *IEEE Transactions on Automatic Control*, Vol. AC-24, No. 6, December 1979, pp. 866–878.

4. E. Gilbert, "Minimal Order Realizations for Continuous-Time Two-Power Input-Output Maps," *IEEE Transactions on Automatic Control*, Vol. AC-28, No. 4, April 1983, pp. 452–464.

5. K. N. Swamy, "On Sylvester's Criterion for Positive-Semidefinite Matrices," *IEEE Transactions on Automatic Control*, Vol. AC-18, No. 3, June 1973, pp. 306.

6. S. Dolecki and D. L. Russell, "A General Theory of Observation and Control," *SIAM Journal on Control and Optimization*, Vol. 15, 1977, pp. 185–220.

24

STABILITY
OF THE TIME-VARYING
LINEAR STATE MODEL

INTRODUCTION

The direction of this chapter points toward an extension of the development of time-invariant stability theory to the time-varying impulse response matrix and the linear time-varying state model. In the BIBO and BIBS case, the definitions remain unchanged; for example, BIBO requires that bounded inputs map to bounded outputs in the L_∞ sense. Unfortunately, the eigenvalue characterizations so useful and so elegant in the time-invariant case do not directly generalize to the time-varying case. Hence, arriving at characterizations of BIBS and BIBO stability involves ideas that are simply absent from the time-invariant case. A number of new definitions are required, e.g., *stability in the sense of Lyapunov* (i.s.l.), as well as some precision in interpretation— e.g., some of the definitions depend on the initial time t_0.

The following examples illuminate the perspective of the chapter. In interpreting them, recall that the complete state response has contributions from both the zero-input and zero-state responses. Unboundedness of either precludes BIBS stability.

EXAMPLE 24.1 [1]

A matrix $A(t)$ with negative eigenvalues does not imply stability of the zero-input state response. For consider the zero-input state dynamical equation

$$\dot{x}(t) = A(t)x(t) \tag{24.1}$$

with

$$A(t) = \begin{bmatrix} -1 - 9\cos^2(6t) + 12\sin(6t)\cos(6t) & 12\cos^2(6t) + 9\sin(6t)\cos(6t) \\ -12\sin^2(6t) + 9\sin(6t)\cos(6t) & -1 - 9\sin^2(6t) - 12\sin(6t)\cos(6t) \end{bmatrix}$$

The eigenvalues of $A(t)$ are negative and independent of time, with values -1 and -10. Would an arbitrary state trajectory remain bounded? This, of course, depends on the state transition matrix, which has the form

$$\Phi(t,0) = 0.2 \begin{bmatrix} \cos(6t) + 2\sin(6t) & 2\cos(6t) - \sin(6t) \\ 2\cos(6t) - \sin(6t) & -\cos(6t) - 2\sin(6t) \end{bmatrix} \begin{bmatrix} e^{2t} & 2e^{2t} \\ 2e^{-13t} & -e^{-13t} \end{bmatrix}$$

For an arbitrarily chosen initial condition, e.g., $x(0) = (1,1)^t$, the term e^{2t} produces an unbounded zero-input response. Thus, the system cannot be BIBS stable.

From example 24.1, then, the stability characterization in terms of the eigenvalue locations of the A-matrix do not extend, at least directly, to the time-varying case. But perhaps, the presence of eigenvalues in the open right-half complex plane implies instability. Here again, however, the speculation fails to pan out.

EXAMPLE 24.2 [2]

The presence of eigenvalues of $A(t)$ in the right-half plane does not imply instability of the state trajectory. Again, consider the zero-input state dynamical equation 24.1, this time with

$$A(t) = \begin{bmatrix} -5.5 + 7.5\sin(12t) & 7.5\cos(12t) \\ 7.5\cos(12t) & -5.5 - 7.5\sin(12t) \end{bmatrix}$$

which has constant eigenvalues of 2 and -13 for all t. Interestingly enough, the state transition matrix has asymptotically stable columns:

$$\Phi(t,0) = \begin{bmatrix} \cos(6t) + 3\sin(6t) & \cos(6t) - 3\sin(6t) \\ 3\cos(6t) - \sin(6t) & -3\cos(6t) - \sin(6t) \end{bmatrix} \begin{bmatrix} 0.5e^{-t} & \frac{1}{6}e^{-t} \\ 0.5e^{-10t} & -\frac{1}{6}e^{-10t} \end{bmatrix}$$

Clearly, then, the presence of the positive eigenvalue at 2 does not cause instability.

These examples foreshadow the complexity of the stability analysis of time-varying systems over that of their time-invariant counterparts. Intuitively, one might imagine time-invariant systems as being analogous to constant lines having zero derivative: their behavior is rather straight and "expected." The behavior of time-varying systems would then be analogous to curved lines, whose derivative is not identically zero and whose behavior is far from "expected." Some of the complexity arises because of initial time and initial condition dependent behavior. To glimpse some of the initial condition dependent behavior, consider Figure 24.1, which shows a ball restricted to lie on various "curved surfaces." One can think of the ball as being subject to gravity and the natural perturbations of its initial state.

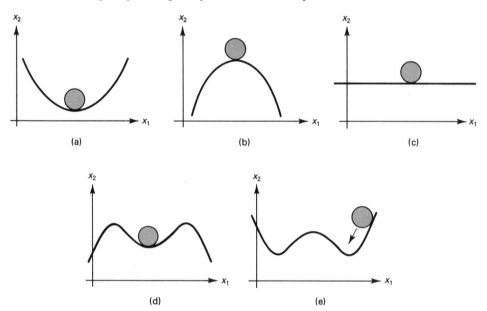

Figure 24.1   Illustrations of stability.

In Figure 24.1(a) any perturbation in the ball's rest state will eventually result in a return to the same rest state, implying that the rest state is stable. In part (b) any perturbation in the rest state sends the ball on a slide to infinity, suggesting an unstable rest state. In (c) perturbations in the rest state result in another rest state, assuming no initial velocity is given to the ball. In effect, every state is a rest state. Part (d) mixes parts (a) and (b): sufficiently small perturbations result in a return to the rest state, but large perturbations result in a journey to infinity. Finally, in part (e) perturbations from a rest state always result in a return to the same or a different rest state, further complicating the analysis.

Despite such diverse situations, one characterization carries over from the time-invariant case more or less intact: that of the impulse response matrix. We examine this in the next section.

BIBO STABILITY VIA THE IMPULSE RESPONSE MATRIX

Suppose the response $y(\cdot)$ of a time-varying linear lumped system satisfies the convolution-like integral

$$y(t) = \int_{-\infty}^{t} H(t,q)u(q)dq \qquad (24.2)$$

where $u(\cdot)$ is a piecewise-continuous input function and $H(t,q)$ is the impulse response matrix, assumed to be piecewise continuous in t and q. Then we have the following theorem, which is an extension of theorem 22.1 to the time-varying case.

Theorem 24.1. A causal system modeled by the convolution-like integral in equation 24.2 is BIBO stable if and only if, for some finite K,

$$|| H(t,\cdot) ||_1 < K \qquad (24.3)$$

for all t.

Proof. In order to be consistent with the definition of a norm, define the projection operators E^T as

$$E^T y(t) = \begin{cases} y(t) & \text{for } t \leqslant T \\ 0 & \text{for } t > T \end{cases} \qquad (24.4)$$

Suppose that $|| H(t,\cdot) ||_1 \leqslant K$ for all t. Consider the L_∞ norm of $E^t y(\cdot)$:

$$|| E^t y(\cdot) ||_\infty = || \int_{-\infty}^{t} H(t,q)u(q)dq ||_\infty \leqslant || H(t,\cdot) ||_1 || u(\cdot) ||_\infty$$

for all t, by the properties of vector-induced norms. Hence, bounded inputs produce bounded outputs.

The converse of the proof is analogous to that of part 2 of theorem 22.1. It is left as an exercise at the end of the chapter.   ∎

Corollary 24.1. A relaxed system modeled by the integral in equation 24.2 is BIBO stable if and only if there is a single finite constant K such that

$$|| y(\cdot) ||_\infty = K || u(\cdot) ||_\infty \qquad (24.5)$$

Proof. If the system is BIBO stable, then K may be taken to be

$$K = \sup_t || H(t,\cdot) ||_1$$

Conversely, given equation 24.5, BIBO stability follows by definition.   ∎

From chapter 18, the impulse response matrix of the time-varying state model is given by

$$H(t,q) = C(t)\Phi(t,q)B(q)1^+(t-q) + D(t)\delta(t-q) \qquad (24.6)$$

Hence, BIBO stability of the impulse response can be determined from this equation.

LYAPUNOV STABILITY:
BASIC DEFINITIONS, EXAMPLES, AND RESULTS

Consider the nonlinear time-varying zero-input state model description

$$\dot{x} = f(x,t), \quad x(t_0) = x_0 \qquad (24.7)$$

where, as usual, $x \in \mathbf{R}^n$ and $f:\mathbf{R}^n \times \mathbf{R} \rightarrow \mathbf{R}^n$. An equilibrium state, denoted x_e, satisfies $f(x_e,t) = 0$ for all t. This is consistent with the ideas expressed in Figure 24.1. Recall that $\phi(\cdot,t_0,x_0)$ represents a solution to equation 24.7. The cornerstone definition of Lyapunov stability theory now follows.

Definition 24.1. The equilibrium state x_e is *stable in the sense of Lyapunov* (i.s.l.) if, for any given initial time t_0 and positive number e, there exists an e- and t_0- dependent positive number $\delta(t_0,e)$ such that

$$\| \phi(\cdot,t_0,x_0) - x_e \|_\infty < e \qquad (24.8)$$

whenever

$$\| x_0 - x_e \| < \delta(t_0,e)$$

Observe carefully the dependence δ on t_0. A later example will show the critical importance of this condition.

To generate a more convenient framework for analysis, one usually transfers the coordinate axes to x_e via the linear change of variable $x - x_e$. Hence, without loss of generality, one may refer to the origin rather than x_e as being stable i.s.l. The following example helps solidify these basic notions.

EXAMPLE 24.3.

Consider the unforced nonlinear series RLC circuit of Figure 24.2.

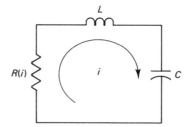

Figure 24.2 Series *RLC* circuit of example 24.3.

Here the resistor $R(i)$ may be a nonlinear function of the loop current i. Recall that the time derivative of the charge q in the capacitor determines the current i. Applying Kirchhoff's voltage law to the circuit permits a straightforward derivation of the nonlinear state model

$$\begin{bmatrix} \dot{q} \\ \dot{i} \end{bmatrix} = \begin{bmatrix} 0 & 1 \\ -(LC)^{-1} & -R(i)/L \end{bmatrix} \begin{bmatrix} q \\ i \end{bmatrix}$$

It is possible to determine zero-input stability information about the system modeled by these equations qualitatively, without solving the equations. The qualitative behavior of the system follows from a consideration of the stored energy in the circuit, given by

$$E(t) = 0.5L(i_L)^2 + 0.5C(v_c)^2 = 0.5Li^2 + \frac{0.5}{C}q^2$$

A given initial charge on the capacitor produces a fixed initial energy. The level of the stored energy then either diminishes, increases, or remains constant, depending on its time derivative

$$\dot{E}(t) = Li\,\dot{i} + C^{-1}q\,\dot{q} = -R(i)\,i^2$$

This time derivative depends on the value of $R(i)$. For example, if $R(i) > 0$ for all i, then any initial energy will decrease to zero as $t \to \infty$, i.e., the system is stable i.s.l. If $R(i) < 0$, then any initial energy will increase without bound as $t \to \infty$ making the circuit not stable i.s.l. Finally, if $R(i) = 0$ for all i, i.e., the pure LC case, the energy will remain constant, alternately shifting from the capacitor to the inductor and conversely. Figure 24.3 illustrates these three possible situations in the phase plane. The Lyapunov function analysis carried out in chapter 22 is a generalization of this energy analysis for systems.

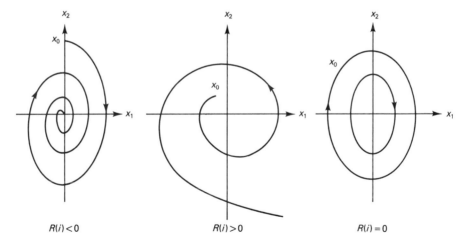

Figure 24.3 Possible responses for circuit of Figure 24.3.

Although illustrative, the preceding definition and example lie within the realm of nonlinear systems. They do not at all exploit the linear structure of $\dot{x}(t) = A(t)x(t)$, i.e., the linearity of the response $\phi(t,t_0,x_0)$ with respect to the initial state x_0. Taking account of this linearity results in the following proposition.

Proposition 24.1.   The zero solution of $\dot{x}(t) = A(t)x(t)$ is stable i.s.l. if and only if for each $t_0 \geqslant 0$ and each $x_0 \in \mathbf{R}^n$, there exists a finite constant $K(t_0)$ such that

$$|| \phi(\cdot,t_0,x_0)1^+(\cdot - t_0) ||_\infty \leqslant K(t_0) ||x_0|| \qquad (24.9)$$

Proof.   We use the method of contradiction to obtain equation 24.9 from definition 24.1 Suppose equation 24.9 does not hold; we show that the zero solution is therefore not stable i.s.l. If equation 24.9 does not hold, then there is some $x_0 \neq 0$ for which no $K(t_0)$ exists that satisfies equation 24.9. Hence,

$$\lim_{t \to \infty} || E^t \phi(\cdot,t_0,x_0)1^+(\cdot - t_0) ||_\infty \to \infty \qquad (24.10)$$

Now, pick any $\delta(t_0,e) > 0$. There exists $\alpha \neq 0$ such that $|| \alpha x_0 || \leqslant \delta(t_0,e)$. Since $\phi(\cdot,t_0,x_0)$ is linear in the initial state, $\phi(\cdot,t_0,\alpha x_0) = \alpha \phi(\cdot,t_0,x_0)$. But then, $|| E^t \phi(\cdot,t_0,\alpha x_0)1^+(\cdot - t_0) ||_\infty = |\alpha| \, || E^t \phi(\cdot,t_0,x_0)1^+(\cdot - t_0) ||_\infty$, and, from equation 24.9,

$$\lim_{t \to \infty} || E^t \phi(\cdot,t_0,\alpha x_0)1^+(\cdot - t_0) ||_\infty \to \infty$$

But then, there does not exist any $e > 0$ that satisfies equation 24.8.

For the converse suppose that for each $t_0 \geqslant 0$ and each $x_0 \in \mathbf{R}^n$, there exists a finite constant $K(t_0)$ that satisfies equation 24.9. Choose $\delta(t_0,e) = e/K(t_0)$ and suppose $||x_0|| \leqslant \delta(t_0,e)$. Then $|| \phi(\cdot,t_0,x_0)1^+(\cdot - t_0) ||_\infty \leqslant K(t_0) ||x_0|| \leqslant K(t_0)[e/K(t_0)] = e$ implying stability of the zero solution of $\dot{x}(t) = A(t)x(t)$ i.s.l.   ∎

This proposition leads to the following theorem, which links the i.s.l. stability of $x(\cdot) \equiv 0$ of $\dot{x}(t) = A(t)x(t)$ to the single initial time $t_0 = 0$.

Theorem 24.2.   The solution $x(\cdot) \equiv 0$ of $\dot{x}(t) = A(t)x(t)$ is stable i.s.l. if and only if there exists a finite constant K such that

$$|| \Phi(\cdot,0)1^+(\cdot) ||_\infty \leqslant K \qquad (24.11)$$

Proof.   For convenience, assume that all norms are taken over the region $t \geqslant t_0 \geqslant 0$. Suppose that equation 24.11 holds. Then, by the usual norm properties, the L_∞–norm of the zero-input state trajectory $\phi(\cdot,t_0,x_0)$ is

$$|| \phi(\cdot,t_0,x_0) ||_\infty = || \Phi(\cdot,t_0)x_0 ||_\infty \leqslant || \Phi(\cdot,0) ||_\infty \, || \Phi(0,t_0) || \, ||x_0||$$

where $|| \Phi(0,t_0) ||$ can be taken as the matrix spectral norm (definition 14.1), and where $\Phi(t,t_0) = \Phi(t,0)\Phi(0,t_0)$ by the semigroup property. Since $|| \Phi(\cdot,0) ||_\infty \leqslant K$, taking $K(t_0) = K|| \Phi(0,t_0) ||_\infty$ yields the inequality of equation 24.9, from which it follows that the aforementioned solution is stable i.s.l.

Proof of the converse is by the method of contradiction. Suppose that equation 24.11 fails to hold. Then some element, say, $\Phi_{km}(\cdot,t_0)$ of $\Phi(\cdot,t_0) = [\Phi_{ij}(\cdot,t_0)]$ has the property that as $t \to \infty$, $|\Phi_{km}(t,t_0)| \to \infty$. Choosing an initial state $\hat{x}_0 = (0,...,x_0^m,0,...,0)^t$ with $x_0^m \neq 0$ then forces the solution $\phi(\cdot,t_0 x_0)$ to be unbounded. Hence, there does not exist any $K(t_0)$ such that $||\phi(\cdot,t_0,\hat{x}_0)||_\infty \leq K(t_0)||\hat{x}_0||$; i.e., the system is not stable i.s.l. It then follows that if $x(\cdot) \equiv 0$ is stable i.s.l., then equation 24.11 results. ∎

Using theorem 24.2, a check or test for the stability i.s.l. of the zero solution of $\dot{x}(t) = A(t)x(t)$ will depend on checking the L_∞ norm of the n basis solutions $\phi_i(\cdot,0,\eta_i)$, $i = 1,...,n$, i.e., the n columns of the state transition matrix $\Phi(t,0)$.

A question crops up upon close scrutiny of the bound in equation 24.11: is this bound truly dependent on t_0? In other words, is there no uniformity in the bound? An example helps clarify the subtlety.

EXAMPLE 24.4 [3]

A first-order zero-input state model

$$\dot{x}(t) = [4t\sin(t) - 2t]\,x(t)$$

has the solution trajectory

$$\phi(t,t_0,x_0) = x_0 \exp[4\sin(t) - 4t\cos(t) - t^2 - 4\sin(t_0) \qquad (24.12)$$
$$+ 4t_0\cos(t_0) + t_0^2]$$

for $t \geq t_0$. Observe that for any t_0 and x_0, the term $-t^2$ will eventually dominate, driving the solution trajectory to zero as $t \to \infty$. Coupled with the analyticity of the state trajectory given in equation 24.12, this observation implies the stability i.s.l. of the model.

Now suppose t_0 takes on successive values denoted by $t_0 = 2n\pi$, and suppose further that equation 24.12 is evaluated π seconds later in each case. Then

$$\phi((2n+1)\pi,2n\pi,x_0) = x_0\exp[(4n+1)\pi(4-\pi)]$$

Letting $\beta = \min|x_0^i|$ over $x_0^i \neq 0$, where $x_0 = (x_0^1,...,x_0^n)^t$, one can place a lower bound on the L_∞ norm, yielding

$$||\phi(\cdot,t_0,x_0)||_\infty \geq \beta \exp[(4n+1)\pi(4-\pi)]$$

Thus, in equation 24.9, $K(t_0) = K(2n\pi)$ must take on successively larger values, indicating that the system, though stable, is not uniformly so.

In general, then, the dependence of $K(t_0)$ on t_0 is complete. However, when the conditions of example 24.4 are not present, i.e., when the behavior is uniform in t_0, the zero solution is said to be uniformly stable i.s.l.

Definition 24.2. The zero solution of $\dot{x}(t) = A(t)x(t)$ is *uniformly stable i.s.l.* if, for any given positive number e, there exists an e-dependent positive number $\delta(e)$ such that

$$||\phi(\cdot,t_0,x_0)1^+(\cdot - t_0)||_\infty < e$$

for all t_0 whenever

$$||x_0|| < \delta(e)$$

It is now straightforward to verify the following analogs of proposition 24.1 and theorem 24.2.

Proposition 24.2. The zero solution of $\dot{x}(t) = A(t)x(t)$ is uniformly stable i.s.l. if and only if, for each $x_0 \in \mathbf{R}^n$, there exists a finite constant K such that

$$||\phi(\cdot,t_0,x_0)1^+(\cdot - t_0)||_\infty \leqslant K||x_0||$$

for all t_0.

Theorem 24.3. The solution $x(\cdot) \equiv 0$ of $\dot{x}(t) = A(t)x(t)$ is uniformly stable i.s.l. if and only if there exists a finite constant K such that

$$||\Phi(\cdot,t_0)1^+(\cdot - t_0)||_\infty \leqslant K$$

for all $t_0 \geqslant 0$.

Proof. The proof of theorem 24.3, as well as that of proposition 24.2, is left as an exercise in the problem section at the end of the chapter; note that the norms are taken over the interval $[t_0,\infty)$.   ∎

ASYMPTOTIC STABILITY i.s.l.

Oftentimes, control engineering entails the stability analysis of an "error signal." Does a particular response signal converge to a known reference signal, the difference being the error? If such convergence does occur, the error signal converges to zero. Provided the error signal stays bounded for all time, such convergence is termed asymptotic stability. In addition to this ubiquitous use, asymptotic stability plays an instrumental role in describing the input-output stability of the time-varying state model.

Definition 24.3. The zero solution of $\dot{x}(t) = A(t)x(t)$ is *asymptotically stable i.s.l.* if and only if (*i*) it is stable i.s.l. and (*ii*) for all $x_0 \in \mathbf{R}^n$, $\phi(t,t_0,x_0) \to 0$ as $t \to \infty$.

Some simple time-invariant system models serve as good, quick illustrations of the difference between asymptotic stability and stability i.s.l. Consider a 2×2 A-matrix whose eigenvalues are $\pm jb$, $b \neq 0$. For any nonzero initial condition, the solution to $\dot{x}(t) = Ax(t)$ is purely sinusoidal. If the eigenvalues are "$a \pm jb$" with $a < 0$, then the sinusoidal response exponentially decreases to zero for large times, implying asymptotic stability.

From definition 24.3, it would appear that there is a straightforward modification of theorem 24.2 as follows.

Theorem 24.4. The zero solution of $\dot{x}(t) = A(t)x(t)$ is asymptotically stable i.s.l. if and only if (*i*) there is a finite number K such that

$$\|\Phi(\cdot,0)1^+(\cdot)\|_\infty \leqslant K$$

and (ii) $\Phi(t,0)\to[0]$ as $t\to\infty$.

Proof. In light of theorem 24.2 and the definition 24.3, asymptotic stability i.s.l. clearly implies conditions (i) and (ii) of the theorem. To prove the converse, observe that condition (i) implies stability i.s.l., again by theorem 24.2. Moreover, since $\phi(t,t_0,x_0) = \Phi(t,t_0)x_0 = \Phi(t,0)[\Phi(0,t_0)x_0]$, condition (ii) implies that $\phi(t,t_0,x_0)\to 0$ as $t\to\infty$. ∎

The notion of *uniform asymptotic stability* conveys the independence of definition 24.3 from t_0. Specifically, the zero solution is *uniformly asymptotically stable* if and only if (i) it is uniformly stable i.s.l. and (ii) for all t_0, $\phi(t,t_0,x_0)\to 0$ as $t\to\infty$. The subtlety here is that in condition (ii) of definition 24.3, asymptotic stability depends on the choice of $\delta(t_0)$ for which $\|x_0\| \leqslant \delta(t_0)$, i.e., if $\|x_0\| \leqslant \delta(t_0)$, then $\phi(t,t_0,x_0)\to 0$ as $t\to\infty$; condition (ii) in the definition of *uniform* asymptotic stability requires that δ be chosen independent of t_0, i.e., for all x_0 in $\mathbf{R}^n$ such that $\|x_0\| \leqslant \delta$ and for all t_0, $\phi(t,t_0,x_0)\to 0$ as $t\to\infty$.

Exercise. Construct an example of a system which is uniformly stable i.s.l. but not uniformly asymptotically stable.

Theorem 24.5. The zero solution of $\dot{x}(t) = A(t)x(t)$, $t \geqslant t_0$, is uniformly asymptotically stable if and only if there is a finite number K such that the L_∞ norm

$$\|\Phi(\cdot,t_0)1^+(\cdot)\|_\infty \leqslant K \tag{24.13}$$

for all $t_0 \geqslant 0$, and (ii) for all $t_0 \geqslant 0$, $\|\Phi(t,t_0)\|_2\to 0$ as $t\to\infty$, where, for each t,

$$\|\Phi(t,t_0)\|_2^2 = \sum_{i=1}^n \sum_{j=1}^n [\Phi_{ij}(t,t_0)]^2$$

Proof. The proof of the theorem is left as an exercise but closely resembles combinations of prior proofs. Note, however, that "uniformly" means that there is a $T(e)$, T depends on e, such that for all $t \geqslant t_0 + T(e)$, $\|\Phi(t,t_0)\|_2 \leqslant e$. ∎

With this meaning of "uniformly," we have the following theorem.

Theorem 24.6 [4]. The zero solution of $\dot{x}(t) = A(t)x(t)$, $t \geqslant t_0$, is uniformly asymptotically stable if and only if there are two positive constants K_1 and K_2 such that

$$\|\Phi(t,t_0)\|_2 \leqslant K_1\exp[-K_2(t-t_0)] \tag{24.14}$$

Proof. If equation 24.14 is true, then uniform asymptotic stability follows immediately. Suppose, then, for the converse, that the zero solution is uniformly asymptotically stable. Construct the bound given in equation 24.14.

Step 1. Since the zero solution is uniformly asymptotically stable, there is a $T > 0$ such that for all $t_0 \geqslant 0$,

$$||\Phi(t_0 + T, t_0)||_2 \leqslant 0.5 \qquad (24.15)$$

Using the semigroup property of the state transition matrix, a sequential application of the Holder inequality, and the bound of equation 24.15, we obtain

$$||\Phi(t_0 + kT, t_0)||_2 \leqslant ||\Phi(t_0 + kT, t_0 + (k-1)T)$$

$$\times \Phi(t_0 + (k-1)T, t_0 + (k-2)T)...\Phi(t_0 + T, t_0)||_2$$

$$\leqslant ||\Phi(t_0 + kT, t_0 + (k-1)T)||_2 ... ||\Phi(t_0 + T, t_0)||_2$$

$$\leqslant 2^{-k} \qquad (24.16)$$

Step 2. From equation 24.13, together with the assumption of uniform asymptotic stability, there is a constant K such that $||\Phi(\cdot, t_0)1^+(\cdot)||_\infty \leqslant K$. With this bound and the bound of equation 24.16, another application of the semigroup property and Holder's inequality yields

$$||\Phi(t, t_0)||_2 \leqslant ||\Phi(t, t_0 + kT)||_2 ||\Phi(t_0 + kT, t_0)||_2 \leqslant K 2^{-k}$$

Step 3. Suppose $t_0 + kT \leqslant t < t_0 + (k+1)T$, which is always possible for some k. Define K_1 in equation 24.14 as $K_1 = 2K$ and $K_2 = T^{-1}ln(2)$, or, more to the point, $\exp(K_2 T) = 2$. Then

$$||\Phi(t, t_0)||_2 \leqslant K 2^{-k} \leqslant K_1 2^{-(k+1)} \leqslant K_1 [\exp(K_2 T)]^{-(k+1)}$$

$$\leqslant K_1 \exp(-(k+1)K_2 T) \leqslant K_1 \exp(-K_2(t - t_0))$$

as was to be shown.   ∎

BIBS STABILITY

Recall the time-varying linear state dynamics given by

$$\dot{x}(t) = A(t)x(t) + B(t)u(t), \quad x(t_0) = x_0 \qquad (24.17a)$$

$$y(t) = C(t)x(t) + D(t)u(t) \qquad (24.17b)$$

for $t \geqslant t_0 \geqslant 0$. An interesting question is, under what conditions does $||x(\cdot)1^+(\cdot)||_\infty$ remain bounded whenever $||u(\cdot)1^+(\cdot)||_\infty$ is bounded? The following theorem proffers one answer to this question. (In the remainder of the development, $1^+(\cdot)$ will be deleted for convenience but understood to be present.)

Theorem 24.7. The state dynamics given by equation 24.17a is BIBS stable if and only if (*i*) the zero solution of $\dot{x}(t) = A(t)x(t)$ is stable i.s.l. and (*ii*) there exists a constant K such that $||\Phi(t, \cdot)B(\cdot)||_1 \leqslant K$ for all t.

Proof. The zero solution is stable i.s.l. if and only if there exists a constant K_1 such that $||\Phi(\cdot, 0)||_\infty \leqslant K_1$, by theorem 24.2. Suppose $||u(\cdot)||_\infty$ is finite, and consider the solution

$$x(t) = \Phi(t, t_0)x_0 + \int_{t_0}^t \Phi(t, q)B(q)u(q)dq$$

to equation 24.17a. Since x_0 is arbitrary, $||x(\cdot)||_\infty$ is bounded if and only if (i) $||\Phi(\cdot,t_0)x_0||_\infty$ is bounded, and (ii)

$$||E^t \int_{t_0}^{t} \Phi(\cdot,q)B(q)u(q)||_\infty$$

is bounded. But condition (i) holds if and only if

$$||\Phi(\cdot,t_0)x_0||_\infty \leqslant K_1 ||x_0||$$

and condition (ii) holds if and only if the L_1 norm

$$||\Phi(t,\cdot)B(\cdot)||_1$$

remains bounded for all t by theorem 24.1. But this is what was to be proven. ∎

BIBO STABILITY

In the time-invariant case, BIBS stability implies BIBO stability. Unfortunately, this relationship fails to carry over to the time-varying case. The difficulty lies with the matrix $C(t)$ which could have unbounded entries generating an unbounded output from a bounded state trajectory. However, if $||C(\cdot)||_\infty$ and $||D(\cdot)||_\infty$ are finite and the state dynamics are BIBS stable, then whenever $||u(\cdot)||_\infty$ is bounded,

$$||y(\cdot)||_\infty \leqslant ||C(\cdot)||_\infty ||x(\cdot)||_\infty + ||D(\cdot)||_\infty ||u(\cdot)||_\infty < \infty$$

This implies BIBO stability. The norms, of course, are taken over the nonnegative real axis where it is implicitly assumed that $t_0 \geqslant 0$. Formalizing this simple derivation, we have the following theorem.

Theorem 24.8.   If the time-varying state model of equation 24.17a is BIBS stable, and if $||C(\cdot)||_\infty$ and $||D(\cdot)||_\infty$ are finite, then the system, 24.17, is also BIBO stable.

As might be imagined, the converse of theorem 24.8 is not true. For example, consider the first-order system defined for $t \geqslant 0$ by

$$\dot{x} = x + u, \quad x(0) = x_0$$
$$y = e^{-2t}x$$

The response satisfies

$$y(t) = e^{-2t}[e^t]x_0 + e^{-2t} \int_{t_0}^{t} e^{(t-q)}u(q)dq$$

$$= e^{-t}x_0 + e^{-t} \int_{t_0}^{t} e^{-q}u(q)dq$$

It is then straightforward to show that for all t,

$$||E^t y(\cdot)||_\infty \leqslant ||x_0|| + ||u(\cdot)||_\infty$$

implying BIBO stability. The catch, however is the effect $C(t) = e^{-2t}$ has on the zero-input state-response: the rate of decrease in $C(t)$ swamps the rate of increase in the unbounded zero-input state response. This results in an unstable state response but a bounded output response.

The notion of bounded-input, bounded-output stability in the context of the state model depends implicitly on the initial time t_0. In fact, the bound on $\|y(\cdot)\|_\infty$ may depend on both $u(\cdot)$ and $x(t_0) = x_0$. With a minor modification to the definition of BIBO stability two further results are possible.

Definition 24.4.   The state model given by equations 24.17 is *input-output stable* if and only if, for any initial condition x_0 and bound K_u, if $\|u(\cdot)\|_\infty \leqslant K_u$, then there is a $K_y < \infty$ such that $\|y(\cdot)\|_\infty \leqslant K_y$.

Again, the preceding norms and those which follow are only taken over the nonnegative real axis. This definition allows us to state and prove a theorem analogous to theorem 24.8.

Theorem 24.9 [4].   If the zero solution of $\dot{x}(t) = A(t)\,x(t)$ is uniformly asymptotically stable i.s.l., and if $\|B(\cdot)\|_\infty$, $\|C(\cdot)\|_\infty$, and $\|D(\cdot)\|_\infty$ are finite, then the system given by equation 24.17 is input-output stable.

Proof.   Suppose $\|B(\cdot)\|_\infty \leqslant K_B$, $\|C(\cdot)\|_\infty \leqslant K_C$, and $\|D(\cdot)\|_\infty \leqslant K_D$. Let $u(\cdot)$ be a bounded input satisfying $\|u(\cdot)\|_\infty \leqslant K_u$. Then by inspection,

$$\|y(\cdot)\|_\infty \leqslant \|C(\cdot)\|_\infty \|x(\cdot)\|_\infty + \|D(\cdot)\|_\infty \|u(\cdot)\|_\infty$$

$$\leqslant K_C \|x(\cdot)\|_\infty + K_D K_u$$

Provided $\|x(\cdot)\|_\infty$ is finite, input-output stability follows.

To show the finiteness of $\|x(\cdot)\|_\infty$, recall that, from theorem 24.6, uniform asymptotic stability implies the existence of positive finite constants K_1 and K_2 such that

$$\|\Phi(t,t_0)\|_2 \leqslant K_1 \exp[-K_2(t-t_0)]$$

for all $t \geqslant t_0 \geqslant 0$. Recall also that

$$x(t) = \Phi(t,t_0)x_0 + \int_{t_0}^{t} \Phi(t,q)B(q)u(q)dq$$

This permits the string of inequalities

$$\|x(\cdot)\|_\infty \leqslant K_1 \|x_0\| + K_B K_u \int_{t_0}^{t} \|\Phi(t,q)\|_2\, dq$$

$$\leqslant K_1 \|x_0\| + K_B K_u \int_{t_0}^{t} K_1 \exp[-K_2(t-q)]dq$$

$$\leqslant K_1 \|x_0\| + \frac{K_B K_u K_1}{K_2}$$

establishing the boundedness of $\|x(\cdot)\|_\infty$.   ∎

The following theorem estabilishes some necessary and sufficient conditions for input-output stability.

Theorem 24.10 [4]. Suppose $||B(\cdot)||_\infty$, $||C(\cdot)||_\infty$, and $||D(\cdot)||_\infty$ are finite and that there exist positive constants e_1 and e_2 such that

$$0 < e_1 \leqslant \inf |\det[B(t)]| \tag{24.18a}$$

and

$$0 < e_2 \leqslant \inf |\det[C(t)]| \tag{24.18b}$$

where "$\inf |\det[B(t)]|$" denotes the infimum value of $|\det[B(t)]|$. Then the zero solution of $\dot{x}(t) = A(t)x(t)$ is uniformly asymptotically stable i.s.l. if and only if the model given by equations 24.17 is input-output stable.

Proof. The "only if" part of the theorem follows from theorem 24.9. For the "if" part, suppose that the model given by equations 24.17 is input-output stable. Then *the zero solution is uniformly stable i.s.l.* (claim 1). This follows because if the input were identically zero, then by hypothesis the response would be bounded regardless of the time t_0 for which $x(t_0) = x_0$. As evidenced in the proof of theorem 24.9, the bound will depend on x_0 but not t_0. Hence, there exists a constant K_1 such that

$$||\Phi(\cdot, t_0)||_\infty \leqslant K_1$$

for all $t_0 \geqslant 0$. It remains to show asymptotic stability.

To do so, we make the following claim: *the L_1 norm of $\Phi(t, \cdot)$ is finite, i.e.,* $||\Phi(t, \cdot)||_1 \leqslant K_2$ *for all* $t \geqslant 0$ (claim 2). To verify this claim, notice that equation 24.18a implies that $B^{-1}(t)$ exists and has a finite L_∞-bound. Hence, for any bounded input $u(\cdot)$, $B^{-1}(t)u(t) = \hat{u}(t)$ is also bounded, and for any bounded input $\hat{u}(\cdot)$ there exists a bounded $u(\cdot)$ such that $B^{-1}(t)u(t) = \hat{u}(t)$. Now, recall that $\Phi(\cdot, q)B(q)1^+(t-q) = H(t,q)$ is the impulse response of the state dynamical equation 24.17a. Thus, if $||\Phi(t, \cdot)||_1$ is not finite for all $t \geqslant 0$, then, using the class of bounded inputs $\{\hat{u}(\cdot)\}$ leads, by theorem 24.1, to the result that the state response is unbounded. Then, by the same arguments as just used, it follows from the bounded invertibility of $C(t)$ that the system response $y(\cdot)$ is also unbounded, contradicting the assumption of input-output stability. Hence, claim 2 is true.

From claims 1 and 2 it follows that for all t,

$$K_1 K_2 \geqslant \int_{t_0}^{t} \left[\max_i \sum_j |\Phi_{ij}(t,q)| \right] ||E^q \Phi(\cdot, t_0)||_\infty \, dq \tag{24.19a}$$

$$\geqslant \max_i \int_{t_0}^{t} \sum_j |\Phi_{ij}(t,q)\Phi_{ji}(q,t_0)| \, dq \tag{24.19b}$$

$$\geqslant [\max_i \sum_j |\Phi_{ij}(t,t_0)|](t-t_0) \tag{24.19c}$$

where equation 24.19b follows from equation 24.19a by the usual norm properties and equation 24.19c follows from equation 24.19b by the semigroup property of the state transition matrix and the resulting independence of the integrand from q.

As $t \to \infty$, then

$$\frac{K_1 K_2}{t - t_0} \geqslant \max_i \sum_j |\Phi_{ij}(t,t_0)| \to 0$$

proving that the zero solution of $\dot{x}(t) = A(t)x(t)$ is uniformly asymptotically stable.   ∎

THE ZERO SOLUTION:
SUFFICIENT CONDITIONS FOR STABILITY AND INSTABILITY

This section briefly sets forth several results which determine sufficient (albeit conservative) conditions for stability and instability of the zero solution of $\dot{x}(t) = A(t)x(t)$. In one sense, this frees one from determining a complete set of basis solutions in order to determine system stability. However, since the results are not necessary, there is an inherent conservativeness to their application. Proofs and elaborate discussions are left to the problems at the end of the chapter and to the references.

The following theorem generates upper and lower bounds on the Euclidean norm of the solution trajectory $\phi(t,t_0,x_0)$ of $\dot{x}(t) = A(t)x(t)$. This opens the door to a series of corollaries which state the aforementioned sufficient conditions.

Theorem 24.11 [4,5].   Let $\lambda_{\min}(t)$ and $\lambda_{\max}(t)$ be the smallest and largest eigenvalues, respectively, of the real symmetric matrix

$$M(t) = A(t) + A^t(t) . \tag{24.20}$$

Then the solution trajectory $\phi(t,t_0,x_0)$ of $\dot{x}(t) = A(t)x(t)$ satisfies

$$||x_0|| \exp[0.5 \int_{t_0}^{t} \lambda_{\min}(q)dq] \leqslant ||\phi(t,t_0,x_0)||_2 \tag{24.21}$$

$$\leqslant ||x_0|| \exp[0.5 \int_{t_0}^{t} \lambda_{\max}(q)dq]$$

In bounding the state trajectory, equation 24.21 permits a number of corollaries which are useful in conservative checks of asymptotic and uniform asymptotic stability.

Corollary 24.11a.   The zero solution of $\dot{x}(t) = A(t)x(t)$ is stable i.s.l. if, for each t_0, there is a bound $K(t_0)$ such that

$$\lim_{t \to \infty} \int_{t_0}^{t} \lambda_{\max}(q)\,dq < K(t_0)$$

Of course, if the bound $K(t_0) = K$ is independent of t_0, then the zero solution is uniformly stable i.s.l. With regard to asymptotic stability, we have the following corollary.

Corollary 24.11b.   The zero solution of $\dot{x}(t) = A(t)\,x(t)$ is asymptotically stable i.s.l. if

$$\lim_{t \to \infty} \int_{t_0}^{t} \lambda_{\max}(q)\,dq = -\infty$$

Two results on the complete instability of the zero solution now follow.

Corollary 24.11c.   The zero solution of $\dot{x}(t) = A(t)\,x(t)$ is unstable i.s.l. if

$$\lim_{t \to \infty} \int_{t_0}^{t} \lambda_{\min}(q)\,dq = \infty$$

Corollary 24.11d.   The zero solution of $\dot{x}(t) = A(t)\,x(t)$ is not asymptotically stable i.s.l. if there exists a constant $K > 0$ such that

$$\lim_{t \to \infty} \int_{t_0}^{t} \lambda_{\min}(q)\,dq > -K$$

CONCLUDING REMARKS

An in-depth discussion extending the basic results on stability for time-varying linear continuous-time state models derived in this chapter is given in Willems [4]; reference [5] also contains a number of interesting results for periodic time-varying linear state models. The discrete-time case is covered especially well in LaSalle [6]. As may be easily inferred, the results in the time-varying case fail to be wholly satisfactory. The problem rests with the need to construct a set of basis solutions, i.e., the columns of the state transition matrix $\Phi(t,\cdot)$. The really strong results express a system property like stability in terms of the structural properties of the system— i.e., the structural properties of $A(t)$, $B(t)$, $C(t)$, and $D(t)$, without recourse to constructing a set of basis solutions. Such results do exist, but only in a conservative sufficient condition capacity. Strengthening the existing characterizations or, even better, discovering new characterizations which are necessary *and* sufficient remains an open research question. Finally some recent results in [7] extend the notion of poles/zeros to time-varying systems and apply the results to the study of asymptotic stability.

PROBLEMS

1. Fill in the details of the proof of theorem 24.1.

2. For each of the following state model matrices, determine the impulse response matrix, and the norm given by equation 24.3, and thus the BIBO stability of the zero-state system response:

(a)

$$A(t) = \begin{bmatrix} 0 & -t \\ 0 & -t \end{bmatrix}, B(t) = \begin{bmatrix} -1 \\ 1 \end{bmatrix}, C(t) = [1 \ 1]$$

(b)

$$A(t) = \begin{bmatrix} -t & 0 \\ 0 & 1 \end{bmatrix}, B(t) = \begin{bmatrix} 1 \\ 0 \end{bmatrix}, C(t) = [1 \ 1]$$

(c)

$$A(t) = \begin{bmatrix} -1 & 0 & 0 \\ 2t & -1 & 0 \\ 1 & 2t & -1 \end{bmatrix}, B(t) = \begin{bmatrix} 0 \\ 1 \\ 0 \end{bmatrix}, C(t) = [0 \ 1 \ 0]$$

(d)

$$A(t) = \begin{bmatrix} -1 & \cos(t) \\ \cos(t) & -1 \end{bmatrix}, B(t) = \begin{bmatrix} 1 \\ 0 \end{bmatrix}, C(t) = [0 \ 1]$$

3. Determine conditions, if any, on the real, scalar-valued, piecewise-continuous functions $f(t)$ for the system

$$A(t) = \begin{bmatrix} 0 & f(t) \\ -f(t) & 0 \end{bmatrix}, B(t) = C(t) = I$$

to be BIBO stable.

4. Repeat problem 3 for

$$A(t) = \begin{bmatrix} 0 & f(t) \\ f(t) & 0 \end{bmatrix}, B(t) = C(t) = I$$

5. Plot the state trajectories given by equation 24.12 of example 24.4 when $t_0 = 0$, 2π, and 4π. What happens when the trajectory is evaluated π seconds later in each case?

6. Fill in the details of the proof of proposition 24.2.

7. Fill in the details of the proof of theorem 24.3.

8. Using theorem 24.4, show that $\dot{x}(t) = Ax(t)$ is asymptotically stable if and only if all the eigenvalues of A are in the open left-half complex plane.

9. Provide the details of the proof of theorem 24.5.

10. The results derived in this chapter are typically stated in terms of the state transition matrix. Justify the statement that the results can be equivalently stated in terms of an arbitrary fundamental matrix. Restate some of the theorems in terms of such a matrix.

11. Show that the zero solution of $\dot{x}(t) = A(t)x(t)$ is unstable if

$$\lim_{t \to \infty} \int_{t_0}^{t} \text{tr}[A(q)] dq = \infty$$

(*Hint*: use the relation $\det[\Phi(t, t_0)] = \exp\left[\int_{t_0}^{t} \text{tr}[A(q)] dq\right]$.)

12. Prove theorem 24.11 after consulting references [4] and [5].

REFERENCES

1. R. E. Vinogradov, "On a Criterion for Instability in the Sense of A. M. Lyapunov for Solutions of Linear Systems of Differential Equations," *Doklady Akademii Nauk USSR*, Vol. 84, No. 2, 1952, pp. 201–204.

2. M. Y. Wu, "A Note on Stability of Linear Time-Varying Systems," *IEEE Transactions on Automatic Control*, Vol. AC-19, No. 2, April 1974, p. 162.

3. J. L. Massera, "On Liapunoff's Conditions of Stability," *Annals of Mathematics*, Vol. 50, 1949, pp. 705–721.

4. Jacques L. Willems, *Stability Theory of Dynamical Systems* (New York: Wiley Interscience, 1970).

5. T. Wazewski, "Sur la limitation des intégrales des systèmes d'equations différentielles linéaires ordinaires," *Studia Mathematica*, Vol. 10, 1958, pp. 48–59.

6. J. P. LaSalle, *The Stability and Control of Discrete Processes* (New York: Springer-Verlag, 1986).

7. Edward W. Kamen, "The Poles and Zeros of a Linear Time-Varying System," *Linear Algebra and Its Applications*, Vol. 98, 1988, pp. 263–289.

25

THE DISTURBANCE
DECOUPLING PROBLEM

INTRODUCTION

Oftentimes a disturbance affects some part of the state space of a system through some input matrix, say, S. Design constraints may require that the effects of the disturbance not appear in some subset of the observable output variables. Under certain conditions, it is possible to construct a state feedback to decouple disturbances from a chosen set of output variables. This chapter sketches the rudiments of this problem and outlines a solution method in the context of linear time-invariant state models. The basic references on this material are [1–4].

NOTATION AND BASIC ALGEBRA

This section utilizes a more advanced and compact notation than in preceding chapters. It is consistent with that of advanced treatments on geometric control, e.g., [1]. Thus, let script or italic letters denote subspaces. For example, $\mathcal{X} = \mathbf{R}^n$ denotes the state space and $\mathcal{U} = \mathbf{R}^m$ the input space. Spaces such as $\mathcal{X}$ have subspaces which can be added together to construct larger spaces. For instance, if $\mathcal{S}_1$

and $\mathcal{S}_2$ are subspaces of $\mathcal{X}$ (i.e., $\mathcal{S}_1 \subset \mathcal{X}$ and $\mathcal{S}_2 \subset \mathcal{X}$) then their sum is a new space

$$\mathcal{S}_3 \triangleq \mathcal{S}_1 + \mathcal{S}_2 = \{s_3 \mid s_3 = \alpha_1 s_1 + \alpha_2 s_2\}$$

where α_1 and α_2 are arbitrary (possibly complex) scalars. Another convenient concept-cum-notation is the multiplication of a space by a matrix. For example, $A\mathcal{S}$ denotes a new subspace defined as

$$A\mathcal{S} \triangleq \{v \mid v = As , \; s \in \mathcal{S}\}$$

This idea provides a simple way of referring to the controllable subspace of a pair (A,B):

$$<A \mid \mathcal{B}> \triangleq \mathcal{B} + A\mathcal{B} + \ldots + A^{n-1}\mathcal{B}$$

denotes the controllable subspace of the pair (A,B), where $\mathcal{B} = \text{Im}(B)$, i.e., the subspace $\mathcal{B} \subset \mathcal{X}$ is the image of the matrix B. This notation incorporates the ideas of subspace addition as well as multiplication of a subspace by a matrix.

A variation of the multiplication of a subspace by a matrix is the operation where $B^{-1}\mathcal{S} = \{x \mid Bx \in \mathcal{S} \subset \mathcal{U}\}$. The notation involved conveniently encompasses the set of all points mapping to $\mathcal{S}$ under B.

Finally, it is often convenient to break $\mathcal{X}$ up into orthogonal subspaces. For example, if $\text{Im}(R) = \mathcal{R} \subset \mathcal{X}$, then $\mathcal{R}^{\perp}$, the orthogonal complement of $\mathcal{R}$ in $\mathcal{X}$, is defined as

$$\mathcal{R}^{\perp} = \{w \mid w^t R = 0\}$$

Thus,

$$\mathcal{X} = \mathcal{R} \oplus \mathcal{R}^{\perp}$$

i.e., $\mathcal{X}$ is a direct sum of vectors in $\mathcal{R}$ and vectors in $\mathcal{R}^{\perp}$. Hence, if $x \in \mathcal{X}$, then there exist a $v \in \mathcal{R}$ and a $w \in \mathcal{R}^{\perp}$ such that

$$x = v + w$$

STRUCTURE OF THE DISTURBANCE DECOUPLING PROBLEM (DDP)

Consider a time-invariant state model

$$\dot{x} = Ax + Bu + Sq$$
$$z = \hat{C}x \tag{25.1}$$

where q is a time-dependent deterministic disturbance vector (assumed to be the solution of some linear ordinary matrix differential equation), S is a $n \times m_q$ matrix specifying how the disturbance affects the state space, and $\hat{C}$ is an $r_z \times n$ matrix relating the state variables to the output variables, denoted z, which are to be "disturbance free" after suitable state feedback. The goal then is to find state feedback

$F:\mathcal{X} \to \mathcal{U}$ such that the effects of the disturbance q are absent from the output variables z.

Up to no. the output variables have standardly been denoted by y. The use of z in this context occurs because often only a subset of the y-entries or some linear combination thereof will have a disturbance-free condition imposed on it.

To develop some insight into the problem, a feedback-compensated state model has the form

$$\dot{x} = (A + BF)x + Sq$$
$$z = \hat{C}x$$

Solving for the model's zero-state response yields

$$z = \hat{C} \int_0^t e^{(A + BF)(t-\tau)} Sq(\tau)d\tau$$

The integral

$$\int_0^t e^{(A + BF)(t-\tau)} Sq(\tau)d\tau$$

has a range that is equal to the controllable subspace $<A + BF \mid \mathcal{S}>$ of the pair $(A + BF, S)$. This space determines the total spread of the disturbance throughout the state space. Hence, the z-variables are free of the disturbance if there exists an $F:\mathcal{X} \to \mathcal{U}$ such that

$$<A + BF \mid \mathcal{S}> \subset N[\hat{C}] \tag{25.2}$$

where $N[\hat{C}]$ is the null space of $\hat{C}$.

Ultimately the task of this chapter is to characterize the existence of such an F in terms of certain structural properties of the system— i.e., of the matrices A, B, S, and $\hat{C}$— and to present an algorithm for its computation. As a brief glimpse of the direction of the development, observe that the space $<A + BF \mid \mathcal{S}>$ is $(A + BF)$-invariant, i.e., if $v \in <A + BF \mid \mathcal{S}>$, then $(A + BF)v \in <A + BF \mid \mathcal{S}>$. This is clear from the development of chapter 20. No doubt there are many such spaces, say, $\mathcal{V}_F = <A + BF \mid \mathcal{S}>$, parameterized by different feedbacks F. It turns out that these $\mathcal{V}_F$'s are independent of specific F's, i.e., they can be defined without regard to a particular F. One then looks at the set of such $\mathcal{V}$'s and asks the question, Is there a largest such space contained in $N[\hat{C}]$? Since these $\mathcal{V}$'s can be described without knowledge of any F, so can this largest space. A test for a solution to the DDP would then reduce to checking whether or not $\mathcal{S}$ is a subspace of this largest space, denoted $\mathcal{V}^*$.

In this chapter we (*i*) show that there exists a largest $(A + BF)$-invariant space, denoted $\mathcal{V}^*$, contained in $N[\hat{C}]$ which depends only on A, B, and $\hat{C}$; (*ii*) show that the DDP is solvable if and only if $\mathcal{S} \subset \mathcal{V}^*$; and (*iii*) develop an algorithm for determining $\mathcal{V}^*$. In the process, we show that $\mathcal{V}^*$ is independent of $\mathcal{S}$. The technique rests squarely upon the notion of (A,B)-invariance [1–4].

(A,B)-INVARIANT SUBSPACES

Definition 25.1   A subspace $\mathcal{V} \subset \mathcal{X} = \mathbf{R}^n$ is said to be (A,B)-invariant if there exists $F:\mathcal{X} \rightarrow \mathcal{U}$ such that $(A + BF)\mathcal{V} \subset \mathcal{V}$.

The following two examples help illuminate the notion of (A,B)-invariance.

EXAMPLE 25.1

Suppose $\dot{x} = Ax + Bu$, where

$$A = \begin{bmatrix} 1 & 1 \\ 0 & -1 \end{bmatrix} \quad B = \begin{bmatrix} 1 \\ 0 \end{bmatrix}$$

The eigenvalue-eigenvector pairs are

$$\left(\lambda_1 = 1, e_1 = \begin{bmatrix} 1 \\ 0 \end{bmatrix} \right)$$

and

$$\left(\lambda_2 = -1, e_2 = \begin{bmatrix} 1 \\ -2 \end{bmatrix} \right)$$

With $F = [0 \ 0]$, it is clear that $(A + BF)e_1 = e_1$ and $(A + BF)e_2 = -e_2$. Hence, $\mathcal{V}_1 = \{v_1 \mid v_1 = \alpha e_1\}$ and $\mathcal{V}_2 = \{v_2 \mid v_2 = \alpha e_2\}$, for all real scalars α are (A,B)-invariant subspaces of $\mathcal{X}$. An important observation is that $\mathcal{V}_3 = \mathcal{V}_1 + \mathcal{V}_2 = \{v_3 \mid v_3 = \alpha_1 v_1 + \alpha_2 v_2\}$ (for all real scalars α_1 and α_2) is also (A,B)-invariant.

EXAMPLE 25.2

Consider again the pair (A,B) of the previous example. Any vector which is a potential eigenvector of $(A + BF)$ is a candidate (A,B)-invariant subspace. For $F = [f_1 \ f_2]$,

$$A + BF = \begin{bmatrix} 1 + f_1 & 1 + f_2 \\ 0 & -1 \end{bmatrix}$$

which has eigenvalues $\lambda_1 = -1 + f_1$ and $\lambda_2 = 1$. The corresponding eigenvectors are

$$e_1 = \begin{bmatrix} 1 \\ 0 \end{bmatrix}$$

and

$$e_2(f_1, f_2) = \begin{bmatrix} -1 \\ \dfrac{-2 - f_1}{1 + f_2} \end{bmatrix}$$

The sets $\{e_1\}$ and $\{e_2(F)\}$ determine all the one-dimensional (A,B)-invariant subspaces, and obviously $\mathbf{R}^2$ is a two-dimensional (A,B)-invariant subspace.

From the preceding definition and examples, it follows that if $\mathcal{V}$ is (A,B)-invariant, then there exists F such that

$$x(t) = e^{(A + BF)t} x_0 \in \mathcal{V}$$

for all t whenever $x_0 \in \mathcal{V}$. Thus, if we can choose F to direct the disturbance $q(\cdot)$ to an (A,B)-invariant subspace contained in $N[\hat{C}]$, it will remain there and hence solve the DDP. Toward this end, denote the set of (A,B)-invariant subspaces associated with the pair (A,B) contained in $\mathcal{X}$ by $\mathcal{J}(A,B;\mathcal{X})$.

The following theorem rids the notion of (A,B)-invariance of its ostensible dependence on a particular F.

Theorem 25.1. Let $\mathrm{Im}(B) = \mathcal{B}$, and suppose $\mathcal{V}$ is a subspace of the state space $\mathcal{X}$. Then $\mathcal{V} \in \mathcal{J}(A,B;\mathcal{X})$ if and only if $A\mathcal{V} \subset \mathcal{V} + \mathcal{B}$.

Proof. Suppose $\mathcal{V}$ is (A,B)-invariant. Then there exists an F such that $(A + BF)v = w$ for each v and some w (depending on v) in $\mathcal{V}$. Hence,

$$Av = w - BFv \in \mathcal{V} + \mathcal{B}$$

Consequently, $A\mathcal{V} \subset \mathcal{V} + \mathcal{B}$.

Conversely, suppose $A\mathcal{V} \subset \mathcal{V} + \mathcal{B}$. The converse follows by showing the existence of F such that $(A + BF)\mathcal{V} \subset \mathcal{V}$.

Step 1. Let $\{v_1,...,v_\mu\}$ be a basis for $\mathcal{V}$. By the hypothesis of the theorem, $A\mathcal{V} \subset \mathcal{V} + \mathcal{B}$. Thus, for each v_i, there exist a $w_i \in \mathcal{V}$ and $u_i \in \mathcal{U} = \mathbf{R}^m$ such that

$$Av_i = w_i - Bu_i$$

where $B:\mathcal{U}\to\mathcal{X}$.

Step 2. Let $\{v_{\mu+1},...,v_n\}$ be any set of vectors such that $\{v_1,...,v_n\}$ is a basis for $\mathbf{R}^n$. Define F by the equation

$$F[v_1,...,v_\mu,v_{\mu+1},...,v_n] = [u_1,...,u_\mu,u_{\mu+1},...,u_n]$$

where the vectors $u_{\mu+1},...,u_n$ are arbitrary. Clearly, F exists and is computable, since the matrix $M = [v_1,...,v_n]$ is nonsingular.

Step 3. By construction, for $i = 1,...,\mu$,

$$(A + BF)v_i = Av_i + BFv_i = Av_i + Bu_i = w_i \in \mathcal{V}$$

Finally, since the desired property holds for the basis $\{v_1,...,v_\mu\}$ it holds for all of $\mathcal{V}$.

∎

It is convenient at this point to name the class of F's for which $(A + BF)\mathcal{V} \subset \mathcal{V}$ for a particular $\mathcal{V}$. Accordingly, we denote by $\underline{F}(\mathcal{V})$ the set $\{F:\mathcal{X}\to\mathcal{U} \mid (A + BF)\mathcal{V} \subset \mathcal{V}\}$. The following proposition interrelates the members of the family of such F's.

Proposition 25.1. F_1 and F_2 are contained in $\underline{F}(\mathcal{V})$ if and only if

$$(F_1 - F_2)\mathcal{V} \subset B^{-1}\mathcal{V} \triangleq \{u \mid Bu \in \mathcal{V}\}$$

Proof. *Part 1.* Let $v \in \mathcal{V}$ and $F_1, F_2 \in \underline{F}(\mathcal{V})$. By definition, (i) $(A + BF_1)v = w_1 \in \mathcal{V}$ and (ii) $(A + BF_2)v = w_2 \in \mathcal{V}$. Subtracting these two expressions yields

$$B(F_1 - F_2)v = w_1 - w_2 \in \mathcal{V}$$

since the difference of two vectors in $\mathcal{V}$ lies in $\mathcal{V}$. Equivalently,

$$(F_1 - F_2)\mathcal{V} \subset B^{-1}\mathcal{V}$$

Part 2. Now suppose $(F_1 - F_2)\mathcal{V} \in B^{-1}\mathcal{V}$, where $F_1 \in \underline{F}(\mathcal{V})$. We show that $F_2 \in \underline{F}(\mathcal{V})$. The hypothesis implies that for each $v \in \mathcal{V}$, there is a $w \in \mathcal{V}$ such that $BF_1v - BF_2v = w$. Adding and subtracting Av yields

$$(A + BF_1)v - (A + BF_2)v = w$$

Since $F_1 \in \underline{F}(\mathcal{V})$, $(A + BF_1)v = w_1 \in \mathcal{V}$, and it follows that

$$(A + BF_2)v = w_1 - w \in \mathcal{V}$$

Hence, $F_2 \in \underline{F}(\mathcal{V})$. ∎

SUPREMAL (*A,B*)-INVARIANT SUBSPACES

The purposes of this section are to show that the set $\underline{\mathcal{I}}(A,B;\mathcal{X})$ is closed under subspace addition and to prove the existence of a largest element of the set.

Proposition 25.2. The class of (*A,B*)-invariant subspaces $\underline{\mathcal{I}}(A,B;\mathcal{X})$ is closed under subspace addition.

Proof. "Closed under subspace addition" means that if $\mathcal{V}_1$ and $\mathcal{V}_2$ are elements of $\underline{\mathcal{I}}(A,B;\mathcal{X})$ then $\mathcal{V}_1 + \mathcal{V}_2 \in \underline{\mathcal{I}}(A,B;\mathcal{X})$. By definition, if $\mathcal{V}_1$ and $\mathcal{V}_2 \in \underline{\mathcal{I}}(A,B;\mathcal{X})$, then $A\mathcal{V}_1 \subset \mathcal{V}_1 + \mathcal{B}$ and $A\mathcal{V}_2 \subset \mathcal{V}_2 + \mathcal{B}$. Now,

$$A(\mathcal{V}_1 + \mathcal{V}_2) = A\mathcal{V}_1 + A\mathcal{V}_2 \subset (\mathcal{V}_1 + \mathcal{V}_2) + \mathcal{B}$$

from which it follows that $(\mathcal{V}_1 + \mathcal{V}_2) \in \underline{\mathcal{I}}(A,B;\mathcal{X})$. But this is no more than to say that $\underline{\mathcal{I}}(A,B;\mathcal{X})$ is closed under subspace addition. ∎

Since $\mathcal{X}$ is finite-dimensional, the property of closure under subspace addition suggests that by adding all the elements of $\underline{\mathcal{I}}(A,B;\mathcal{X})$ together, one could construct a largest element. If such a largest element, say, $\mathcal{V}^*$, exists, then if $\mathcal{V}$ is any other member of the class, $\mathcal{V} \subset \mathcal{V}^*$. Further, this largest member must be unique, for if $\mathcal{V}_1^*$ and $\mathcal{V}_2^*$ are two supremal elements, then $\mathcal{V}_1^* \subset \mathcal{V}_2^*$ and $\mathcal{V}_2^* \subset \mathcal{V}_1^*$, implying that $\mathcal{V}_1^* = \mathcal{V}_2^*$.

Theorem 25.2. Let $\underline{\mathcal{I}}(\mathcal{X})$ be a nonempty family of subspaces of $\mathcal{X}$ that is closed under the operation of subspace addition. Then $\underline{\mathcal{I}}(\mathcal{X})$ contains a supremal or largest element.

Proof. Since $\underline{J}(\mathfrak{X})$ is a family of finite-dimensional subspaces contained in the n-dimensional space $\mathfrak{X}$, there exists at least one element of largest dimension. Denote this element by $\mathcal{V}^*$. Let $\mathcal{V} \in \underline{J}(\mathfrak{X})$. Since $\underline{J}(\mathfrak{X})$ is closed under subspace addition, $\mathcal{V} + \mathcal{V}^* \in \underline{J}(\mathfrak{X})$. By hypothesis, $\dim(\mathcal{V}^*) \geqslant \dim(\mathcal{V} + \mathcal{V}^*)$. By the properties of subspaces, $\dim(\mathcal{V} + \mathcal{V}^*) \geqslant \dim(\mathcal{V}^*)$. Therefore,

$$\dim(\mathcal{V}^*) \geqslant \dim(\mathcal{V} + \mathcal{V}^*) \geqslant \dim(\mathcal{V}^*)$$

for all $\mathcal{V} \in \underline{J}(\mathfrak{X})$. Hence, $\mathcal{V} \subset \mathcal{V}^*$ for all $\mathcal{V} \in \underline{J}(\mathfrak{X})$.

The preceding notions need a bit of refinement before being applied to the solution of the DDP. Let $\mathcal{H} \subset \mathfrak{X}$ be a subspace, and define the class

$$\underline{J}(A,B;\mathcal{H}) \triangleq \{\mathcal{V} \subset \mathfrak{X} \mid \mathcal{V} \subset \mathcal{H} \text{ and } \mathcal{V} \in \underline{J}(A,B;\mathfrak{X})\}$$

That is, $\underline{J}(A,B;\mathcal{H})$ is the set of (A,B)-invariant subspaces of $\mathfrak{X}$ contained in $\mathcal{H}$. Proposition 25.2 and theorem 25.2 have the following two corollaries, respectively.

Corollary 25.2a. $\underline{J}(A,B;\mathcal{H})$ is closed under subspace addition.

Corollary 25.2b. $\underline{J}(A,B;\mathcal{H})$ contains a supremal element denoted by

$$\mathcal{V}^* = \sup \underline{J}(A,B;\mathcal{H})$$

Before proceeding to a solution of the DDP, the following section describes a method for computing $\mathcal{V}^*$.

COMPUTATION OF $\mathcal{V}^*$

Two things are required for the computation of $\mathcal{V}^*$: a theoretical procedure and a numerical algorithm [5]. The following theorem provides the former.

Theorem 25.3. Let $\dot{x} = Ax + Bu$ and $\mathcal{H}$ be a subspace of $\mathfrak{X}$, e.g., $\mathcal{H} = N[\hat{C}]$. Define a sequence of subspaces, denoted $\mathcal{V}^j$, according to

$$\mathcal{V}^o = \mathcal{H}$$
$$\mathcal{V}^j = \mathcal{H} \cap A^{-1}(\mathcal{B} + \mathcal{V}^{j-1})$$

for $j = 1,2,3,\ldots$ Then $\mathcal{V}^j \subset \mathcal{V}^{j-1}$ and for some $k \leqslant \dim(\mathcal{H})$,

$$\mathcal{V}^k = \mathcal{V}^* = \sup \underline{J}(A,B;\mathcal{H})$$

Proof. The proof of this result has two parts: (*i*) show that $\mathcal{V}^j \subset \mathcal{V}^{j-1}$, i.e., $\mathcal{V}^0, \mathcal{V}^1, \ldots, \mathcal{V}^k$ is a nonincreasing sequence of subspaces, and (*ii*) show that for some $k \leqslant \dim(\mathcal{H})$, $\mathcal{V}^k = \sup \underline{J}(A,B;\mathcal{H})$.

Part 1. This part is shown by induction. Since $\mathcal{V}^0 = \mathcal{H}$,

$$\mathcal{V}^1 = \mathcal{H} \cap A^{-1}(\mathcal{B} + \mathcal{V}^0) \subset \mathcal{H} = \mathcal{V}^0$$

The second stage of the induction utilizes the hypothesis, $\mathcal{V}^j \subset \mathcal{V}^{j-1}$, with the goal of demonstrating that $\mathcal{V}^{j+1} \subset \mathcal{V}^j$. But this follows readily:

$$\mathcal{V}^{j+1} \triangleq \mathcal{K} \cap A^{-1}(\mathcal{B} + \mathcal{V}^j)$$

$$\subset \mathcal{K} \cap A^{-1}(\mathcal{B} + \mathcal{V}^{j-1}) \triangleq \mathcal{V}^j$$

Part 2. This part has several stages. The first stage shows that the foregoing sequence terminates for some $k \leqslant \dim(\mathcal{K})$. The second stage shows that $\mathcal{V}^k \in \underline{\mathcal{J}}(A,B;\mathcal{K})$, and the third stage, $\mathcal{V}^k = \sup \underline{\mathcal{J}}(A,B;\mathcal{K})$.

To show that the sequence terminates, observe that each $\mathcal{V}^j$ is finite-dimensional and contained in $\mathcal{K}$. Hence, there must exist a $k \leqslant \dim(\mathcal{K})$ such that $\mathcal{V}^j = \mathcal{V}^k$ for all $j \geqslant k$.

To show that $\mathcal{V}^k \in \underline{\mathcal{J}}(A,B;\mathcal{K})$, recall that by definition $\mathcal{V} \in \underline{\mathcal{J}}(A,B;\mathcal{K})$ if and only if $A\mathcal{V} \subset \mathcal{V} + \mathcal{B}$ and $\mathcal{V} \subset \mathcal{K}$. But this is true if and only if $\mathcal{V} \subset A^{-1}(\mathcal{V} + \mathcal{B})$ and $\mathcal{V} \subset \mathcal{K}$, which is in turn true if and only if $\mathcal{V} \subset \mathcal{K} \cap A^{-1}(\mathcal{V} + \mathcal{B})$. But then, by the argument of the previous paragraph,

$$\mathcal{V}^k = \mathcal{V}^{k+1} = \mathcal{K} \cap A^{-1}(\mathcal{B} + \mathcal{V}^k)$$

Hence, $\mathcal{V}^k$ is (A,B)-invariant.

Finally, to show that $\mathcal{V}^k = \sup \underline{\mathcal{J}}(A,B;\mathcal{K})$, let $\mathcal{V} \in \underline{\mathcal{J}}(A,B;\mathcal{K})$ be arbitrary. $\mathcal{V}^k$ is supremal if $\mathcal{V} \subset \mathcal{V}^k$. This follows by a simple induction. Clearly if $\mathcal{V} \in \underline{\mathcal{J}}(A,B;\mathcal{K})$, $\mathcal{V} \subset \mathcal{K} = \mathcal{V}^0$. Suppose now, for induction, that $\mathcal{V} \subset \mathcal{V}^{j-1}$. By definition,

$$\mathcal{V}^j \triangleq \mathcal{K} \cap A^{-1}(\mathcal{B} + \mathcal{V}^{j-1})$$

Now by hypothesis, $\mathcal{V} \subset \mathcal{K}$, $\mathcal{V} \subset \mathcal{V}^{j-1}$, and $\mathcal{V} \in \underline{\mathcal{J}}(A,B;\mathcal{K})$. Thus, $A\mathcal{V} \subset \mathcal{V} + \mathcal{B} \subset \mathcal{V}^{j-1} + \mathcal{B}$, or, equivalently, $\mathcal{V} \subset A^{-1}(\mathcal{B} + \mathcal{V}^{j-1})$. Hence, $\mathcal{V} \subset \mathcal{V}^j$ for all j. But then, $\mathcal{V} \subset \mathcal{V}^k$ for any arbitrary k, and so $\mathcal{V}^k = \sup(A,B;\mathcal{K})$, i.e., $\mathcal{V}^k$ is supremal.   ∎

The theoretical sequencing treatment of theorem 25.3 sets up the structure for a numerical algorithm. A characterization of the various spaces $\mathcal{V}^j$ depends on the computation of basis vectors, which in turn depend on the orthogonality properties of certain matrices as described in the following lemma.

Lemma 25.1. Let A be an $n \times n$ matrix, and R be a matrix with independent columns. Also, let $\mathcal{R} = \mathrm{Im}(R)$. Suppose W is a maximal-rank matrix such that $W^t R = [0]$, i.e., the columns of W are a basis for $\mathcal{R}^\perp$, the subspace of $\mathcal{X}$ that is orthogonal to the subspace $\mathcal{R}$. Then

$$A^{-1}\mathcal{R} = N[W^t A]$$

i.e., the subspace $A^{-1}\mathcal{R}$ equals the null space of $W^t A$.

Proof. *Part 1.* Let $x \in A^{-1}\mathcal{R}$. By definition, there is a $y \in \mathcal{R}$ such that $Ax = y$. Since $y \in \mathcal{R}$, $W^t y = 0$, i.e., $W^t y = W^t Ax = 0$ for all $x \in A^{-1}\mathcal{R}$. Consequently, $x \in N[W^t A]$.

Part 2. Conversely, let $x \in N[W^t A]$. Then $W^t Ax = 0$. But then, $Ax \in \mathcal{R} = N[W^t]$ or $x \in A^{-1}\mathcal{R}$.

Based on lemma 25.1, we now present the following algorithm for computing $\mathcal{V}*$:

Step 1. Establish a basis for $\mathcal{V}^0 = \mathcal{K} = N[\hat{C}]$. Let V_0 denote a maximal-rank matrix that satisfies the equation

$$\hat{C} V_0 = [0]$$

Remark. A singular value decomposition of $\hat{C}$ will immediately yield a V_0 whose columns are an orthonormal basis for $\mathcal{V}^0$.

Step 2. Construct a basis for $(\mathcal{B} + \mathcal{V}^0)$. To accomplish this, form the matrix $[B \; V_0]$ and eliminate redundant columns to form $[B \; V_0]'$, whose columns are a basis for $(\mathcal{B} + \mathcal{V}_0)$.

Remark. If one is using an SVD program, forming $[B \; V_0]'$ is unnecessary, as will be plain from step 3.

Remark. The aim at this point is to determine a basis for $A^{-1}(\mathcal{B} + \mathcal{V}^0) \triangleq N[W_1^t A]$, where W_1 is a matrix whose columns are a basis for $(\mathcal{B} + \mathcal{V}^0)^\perp$, the space orthogonal to $(\mathcal{B} + \mathcal{V}^0)$.

Step 3. Compute the matrix W_1 *mentioned in the preceding remark.* W_1 must satisfy

$$([B \; V_0]')^t \, W_1 = \begin{bmatrix} B^t \\ V_0^t \end{bmatrix} W_1 = [0]$$

Step 4. Construct a matrix V_1 *whose columns are a basis for* $\mathcal{V}^1$. Observe that $\mathcal{V}^1 = \mathcal{K} \cap A^{-1}(\mathcal{B} + \mathcal{V}^0)$. Hence, V_1 must be a maximal-rank matrix satisfying (*i*) $\hat{C} V_1 = [0]$ and (*ii*) $[W_1^t A] V_1 = [0]$, i.e.,

$$\begin{bmatrix} \hat{C} \\ W_1^t A \end{bmatrix} V_1 = [0]$$

Step 5. Repeat steps 3 and 4 until $rank[V_j] = rank[V_{j-1}]$. Note that W_j is a maximal-rank matrix satisfying

$$\begin{bmatrix} B^t \\ V_{j-1}^t \end{bmatrix} W_j = [0]$$

The columns of W_j are orthogonal to every basis for the space $(\mathcal{B} + \mathcal{V}^{j-1})$. Hence, the columns of W_j are a basis for $(\mathcal{B} + \mathcal{V}^{j-1})^\perp$. Note that V_j is a matrix whose columns are a basis for $\mathcal{V}^j$. Hence, V_j is a maximal-rank matrix satisfying

$$\begin{bmatrix} \hat{C} \\ W_j^t A \end{bmatrix} V_j = [0]$$

EXAMPLE 25.3 [1]

Compute a basis for $\mathcal{V}*$ when

$$A = \begin{bmatrix} 0 & 1 & 0 & 0 & 0 \\ 0 & 0 & 1 & 0 & 0 \\ 0 & 0 & 0 & 0 & 0 \\ 0 & 0 & 0 & 0 & 1 \\ 0 & 0 & 0 & 0 & 0 \end{bmatrix}, \quad B = \begin{bmatrix} 0 & 0 \\ 0 & 0 \\ 1 & 0 \\ 0 & 1 \\ 0 & 0 \end{bmatrix}, \quad \hat{C} = \begin{bmatrix} 1 & 0 & 0 & 0 & 0 \\ 0 & 0 & 0 & 1 & 0 \end{bmatrix}$$

Step 1. We compute a basis for $\mathcal{V}^0 = \mathcal{K} = N[\hat{C}]$. V_0 is a maximal-rank matrix such that $\hat{C}V_0 = [0]$. By inspection, V_0 can be taken as

$$V_0 = \begin{bmatrix} 0 & 0 & 0 \\ 1 & 0 & 0 \\ 0 & 1 & 0 \\ 0 & 0 & 0 \\ 0 & 0 & 1 \end{bmatrix}$$

Clearly, rank$[V_0] = 3$.

Step 2. To construct a basis for $(\mathcal{B} + \mathcal{V}^0)$, we form

$$[V_0 \; B] = \begin{bmatrix} 0 & 0 & 0 & 0 & 0 \\ 1 & 0 & 0 & 0 & 0 \\ 0 & 1 & 0 & 1 & 0 \\ 0 & 0 & 0 & 0 & 1 \\ 0 & 0 & 1 & 0 & 0 \end{bmatrix}$$

By inspection,

$$[V_0 \; B]' = \begin{bmatrix} 0 & 0 & 0 & 0 \\ 1 & 0 & 0 & 0 \\ 0 & 1 & 0 & 0 \\ 0 & 0 & 1 & 0 \\ 0 & 0 & 0 & 1 \end{bmatrix}$$

Step 3. We compute W_1 which is a maximal-rank matrix satisfying

$$([V_0 \; B]')^t \, W_1 = \begin{bmatrix} 0 & 1 & 0 & 0 & 0 \\ 0 & 0 & 1 & 0 & 0 \\ 0 & 0 & 0 & 1 & 0 \\ 0 & 0 & 0 & 0 & 1 \end{bmatrix} W_1 = \begin{bmatrix} 0 \\ 0 \\ 0 \\ 0 \end{bmatrix}$$

By inspection, $W_1^t = [1 \; 0 \; 0 \; 0 \; 0]$.

Step 4. Find V_1, a maximal-rank matrix satisfying

$$\begin{bmatrix} \hat{C} \\ W_1^t A \end{bmatrix} V_1 = [0]$$

Specifically

$$\begin{bmatrix} \hat{C} \\ W_1^t A \end{bmatrix} V_1 = \begin{bmatrix} 1 & 0 & 0 & 0 & 0 \\ 0 & 0 & 0 & 1 & 0 \\ 0 & 1 & 0 & 0 & 0 \end{bmatrix} V_1 = [0]$$

Again by inspection,

$$V_1 = \begin{bmatrix} 0 & 0 \\ 0 & 0 \\ 1 & 0 \\ 0 & 0 \\ 0 & 1 \end{bmatrix}$$

and $\operatorname{rank}(V_1) = 2$.

Step 5. We repeat steps 2 and 4 until we find a j such that $\operatorname{rank}[V_j] = \operatorname{rank}[V_{j-1}]$. A matrix whose columns span $(\mathcal{B} + \mathcal{V}^1)$ is

$$[V_1 \ B] = \begin{bmatrix} 0 & 0 & 0 & 0 \\ 0 & 0 & 0 & 0 \\ 1 & 0 & 1 & 0 \\ 0 & 0 & 0 & 1 \\ 0 & 1 & 0 & 0 \end{bmatrix}$$

By inspection, a matrix whose columns are a basis for $(\mathcal{B} + \mathcal{V}^1)$ is

$$[V_1 \ B]' = \begin{bmatrix} 0 & 0 & 0 \\ 0 & 0 & 0 \\ 1 & 0 & 0 \\ 0 & 1 & 0 \\ 0 & 0 & 1 \end{bmatrix}$$

We continue by constructing W_2, which is a maximal-rank matrix satisfying

$$([V_1 \ B]')^t W_2 = \begin{bmatrix} 0 & 0 & 1 & 0 & 0 \\ 0 & 0 & 0 & 1 & 0 \\ 0 & 0 & 0 & 0 & 1 \end{bmatrix} W_2 = [0]$$

In particular

$$W_2^t = \begin{bmatrix} 1 & 0 & 0 & 0 & 0 \\ 0 & 1 & 0 & 0 & 0 \end{bmatrix}$$

Using W_2^t, we now compute V_2, a maximal-rank matrix satisfying

$$\begin{bmatrix} \hat{C} \\ W_2^t A \end{bmatrix} V_2 = \begin{bmatrix} 1 & 0 & 0 & 0 & 0 \\ 0 & 0 & 0 & 1 & 0 \\ 0 & 1 & 0 & 0 & 0 \\ 0 & 0 & 1 & 0 & 0 \end{bmatrix} V_2 = [0]$$

Hence, $V_2 = [0\ 0\ 0\ 0\ 1]^t$.

Repeating steps 3 and 4 again, we construct $V_3 = V_2$, so that $\mathcal{V}^* = V_2$.

SOLUTION TO THE DDP

Recall the state space model introduced at the beginning of the chapter, i.e.

$$\dot{x} = Ax + Bu + Sq$$
$$z = \hat{C}x$$

where $q(\cdot)$ is the disturbance vector. Also recall that the system is said to be disturbance decoupled if there exists $F:\mathcal{X}\to\mathcal{U}$ such that

$$<A + BF\,|\,\mathcal{S}> \subset \mathcal{K} = N[\hat{C}]$$

Solvability of this problem is characterized by the theorem below.

Theorem 25.4. The state space system given by equation 25.1 is disturbance-decoupled (see equation 25.2) if and only if $\mathcal{S} \subset \mathcal{V}^*$, where $\mathcal{V}^* = \sup \underline{\mathcal{J}}(A,B;\mathcal{K})$.

Proof. *Part 1.* Suppose $\mathcal{V}^* = \sup \underline{\mathcal{J}}(A,B;\mathcal{K})$. Then there exists an $F \in \underline{F}(\mathcal{V}^*)$ such that $(A + BF)\mathcal{V}^* \subset \mathcal{V}^*$. But since $\mathcal{S} \subset \mathcal{V}^*$, it follows that $<A + BF\,|\,\mathcal{S}> \subset <A + BF\,|\,\mathcal{V}^*> = \mathcal{V}^* \subset \mathcal{K} = N[\hat{C}]$. Therefore, the system is disturbance-decoupled and F solves the DDP.

Part 2. Suppose F solves the DDP, i.e., the system is disturbance-decoupled. Define

$$\mathcal{V} = <A + BF\,|\,\mathcal{S}> \subset \mathcal{K} = N[\hat{C}]$$

where the inclusion follows since F solves the DDP. Clearly, $\mathcal{V}$ is (A,B)-invariant and contains $\mathcal{S}$. Thus $\mathcal{S} \subset \mathcal{V} \subset \mathcal{V}^*$ by the supremal nature of $\mathcal{V}^*$. ∎

Clearly, any $F \in \underline{F}(\mathcal{V}^*)$ will solve the DDP. But what other properties of F are important for an intelligent choice? Certainly, the stability of $(A + BF)$ is crucial. Details of this consideration find exposition in chapter 5 of [1].

At this point, it should be rather obvious that testing for a solution relies on constructing $\mathcal{V}^*$ and checking to see whether $\mathcal{S} \subset \mathcal{V}^*$. Once $\mathcal{V}^*$ is known, it is possible to characterize the entire set of S's for which there exists a solution. Of course, if there does exist a solution, one must compute an F which achieves the decoupling. This is done through the use of theorem 25.1.

EXAMPLE 25.4

Reconsider again the matrices A, B, and $\hat{C}$ of example 25.3. For these matrices

$$\mathcal{V}^* = \{(0\ 0\ 0\ 0\ 1)^t\}$$

It is necessary, of course, that $S = [0\ 0\ 0\ 0\ \alpha]^t$ for arbitrary nonzero α. Let $v_1 = (0\ 0\ 0\ 0\ 1)^t$. Since $\mathcal{A}\mathcal{V}^* \subset \mathcal{V}^* + \mathcal{B}$, there is a $w_1 \in \mathcal{V}^*$ and a $u_1 \in \mathcal{U}$ such that

$$Av_1 = w_1 - Bu$$

Thus,

$$Av_1 = \begin{bmatrix} 0 \\ 0 \\ 0 \\ 1 \\ 0 \end{bmatrix} = \begin{bmatrix} 0 \\ 0 \\ 0 \\ 0 \\ 1 \end{bmatrix} \alpha_1 - \begin{bmatrix} 0 & 0 \\ 0 & 0 \\ 1 & 0 \\ 0 & 1 \\ 0 & 0 \end{bmatrix} \begin{bmatrix} \alpha_2 \\ \alpha_3 \end{bmatrix} = \begin{bmatrix} 0 & 0 & 0 \\ 0 & 0 & 0 \\ 0 & -1 & 0 \\ 0 & 0 & -1 \\ 1 & 0 & 0 \end{bmatrix} \begin{bmatrix} \alpha_1 \\ \alpha_2 \\ \alpha_3 \end{bmatrix}$$

for appropriate scalars α_1, α_2, and α_3. Solving this equation using a pseudo left inverse yields $\alpha_1 = 0$, $\alpha_2 = 0$, $\alpha_3 = -1$, in which case $u_1 = (0\ -1)^t$. One must then choose F so that

$$F[v_1 | v_2, v_3, v_4, v_5] = [u_1 | u_2, u_3, u_4, u_5]$$

where the v_i's form a basis for R^5 and $u_2,...,u_5$ are arbitrary. A simple choice for F is given by

$$F \begin{bmatrix} 0 & 1 & 0 & 0 & 0 \\ 0 & 0 & 1 & 0 & 0 \\ 0 & 0 & 0 & 1 & 0 \\ 0 & 0 & 0 & 0 & 1 \\ 1 & 0 & 0 & 0 & 0 \end{bmatrix} = \begin{bmatrix} 0 & 0 & 0 & 0 & 0 \\ -1 & 0 & 0 & 0 & 0 \end{bmatrix}$$

so that

$$F = \begin{bmatrix} 0 & 0 & 0 & 0 & 0 \\ 0 & 0 & 0 & 0 & -1 \end{bmatrix}$$

which solves the DDP.

CONCLUDING REMARKS

The ideas set out in this chapter have significant extensions [6–9]. The reader should consult these references for further information. Reference [6] provides a good discussion of the nonlinear problem. For an extension to interconnected systems, see [10–12].

PROBLEMS

1. Let two matrices M_1 and M_2 be given by

$$M_1 = \begin{bmatrix} 1 & 0 \\ 0 & 1 \\ 1 & 1 \\ 0 & 1 \end{bmatrix}$$

and

$$M_2 = \begin{bmatrix} 1 & 0 \\ 1 & 0 \\ 0 & 0 \\ -1 & 1 \end{bmatrix}$$

Find a basis for the space

$$\mathcal{S} = Im(M_1) + Im(M_2)$$

2. If a basis for the space $\mathcal{S}$ is $\{v_1, v_2\}$ where $v_1 = (1\ 0\ 0)^t$ and $v_2 = (1\ 0\ -1)^t$, then determine a basis for the space $A\mathcal{S}$, where

$$A = \begin{bmatrix} 1 & 1 & 1 \\ 1 & 0 & 1 \\ 0 & -1 & 0 \end{bmatrix}$$

3. Determine a basis for the space $B^{-1}\mathcal{S}$, where

$$B = \begin{bmatrix} 1 & 0 & 0 \\ 0 & 1 & -1 \\ 0 & -1 & 1 \\ 1 & 2 & -2 \end{bmatrix}$$

and a basis for $\mathcal{S}$ is $\{(1,1,1)^t\}$.

4. For the matrix B defined in problem 3, determine a basis for $\mathcal{B}^{\perp}$ in R^4. If $x = (1,1,1,1)^t$ find the components of x in $\mathcal{B}$ and in $\mathcal{B}^{\perp}$.

5. Consider equation 25.1 with

$$A = \begin{bmatrix} \lambda_1 & 0 & 0 \\ 0 & \lambda_2 & 0 \\ 0 & 0 & \lambda_3 \end{bmatrix}, \quad B = \begin{bmatrix} 0 \\ 1 \\ 1 \end{bmatrix}, \quad S = \begin{bmatrix} 1 \\ 0 \\ 0 \end{bmatrix}$$

By inspection, determine the class of $\hat{C}$-matrices for which the system is disturbance-decoupled.

6. Let $F = [f_1\ f_2]$ and consider the system

$$A = \begin{bmatrix} 0 & 0 \\ 0 & 0 \end{bmatrix}, \quad B = \begin{bmatrix} 0 \\ 1 \end{bmatrix}$$

Investigate $(A + BF)$, and determine the structure of various (A,B)-invariant subspaces.

7. Provide detailed proofs of corollaries 25.2a and 25.2b.

8. Consider the state model

$$\dot{x} = Ax + Bu + Sq(t)$$

$$z = \hat{C}x$$

where

$$A = \begin{bmatrix} -1 & 0 & 0 & 0 & 0 \\ -3 & 0 & 0 & 1 & -1 \\ -1 & -2 & 0 & 2 & -2 \\ -0.5 & -1 & 0.5 & 1 & -1 \\ 3.5 & -1 & 0.5 & -1 & 0 \end{bmatrix}, \quad B = \begin{bmatrix} 0 & 0 \\ -1 & 0 \\ 0 & 0 \\ 0 & 1 \\ 1 & 1 \end{bmatrix}, \quad S = \begin{bmatrix} 0 & 1 \\ 1 & 0 \\ 0 & 1 \\ 0 & 1 \\ -1 & 1 \end{bmatrix}$$

and

$$\hat{C} = [0 \ -1 \ 0 \ 1 \ -1]$$

(a) Find a basis for $\mathcal{V}^*$.

(b) Is the DDP solvable? If so, find a state feedback F which solves the problem.

9. Consider again a system of the form

$$\dot{x} = Ax + Bu + Sq$$
$$z = \hat{C}x$$

where

$$A = \begin{bmatrix} 2 & 0 & 3 & 1 & 0 & 0 \\ 1 & 0 & 1 & 0 & -1 & 0 \\ 0 & 0 & 1 & 0 & 0 & 0 \\ -1 & 0 & -2 & 0 & 0 & 0 \\ 2 & 1 & 2 & 0 & 0 & 0 \\ 0 & 0 & 0 & 0 & 0 & 1 \end{bmatrix}, \quad B = \begin{bmatrix} 0 & 1 & 1 \\ 0 & 0 & 0 \\ 0 & 0 & -1 \\ 0 & -1 & 0 \\ -1 & 0 & 0 \\ 0 & 0 & 0 \end{bmatrix}$$

and

$$\hat{C} = [1 \ 0 \ 0 \ 0 \ 1 \ 0]$$

(a) Find a basis for $\mathcal{V}^*$.

(b) Describe the class of all S's for which the DDP is solvable.

10. Consider the matrices

$$A = \begin{bmatrix} -1 & -1 & 0 & -2 & 0 \\ 0 & 0 & 0 & 1 & 0 \\ 1 & 1 & 1 & 1 & 0 \\ 1 & 1 & 0 & 1 & 0 \\ -1 & -1 & -1 & -1 & 0 \end{bmatrix}, \quad B = \begin{bmatrix} 1 & 0 \\ 0 & 0 \\ -1 & 0 \\ 0 & 0 \\ 0 & 1 \end{bmatrix}, \quad S = \begin{bmatrix} -1 \\ 1 \\ 0 \\ 1 \\ 0 \end{bmatrix},$$

and

$$\hat{C} = \begin{bmatrix} -1 & 0 & -1 & -1 & 0 \\ 0 & 1 & 0 & -1 & 0 \\ 0 & 0 & 0 & 0 & -1 \end{bmatrix}$$

(a) Find $\mathcal{V}^* = \sup \mathcal{I}(A,B;\ker(D))$.

(b) Show that the system can be disturbance-decoupled.

(c) Find a matrix F which will do the job.

(d) Show that if F_1, and F_2 are any two matrices which will decouple the disturbance from the output, then $F_1 | \mathcal{V}^* = F_2 | \mathcal{V}^*$.

(e) Lastly show that $(A + BF)$ has unstable eigenvalues and, in fact, $(A + BF) | \mathcal{V}^*$ has an eigenvalue in the right half complex plane.

REFERENCES

1. W. M. Wonham, *Linear Multivariable Control: a Geometric Approach* (New York: Springer-Verlag, 1979).

2. W. M. Wonham, "Decoupling and Pole Assignment in Linear Multivariable Systems: a Geometric Approach," *SIAM Journal of Control*, Vol. 8, No. 1, 1970, pp. 317–337.

3. G. Basile and G. Marro, "Controlled and Conditioned Invariant Subspaces in Linear System Theory," *Journal of Optimization Theory and Applications*, Vol. 3, No. 5, 1969, pp. 306–315.

4. G. Basile and G. Marro, "On the Observability of Linear Time-invariant Systems with Unknown Inputs," *Journal of Optimization Theory and Applications*, Vol. 3, No. 6, 1969, pp. 410–415.

5. B. C. Moore and A. J. Laub, "Computation of Supremal (A,B)-invariant and (A,B)-controllability Subspaces," *IEEE Transactions on Automatic Control*, Vol. AC-23, No. 5, 1978, pp. 783–792.

6. A. Isidori, *Nonlinear Control Systems: An Introduction* (New York: Springer-Verlag, 1985).

7. J. C. Willems and C. Commault, "Disturbance Decoupling by Measurement Feedback with Stability or Pole Placement," *SIAM Journal of Control and Optimization*, Vol. 19, 1981, pp. 490–504.

8. Arno Linnemann, "Numerical Aspects of Disturbance Decoupling by Measurement Feedback," *IEEE Transactions on Automatic Control*, Vol. AC-32, No. 10, October 1987, pp. 922–926.

9. V. A. Arenentano, "Almost Disturbance Decoupling by a Proportional Derivative State Feedback Law," *Automatica*, Vol. 22, No. 4, 1986, pp. 449–456.

10. Dale Sebok and R. DeCarlo, "Preliminary Results on Decentralized Disturbance Decoupling with Internal Stability," *Proceedings of the 24th IEEE Conference on Decision and Control*, Fort Lauderdale, FL, December, 1985, pp. 1493–1494.

11. Vitor M. P. Leite, "Disturbance Decoupling in Decentralized Linear Systems by Non-dynamic Feedback of State or Measurement," *International Journal of Control*, Vol. 42, No. 4, 1985, pp. 913–937.

12. Vitor M. P. Leite, "Further Aspects of Disturbance Decoupling in Decentralized Linear Systems," *International Journal of Control*, Vol. 42, No. 4, 1985, pp. 939–948.

INDEX